				IIIA	IVA	VA	VIA	VIIA	VIIIA
									2 He 4.00260
				5 B 10.81	6 C 12.011	7 N 14.0067	8 O 15.9994	9 F 18.998403	10 Ne 20.179
		IB	IIB	13 Al 26.98154	14 Si 28.0855	15 P 30.97376	16 S 32.06	17 Cl 35.453	18 Ar 39.948
28 Ni 58.69	29 Cu 63.546	30 Zn 65.38	31 Ga 69.72	32 Ge 72.59	33 As 74.9216	34 Se 78.96	35 Br 79.904	36 Kr 83.80
46 Pd 106.42	47 Ag 107.868	48 Cd 112.41	49 In 114.82	50 Sn 118.69	51 Sb 121.75	52 Te 127.60	53 I 126.9045	54 Xe 131.29
78 Pt 195.08	79 Au 196.9665	80 Hg 200.59	81 Tl 204.383	82 Pb 207.2	83 Bi 208.9804	84 Po (209)	85 At (210)	86 Rn (222)

63 Eu 151.96	64 Gd 157.25	65 Tb 158.9254	66 Dy 162.50	67 Ho 164.9304	68 Er 167.26	69 Tm 168.9342	70 Yb 173.04	71 Lu 174.967
95 Am (243)	96 Cm (247)	97 Bk (247)	98 Cf (251)	99 Es (252)	100 Fm (257)	101 Md (258)	102 No (259)	103 Lr (260)

Understanding Chemistry

Understanding Chemistry

Robert J. Ouellette
The Ohio State University

Macmillan Publishing Company
New York

Collier Macmillan Publishers
London

Macmillan Publishing Company
866 Third Avenue, New York, New York 10022

Collier Macmillan Canada, Inc.

Library of Congress Cataloging in Publication Data

Ouellette, Robert J.,
  Understanding chemistry.

  Includes index.
  1. Chemistry.   I. Title.
QD31.2.O86   1987   540   86-12606
ISBN 0-02-389750-3

Printing: 1 2 3 4 5 6 7 8   Year: 7 8 9 0 1 2 3 4 5 6

Preface

Many college curricula have chemistry as a required course. The standard first year college chemistry course is based on the student's having completed one year of high school chemistry and a minimum of one year of high school algebra. Some students lack the prerequisite high school chemistry course, have had an inadequate course, or had the course more than two years ago. For such students to be successful in the full year college course it is often necessary to take a one semester course designed to introduce or review the basic concepts of chemistry.

Understanding Chemistry has developed from the 25 years of experience of the author in teaching a beginning or preparatory chemistry course. This book is written with the assumption that the students have had no previous chemistry course and that their mathematical background is limited. Detailed applications of chemistry, highly theoretical aspects of the science, and special topic areas are omitted. Only the central features of chemistry are presented and at a level suitable for the students with the background described.

Learning objectives are stated at the start of each chapter. All key terms are boldfaced in the chapter and are given again in a list with definitions at the end of the chapter. A chapter summary is provided.

Problem solving is the major emphasis of this book since the solution of problems creates the most difficulty for students in the first year chemistry course. There are 250 solved Examples in the text, many of which refer to additional questions at the end of the chapter. The 1800 Exercises and Additional Exercises at the ends of the chapters are grouped by concept and in order of chapter presentation. In addition they are arranged in columns of matched pairs that contain similar material. Because many exercises have multiple parts, the student has about 5000 problems to use to develop problem-solving skills. This large number of problems offers the teacher a variety of assignments that may change with each course offering.

Significant figures are considered in all Examples. The typical answer obtained by using a calculator is given and then is adjusted to provide the correct answer containing the proper number of significant figures.

Answers to the Exercises in the left column are given at the end of the book. Solutions to both columns of Exercises are available in a separate Solutions Manual. Solutions to the Additional Exercises are provided in the Teacher's Manual.

Supporting materials for the student include the Study Guide and the Solutions Manual. There are 275 Examples with explanations in the Study Guide and 585 Drill Problems. The study guide is organized according to the learning objectives for each chapter of the text. Sometimes the topics are explained in a somewhat different way than in the text. However, in no case are the two approaches contradictory. An alternate explanation or viewpoint is intended to help the student learn the material. The majority of the material follows closely the explanation and examples given in the text. Therefore the Study Guide provides a reinforcement of what the student must learn.

The Solutions Manual is meant to be used after the text and the Study Guide have been studied carefully. A serious and honest intellectual commitment must be made to solve a problem before looking up the solution. One must be an active participant in learning—not a casual reader of both the text questions and the solutions. It is necessary to put ideas down on paper even though the answer might not be right. Then the Solutions Manual can be used to determine the sources of errors in the approach to the problem.

Supporting materials for the teacher include a Teacher's Manual and a Test Bank. These materials are available to the teacher from the publisher. They were developed simultaneously with the text. To provide an integrated approach, the Teacher's Manual was revised after the manuscript for this book was copyedited.

The author acknowledges the contributions of Mary Bailey in her preparation of the Study Guide and the Solutions Manual and Judith Casey in the preparation of the Teacher's Manual. They read the manuscript in its many versions and contributed to its development while writing their manuscripts. As a result the supporting materials are strongly integrated with the book. The classroom experience of both of my colleagues will benefit the student in learning chemistry and the teacher in presenting the science.

The author wishes to thank the many reviewers selected by Macmillan: Professors David Becker, Wade Freeman, Edward Garvin, August Frederick Goellner, Philip Jaffe, R. C. Phillips, Irvin Plitzuweit, George Schenk, Carl Shonk, and Vernon Thielmann.

The author has received substantial developmental support from the Macmillan Publishing Company throughout the project. Separate series of academic reviews were provided for both the first and second drafts of this manuscript. The manuscript was then revised a third time based on the comments of the Development Editor, Madalyn Stone. A copyeditor, versed in chemistry, meticulously verified the data as well as the consistency of the material. Finally another individual proofed the galleys after the author had corrected them. The author is grateful for the confidence of Peter Gordon, the Chemistry Editor, as exemplified by the support that he provided. The foresight of Peter Gordon in supporting the simultaneous development of all supplementary materials represents an unusual commitment to the project and allowed the author to develop a coordinated teaching package.

It was again my good fortune and pleasure to have Elisabeth Belfer as Production Supervisor. The author appreciates the immense help that she provided in the critical months required to bring the material from manuscript to a completed book.

The manuscript was typed in its several versions by Joy Snyder and Paula Wise. The days of word processors make the task of revisions easier than in the past, but quality work at the keyboard in formating chemistry is still required and was competently provided by both individuals.

R. J. O.

Contents

6 Chemical Equations and Reactions 122

7 Stoichiometry 141

11

Nomenclature of Inorganic Compounds   218

12

The Gaseous State   239

13 Liquids and Solids   266

14 Solutions   292

15 Reaction Rates and Equilibrium   320

16 Acids and Bases   345

20   Functional Groups in Organic Chemistry   453

Appendix / Mathematical Review   489

1 Introduction to Chemistry

Learning Objectives

After studying Chapter 1 you should be able to

1. Distinguish between the terms composition and structure.
2. Distinguish between the terms reactant and product.
3. Distinguish among the terms hypothesis, theory, and law.

1.1 Matter and Our World

Any physical material that you can see, smell, or touch consists of matter. Your body is made of matter. Among the examples of matter in your environment are the oxygen in the air that you breathe, the water you drink, and the aspirin that you take for a headache. **Matter** is anything that has mass and occupies space. Mass is probably not a term that you commonly use. By **mass** we mean a quantity of matter. Normally you describe a quantity of material in terms of its weight, but there is a subtle difference between mass and weight. **Weight** is the result of the gravitational attraction of the earth for the object. Without gravity an object would not have weight, but it would still have mass. For example, although the astronauts in the space shuttle are weightless in space, they still have mass. Since the force of gravity is about the same anywhere on the earth, two objects that have the same mass have the same weight. Therefore, mass and weight are often used interchangeably, but strictly speaking only mass is a measure of the quantity of matter.

Matter is constantly undergoing change in its identity. The germination of flowers in the spring, the growth of a plant in the summer, the changes in the colors of tree leaves in the autumn, and the decay of vegetation in the winter all involve changes in matter. From conception, through your life, and ultimately in death, the biological and physiological processes of your body involve changes in matter.

Humans have long been curious about the changes in matter. What is responsible for these changes? Can we understand the reasons for these changes? In early history, black magic and superstitions dominated our ancestors' thoughts about their world. Today scientists have a better understanding of why changes in matter occur and can, to some extent, control certain events.

1.2 What Is Chemistry?

Chemistry is a science of the composition, structure, and reactions of matter. To understand this definition, we must understand the words used. You have already learned what matter is. Although you have some idea of what science is, let's define it. **Science** is the observation, description, experimental investigation, and explanation of natural phenomena. For the remainder of this section we will examine what the chemist means by the words *composition, structure,* and *reaction.*

In chemistry **composition** is defined as the identity and amount of the components of matter. To understand the term composition, let's consider the components of a bicycle. A bicycle is composed of a set of handlebars, a frame, a seat, two wheels, and several other items. This description, which is a statement of composition, gives the type and number of items that are contained in a bicycle.

Composition can be given not only in terms of the number of items but also in terms of their mass. You often purchase produce by weight rather than by number of items. Composition can be based on weight. Consider a bag of produce containing two grapefruits and four oranges. The composition of the bag can be stated in terms of the weight of the fruit. If the weight of an orange is 7 oz and the weight of a grapefruit is 14 oz, then the total weight is 56 oz.

$$
\begin{aligned}
4 \text{ oranges} &= 4 \times (7 \text{ oz}) = 28 \text{ oz} \\
2 \text{ grapefruits} &= 2 \times (14 \text{ oz}) = \underline{28 \text{ oz}} \\
\text{total weight} &= 56 \text{ oz}
\end{aligned}
$$

The percent composition of the oranges based on weight can be calculated using the definition of percent, which is the ratio of a part divided by the whole with the quotient multiplied by 100.

$$\text{percent} = \frac{\text{part}}{\text{whole}} \times 100$$

$$\text{percent of oranges by weight} = \frac{4 \times (7\text{ oz})}{56\text{ oz}} \times 100 = 50\%$$

Example 1.1

A 110 lb woman may eat 12 oz of carbohydrates, 2 oz of fats, and 16 oz of proteins one day. What is the percent of carbohydrates in her diet?

Solution The total amount of the components of the diet is not given. Thus you must first calculate the total weight of the materials.

total weight = 12 oz + 2 oz + 16 oz = 30 oz

Now you may calculate the percent of the desired material, carbohydrates, in the diet using the definition of percent.

$$\text{percent of carbohydrates} = \frac{12\text{ oz}}{30\text{ oz}} \times 100 = 40\%$$

Example 1.2

A 5.0 grain capsule of a medicine contains 3.8 grains of active ingredients, and the remainder is an ''inactive'' ingredient. What percent of the capsule is an ''inactive'' ingredient?

Solution The amount of ''inactive'' ingredient is not given, and it must first be calculated.

''inactive'' ingredient = 5.0 grains − 3.8 grains = 1.2 grains

Now calculate the percent of the ''inactive'' ingredient.

$$\text{percent of ''inactive'' ingredient} = \frac{1.2\text{ grains}}{5.0\text{ grains}} \times 100 = 24\%$$

[*Additional examples may be found in 1.5–1.12 at the end of the chapter.*]

Based on the accumulated knowledge of years of experiments, chemists now describe the composition of matter in terms of the number and type of atoms and molecules characteristic of that substance. **Atoms** are the simplest units of matter, and **molecules** are units of matter consisting of combinations of atoms. For example, water consists of water molecules that are composed of two hydrogen atoms and one oxygen atom.

In addition to knowing the number and type of the atoms of a substance, such as water, chemists use mass to describe the composition of matter. The mass of atoms is given in atomic mass units, abbreviated amu (Table 1.1). Don't be alarmed by this unfamiliar unit. You can use atomic mass units the same way that you used ounces and grains in the two preceding examples.

Table 1.1 Atomic Weights and Symbols of Selected Elements

Element	Atomic weight (amu)	Chemical symbol
hydrogen	1	H
carbon	12	C
nitrogen	14	N
oxygen	16	O
fluorine	19	F
phosphorus	31	P
sulfur	32	S

The hydrogen atom has a mass of 1 atomic mass unit (amu). The oxygen atom, which is 16 times heavier than the hydrogen atom, has a mass of 16 amu. A water molecule, which consists of two hydrogen atoms and an oxygen atom, has a mass of 18 amu.

$$
\begin{aligned}
\text{mass of 2 hydrogen atoms} &= 2 \times (1 \text{ amu}) = \underline{2 \text{ amu}} \\
\text{mass of 1 oxygen atom} &= 1 \times (16 \text{ amu}) = \underline{16 \text{ amu}} \\
\text{mass of 1 water molecule} &= 18 \text{ amu}
\end{aligned}
$$

In terms of mass, the water molecule is then 11% hydrogen and 89% oxygen.

$$
\% \text{ hydrogen} = \frac{2 \times (1 \text{ amu})}{18 \text{ amu}} \times 100 = 11\%
$$

$$
\% \text{ oxygen} = \frac{1 \times (16 \text{ amu})}{18 \text{ amu}} \times 100 = 89\%
$$

Of course, the total of the percents of all components must equal 100%; in this case

$$
\begin{aligned}
\text{total percent} &= \% \text{ oxygen} + \% \text{ hydrogen} = 100\% \\
\text{total percent} &= 89\% + 11\% = 100\%
\end{aligned}
$$

Example 1.3

Carbon dioxide consists of molecules that have one carbon atom and two oxygen atoms. The masses of carbon and oxygen atoms are 12 and 16 amu, respectively. Calculate the composition of carbon dioxide in terms of the mass of the atoms.

Solution First determine the total mass of the carbon dioxide molecule.

$$
\begin{aligned}
\text{mass of 1 carbon atom} &= 1 \times (12 \text{ amu}) = 12 \text{ amu} \\
\text{mass of 2 oxygen atoms} &= 2 \times (16 \text{ amu}) = \underline{32 \text{ amu}} \\
\text{mass of 1 carbon dioxide molecule} & \phantom{= 2 \times (16 \text{ amu}) =} \;\; 44 \text{ amu}
\end{aligned}
$$

The percent of carbon in the carbon dioxide molecule based on mass is determined in a manner similar to any percent composition problem.

$$
\% \text{ carbon} = \frac{12 \text{ amu}}{44 \text{ amu}} \times 100 = 27\%
$$

Because there are only two components in the molecule, the percent of oxygen can be determined by subtracting the percent of carbon from 100%.

% oxygen = 100% − % carbon

% oxygen = 100% − 27% = 73%

[*Additional examples may be found in 1.13–1.16 at the end of the chapter.*]

Composition by itself does not tell us all that we really need to know about matter. For example, could an aborigine in the South American jungle visualize a bicycle if you provided a picture of each of the components of the bicycle (Figure 1.1)? The answer is no, since the aborigine needs to know how the parts of the bicycle fit together.

In chemistry **structure** may be defined as the arrangement of the atoms of a substance. In other words, structure describes how atoms are put together to make molecules. As the science of chemistry is developed in this book, you will learn that atoms in molecules are bonded to each other. **Bonded** means held or fastened together. The manner in which atoms are bonded to each other then creates a structure. For example, in water the two hydrogen atoms are bonded to a central oxygen atom but not to each other (Figure 1.2). The capital letters H and O are used by

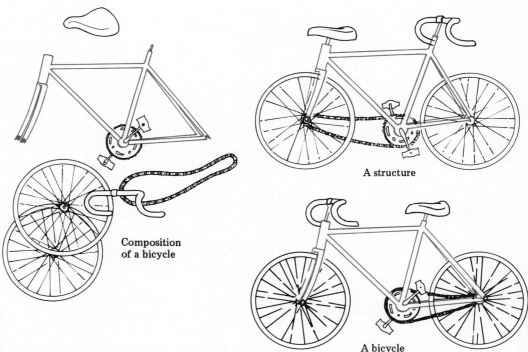

Figure 1.1

A structure

Composition
of a bicycle

A bicycle

The composition and structure of a bicycle
The parts of a bicycle are its composition. To show the structure of a bicycle, the parts are put together, but if the parts are improperly assembled, the result is not a bicycle.

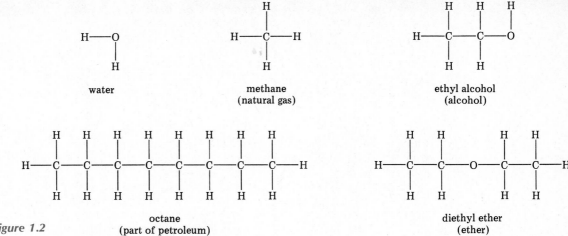

water   methane
(natural gas)   ethyl alcohol
(alcohol)

octane
(part of petroleum)   diethyl ether
(ether)

Figure 1.2

Structures of some common substances
C is the symbol for the carbon atom. O is the symbol for the oxygen atom. H is the symbol for the hydrogen atom. The dashes between the symbols represent bonds that hold the molecule together and create the structure.

chemists to represent (or symbolize) the hydrogen and oxygen atoms, respectively. The dashes between the symbols or the atoms represent the bonds. Structures for some common substances are also shown in Figure 1.2.

Example 1.4

Describe the composition of ethyl alcohol (Figure 1.2) in terms of the atoms in the molecule.

Solution The representation of ethyl alcohol in Figure 1.2 tells us that the molecule has two carbon atoms, six hydrogen atoms, and one oxygen atom.

Example 1.5

Describe the structure of methane (Figure 1.2).

Solution The representation of methane in Figure 1.2 tells us that methane has four hydrogen atoms, each of which is bonded to a central carbon atom.

Example 1.6

Describe the structure of ethyl alcohol (Figure 1.2) based on the bonds of the carbon atom.

Solution From the structure in Figure 1.2, you can see that the two carbon atoms are bonded to each other. One carbon atom is also bonded to three hydrogen atoms. The other carbon atom is bonded to two hydrogen atoms and one oxygen atom.

[*Additional examples may be found in 1.17–1.20 at the end of the chapter.*]

A **reaction** involves changes in the composition and structure of matter. There are many examples of chemical reactions occurring in your everyday life. Examples that you may not have considered include sight perception, muscle movement, and the metabolism of food. Although you might think of these as physiological processes, they all involve chemical reactions. The metabolic processes in your body convert glucose into carbon dioxide and water. The body uses oxygen in this reaction. Chemists represent the reaction with a **chemical equation.**

glucose + oxygen ⟶ carbon dioxide + water
(reactant) (reactant)   (product)   (product)

The words to the left of the arrow of a chemical equation represent **reactants,** the materials that are used in the reaction and will undergo change. The words to the right of the arrow represent **products,** the substances that result from or are produced in a chemical reaction. The arrow means "is converted into." In this chemical reaction, substances are transformed or converted into other substances that have different compositions and structures.

Example 1.7

Photosynthesis in plants can be represented by the following equation. What are the reactants and the products in this chemical reaction?

carbon dioxide + water ⟶ glucose + oxygen

Solution The words on the left of the arrow represent the substances carbon dioxide and water, the reactants. The words on the right side of the arrow represent the substances glucose and oxygen, the products.

[*Additional examples may be found in 1.21–1.24 at the end of the chapter.*]

In this section you have been given an overview of the science of chemistry by discussing the terms composition, structure, and reaction. Throughout the remainder of the text the meaning of these terms will be refined. The accumulated knowledge of the science of chemistry is the result of a long history of successful and unsuccessful approaches to studying matter. In the next section a brief history of the development of chemistry is presented.

1.3 A Brief History of Chemistry

Human beings have always been concerned with and fascinated by their surroundings. Initially, we had to adapt to our environment to survive. By observing the processes that occur, we learned to control them or predict the path they would take, thereby improving our adaptability. Basically, this is still our chief concern because we can survive only if we can adapt to future events. Our early ancestors may have considered and examined nature in an attempt to control it, but they were not concerned with the reasons behind the observed events.

The early development of chemistry was essentially utilitarian, concerned with how the materials of the earth could be put to use to improve the quality of life. Basic chemical industry as it appeared between 4000 and 3000 B.C. involved the very practical processes of production of metal, glass, pottery, pigments, dyes, and perfume. Gold, which occurs free in nature, was used by the ancients for ornamental purposes, much as it is today. The metallurgy of copper, lead, tin, iron, and silver developed before 3000 B.C. Ornamental glass and glazed pottery were produced in China, India, and Egypt. Such early chemistry, as used and developed by these societies, was more a practical art than a science.

Chemistry as a science developed as an offshoot of philosophy, which means the love or pursuit of knowledge. The philosophy of Greece about the time of Socrates (500 B.C.) is generally regarded as the beginning of scientific thought. Although other civilizations had recorded and used chemical processes, they did not ask penetrating questions about them nor did they attempt to explain natural phenomena. The Greeks posed many questions about the nature of the world and approached science as rationalists; they believed that reason is the prime source of knowledge and is the only valid basis for action or belief. There are some similarities between what some Greek philosophers thought about matter and what is now well established. For example, the concept of the atom was suggested based on philosophical grounds. However, their understanding of matter was often wrong because the conclusions were the result of contemplative philosophy and were not verified by experimental methods. The development of science as we know it depends on continuing observations and experiments.

The alchemical laboratory of the ninth to sixteenth centuries was the center of limited experimentation. Although the purpose of the alchemist was to change common metals into gold and to discover the elixir of life, the net result was the development of experimental methods that would lead to the science of chemistry.

The iatrochemists of the sixteenth and seventeenth centuries sought to merge the fields of medicine and chemistry under one discipline. They believed that man is a composite of chemicals and that health depends on the proper proportions of the elements. This belief, although somewhat simplistic, is essentially correct. The experimental method continued to be developed in the laboratory of the physician–chemist, and the standards of the developing field of chemistry were improved.

In all human endeavors there are certain individuals who make contributions that are recognized as historically significant. Chemistry is no different. Many have labored in the chemistry laboratory to uncover the facts of the science but only certain of those facts led to significant advances in our understanding of nature. Robert Boyle, Antoine Lavoisier, and John Dalton are three individuals whose talents led to discoveries that contributed to our understanding of chemistry.

The study of the properties of gases by Robert Boyle (1627–1691) is a classic example of the experimental methods of science. In Chapter 12, we will discuss Boyle's observations on the behavior of a volume of a gas under changes in pressure. *The Skeptical Chymist,* a book written by Boyle in 1661, established chemistry as a field of study relying on experiments and concerned with the search for explanations of natural phenomena consistent with facts gathered in the laboratory. The definition of an element as stated by Boyle is considered to be the first concise scientific expression of the nature of this fundamental unit of matter.

Antoine Lavoisier (1743–1794) studied and provided an explanation of the combination of matter with oxygen, a component of air. These processes include combustion, in which materials such as wood burn rapidly, and corrosion, in which materials such as metals slowly combine with oxygen. The quantitative work of Lavoisier established methods of studying chemical reactions. Conclusions must be supported by experiment, and experiments must be repeated to establish that the results are correct.

John Dalton (1766–1844) proposed that atoms can account for the quantitative observations that had been made on the composition of compounds and chemical reactions. His atomic theory, which is described in Chapter 4, is used in the remainder of the text.

Prior to World War I, chemistry was dominated by European scientists; however, since that time the United States has become the preeminent country in chemistry, and scientists from many countries come to the United States to study. Some European scientists who have made major contributions to our understanding of chemistry include Niels Bohr (1885–1962), a Dane who proposed a theory of the structure of the atom, and Otto Hahn (1879–1968), a German who discovered nuclear fission.

Many Americans have made significant contributions to chemistry as well. Gilbert Lewis (1875–1946) developed a concept of the chemical bond that is presented in Chapter 10. Linus Pauling (1901–) contributed to the theories of bonding as well as to our knowledge of the structure of proteins. Melvin Calvin (1911–) developed our current understanding of photosynthesis, a process in which solar energy is used to produce plant matter. Glenn Seaborg (1912–) prepared nine new elements in nuclear reactions and contributed to our understanding of the structure of the nucleus, which is a part of the atom. Paul Flory (1910–1985) determined ways of studying polymers, which are the basis of the plastics industry.

1.4 The Scientific Method

Observations and Laws

All sciences depend on observations. Accumulation of quantitative observations of certain regularities in nature leads to the statement of laws of nature. A **law** is simply the conclusion or an explicit statement of fact that is already inherent in the information obtained by observations. No new understanding of nature results by stating a law; the law merely summarizes what has been observed.

An example of a scientific law is the law of conservation of mass. This law states that in a chemical reaction the mass of the products formed is equal to the mass of the reactants that are used. In other words, the mass has been conserved. This law is important to the study of chemistry and is discussed in Chapter 3.

Models

Science is more than just experiments and the statement of laws. Science is characterized by a curiosity and interest in how and why matter behaves as it does. To discover these facts models are used. A **model** is a speculation or interpretation of what is responsible for a phenomenon. For example, we could use a model to discover the shape of an object in a box (Figure 1.3). If the box is tilted front to back, the observer may hear and feel the rolling object. The observer might use a sphere as a model for the object. As a test of this model, the observer might tilt the box from side to side and discover that the object appears to slide rather than roll. This evidence would show the sphere model to be faulty because a spherical object would still roll side to side. Thus the observer might consider a cylindrical object as a model for the item in the box. However, the actual object could be a solid cylinder, a hollow cylinder, or a cylinder with rounded ends. Thus depending on the number of observations made, the model may or may not bear a resemblance to the real object, except in the functions already perceived.

The structure of matter is submicroscopic in scale and cannot be seen with optical microscopes. However, based on many observations of chemical properties and reactions, the chemist has developed models of atoms and molecules for the structure of matter. Think of the chemist as a detective seeking to unravel a mystery of structure by accumulating circumstantial evidence. We cannot ''see'' the atoms and molecules, but there is an overwhelming body of

Figure 1.3

The development of a model
The observer in this experiment postulates a model
for the object in the box based on the sensation of
motion caused by tilting the box.

circumstantial evidence on chemical structure that suggests that the concept of atoms and molecules is correct. However, unlike a mystery, the case is never solved. Each new bit of evidence is used to recheck our model.

In order to communicate our ideas to others we may wave our hands about or draw a sketch on paper. Models are also expressed in more concrete terms. In some cases a mathematical equation is an appropriate expression of a model. However, pictures also can be used to depict models. Picture models are easy to understand and are used frequently in this text. These models must not be viewed as more important than the experimental facts. The experimental facts reflect nature; the model is a human creation. A scientific model is not a small-scale working version of the real object, as model planes are miniature representations of actual planes. Scientific models are attempts to illustrate what we cannot see. These models are imagined; they are not photographs of the real object.

Hypothesis and Theory

The terms hypothesis and theory describe models that are used to explain laws. A **hypothesis** is a tentative model, whereas a **theory** describes a model that has been tested many times. The dividing line between the two is arbitrary and cannot be precisely defined. When a substantial majority of scientists accept a model, it is called a theory. John Dalton proposed a model in 1808 to account for the then known laws of chemistry. As a result of many years of testing his model was accepted and is now called Dalton's atomic theory.

The use of hypotheses and theories is a part of the scientific method (Figure 1.4). The **scientific method** involves making observations, recording them, and using them to solve a problem. In the scientific method the facts are collected by experimentation and then analyzed to find trends. A hypothesis is formulated to account for the data. Additional experiments are planned to test the hypothesis, and the hypothesis is modified or discarded as necessary to account for any additional observations. Finally, after the hypothesis has survived many tests, a theory is stated.

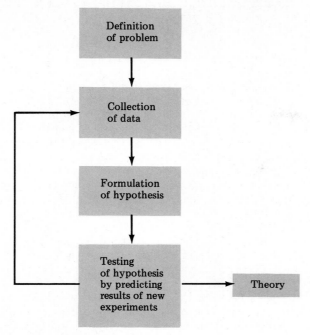

Figure 1.4

The scientific method
This flow diagram shows the development of a theory using the scientific method. The formulation of the first hypothesis is based on the data initially collected. The testing of the hypothesis requires new experiments, which yield additional data. It may be necessary to reformulate a hypothesis many times before a theory is stated.

Example 1.8 _____

Your friend lends you her calculator. When you push the ''on'' button, the calculator does not work. Describe a scientific approach to discovering why the calculator does not work.

Solution   First you should push the ''on'' button one more time to make sure that your first observation is correct. If the observation is verified, you might hypothesize that the calculator does not have a battery. A check of the battery compartment will confirm or disprove this hypothesis. If the calculator has a battery, you could hypothesize that the battery is ''dead.'' At this point you could replace the battery with a new battery and try the calculator again. If it still does not work, then you might conclude that something is wrong with the display or another part of the calculator.

[*Additional examples may be found in 1.25–1.28 at the end of the chapter.*]

1.5 How to Study Chemistry

In this book you will encounter numerous theories, laws, symbols, and terms that are unfamiliar. The very language that is used may be quite different from the language used in your other college courses. There is no short-cut to learning the language used in chemistry. You must first start memorizing facts as they are encountered. In a sense, a chemistry course is much like a language course. You cannot learn a language without memorizing words and rules of syntax.

There is much interdependent material in chemistry. If you do not study and learn the facts for even a few days your overall performance in the course will be affected. You will not be able to read several chapters the night before an exam and be successful.

Merely reviewing chapter material, even on a regular basis, is not enough for success in chemistry. It is necessary to practice using the terms, symbols, and equations. Problems must be done regularly to develop proficiency in methods of approach. Only by working out many problems can a full understanding of chemistry be acquired.

In this book there are many examples of problems and their solutions. Try to do each as you proceed through the book. If you find it necessary to depend on the author's solutions, return to the previous subject matter and learn why you were unable to solve the problem. At the end of the chapter you will find other problems of the same type. Try one or more of these to see if you have improved your ability.

Assigned chapters should be read at least twice. Read the material the first time for an overall impression and to determine what areas you need to study carefully. In your second reading acquire as complete an understanding as possible. If you study before the lecture you will be in the best position to learn from the class session. You can listen in a more relaxed manner and actively think of what is being discussed. Furthermore you will be prepared to listen closely to the subject matter that is the most troublesome for you. If you do not study before your class session, the material presented in class will be new to you. Consequently you will be transcribing words that on the whole have little meaning to you. Learning from notes that were taken without any understanding is a futile process. Prepare for every class and chemistry will not be the difficult subject that it is reputed to be.

Summary

Every material or object that we observe consists of matter that has mass and occupies space. Chemistry is the study of matter, its composition, structure, and reactions.

The science of chemistry is one part of our cultural heritage. It is based on the accumulation of studies of many individuals. However, some individuals such as Robert Boyle, Antoine Lavoisier, and John Dalton made major contributions in an era regarded as the start of modern chemistry.

In order to develop a science, models that represent ideas are used. These models take the form of mathematical equations or pictures in order to communicate ideas to others. The scientific method is an approach to the study of science that is based on observation and experiment to obtain facts that are stated as laws. The facts in turn are studied to determine what is responsible for a natural phenomenon. A tentative model used to explain a natural phenomenon is called a hypothesis, whereas a model that is well established is called a theory. Dalton's concept of atoms and molecules is a theory used to explain chemical reactions.

New Terms

Atoms are the simplest units of matter.
Bonding describes how the atoms in a molecule are held or fastened together.
A **chemical equation** represents reactants and products in a chemical reaction.

Chemistry is a science of the composition, structure, and reactions of matter.
Composition means the identity and amount of the components of a sample of matter.
A **hypothesis** is a tentative model used to explain a law.

A **law** is an explicit statement of fact obtained by observation or experimentation.

Mass is a quantity of matter.

Matter is anything that occupies space and has mass.

A **model** is an idea that may correspond to what is responsible for a natural phenomenon.

Molecules consist of combinations of atoms.

Products are the substances that are produced in a chemical reaction.

Reactants are the substances that enter into a chemical reaction.

A **reaction** involves changes in the composition and structure of matter.

Science is the observation, description, experimental investigation, and explanation of natural phenomena.

The **scientific method** is a sequence of steps involving observation, experimentation, and formulation of laws and theories that lead to scientific knowledge.

Structure means the arrangement of the components of a substance.

A **theory** is a tested model that is used to explain a law.

Weight is a result of the gravitational attraction between an object and the earth.

Exercises

Terminology

1.1 Distinguish between the terms composition and structure.

1.2 Distinguish between the terms reactants and products.

1.3 What is the difference between a theory and a law?

1.4 What is the difference between a hypothesis and a theory?

Percent Composition

1.5 In a high school class of 60 students, 15 continue their education in college. What percent of the class did not go to college?

1.6 In a class of 80 students there were 8 A, 16 B, 40 C, 12 D and 4 E grades. What is the percent distribution of each grade?

1.7 A 150 lb marathon runner has a 6% body fat content. How many pounds of fat does the runner have?

1.8 The average human body has 9% bone. How many pounds of bone does a 60 lb child have?

1.9 Approximately 2% of the human body is calcium. How many pounds of calcium does a 100 lb woman have?

1.10 Approximately 1.1% of the human body is phosphorus. How many pounds of phosphorus does a 200 lb man have?

1.11 A brass is 65% copper and 35% zinc. How much copper is required to make 20 oz of brass?

1.12 White gold is 60% gold and 40% platinum. How many ounces of platinum are in 2.0 oz of white gold?

1.13 Carbon monoxide, a poisonous gas, consists of molecules containing one carbon atom and one oxygen atom. The atomic weights of carbon and oxygen atoms are 12 amu and 16 amu, respectively. Calculate the percent composition of carbon monoxide.

1.14 Methane is a gas used in the home for cooking and heat. It consists of molecules containing one carbon atom and four hydrogen atoms. The atomic weights of carbon and hydrogen atoms are 12 amu and 1 amu, respectively. Calculate the percent composition of methane.

1.15 Propane is used in camp cooking stoves. It consists of molecules containing three carbon atoms and eight hydrogen atoms. The atomic weights of carbon and hydrogen atoms are 12 amu and 1 amu, respectively. Calculate the percent composition of propane.

1.16 Ethyl alcohol consists of molecules containing two carbon atoms, six hydrogen atoms, and one oxygen atom. The atomic weights of carbon, hydrogen, and oxygen atoms are 12 amu, 1 amu, and 16 amu, respectively. Calculate the percent composition of ethyl alcohol.

Structure of Molecules

1.17 Hydrogen sulfide, the foul-smelling gas of rotten eggs, consists of molecules containing two hydrogen atoms and one sulfur atom. The symbols for hydrogen and sulfur are H and S, respectively. The molecule consists of a central sulfur atom bonded to two hydrogen atoms. The two hydrogen atoms are not bonded to each other. Draw the molecule.

1.18 Ammonia consists of molecules containing three hydrogen atoms and one nitrogen atom. The symbols for hydrogen and nitrogen are H and N, respectively. The molecule consists of a central nitrogen atom bonded to three hydrogen atoms. The three hydrogen atoms are not bonded to each other. Draw the molecule.

1.19 Methyl alcohol, a poisonous alcohol that can cause permanent blindness even in small quantities, consists of molecules containing one carbon atom, four hydrogen atoms, and one oxygen atom. Given that the carbon atom forms four bonds, an oxygen atom forms two bonds, and a hydrogen atom forms one bond, draw the structure of methyl alcohol.

1.20 The molecule of butane, a substance used in butane cigarette lighters, consists of four carbon atoms and ten hydrogen atoms. The carbon atoms are connected in a series similar to that shown for octane in Figure 1.2. Given the facts that a carbon atom forms four bonds and a hydrogen atom forms one bond, draw the structure of butane.

Chemical Reactions

1.21 Sulfur burns in oxygen to form sulfur dioxide. Write a word equation for the reaction.

1.23 Ammonia is produced by the reaction of nitrogen and hydrogen gas as given by the following equation. Identify each substance in the equation as a reactant or product.

hydrogen + nitrogen $\longrightarrow$ ammonia

1.22 At high temperatures nitrogen combines with oxygen to form nitric oxide. Write a word equation for the reaction.

1.24 Water can be decomposed by an electric current. The reaction is given by the following equation. Identify each substance in the equation as a reactant or product.

water $\longrightarrow$ oxygen + hydrogen

The Scientific Method

1.25 You are visiting a friend and decide to turn on the television set while he is out of the room. After pushing the on button and waiting for a few seconds the screen remains dark. Describe a scientific way to approach this problem.

1.27 You perform an experiment and do not obtain the results described in your textbook. What do you do next?

1.29 How many examples of exceptions to the results predicted by a theory are allowed before the theory must be replaced?

1.26 You live in an apartment with three other students. Just prior to leaving for your class, you discover that your calculator is "missing." Describe a scientific way to approach this problem.

1.28 Explain the reason for repeating experiments several times before developing a law.

1.30 Why is it important that scientific models be published in scientific journals or be presented in scientific meetings?

Additional Exercises

1.31 A flock of 50 sheep has 2 black sheep. What is the percent of black sheep in the flock?

1.33 Approximately 0.2% of the human body is potassium. How many pounds of potassium are there in a 30 lb child?

1.35 Hydrazine, used as a fuel in early rockets, consists of molecules containing two nitrogen atoms and four hydrogen atoms. Calculate the percent composition of hydrazine.

1.37 Ethylene glycol, a substance used in antifreezes, consists of molecules containing two carbon atoms, six hydrogen atoms, and two oxygen atoms. Calculate the percent composition of ethylene glycol.

1.39 Examine the following structure for dimethyl ether and compare it to the structure of ethyl alcohol given in Figure 1.2. What can be said about the composition of the two substances? What can be said about the structures of the two substances?

1.32 A 110 lb woman has a blood volume of 3.6 qt. Her plasma volume is 2.3 qt, and her red cell volume is 1.3 qt. What percent of her blood volume is the red cell volume?

1.34 Approximately 18% of the human body is carbon. How many pounds of carbon are there in a 150 lb man?

1.36 Hydrogen peroxide, used as an oxidizer in early rockets, consists of molecules containing two oxygen atoms and two hydrogen atoms. Calculate the percent composition of hydrogen peroxide.

1.38 Formaldehyde, the substance used to preserve biological specimens, consists of molecules containing one carbon atom, two hydrogen atoms, and one oxygen atom. Calculate the percent composition of formaldehyde.

1.40 Ethyl mercaptan, a foul-smelling compound added to natural gas to detect leaks, consists of molecules containing two carbon atoms, six hydrogen atoms, and one sulfur atom. Sulfur, like oxygen, can form two bonds. The structure of ethyl mercaptan resembles the structure of ethyl alcohol. Draw the structure of ethyl mercaptan.

```
    H     H
    |     |
H — C — O — C — H
    |     |
    H     H
```

1.41 Carbon dioxide was called a product in Section 1.2, but was called a reactant in Example 1.7. How can a substance be considered both a reactant and a product?

1.43 A student in another science class tells you that scientific laws can be stated only after a theory has been established. What is your response?

1.42 Can there be more than one product in a chemical reaction? Can there be more than one reactant in a chemical reaction?

1.44 It has been said that science progresses most rapidly when scientists can suggest several hypotheses to explain a natural phenomenon. Explain why this statement is valid.

2

Numbers, Measurements, and Units

Learning Objectives

After studying Chapter 2 you should be able to

1. Use the factor unit method in problem solving.

2. Distinguish between precision and accuracy.

3. Determine the number of significant figures in a number.

4. Express numbers in scientific notation.

5. Express the results of mathematical operations to the proper number of significant figures, rounding off where appropriate.

6. Convert a measurement in one metric unit into another metric unit or into an English unit.

2.1 Numbers

The word "numbers" in the title of the chapter may have caused an increase in the blood pressure of some of you who have the dreaded affliction known as math anxiety. Why do we have to talk about numbers? To answer that question, consider the words "increase" and "blood pressure" in the first sentence. Increase is a qualitative adjective and blood pressure can be considered qualitatively as being high or low. However, how much did the blood pressure increase? Has the blood pressure gone so high as to be life threatening? Only a quantitative measure of blood pressure using numbers and appropriate units will allow a doctor to make that judgement.

Understanding chemistry is based on quantitative observations and expressing the results in terms of numbers. Only with numbers can you understand how small an atom is or how much product will be produced in a reaction. The study of chemistry requires that you gain some mastery with numbers and learn how to use them in solving problems. For this reason, the emphasis of this chapter is on numbers, measurements, and units of measure. You will learn how to solve problems using conversion factors, how to express numbers with the proper number of significant figures, and how to express numbers in scientific notation.

2.2 Measurements

At one time or another you most likely have measured quantities such as distance, volume, and weight or have depended on the measurements of these quantities by others. **Measurements** are made by use of or comparison to a standard measuring device. The length of a long jump at a track meet can be done with a tape measure in feet. The weight of prepackaged hamburger in a supermarket showcase was measured on a scale in pounds. The volume of gasoline delivered at a gas station is measured by the pump in gallons. Where did these standards of feet, pounds, and gallons come from? The choice of standards is arbitrary. Our civilization, as did all previous societies, developed the dimensions that were convenient. The only significant difference between present-day measurements and those of past civilizations is that today's are more accurate. Accuracy is a concept we will discuss later in the chapter.

Countries have differed considerably in their use of units. In the eighteenth century the French used 9216 grains per pound whereas the English used 7000 grains. Units of measure may vary within one country. You use the Avoirdupois pound, which is divided into 16 ounces; each ounce is divided into 16 drams, and each dram into $27\frac{11}{32}$ grains. The Troy pound, used for weighing precious metals such as gold, is slightly smaller than the Avoirdupois pound. It is divided into 12 Troy ounces; each Troy ounce is divided into 20 pennyweights, and each pennyweight contains 24 grains.

With increased trade and commerce between countries and with advancing technology, it became necessary to develop more sophisticated and universal measurement systems. The two major measurement systems are the English and metric systems, with the latter being the more widely used. The English system of measurement is still used in the United States. The metric system is used by most other countries and by scientists in all countries, including the United States. The advantage of the metric system over the English system is based on the multiple units and subunits of each system. There is no regularity in the various units of the English system. The metric system is a decimal system in which the units differ from each other by powers of ten. The metric system is discussed in Section 2.8.

2.3 Conversion Factors and Problem Solving

Because different units of measurements may be used, it is often necessary to convert one measured quantity into another equivalent quantity with different units. This is accomplished by a **conversion factor,** which is a multiplier consisting of two or more units used to multiply one quantity to convert it into a second quantity having different units. The use of conversion factors is called the **factor unit method** because the factor is numerically equivalent to one. All calculations using conversion factors can be summarized by the relationship

information given × conversion factor = information sought

The factor unit method is widely used to solve problems. The following is a step by step approach to this problem-solving technique.

1. Examine the data given and note the units associated with all numbers. Determine what is asked for, that is, what units are desired.
2. Write down the vital data given along with the units to the left of the equal sign, and the desired unknown along with the desired units to the right of the equal sign.
3. Develop the conversion factors with their units that when multiplied by the known data, will give the desired unknown. There are two ways to write a conversion factor—in one form or its inverse (reciprocal).
4. Check your work to see that the units are equal on both sides of the equation.
5. Carry out the arithmetic and check the answer for mathematical reasonableness.

To illustrate the factor unit method, let us calculate the length in feet (ft) of a 100 yard (yd) football field. You know the answer without using the factor unit method but the best way to learn the method is to use it with familiar quantities. The number of feet in the field is the unknown quantity. Your given quantity is 100 yd in the field. Therefore, write

100 yd × factor = ? ft

In addition, we know that 1 yd is equal to 3 ft.

1 yd = 3 ft

We can write a factor F_1 that is a mathematical way of saying that 3 ft equals 1 yd.

$$F_1 = \frac{3 \text{ ft}}{1 \text{ yd}}$$

We also can write a factor F_2 that is a mathematical way of saying that 1 yd equals 3 ft.

$$F_2 = \frac{1 \text{ yd}}{3 \text{ ft}}$$

Since both F_1 and F_2 express the same information, how do you know which to use in solving the problem of finding the equivalent of 100 yd in feet? Choose the factor that eliminates the unit that you do not want and results in the desired unit. If the unit to be eliminated is in the numerator, choose the conversion factor that has that unit in the denominator. If the unit to be eliminated is in the denominator, choose the conversion factor that has that unit in the numerator. Since the length in yards is in the numerator and you are interested in feet, F_1 is the proper conversion factor. Multiplication of 100 yd per field by F_1 gives an answer in feet since the yards cancel.

$$100 \text{ yd} \times F_1 = 100 \text{ yd} \times \frac{3 \text{ ft}}{1 \text{ yd}} = 300 \text{ ft}$$

The answer is mathematically reasonable because you know that a distance expressed in feet is numerically larger than the same distance expressed in yards.

Example 2.1

The phosphorus content of a cow manure used as a fertilizer is 2 pounds per ton (lb/ton) of manure. Calculate the amount of phosphorus in a pound of manure.

Solution The given datum is 2 lb phosphorus per ton or in a ratio

$$\frac{2 \text{ lb phosphorus}}{1 \text{ ton manure}}$$

In order to calculate the desired quantity, you need to convert to pounds of manure. Writing down the given and desired quantities on the proper sides of an equal sign, we have

$$\frac{2 \text{ lb phosphorus}}{1 \text{ ton manure}} \times \text{factor} = \frac{? \text{ lb phosphorus}}{1 \text{ lb manure}}$$

The possible conversion factors to use involve the quantities 2000 lb and 1 ton

$$F_1 = \frac{1 \text{ ton}}{2000 \text{ lb}} \quad \text{or} \quad F_2 = \frac{2000 \text{ lb}}{1 \text{ ton}}$$

Only the first of these two factors will give the proper units when multiplied by the given quantity.

$$\frac{2 \text{ lb phosphorus}}{1 \text{ ton manure}} \times \frac{1 \text{ ton manure}}{2000 \text{ lb manure}} = \frac{0.001 \text{ lb phosphorus}}{1 \text{ lb manure}}$$

It may be necessary to use two or more conversion factors to solve a problem. Consider calculating the number of inches (in.) in the 2.5 mile race track at Indianapolis. In a mile there are 5280 ft, and in a foot there are 12 in. Thus we can convert the distance in miles into feet and then convert the resulting answer in feet into inches.

$$2.5 \text{ mile} \times \frac{5280 \text{ ft}}{1 \text{ mile}} = 13,200 \text{ ft}$$

$$13,200 \text{ ft} \times \frac{12 \text{ in.}}{1 \text{ ft}} = 158,400 \text{ in.}$$

The conversion could be done using one equation containing two factors.

$$2.5 \text{ mile} \times \frac{5280 \text{ ft}}{1 \text{ mile}} \times \frac{12 \text{ in.}}{1 \text{ ft}} = 158,400 \text{ in.}$$

Example 2.2

How many boxes of individual fruit cups are needed to give 12 children 2 fruit cups each? There are 4 fruit cups in a box.

Solution There are two types of conversion factors that are required to solve this problem. One type of conversion factor gives the relationship between the children and the number of fruit cups required.

$$F_1 = \frac{1 \text{ child}}{2 \text{ fruit cups}} \qquad F_2 = \frac{2 \text{ fruit cups}}{1 \text{ child}}$$

The other type of conversion factor gives the relationship between boxes and the number of fruit cups.

$$F_3 = \frac{1 \text{ box}}{4 \text{ fruit cups}} \qquad F_4 = \frac{4 \text{ fruit cups}}{1 \text{ box}}$$

The first part of the problem involves finding the desired number of fruit cups for the given number of children. This quantity can be determined by using F_2 as the proper conversion factor.

$$12 \text{ children} \times \frac{2 \text{ fruit cups}}{1 \text{ child}} = 24 \text{ fruit cups}$$

Next determine the number of boxes that will give you 24 fruit cups using F_3 as the conversion factor.

$$24 \text{ fruit cups} \times \frac{1 \text{ box}}{4 \text{ fruit cups}} = 6 \text{ boxes}$$

The problem could be set up combining both steps.

$$12 \text{ children} \times \frac{2 \text{ fruit cups}}{1 \text{ child}} \times \frac{1 \text{ box}}{4 \text{ fruit cups}} = 6 \text{ boxes}$$

Note that after cancelling units, the desired unit remains. If you had decided to multiply $12 \times 2 \times 4$ to obtain 96, you would not obtain the correct answer, as is clear from the incorrect units that result.

$$12 \text{ children} \times \frac{2 \text{ fruit cups}}{1 \text{ child}} \times \frac{4 \text{ fruit cups}}{1 \text{ box}} = 96 \frac{(\text{fruit cups})^2}{\text{box}}$$

[*Additional examples may be found in 2.5–2.10 at the end of the chapter.*]

2.4 Precision and Accuracy

No measurement in chemistry or any science that is obtained with a measuring device is exact. The uncertainty or degree of inexactness of a measurement depends on both the measuring device and the individual making the measurements. The scientist strives to reduce the uncertainty of data and obtain data that is accurate and precise. The terms precision and accuracy may mean the same thing to you, but their meanings are very different in the sciences. **Precision** means the degree of reproducibility of a measurement. **Accuracy** means the degree to which a measurement represents the true value of what is measured.

The precision of a measurement is determined by repeated individual measurements and depends on the measuring device being used as well as the skill of the individual making the measurements. For example, you may determine your temperature several times within a few minutes, and the repeated readings might be 98.6, 98.5, and 98.7°F. Your temperature is best represented by the average of these readings, 98.6°F. The temperature readings differ from this average value by about 0.1°F, which is the precision of the measurement. The smaller the difference between the individual measurements and the average value, the higher is the precision.

Accuracy depends on the quality of the measuring device. For example, if your thermometer is improperly produced by the manufacturer, you cannot accurately determine your temperature. If the thermometer always reads 1°F low and you measure your temperature as 98.6°F, then your temperature is actually 99.6°F. The measured value differs from the true value by 1.0°F and is inaccurate by that amount. Note that you cannot improve the accuracy of the measured value by repeated measurements. The repeated measurements can only increase the precision of the measurement.

An illustrative comparison of the terms precision and accuracy is shown in Figure 2.1, which shows targets used for a rifle shooting match. The goal is to come closest to the center of the target the most number of times. Accuracy is represented by the closeness of the average of the shots to the center of the target. Precision is reflected by the close grouping of shots. Thus both shooter 1 and 3 are precise in the use of their respective rifles, but only 3 is accurate too. In the case of shooter 1, the low accuracy is probably the result of an improperly adjusted rifle sight. In the case of shooter 2, the rifle is adjusted to give accurate results on the average, but the individual is not as skilled or precise as either shooter 1 or 3.

All measuring devices have limits to their accuracy. For example, some watches have second hands, whereas others only record minutes. You need a second hand to determine your pulse rate accurately. The minute hand, however, is accurate enough to help you to get to your chemistry class on time.

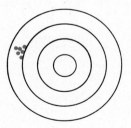 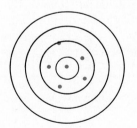 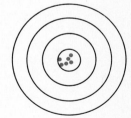

shooter 1
good precision
poor accuracy

shooter 2
poor precision
good accuracy

shooter 3
good precision
good accuracy

Figure 2.1

Accuracy and precision
Precision, as shown on the target, is measured by the close grouping of individual shots regardless of their location. Accuracy is reflected by the average position of the shots with respect to the center of the target.

Example 2.3

How could you determine that shooter 1 of Figure 2.1 is capable of accurately hitting the target?

Solution   Shooter 1 could be given the rifle of shooter 3. With a more accurate rifle, shooter 1 should retain his precision but should be more accurate as well. It is also possible to verify that the rifle originally used by shooter 1 is inaccurate by allowing shooter 3 to use that rifle.

Example 2.4

A crucible used in a chemistry laboratory is known to weigh 24.3162 grams (g). The students in a class determine the weight of the crucible by using simple balances, repeating the determination several times. Consider the following information obtained by two students using different balances. How would you describe their data?

	Trial 1	Trial 2	Trial 3	Trial 4	Trial 5
student 1	24.8 g	24.9 g	24.8 g	24.7 g	24.8 g
student 2	24.0 g	24.6 g	24.4 g	24.2 g	24.3 g

Solution   The average of the data of student 1 is 24.8 g. This value differs from the exact value by 0.5 g and is therefore not very accurate. However, the agreement among the individual trials is good and is indicative of good precision. The average of the data of student 2 is 24.3 g and is an accurate determination of the weight of the crucible. However, the range of values for individual trials is quite large, and therefore the data is not precise.

[*Additional examples may be found in 2.11–2.14 at the end of the chapter.*]

2.5   Significant Figures

If you count the number of students present in a class you may determine that there are exactly 26 students. There is no uncertainty in the number; it is an **exact number.** There also is no uncertainty in the number of feet in 1 yd; it is exactly 3 ft. *Exact numbers arise from either a direct count or a defined equivalence.*

Measured numbers are never exact. There is always an uncertainty that is either due to limitations in the measuring device or in how the measurement is made by a particular individual. The number of digits in a number that give reliable information is called the number of significant figures. The **significant figures** in a number include all the numbers known with certainty plus the first digit to the right that is estimated and has an uncertain value.

The degree of reliability of any number is given by the number of significant figures. Whenever you make a measurement, it is necessary to make sure that the number that you write contains the same number of significant figures as the reliability of the measurement. No measurement should ever be expressed or written as more reliable than the actual measurement.

In any measurement of a physical quantity with a continuous reading device such as a balance, there is some uncertainty due to estimation required by the observer. For example, the mass determined in Figure 2.2a can be estimated as 7.8 g. The quantity 7.8 g contains one certain figure and one estimated figure for a total of two significant figures. If the object is placed on a more accurate balance (Figure 2.2b), the mass is seen to be 7.8 g. However, the mass can now be estimated to the nearest hundredth of a gram as 7.82 g. In this case there are three significant figures, and the 2 is the uncertain figure.

Zeros are sometimes significant figures and at other times are merely used to place the number on the measurement scale being used. Thus zeros in a quantity may or may not be significant figures. Suppose two different samples are weighed on a balance that can give the mass accurately to 0.001 g. If the mass of one sample is found to be 5.000 g, the sample is between 4.999 and 5.001 g, and the number of significant figures is four. Thus, the three zeros following the decimal point are significant figures. If the other sample has a mass of 0.015 g, the sample actually has a mass between 0.014 and 0.016 g, and the number 0.015 contains only two significant figures. Neither the zero before the decimal point nor the zero in the tenths place is significant.

Numbers containing zeros, such as 10,000, are difficult to interpret; their significance may depend on the source of the quantity. The number might contain five significant figures and therefore represent a quantity between 9999 and 10,001. If the quantity refers to the number of people in a crowd, the actual number might be between 9000 and 11,000 based on a rough estimate to the nearest 1000 people. For this reason it is necessary to use scientific notation (see Section 2.6) involving powers of ten to give the correct number of significant figures. For the quantity 10,000 known to the nearest unit, the number would be expressed as 1.0000×10^4. To denote the number of people in a crowd known only to the nearest thousand, we would use 1.0×10^4.

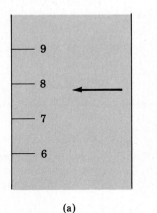
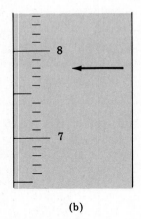

Figure 2.2   (a)   (b)

Significant figures in determining a mass
In case (a), the scale of the balance only gives the mass in grams to the nearest gram. The mass is somewhat less than 8 g and is clearly more than 7 g. You should report the mass as 7.8 g by estimating to the nearest 0.1 g. The value of the mass has two significant figures. In case (b), the balance gives greater accuracy. The mass is slightly more than 7.8 g and clearly less than 7.9 g. You should report the mass as 7.82 g, a quantity that has three significant figures.

Table 2.1   Rules for Significant Figures

1. All digits 1 through 9 inclusive are significant. Thus 14.9 contains three significant figures and 125.62 contains five significant figures.

2. Zero is significant if it appears between two nonzero digits. Thus 306, 30.6, 3.06, and 0.306 all contain three significant figures.

3. A terminal zero to the right of a decimal in a number greater than 1 is significant if it is expressing a reliable measurement. Thus 279.0, 27.90, and 2.790 all contain four significant figures.

4. A terminal zero to the right of a decimal point in a number less than 1 is significant if it expresses reliable information. Thus 0.2790 contains four significant figures if indeed the value has been shown to be zero in the ten-thousandth place.

5. A zero that is used only to fix the decimal in a number less than 1 is not significant. Thus 0.456, 0.0456, 0.00456, and 0.000456 all contain only three significant figures.

6. Terminal zeros in an integer are not significant. Thus 450 contains only two significant figures.

Unless the power of ten notation is used, we normally assume that zeros at the end of a number indicate only the relative size of the number and are not significant figures. Thus 10,000 is viewed as containing only one significant figure. The rules of significant figures are summarized in Table 2.1.

Example 2.5

What is the number of significant figures in each of the following numbers? (Give the rules from Table 2.1 used for your answer.)

a. 5041   b. 5401   c. 5410   d. 54100
e. 0.5401   f. 0.5410   g. 0.05401   h. 0.05410

Solution

a. 4 (1, 2)   b. 4 (1, 2)   c. 3 (1, 6)   d. 3 (1, 6)
e. 4 (1, 2)   f. 4 (1, 4)   g. 4 (1, 2, 5) h. 4 (1, 3)

[*Additional examples may be found in 2.15–2.18 at the end of the chapter.*]

2.6 Scientific Notation

Many numbers that are used in chemistry are either so large or so small that the number of zeros that must be written is inconvenient. For example, 18.02 g of water contains approximately 602200000000000000000000 molecules of water. Each water molecule weighs approximately 0.0000000000000000000002988 g.

We can write both large numbers and small numbers and indicate the number of significant figures in a numerical quantity by using scientific notation. In **scientific notation,** a number is expressed as a product of a coefficient multiplied by 10 raised to a power. The coefficient is a number equal to or greater than 1 but less than 10.

$$\text{coefficient} \rightarrow a \times 10^{b} \leftarrow \text{exponent}$$

To change a number greater than 1 into its equivalent in scientific notation, move the decimal point to the left until the number is greater than 1 but is less than 10. This quantity is the coefficient. The number of places that the decimal was moved is the *positive* power of ten.

2 4 5 2 becomes $2.452 \times 10^{3} \leftarrow$ number of places decimal
was moved to the left
coefficient

The number of molecules of water in an 18.02 g sample is 6.022×10^{23}. The 6.022 is the proper coefficient; it is larger than 1 but less than 10. Neither 0.6022 nor 60.22 is acceptable. The 10^{23} indicates the number of places required to move the decimal to the left to obtain the coefficient 6.022.

To change a number less than 1 into its equivalent in scientific notation, move the decimal point to the right until the number is greater than 1 but is less than 10. This quantity is the coefficient. The number of places that the decimal was moved is the *negative* power of ten.

0 . 0 0 0 2 4 becomes $2.4 \times 10^{-4} \leftarrow$ number of places decimal
was moved to the right
coefficient

The weight of the water molecule is 2.988×10^{-23} g. The 10^{-23} indicates the number of places required to move the decimal to the right to obtain the coefficient 2.988. Some additional examples of scientific notation are given in Table 2.2.

To convert a number in scientific notation with a positive power of ten into an ordinary number, move the decimal point to the right the number of places given by the exponent. To convert a number in scientific notation with a negative power of ten into an ordinary number, move the decimal point to the left the number of places given by the exponent.

Example 2.6

Do the numbers 2462.89 and 2.46289×10^{4} represent the same value?

Solution Either the first number can be converted into scientific notation or the second number may be converted into an ordinary number.

In order to convert 2462.89 into scientific notation you move the decimal point 3 places to the left to obtain a number that is greater than 1 but less than 10.

$$2462.89 = 2.46289 \times 10^{3}$$

This number does not have the same exponent and is not the same value as 2.46289×10^{4}.

In order to convert 2.46289×10^{4} into an ordinary number, the decimal point must be moved four places to the right.

$$2.46289 \times 10^{4} = 24628.9$$

Thus by either method you can show that 2462.89 and 2.46289×10^{4} do not represent the same value.

[*Additional examples may be found in 2.19–2.22 at the end of the chapter.*]

Table 2.2 **Scientific Notation**

Numbers larger than 1		Numbers smaller than 1	
Number	Scientific notation	Number	Scientific notation
1,111,100	1.1111×10^6	0.0000077	7.7×10^{-6}
222,200	2.222×10^5	0.0000666	6.66×10^{-5}
33,300	3.33×10^4	0.00050	5.0×10^{-4}
4,400	4.4×10^3	0.004444	4.444×10^{-3}
555	5.55×10^2	0.033333	3.3333×10^{-2}
66	6.6×10^1	0.22	2.2×10^{-1}
7.7	7.7×10^0		

2.7 Mathematical Operations and Significant Figures

Whenever two or more quantities representing measurements and expressed to their proper number of significant figures are added, subtracted, multiplied, or divided, the answer cannot be expressed to any more reliability than the least significant quantity used. The rules governing the significant figures derived from mathematical operations are given in the next three subsections.

Addition and Subtraction

When numbers are added or subtracted, the answer must not contain any significant figures beyond the place value common to all of the numbers. Thus, the sum of the numbers 12.2 and 13.31 is 25.5 and not 25.51 as would appear on the display of your calculator. Only the tenths place is common to both numbers. The hundredths place is not given in 12.2 and is not known. The same rule applies to subtraction. The difference of the numbers 13.31 and 12.2 is 1.1 and not 1.11. In each case the number of significant figures reflects the reliability of the resultant number based on the reliability of the quantities used in obtaining it.

Example 2.7

What is the sum of $25.1 + 15 + 14.15$? Express the answer to the proper number of significant figures.

Solution The place common to all numbers is the units place. A line can be placed at this point to avoid using the numbers to the right in the answer.

$$
\begin{array}{r}
2\,5\,|\,.\,1 \\
1\,5\, \\
1\,4\,|\,.\,1\,5 \\
\hline
5\,4\,|\,.\,2\,5
\end{array}
$$

The correct sum is 54 based on the quantity 15, which is known only to the units place.

Example 2.8

What is the difference of 16.29 and 3.168?

Solution The place value common to both numbers is the hundredths place. Draw a line at this point.

```
  1 6 . 2 9 |
−    3 . 1 6 | 8
  1 3 . 1 2 | 2
```

The correct answer expressed to the hundredth place is 13.12.

[Additional examples may be found in 2.23–2.26 at the end of the chapter.]

Multiplication and Division

In multiplication or division of numbers, the answer must not contain more significant figures than the least number of significant figures used in the operations. Thus, the product of 201×3 obtained by using a calculator is 603, but the answer can be expressed only to one significant figure since 3 has only one significant figure. The proper answer in scientific notation is 6×10^2. Similarly, the quotient 603/3 is not 201 but in scientific notation is 2×10^2.

Example 2.9

What is the product of 304×11?

Solution

$$304 \times 11 = 3344$$

However, the answer can be expressed to only two significant figures because there are two significant figures in 11. The correct answer is 3.3×10^3.

Example 2.10

What is the quotient 176.1/2.5?

Solution

$$\frac{176.1}{2.5} = 70.44$$

Division yields the number 70.44. However, since the divisor contains only two significant figures, the quotient must be expressed in scientific notation as 7.0×10^1.

[Additional examples may be found in 2.27–2.30 at the end of the chapter.]

Rounding Off

In the preceding examples of this section, the nonsignificant figures were discarded because the examples were chosen so that the first nonsignificant place was less than 5. The rules for rounding off nonsignificant figures are

1. If the first nonsignificant figure is less than 5, it and all other following nonsignificant figures are dropped.
2. If the first nonsignificant figure is more than 5 or is 5 followed by digits other than all zeros, all nonsignificant figures are dropped and the last significant figure is increased by one.
3. If the first nonsignificant figure is only 5 or is 5 followed by only zeros, all nonsignificant figures are dropped and the last significant figure is increased by one if it is odd but remains the same if it is even.

Hence if 34.51 must be expressed to three significant figures, the answer is 34.5 as the 1 in the hundredths place is less than 5. However, if the number 34.51 must be expressed to two significant figures, the answer must be rounded off to 35 as the nonsignificant figures are 5 followed by digits other than zero. If the number is expressed to one significant figure, that quantity is 3×10^1.

Example 2.11

What is the sum of $10.7 + 17.43 + 3.56$?

Solution

$$\begin{array}{r}
10.7 \\
17.4|3 \\
3.5|6 \\
\hline
31.6|9
\end{array}$$

Rounding off to the tenths place as required by the quantity 10.7 gives the correct answer, 31.7.

Example 2.12

What is the product of 5.01×21?

Solution

$$5.01 \times 21 = 105.21$$

Since 21 contains two significant figures, only two are allowed in the answer. The five cannot be dropped but must be used to round off to the correct answer, 1.1×10^2.

[*Additional examples may be found in 2.31–2.34 at the end of the chapter.*]

Most problems in this text require the use of a calculator. Calculators do not determine the number of significant figures; you will have to decide how to round off and discard many digits

Figure 2.3
Significant figures and
the calculator
If you divide 125 by
55.847, the calculator
displays 2.2382581.
The correct answer—
rounded off to three
significant figures— is
2.24.

to obtain the proper answer. For example, if you divide 125 by 55.847, the calculator displays 2.2382581 (Figure 2.3). However, there are only three significant figures in 125; therefore the answer must be 2.24. As a reminder that calculators often give many more digits than are required, the solution to the examples in **this text will give the result displayed on the calculator followed by the correct answer rounded to the proper number of significant figures.**

When doing calculations involving two or more multiplication or division steps, it is usually important to retain additional digits in intermediate answers beyond the correct number of significant figures. Rounding at each of several steps in a calculation can cause an error in the final result. Rounding should be done only on the final answer. With a calculator you can perform several multiplications or divisions in succession and round the final answer to the correct number of significant figures based on the quantities used in the calculation. For example, if you must obtain the product of $1.23 \times 4.003 \times 3.22$, the answer must have three significant figures. When you multiply 1.23 by 4.003, the calculator displays 4.92369. Then multiplying by 3.22 produces 15.8542818. At this point round off the answer to 15.9.

In order to give intermediate answers for each step in the solution to the examples in this text, one digit beyond the required number of significant figures is retained. The extra digit is underlined. Correct rounding of the final answer is based on the proper number of significant figures. In the multiplication of 1.23 by 4.003, the calculator display is recorded as 4.924. Multiplication of 4.924 by 3.22 gives 15.85528. Note that this quantity is different from the 15.8542818 obtained by successive multiplications on the calculator. However, in either number, rounding to three significant figures gives the same result, 15.9.

2.8 The Metric System

In the sciences, as well as in most nations of the world, the metric system is the standard of measurement. The metric system is simple and convenient to use since all units are based on multiples of 10. Conversions are easy because they involve only moving the decimal point as the units are changed. The metric system was established in France in 1790. The metric equivalents of some English units are given in Table 2.3. Most conversion factors involving metric and English units are not exact. Thus such factors must be considered in determining the number of significant figures in a calculation. For exactly 1 kg there are approximately 2.2 lb expressed to two significant figures. Any problem using such a factor can be calculated to no more than two significant figures.

Table 2.3   **A Comparison of Metric and English Measurements**

Dimension	Metric unit	English unit
length	1.00 meter	39.4 inches
	2.54 centimeters (exact)	1.00 inch
	1 kilometer	0.6 mile
volume	1.00 liter	1.06 quarts
	0.946 liter	1.00 quart
mass	454 grams	1.00 pound
	1.0 kilogram	2.2 pounds

Example 2.13

A linebacker weighs 231 lb. What is his mass in kilograms?

Solution A pound is a smaller quantity than a kilogram. Several pounds (lb) are needed to make a kilogram (kg). Therefore, the mass will be numerically less than 231. The proper conversion factor is 1 kg/2.2 lb.

$$231 \text{ lb} \times \frac{1 \text{ kg}}{2.2 \text{ lb}} = 105 \text{ kg}$$

In the conversion unit the kilogram is an exact quantity. However the equivalent in pounds is accurate to only two significant figures. Thus the answer should be expressed as 1.0×10^2 kg.

[*Additional examples may be found in 2.39–2.50 at the end of the chapter.*]

The standard units of the metric system are the second (s) for time, the meter (m) for length, the gram (g) for mass, and the liter (L) for volume. Fractions and multiples of the standard units of the metric system use prefixes to indicate the size of the unit relative to the standard unit. A list of prefixes appears in Table 2.4. You will use these units and their equivalents in applying the factor unit method of solving problems. The abbreviations for the prefixes kilo, deci, centi, milli, and micro should be memorized. All conversion factors using only metric units are exact. Thus metric unit conversion factors do not affect the number of significant figures in a conversion of units problem.

We are familiar with a kind of metric system in our currency. For example, we know that $1.50 is the same thing as 150 cents or that $0.25 is 25 cents. A cent is a hundredth of a dollar, just as a centimeter is a hundredth of a meter.

Table 2.4 **Prefixes Used in the Metric and SI Systems**

Multiplier	Prefix	Symbol
$1\ 000\ 000\ 000\ 000\ 000\ 000 = 10^{18}$	exa	E
$1\ 000\ 000\ 000\ 000\ 000 = 10^{15}$	peta	P
$1\ 000\ 000\ 000\ 000 = 10^{12}$	tera	T
$1\ 000\ 000\ 000 = 10^9$	giga	G
$1\ 000\ 000 = 10^6$	mega	M
$1\ 000 = 10^3$	kilo	k
$100 = 10^2$	hecto	h
$10 = 10^1$	deka	da
$1 = 10^0$		
$0.1 = 10^{-1}$	deci	d
$0.01 = 10^{-2}$	centi	c
$0.001 = 10^{-3}$	milli	m
$0.000\ 001 = 10^{-6}$	micro	μ
$0.000\ 000\ 001 = 10^{-9}$	nano	n
$0.000\ 000\ 000\ 001 = 10^{-12}$	pico	p
$0.000\ 000\ 000\ 000\ 001 = 10^{-15}$	femto	f
$0.000\ 000\ 000\ 000\ 000\ 000 = 10^{-18}$	atto	a

Example 2.14

How many cents are there in $2.27?

Solution Although you can automatically move the decimal point two places to the right to obtain the correct answer, 227 cents, the factor unit method may be used to prepare for problems that involve less familiar units.

$$\$2.27 \times \text{factor} = ? \text{ cents}$$

The possible factors are

$$F_1 = \frac{1 \text{ dollar}}{100 \text{ cents}} \quad \text{and} \quad F_2 = \frac{100 \text{ cents}}{1 \text{ dollar}}$$

The solution requires F_2 in order to cancel the units and convert dollars into cents.

$$\$2.27 \times \frac{100 \text{ cents}}{1 \text{ dollar}} = 227 \text{ cents}$$

Table 2.5 Radii of Objects in the Metric System

Object	Radius (m)
nucleus	1×10^{-15}
atom	1×10^{-10}
amoeba	1×10^{-4}
Earth	1×10^{7}
Sun	1×10^{9}
solar system	1×10^{13}
Milky Way	1×10^{21}

In 1960 the International Bureau of Weights and Measures proposed the International System of units, which is a revision and extension of the metric system. These units are called SI after the French words *Systeme Internationale*. For most subjects covered in this text, the metric system and SI systems are nearly identical and only the metric system will be used.

2.9 The Meter

The meter (m) is the basic unit of length in the metric system. The approximate dimensions of a number of objects are given in Table 2.5. All dimensions are given in meters, but any of these distances could be changed to other metric units of distance by changing the power of ten appropriately.

In chemistry, there is seldom a need to measure length directly in meters. A meter rule is divided into 100 equal parts called centimeters (cm). This quantity is more commonly used. Each centimeter consists of ten smaller units called millimeters (mm). There are 1000 mm in 1 m. A list of metric units for length is given in Table 2.6.

Table 2.6 Metric Units of Length

Unit	Symbol	Meter equivalent	Exponential meter equivalent
kilometer	km	1000 m	1×10^{3} m
meter	m	1 m	1×10^{0} m
decimeter	dm	0.1 m	1×10^{-1} m
centimeter	cm	0.01 m	1×10^{-2} m
millimeter	mm	0.001 m	1×10^{-3} m
micrometer	μm	0.000001 m	1×10^{-6} m
nanometer	nm	0.000000001 m	1×10^{-9} m
picometer	pm	0.000000000001 m	1×10^{-12} m

yardstick

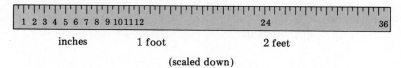

inches   1 foot   2 feet

(scaled down)

meterstick

centimeters

(scaled down)

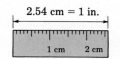   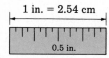

Figure 2.4

The meter and a comparison of length with the English system

An idea of the size of units of length can be shown in common examples. The kilometer is approximately 0.6 mile. A centimeter is approximately 5/16 in., or in another way, there are 2.54 cm in 1 in. (Figure 2.4). A dime is about 1 mm thick.

Example 2.15

Convert 6721 mm into meters.

Solution Two factors can be derived from the fact that 1 m = 1000 mm.

$$F_1 = \frac{1 \text{ m}}{1000 \text{ mm}} \quad \text{and} \quad F_2 = \frac{1000 \text{ mm}}{1 \text{ m}}$$

The solution to the conversion requires F_1 in order to cancel the units of millimeters and yield the meter unit.

$$6721 \text{ mm} \times \frac{1 \text{ m}}{1000 \text{ mm}} = 6.721 \text{ m}$$

Example 2.16

The radius of an atom of gold is 0.144 nanometer (nm). Express this radius in centimeters.

Solution From Table 2.2 you find that the prefix nano means the basic unit times 10^{-9}. Thus a nanometer is 10^{-9} m, and two conversion factors can be written.

$$F_1 = \frac{1 \text{ nm}}{1 \times 10^{-9} \text{ m}} \qquad F_2 = \frac{1 \times 10^{-9} \text{ m}}{1 \text{ nm}}$$

However, we need the answer in centimeters. We know that a centimeter is 1×10^{-2} m, and two conversion factors may be written.

$$F_3 = \frac{1 \text{ cm}}{1 \times 10^{-2} \text{ m}} \qquad F_4 = \frac{1 \times 10^{-2} \text{ m}}{1 \text{ cm}}$$

The two factors needed to convert nanometers into centimeters may be combined in order to cancel the units. The factor F_2 converts the given quantity into meters; the factor F_3 converts the resultant meters into centimeters.

$$0.144 \text{ nm} \times \frac{1 \times 10^{-9} \text{ m}}{1 \text{ nm}} \times \frac{1 \text{ cm}}{1 \times 10^{-2} \text{ m}} = 0.144 \times 10^{-7} \text{ cm}$$

The answer expressed in scientific notation is obtained by moving the decimal point one place to the right, which means that the exponent is changed to -8. The answer is 1.44×10^{-8} cm.

[*Additional examples may be found in 2.61 and 2.62 at the end of the chapter.*]

2.10 The Liter

Volume refers to the amount of space occupied by matter. Because the meter is the standard for length in the SI system, the cubic meter (m^3) is the standard for volume in this system. However a cubic meter is a large quantity, and the cubic centimeter (cm^3) is more often used in measuring volumes in chemistry. An older abbreviation for cubic centimeter, cc, survives in medical usage.

The liter has been chosen as a volume standard in the metric system. This unit is intermediate between cubic centimeters and cubic meters. The liter is the volume of a cube that is 10 cm or 1 dm on each side (Figure 2.5). A liter (L) then has a volume of 1000 cm^3 or 1 dm^3. Fractions of a liter can be expressed by use of the proper prefix. For example, 1/1000 L is 1 milliliter (mL). Note that since a liter is 1000 cm^3, the milliliter and the cubic centimeter are the same volume (Table 2.7). These units are used interchangeably.

Table 2.7   **Metric Units of Volume**

Unit	Symbol	Liter equivalent	Exponential liter equivalent
kiloliter	kL	1000 L	1×10^3 L
liter	L	1 L	1×10^0 L
deciliter	dL	0.1 L	1×10^{-1} L
centiliter	cL	0.01 L	1×10^{-2} L
milliliter	mL	0.001 L	1×10^{-3} L
cubic centimeter	cm^3	0.001 L	1×10^{-3} L
microliter	μL	0.000001 L	1×10^{-6} L

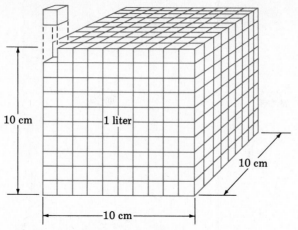

Figure 2.5

The liter and cubic centimeters
Each small cube is 1 cm on each side. The volume of
a small cube is 1 cm³. There are 1000 small cubes in
the large cube. The liter contains 1000 cm³. Each
cubic centimeter is equal to 1 milliliter (mL).

There are several devices used for measuring volumes of liquids in the chemistry laboratory (Figure 2.6). Both the beaker and the graduated cylinder are used to store and dispense approximate volumes. They are used to measure in much the same way as a measuring cup is used in cooking and are generally not very accurate.

A pipet or buret is used to make accurate measurements in chemistry. The pipet can measure a definite amount when filled to its calibration mark. The buret is used to measure out variable amounts of liquids. Both pipets and burets are available in many sizes.

The volumetric flask when filled to its calibration mark is used to contain a specified volume of liquid. It is often used to store accurately prepared samples of liquids.

A syringe is similar to a buret since it can dispense a variable volume of a liquid. Of course, the syringe can also be used to inject small volumes of a liquid.

Example 2.17

An adult of average activity inhales 1.0×10^4 L of air a day. What is the equivalent volume in milliliters?

Solution In order to convert liters into milliliters the conversion factor must have liters in the denominator and milliliters in the numerator.

$$1.0 \times 10^4 \, \cancel{L} \times \frac{1000 \text{ mL}}{1 \, \cancel{L}} = 1.0 \times 10^7 \text{ mL}$$

Note that the units of liters cancel and the units of milliliters result. In addition, the number of milliliters contained in a volume is far greater than the number of liters, as required by the relative sizes of the units.

[*Additional examples may be found in 2.57 and 2.58 at the end of the chapter.*]

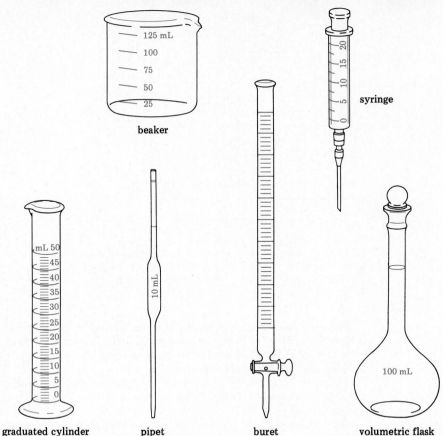

Figure 2.6   graduated cylinder   pipet   buret   volumetric flask

Laboratory equipment used to measure volume

2.11 The Gram

The kilogram (kg) is the standard unit of mass in the SI system. A list of masses in kilograms for a range of objects is given in Table 2.8. The unit of mass most commonly used in chemistry is the gram (g). Smaller quantities of matter used in chemistry are usually expressed in milligrams (mg). The milligram unit is used to designate the contents of drugs in capsules and pills. Micrograms, nanograms, and even picograms are used to measure some highly potent biological chemicals. For example, some hormone levels in the blood are detected in picogram quantities. A list of units of mass is given in Table 2.9.

Example 2.18

The average mass of a DNA molecule is 1.0×10^{-15} g. What is the mass in nanograms?

Solution   Nanograms represent a smaller unit of matter than grams. Thus the answer will be numerically larger (the exponent will be less negative).

Table 2.8 Masses of Objects in the Metric System

Object	Approximate mass (kg)
nucleus of atom	1×10^{-27}
amino acid molecule	1×10^{-25}
hemoglobin molecule	1×10^{-22}
amoeba	1×10^{-8}
raindrop	1×10^{-6}
ant	1×10^{-2}
human child	1×10^{1}
U. S. rocket	1×10^{6}
Earth	1×10^{24}
Sun	1×10^{30}

$$1.0 \times 10^{-15} \, g \times \frac{1 \times 10^{9} \, ng}{1 \, g} = 1.0 \times 10^{-6} \, ng$$

[*Additional examples may be found in 2.63 and 2.64 at the end of the chapter.*]

A kilogram is the mass of 1 L of water at 4°C. (The Celsius temperature scale is discussed in Chapter 3.) Since 1 L of water contains 1000 mL and has a mass of 1000 g, it follows that 1 mL of water has a mass of 1 g.

The pound is equivalent to 453.6 g. The metric equivalent of mass has been printed for years on cans of vegetables and fruits. A kilogram is equal to approximately 2.2 lb.

While you are accustomed to using scales to measure weights, scientists use balances to determine mass. A balance is used to measure the mass of an object by comparing it to the masses of standards. The platform balance (Figure 2.7) in common use in school laboratories compares the mass of an unknown object on the left side with known masses on the right side. When the two platforms are level, the masses must be identical. More sensitive instruments, called single pan analytical balances, measure mass electronically.

Table 2.9 Metric Units of Mass

Unit	Symbol	Gram equivalent	Exponential gram equivalent
kilogram	kg	1000 g	1×10^{3} g
gram	g	1 g	1×10^{0} g
decigram	dg	0.1 g	1×10^{-1} g
centigram	cg	0.01 g	1×10^{-2} g
milligram	mg	0.001 g	1×10^{-3} g
microgram	μg	0.000001 g	1×10^{-6} g
nanogram	ng	0.000000001 g	1×10^{-9} g
picogram	pg	0.000000000001 g	1×10^{-12} g

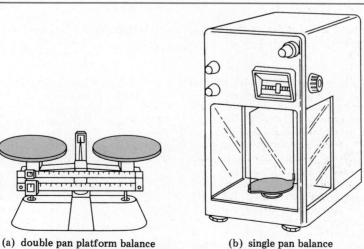

Figure 2.7   (a) double pan platform balance   (b) single pan balance

Laboratory equipment to determine the mass of matter

Summary

The emphasis of this chapter is on numbers, measurements, and units of measure. Every measurement involves an uncertainty in both precision and accuracy. Repetition of measurements improves the precision of the measurement. Accuracy depends on the instrument with which the measurement is made. It is necessary to express a measured quantity so that the uncertainty is clear.

The certainty of a measurement is indicated by using the proper number of significant figures. All arithmetic operations on measured quantities must be carried out with and rounded off to the proper number of significant figures. Scientific notation is used to express large or small numbers.

The metric system is used for measurements in chemistry. Measurements in one type of metric unit can be converted into another metric unit by using conversion factors that are powers of ten.

New Terms

Accuracy represents the degree to which a measurement represents the true value of what is measured.

A **conversion factor** is a multiplier having two or more units that is used to convert a quantity in one unit into its equivalent in another unit.

Exact numbers express a direct count or a defined equivalence.

Measurements are comparisons to a standard measuring device.

Precision means the degree of reproducibility of a measurement.

In **scientific notation** a number is expressed as a product of a coefficient multiplied by 10 raised to a power.

Exercises

Terminology

2.1 Clearly distinguish between precision and accuracy.

2.3 What is a conversion factor?

2.2 What is meant by the significant figures in a number?

2.4 What is meant by scientific notation?

Miscellaneous Conversions

2.5 A Tennessee walking horse is 15 hands high. A hand is 4 in. How high is a Tennessee walking horse in feet?

2.7 A cuffisco, used in Sicily to measure oil, is 5.6 gal. In Tangiers kulas, which equal 4.0 gal, are used for oil. How many kulas are present in 3.0 cuffiscos?

2.9 A formerly used system of measuring volumes had the following equivalents: 1 tun = 4 hogsheads, 1 hogshead = 0.500 butt, 1 butt = 126 gal. How many tun are in 63 gal?

2.6 A heavyweight boxer from Great Britain weighs 13 stones. A stone is 14 lb. What is the weight of the boxer in pounds?

2.8 The barrel used for petroleum is equal to 42 U.S. gal. How many barrels are needed to provide 21,000,000 gal of petroleum?

2.10 A formerly used measure of distance had the following equivalents: 2 fardells = 1 nooke, 4 nookes = 1 yard, 4 yards = 1 hide. How many fardells are in 0.5 hide?

Precision and Accuracy

2.11 You must hire an individual to do testing of water samples for purity. A woman tells you that she is capable of doing precise work. A man tells you that he is capable of doing accurate work. Assuming that both individuals understand the meaning of the terms, which person would you hire?

2.13 Consider the data obtained for the length of an object as measured by three students. You know that the length is 14.54 cm. How do you evaluate the work of the three students with regard to precision and accuracy?

2.12 You use the same balance in laboratory each week and you receive low grades on your reports in spite of excellent agreement between multiple determinations of the same quantity. What should you do?

2.14 Consider the data obtained for the mass of an object as measured by three students. You know that the mass is 8.54 g. How do you evaluate the work of the three students with regard to precision and accuracy?

	Trial 1	Trial 2	Trial 3	Trial 4
student 1	14.4 cm	14.6 cm	14.5 cm	14.5 cm
student 2	14.2 cm	14.1 cm	14.1 cm	14.2 cm
student 3	14.1 cm	14.8 cm	14.9 cm	14.2 cm

	Trial 1	Trial 2	Trial 3	Trial 4
student 1	8.4 g	8.6 g	8.5 g	8.5 g
student 2	8.1 g	8.8 g	8.9 g	8.2 g
student 3	8.2 g	8.1 g	8.1 g	8.2 g

Significant Figures

2.15 How many significant figures does each of the following numbers have?
- (a) 147.89
- (b) 0.0375
- (c) 2146.8
- (d) 0.000408
- (e) 21.6489
- (f) 0.0000039
- (g) 1.230
- (h) 24.0500

2.16 How many significant figures does each of the following numbers have?
- (a) 147.8
- (b) 0.00375
- (c) 214.08
- (d) 0.00509
- (e) 21.6089
- (f) 0.0000209
- (g) 1.200
- (h) 24.0501

2.17 How many significant figures does each of the following numbers have?
- (a) 5.02×10^4
- (b) 1.256×10^3
- (c) 3.15×10^{-4}
- (d) 1.2×10^{-6}
- (e) 2.100×10^{-5}
- (f) 2.01×10^{-30}

2.18 How many significant figures does each of the following numbers have?
- (a) 2.002×10^6
- (b) 2.3360×10^2
- (c) 4.105×10^{-7}
- (d) 2.0×10^{-8}
- (e) 5.90×10^{-5}
- (f) 3.05×10^{-23}

Scientific Notation

2.19 Convert the following into scientific notation.
- (a) 244.89
- (b) 0.0476
- (c) 4143.8
- (d) 0.000455
- (e) 41.6469
- (f) 0.000056
- (g) 33.40
- (h) 33.0540

2.20 Convert the following into scientific notation.
- (a) 57.8
- (b) 0.000421
- (c) 214.08
- (d) 0.00604
- (e) 71.3022
- (f) 0.0000407
- (g) 48.00
- (h) 38.0801

2.21 Write the numerical equivalent of each of the following numbers expressed in scientific notation.
- (a) 5.02×10^4
- (b) 1.256×10^3
- (c) 3.15×10^{-4}
- (d) 1.2×10^{-6}
- (e) 2.100×10^{-5}
- (f) 2.01×10^{-3}

2.22 Write the numerical equivalent of each of the following numbers expressed in scientific notation.
- (a) 4.002×10^6
- (b) 6.3480×10^3
- (c) 8.703×10^{-3}
- (d) 5.0×10^{-4}
- (e) 7.800×10^{-5}
- (f) 2.45×10^{-2}

Mathematical Operations

2.23 Perform each of the following calculations and express the answer to the proper number of significant figures.
- (a) $340 + 3.4 + 0.4589$
- (b) $124.2 + 4.2 + 32.22$
- (c) $3.99 + 121.455 + 0.05$
- (d) $0.123 + 3.43 + 101.2$
- (e) $0.0011 + 2.45 + 22.0$
- (f) $199.88 + 0.198 + 3465$

2.24 Perform each of the following calculations and express the answer to the proper number of significant figures.
- (a) $0.0035 + 4.49 + 44.2$
- (b) $145.88 + 0.095 + 4711$
- (c) $0.1556 + 0.0003 + 0.118$
- (d) $0.2524 + 0.554 + 0.04$
- (e) $457 + 6.2 + 0.4165$
- (f) $256.3 + 7.3 + 12.38$

2.25 Perform each of the following calculations and express the answer to the proper number of significant figures.
- (a) $5.0035 - 4.49$
- (b) $445.56 - 0.195$
- (c) $0.1246 - 0.0003$
- (d) $0.3544 - 0.04$
- (e) $457 - 6.2$
- (f) $156.3 - 7.34$

2.26 Perform each of the following calculations and express the answer to the proper number of significant figures.
- (a) $6.00 - 4.4933$
- (b) $745.562 - 0.21$
- (c) $0.420 - 0.0002$
- (d) $0.351 - 0.04211$
- (e) $257.00 - 3.1$
- (f) $257.311 - 3.12$

2.27 Perform each of the following calculations and express the answer to the proper number of significant figures.
- (a) 6.0035×2.49
- (b) 145.22×0.195
- (c) 0.1246×0.02
- (d) 9.3522×0.050
- (e) 4.257×0.2
- (f) 15.23×1.3

2.28 Perform each of the following calculations and express the answer to the proper number of significant figures.
- (a) 6.0×4.4933
- (b) 745.562×0.212
- (c) 0.42×0.0102
- (d) 0.251×0.4211
- (e) 347.00×2.1
- (f) 211.311×2.12

2.29 Perform each of the following calculations and express the answer to the proper number of significant figures.
- (a) $7.0035/1.49$
- (b) $145.22/0.292$
- (c) $0.1122/0.02$
- (d) $3.3522/0.020$
- (e) $2.257/0.2$
- (f) $22.23/1.4$

2.30 Perform each of the following calculations and express the answer to the proper number of significant figures.
- (a) $6.0/4.4933$
- (b) $225.562/0.518$
- (c) $0.31/0.0102$
- (d) $0.151/0.4211$
- (e) $245.11/4.1$
- (f) $332.311/5.12$

Rounding Off Numbers

2.31 Round off the following numbers to the number of significant figures indicated in parentheses.
  (a) 147.89 (4)   (b) 0.0375 (2)
  (c) 2146.8 (4)   (d) 0.000408 (2)
  (e) 21.6489 (4)   (f) 0.0000039 (1)
  (g) 1.235 (3)   (h) 24.0500 (3)
  (i) 210.06 (3)

2.32 Round off the following numbers to the number of significant figures indicated in parentheses.
  (a) 247.855 (3)   (b) 0.00375 (1)
  (c) 214.08 (4)   (d) 0.00909 (2)
  (e) 44.6089 (3)   (f) 0.0000209 (1)
  (g) 1.254 (3)   (h) 33.0501 (4)
  (i) 215.52 (3)

2.33 Round off the following numbers to the number of significant figures indicated in parentheses.
  (a) 5.55×10^4 (2)   (b) 1.256×10^3 (3)
  (c) 3.15×10^{-4} (2)   (d) 1.6×10^{-6} (1)
  (e) 5.500×10^{-5} (3) (f) 7.26×10^{-30} (2)

2.34 Round off the following numbers to the number of significant figures indicated in parentheses.
  (a) 4.005×10^6 (3)   (b) 5.3470×10^2 (3)
  (c) 4.105×10^{-7} (2)   (d) 2.004×10^{-8} (1)
  (e) 5.902×10^{-5} (3) (f) 3.054×10^{-23} (2)

Metric Abbreviations

2.35 What are the metric abbreviations for each of following terms?
  (a) milliliters   (b) picograms   (c) nanometers
  (d) decimeters   (e) centiliters   (f) kilograms
  (g) micrograms   (h) micrometers (i) picoliters

2.36 What are the metric abbreviations for each of following terms?
  (a) milligrams   (b) picometers   (c) nanograms
  (d) deciliters   (e) centigrams   (f) kilometers
  (g) microliters   (h) milligrams   (i) centimeters

2.37 Write the terms represented by each of the following metric abbreviations.
  (a) mg   (b) pm   (c) cL
  (d) dm   (e) μg   (f) nL
  (g) kg   (h) mL   (i) km

2.38 Write the terms represented by each of the following metric abbreviations.
  (a) dg   (b) μm   (c) pL
  (d) km   (e) cL   (f) ng
  (g) cg   (h) nm   (i) kL

English–Metric Conversions

2.39 Make the following conversions using conversion factors.
  (a) 41 cm into inches
  (b) 14.5 in. into centimeters
  (c) 245 g into pounds
  (d) 212 lb into kilograms
  (e) 452 cm^3 into quarts
  (f) 2.1 qt into liters
  (g) 12.1 yd into meters
  (h) 5.5 m into feet

2.40 Make the following conversions using conversion factors.
  (a) 21 in. into centimeters
  (b) 14.5 in. into millimeters
  (c) 2.8 lb into grams
  (d) 1.89 kg into pounds
  (e) 232 mL into quarts
  (f) 4.6 L into quarts
  (g) 11.2 m into yards
  (h) 4.2 ft into meters

2.41 A speed sign on the approach to a Mexican town reads 40 km/hr. Should you drive your American-made car at 25 or 65 mph?

2.42 A car is traveling at the speed limit of 55 mph. What is the speed in kilometers per hour?

2.43 If 20 kg is the baggage allowance on a small plane, how many pounds are you allowed to carry?

2.44 An individual has a mass of 75 kg. What is the equivalent weight in pounds?

2.45 The stapes, a bone in the middle ear, is 0.10 in. long. What is this equivalent in millimeters?

2.46 The femur, a bone in the leg, is about 27.5 in. long. What is this length in centimeters.

2.47 A person on a diet succeeds in losing 2.0 lb/week. If the diet continues for 12 weeks, how many kilograms will be lost?

2.48 The amount of phosphorus in a 60 kg person is approximately 0.60 kg. What is the mass of phosphorus in pounds?

2.49 A medication label reads "Administer 2 mg/kg of body weight." How much medication should be given to a 110 lb patient?

2.50 A premature infant has a mass of 2 kg. What is the child's weight in pounds?

Metric Conversions

2.51 Indicate which of the following is the larger quantity.
- (a) 1 mL or 1 dL
- (b) 1 pm or 1 nm
- (c) 1 g or 1 kg
- (d) 1 μm or 1 pm
- (e) 1 cg or 1 mg
- (f) 1 dL or 1 cL
- (g) 1 ng or 1 cg
- (h) 1 μL or 1 mL
- (i) 1 cm or 1 km

2.52 Indicate which of the following is the larger quantity.
- (a) 1 ng or 1 μg
- (b) 1 cL or 1 mL
- (c) 1 cm or 1 dm
- (d) 1 kL or 1 pL
- (e) 1 dm or 1 cm
- (f) 1 g or 1 dg
- (g) 1 μm or 1 cm
- (h) 1 dg or 1 ng
- (i) 1 cL or 1 kL

2.53 Make the following conversion using conversion factors.
- (a) 59 mm into centimeters
- (b) 153 cm into kilometers
- (c) 348 mL into liters
- (d) 5.328 L into milliliters
- (e) 248 mg into grams
- (f) 0.056 kg into grams

2.54 Make the following conversion using conversion factors.
- (a) 566 cm into decimeters
- (b) 353 cm into meters
- (c) 545 mL into deciliters
- (d) 2.125 L into centiliters
- (e) 248 cg into grams
- (f) 0.156 g into milligrams

2.55 Make the following conversions using conversion factors.
- (a) 0.011 mL into microliters
- (b) 243 ng into micrograms
- (c) 356.2 pm into nanometers
- (d) 456 μg into decigrams
- (e) 0.0012 μg into picograms
- (f) 0.0034 μm into nanometers

2.56 Make the following conversions using conversion factors.
- (a) 542 μL into milliliters
- (b) 123 μg into nanograms
- (c) 12 nm into picometers
- (d) 149 dg into micrograms
- (e) 14.4 pg into micrograms
- (f) 3.2 nm into micrometers

2.57 Your stomach releases 2.5 L of gastric juice each day. What is the equivalent volume in milliliters?

2.58 A drop of blood has a volume of 0.05 mL. How many drops of blood are there in an average adult body which has 5 L of blood?

2.59 The level of vitamin C in blood is about 0.2 mg/100 mL of serum. What is the amount of vitamin C in grams per liter?

2.60 An analysis of blood indicates that there are 9.5 mg of calcium in 100 mL of blood. How many grams of calcium are there in 6.0 L of blood?

2.61 A frog's egg is about 1.5×10^3 μm in diameter. What is the equivalent diameter in nanometers?

2.62 A light microscope has a resolving power of about 0.2 μm. What is the equivalent resolving power in nanometers?

2.63 An analysis of blood indicates that there is 65 ng of testosterone in 100 mL of blood. How many grams of testosterone are there in 6.0 L of blood?

2.64 A cigarette contains 2.5×10^{-5} g of nickel. How many micrograms of nickel are in the cigarette?

Additional Exercises

2.65 A fathom, which is used to measure the depth of water, is 6.0 ft. What is the distance of 1 fathom in meters?

2.66 A furlong, which is used to measure the length of a race track, is $\frac{1}{8}$ mile. What is the distance of 1 furlong in meters?

2.67 How many significant figures does each of the following numbers have?
- (a) 210.0
- (b) 0.001
- (c) 0.0040
- (d) 589.300
- (e) 210.02
- (f) 0.0022
- (g) 0.00400
- (h) 478.400

2.68 How many significant figures does each of the following numbers have?
- (a) 4.50×10^{10}
- (b) 4×10^{40}
- (c) 4.515×10^{12}
- (d) 4.25×10^{26}

2.69 Convert the following into scientific notation.
- (a) 820.0
- (b) 0.003
- (c) 0.0060
- (d) 154.200
- (e) 510.22
- (f) 0.0067
- (g) 0.00500
- (h) 773.700

2.70 Write the numerical equivalent of each of the following numbers expressed in scientific notation.
- (a) 4.50×10^2
- (b) 4.0×10^4
- (c) 1.515×10^2
- (d) 3.66×10^4

2.71 Round off the following numbers to the number of significant figures indicated in parenthesis.
- (a) 0.00150 (1)
- (b) 0.0415 (2)
- (c) 589.355 (4)
- (d) 0.00325 (2)
- (e) 0.00315 (2)
- (f) 478.457 (4)

2.72 Round off the following numbers to the number of significant figures indicated in parenthesis.
- (a) 2.60×10^{10} (1)
- (b) 4.625×10^{40} (3)
- (c) 4.5150×10^{12} (3)
- (d) 4.2500×10^{26} (3)

2.73 Perform each of the following calculations and express the answer to the proper number of significant figures.
(a) $0.2345 + 0.0001 + 0.452$
(b) $0.4565 + 0.678 + 0.02$
(c) $4.37 + 129.567 + 0.04$
(d) $0.387 + 4.12 + 121.5$
(e) $4.374 - 0.04$ (f) $0.387 - 0.0012$
(g) $6.37 - 0.0422$ (h) $0.5823 - 0.022$

2.74 Perform each of the following calculations and express the answer to the proper number of significant figures.
(a) 6.274×0.042 (b) 0.387×0.001
(c) 6.37×0.0420 (d) 0.5823×0.122
(e) $1.174/0.082$ (f) $8.387/0.011$
(g) $6.37/0.04200$ (h) $1.5823/0.122$

2.75 Calculate the following to the correct number of significant figures.

$$(0.00015 \times 54.6) + 1.002 = \,?$$

2.76 Calculate the following to the correct number of significant figures.

$$\frac{42.7 + 0.259}{28.4445} = \,?$$

2.77 Calculate the following to the correct number of significant figures.

$$(3.53 \div 0.084) - (14.8 \times 0.046) = \,?$$

2.78 Calculate the following to the correct number of significant figures.

$$(5.15 + 82.3) \times (0.024 + 3.000) = \,?$$

2.79 A sprinter can run 100 yd in 9.1 s. How long should it take the sprinter to run 100 m?

2.80 The heart of an infant may weigh 1 oz. What is this mass in grams?

2.81 Wine used to be sold in bottles that were $\frac{4}{5}$ qt but is now sold in 750 mL bottles. Which bottle contains the larger volume of wine?

2.82 The dust deposited from city air might be 2 tons per square mile per day. Express this dust level in milligrams per square meter per day.

2.83 Assume that your heart beats at a rate of 70 beats/min. If 60 mL of blood is pushed into the aorta by each beat, what volume of blood is circulated each day?

2.84 A 160 lb adult has 5 L of blood. Each milliliter of blood contains 5×10^6 red blood cells. How many red blood cells does the adult have?

2.85 The long jump record is 29 ft 2.5 in. What is this length in meters?

2.86 The diameter of a nickel is 0.8 in. What is the diameter in centimeters?

2.87 A tennis ball weighs 62 g. What is the weight in ounces?

2.88 A soccer ball weighs 15 oz. What is the weight in grams?

2.89 The ozone in a city one day is 100 μg/m^3. What is the amount in nanograms per liter?

2.90 A brand of cigarettes contains 15 mg of tar. If a person smokes 20 cigarettes each day, how many grams of tar are inhaled in 30 days?

2.91 The unaided human eye has a resolving power of about 0.1 mm. What is the equivalent resolving power in micrometers?

2.92 In some polluted city air the amount of lead is 3.1×10^{-6} g/m^3. What is the amount of lead in nanograms per cubic centimeter?

2.93 The blood glucose level of a fasting patient drops to 65 mg/dL. What is the number of grams of glucose in 1.0 L of blood?

2.94 The blood urea nitrogen (BUN) of an individual is 1.2 mg/dL. What is the number of grams of BUN per milliliter of blood?

2.95 The cholesterol content of the blood of a male patient is at the rather high value of 325 mg/dL. What is the number of grams of cholesterol in a liter of blood?

2.96 The triglyceride content of the blood of an individual is 121 mg/dL. What is the number of grams of triglycerides in 1.00 mL of blood?

2.97 Seawater is estimated to contain 4.0 pg of gold per 1.0 g of water. If the total mass of the oceans is 1.6×10^{12} Tg, how many grams of gold are present in the oceans of the world?

2.98 Seawater is estimated to contain 1.4 mg of magnesium per 1.0 g of water. The total mass of the oceans is 1.6×10^{12} Tg. How many kilograms of magnesium are present in the oceans of the world?

2.99 Seawater contains 19.4 g of chloride ion in 1.00 L. How many milligrams of chloride ions are in 1.00 mL of seawater?

2.100 Seawater contains 10.5 g of sodium ion in 1.00 kL. How many grams of sodium ions are in 1.00 mL of seawater?

2.101 The diameter of an iridium atom is 2.7 Å. Express this diameter in picometers. (1 Å $= 1 \times 10^{-8}$ cm)

2.102 The distance between two centers of the oxygen atoms in an oxygen molecule is 1.21 Å. What is the difference in nanometers? (1 Å $= 1 \times 10^{-8}$ cm)

3 Classification and Properties of Matter

Learning Objectives

After studying Chapter 3 you should be able to

1. Calculate the density and specific gravity of a sample of matter and use density or specific gravity to calculate the mass or volume of a sample of matter.

2. Convert Fahrenheit temperature to Celsius and vice versa.

3. Convert Celsius temperature to Kelvin and vice versa.

4. Perform calculations relating the heat capacity, change in temperature, and mass of a sample of matter to the amount of heat energy gained or lost.

5. Compare the properties of the three states of matter.

6. Distinguish between physical properties and chemical properties.

7. Distinguish between physical changes and chemical changes.

8. Recognize the experimental consequences of the law of conservation of mass.

9. Distinguish between pure substances and homogeneous and heterogeneous mixtures.

10. Differentiate between elements and compounds.

11. Identify common elements using names and symbols.

12. Recognize the experimental consequences of the law of definite proportions.

3.1 Classification in Science

Scientists group or classify items of interest according to their similarities in order to facilitate communication and organize knowledge. A classification may be simple and consist of large numbers of items in few classes or complex and consist of many classes. Classification in science allows us to make generalizations about the members of each class. The generalizations provide a focus for scientific hypotheses and theories that help us understand nature.

In chemistry the process of simplification by classification is important to understand matter, its composition, structure, and reactions. The classification used depends on the objective of the study. In this chapter matter will be classified in broad terms based on properties and composition. A more detailed classification of matter based on structure is developed in the next chapter. As further details of classification are presented throughout this text, you should be sure that you understand the classification scheme and the terminology used in the classification method.

3.2 Density of Matter

We sometimes talk about one substance being ''heavier'' than another. Metals such as lead are ''heavy'' and will not float in water, whereas a wood such as maple wood is ''light'' and will float in water. However, what we are comparing is not really the weight of the materials, but their densities. **Density** is the mass per unit volume of matter and is a characteristic property of matter.

Since density is a ratio of the mass to the volume, the property does not depend on the amount of material. It is for this reason that density is useful in describing and classifying matter. For example, a 1.0 mL sample of water has a mass of 1.0 g. Thus water has a density of 1.0 g/mL. A 10 mL sample of water has a mass of 10 g. The ratio of mass to volume is still 1 g/mL.

$$\text{density} = \frac{\text{mass}}{\text{volume}} = \frac{1.0 \text{ g}}{1.0 \text{ mL}} = \frac{10 \text{ g}}{10 \text{ mL}} = 1.0 \text{ g/mL}$$

The densities of liquids and solids are expressed in grams per milliliter (g/mL) or grams per cubic centimeter (g/cm^3) because the resultant values avoid small fractions and extremely large numbers. For the same reason the densities of gases are expressed in grams per liter (g/L). A list of the densities of some common substances at 25°C is given in Table 3.1.

Table 3.1   **Densities of Some Liquids, Solids, and Gases at 25°C**

Liquid	Density (g/mL)	Solid	Density (g/cm^3)	Gas	Density (g/L)
alcohol	0.80	iron	7.86	carbon dioxide	1.80
bromine	3.12	gold	19.3	carbon monoxide	1.14
ether	0.71	lead	11.3	hydrogen	0.08
olive oil	0.92	rock salt	2.2	helium	0.16
turpentine	0.87	sugar	1.59	methane	0.66
water	1.00	uranium	19.0	oxygen	1.31
mercury	13.53	wood, maple	0.49	nitrogen	1.14

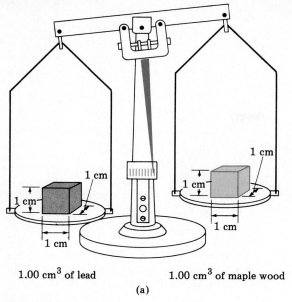
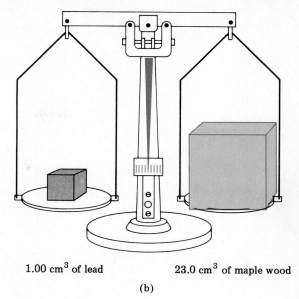

1.00 cm³ of lead	1.00 cm³ of maple wood	1.00 cm³ of lead	23.0 cm³ of maple wood
	(a)		(b)

Figure 3.1

Mass and density of matter
In experiment (a), the equal volumes of lead and maple wood do not have the same mass. The density of lead is greater than the density of wood. In experiment (b), the mass of lead is equal to the mass of wood. Since the density of maple wood is less than the density of lead, the sample of maple wood has a larger volume.

One of the ways of discussing and classifying matter is to use density. For example metals have high densities, whereas various woods have low densities. If the masses of 1.00 cm³ samples of lead and maple wood are determined, the mass of lead is 11.3 g, whereas that of the wood is 0.490 g. Thus the densities of lead and maple wood are 11.3 and 0.490 g/cm³, respectively. These values are in accord with our perception that lead is "heavier" than wood. However, this perception more exactly stated is that for comparable volumes there is a greater mass of lead than of wood (Figure 3.1). Another way of illustrating the greater density of lead compared to maple wood is to compare equal masses of both materials. Consider 11.3 g of lead and 11.3 g of maple wood. The 11.3 g of lead would occupy 1.00 cm³ but the 11.3 g of maple wood would occupy 23.0 cm³. For equal masses of several substances, the more dense material will occupy the smaller volume.

Example 3.1

A 1.50 cm³ sample of aspirin has a mass of 1.74 g. What is the density of aspirin?

Solution Density is defined as mass per unit volume, and for the units given, the density is grams per cubic centimeter (g/cm³). Therefore, the mathematical operation is

$$\frac{\text{mass}}{\text{volume}} = \frac{1.74 \text{ g}}{1.50 \text{ cm}^3} = 1.16 \text{ g/cm}^3 = \text{density}$$

[*Additional examples may be found in 3.3–3.8 at the end of the chapter.*]

Example 3.2

Alcohol has a density of 0.80 g/mL. What is the mass of 35 mL of alcohol?

Solution The mass of the sample may be obtained by multiplying the density by the volume. In this way the units of volume cancel and the unit of mass remains.

$$0.80 \; \frac{g}{mL} \times 35 \; mL = 28 \; g$$

[*Additional examples may be found in 3.9–3.12 at the end of the chapter.*]

Example 3.3

The density of sulfuric acid is 1.82 g/mL. What volume of sulfuric acid is required to provide 72.8 g of the material?

Solution The given unit for the mass of sulfuric acid is grams and the desired unit is the volume which can be calculated in milliliters. By multiplying by a conversion factor (mL/g) the unit of grams will cancel and the unit of milliliters will result. The necessary conversion factor is the reciprocal of the density.

$$72.8 \; g \times \frac{1 \; mL}{1.82 \; g} = 40.0 \; mL$$

[*Additional examples may be found in 3.13 and 3.14 at the end of the chapter.*]

In order to determine the density of a substance, you have to measure both the mass and volume. For a liquid, the mass could be measured by adding the liquid to a weighed vessel, such as a graduated cylinder, and determining the increase in weight. This procedure also allows you to measure the volume of the liquid contained in the graduated cylinder. In the case of a solid object, the mass can also be measured by using a balance. The volume of the solid object can be determined by placing it in a liquid. The object must be more dense than the liquid so that the object sinks. Under these conditions, the object displaces a volume of the liquid that is equal to its volume. The volume of the liquid displaced can be measured using a graduated cylinder (Figure 3.2). The difference between the initial volume of the liquid and the total volume after the solid is immersed is equal to the volume of the object.

Example 3.4

A 33.9 g sample of lead is placed into a graduated cylinder containing 20.0 mL of water and the lead sinks. The level of the water rises to 23.0 mL. What is the density of the lead?

Solution Since the lead sinks, the volume of the water displaced is equal to the volume of the lead sample.

volume of lead = 23.0 mL − 20.0 mL = 3.0 mL

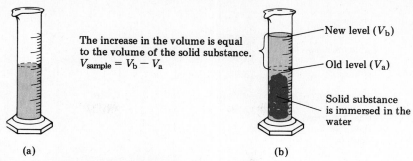

Figure 3.2   (a)   (b)

The increase in the volume is equal to the volume of the solid substance. $V_{sample} = V_b - V_a$

New level (V_b)

Old level (V_a)

Solid substance is immersed in the water

Determination of the density of a solid

The volume of an object can be determined by the difference in volume of (b) and (a), which is the result of the displacement of the liquid by the object. The mass of the object is then divided by this volume difference to determine the density.

$$\text{density} = \frac{33.9 \text{ g}}{3.0 \text{ mL}} = 11.3 \text{ g/mL}$$

However, the density must be reported as 11 g/mL because there are only two significant figures in the volume of lead.

Example 3.5

A graduated cylinder weighing 41.6546 g is used to determine the density of carbon tetrachloride, which is a liquid. After 25.0 mL of the carbon tetrachloride is added, the graduated cylinder and its contents weighs 81.5557 g. What is the density of the carbon tetrachloride?

Solution   The mass of the carbon tetrachloride is determined by difference.

mass of carbon tetrachloride = 81.5557 g − 41.6546 g = 39.9011 g

The density is given by the ratio of the mass to the volume.

$$\text{density} = \frac{39.9011 \text{ g}}{25.0 \text{ mL}} = 1.596 \text{ g/mL}$$

The density calculated to three significant figures based on the volume is 1.60 g/mL.

3.3   Specific Gravity

Water, which has a density of 1 g/mL, has been chosen as a reference material to compare the densities of other materials. **Specific gravity** is the ratio of the density of a substance to the density of water at the same temperature.

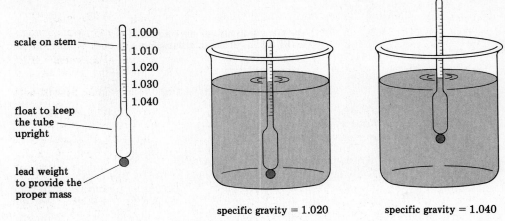

scale on stem

1.000
1.010
1.020
1.030
1.040

float to keep
the tube
upright

lead weight
to provide the
proper mass

Figure 3.3

specific gravity = 1.020   specific gravity = 1.040

Use of a hydrometer
The hydrometer can be used to determine the specific gravity of a liquid. The more dense
the liquid, the higher the hydrometer will float.

$$\text{specific gravity} = \frac{\text{density of substance (g/mL)}}{\text{density of water (g/mL)}}$$

Because the specific gravity is a ratio of two numbers having identical units, the specific gravity is without dimensions. Of course, since the density of water is 1 g/mL, the specific gravity of a substance is numerically equal to its density in grams per milliliter. Since the density of alcohol is 0.80 g/mL, its specific gravity is 0.80.

Specific gravity can be measured with a hydrometer. A hydrometer is illustrated in Figure 3.3. The hydrometer sinks until it displaces a mass of liquid equal to its own mass. Thus, the higher the density of the liquid, the higher the hydrometer floats.

A common test used in hospitals and clinics is the determination of the specific gravity of urine with a hydrometer. The normal range of specific gravities of urine is 1.009–1.030 with an average of about 1.018. If an abnormal amount of glucose is present in the urine, the specific gravity is higher and the hydrometer floats higher in the sample.

Example 3.6

The specific gravity of concentrated hydrochloric acid is listed as 1.19. A 25.0 mL sample of the acid is measured using a pipet. What is the mass in grams of the concentrated hydrochloric acid?

Solution The density of the acid is 1.19 times that of water or 1.19 g/mL.

$$\text{specific gravity} = 1.19 = \frac{\text{density of acid}}{1.00 \text{ g/mL}}$$

density of acid = 1.19×1.00 g/mL = 1.19 g/mL

Now we can determine the mass of the 25.0 mL sample using the definition of density.

$$25.0 \text{ mL} \times 1.19 \frac{g}{mL} = 29.75 \text{ g}$$

The correct answer expressed to three significant figures must be rounded to 29.8 g.

[Additional examples may be found in 3.17–3.20 at the end of the chapter.]

3.4   Temperature Units

Temperature is a measure of the degree of "hotness" of a material. The temperature does not depend on the quantity of the material. Many properties of matter such as density depend on temperature. For this reason, the temperature of a substance must be known before any of its properties can be discussed. Almost all substances expand when heated and contract when cooled. This property is used in the mercury thermometer to measure temperature. The mercury expands when heated and rises up the stem from the bulb at the base of the thermometer. Electronic thermometers with digital readouts operate on somewhat different principles. However, in all thermometers, the temperature recording results from a transfer of heat energy from the object to the thermometer or vice versa. Heat flow occurs from hot objects to cold objects.

Two commonly used temperature scales are the Celsius scale and the Fahrenheit scale; the former is used by scientists and by most people throughout the world, and the latter is used in the United States. On each of these scales the unit is a degree that is given by a ° symbol. The symbol is placed as a superscript to the right of a number followed by C or F to represent Celsius and Fahrenheit, respectively. Thus 32°F means 32 degrees Fahrenheit.

In addition to the Celsius scale, scientists use the Kelvin scale. The units of temperature in this scale are not called degrees but rather kelvins. Thus no degree sign is used for a value such as 273 K.

On the **Celsius scale,** the freezing point and boiling point of water are defined as 0°C and 100°C, respectively. There are 100 degrees between these two points. On the **Fahrenheit** scale, the freezing point of water is 32°F and the boiling point of water is 212°F. There are 180 degrees between the freezing point and boiling point of water. Thus, a 9° interval on the Fahrenheit scale corresponds to a 5° interval on the Celsius scale. The Celsius and Fahrenheit scales are compared in Figure 3.4.

$$\frac{180 \text{ Fahrenheit units}}{100 \text{ Celsius units}} = \frac{9 \text{ Fahrenheit units}}{5 \text{ Celsius units}}$$

The Kelvin scale, an important scale for certain scientific measurements, is identical to the Celsius scale in the size of the unit intervals, but differs in the values assigned to the reference points, the freezing and boiling points of water. On the **Kelvin** scale, which is also called the absolute scale, the zero point is 273.15 kelvins below the zero point on the Celsius scale. No temperature lower than this value is possible. The freezing point of water is 273.15 K, and the boiling point is 373.15 K. For all but the most exact work the 0.15 may be disregarded.

Mathematical expressions that can be used to change values from one scale to another are given below. Note the use of the reciprocal factors $\frac{5}{9}$ and $\frac{9}{5}$ in the equations relating °F and °C. These factors are based on the relative intervals of the Fahrenheit and Celsius scales.

$$°F = (\tfrac{9}{5} \times °C) + 32.0°F$$
$$°C = \tfrac{5}{9} \times (°F - 32.0°F)$$
$$°C = K - 273°C$$
$$K = °C + 273$$

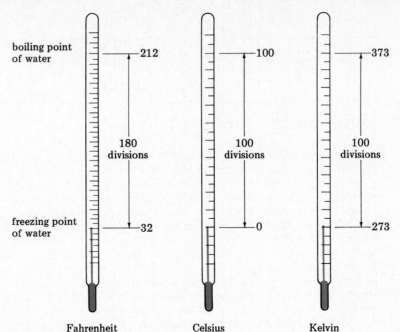

Figure 3.4

boiling point of water — 212 (Fahrenheit) / 100 (Celsius) / 373 (Kelvin)

180 divisions / 100 divisions / 100 divisions

freezing point of water — 32 (Fahrenheit) / 0 (Celsius) / 273 (Kelvin)

Fahrenheit   Celsius   Kelvin

A comparison of temperature scales
The Fahrenheit and Celsius scales have degree intervals of different sizes. The Celsius and Kelvin scales have the same size intervals but use different reference temperatures.

Example 3.7

The normal body temperature is 98.6°F. What temperature is considered normal by a European doctor who uses the Celsius scale?

Solution You may substitute directly into the second formula given.

$$°C = \tfrac{5}{9} \times (°F - 32.0°F)$$
$$= \tfrac{5}{9} \times (98.6°F - 32.0°F)$$
$$= \tfrac{5}{9} \times (66.6°F) = 37.0°C$$

[*Additional examples may be found in 3.21 and 3.22 at the end of the chapter.*]

Example 3.8

Many enzymes that are important to our well-being lose their biological ability to catalyze reactions if the temperature exceeds 45°C. What is this temperature in degrees Fahrenheit?

Solution You may substitute directly into the following formula and solve for the required degrees Fahrenheit.

$$\degree F = (\tfrac{9}{5} \times \degree C) + 32.0\degree F$$
$$= (\tfrac{9}{5} \times 45\degree C) + 32.0\degree F$$
$$= 81\degree F + 32.0\degree F = 113\degree F$$

[Additional examples may be found in 3.23–3.28 at the end of the chapter.]

Example 3.9

The boiling point of liquid nitrogen is 77 K. What is the boiling point on the Celsius scale?

Solution   The temperature is far below the freezing point of water, which is 273 K or 0°C. Thus the boiling point on the Celsius scale is negative.

$$\degree C = K - 273\degree C$$
$$\degree C = 77 - 273\degree C = -196\degree C$$

[Additional examples may be found in 3.29 and 3.30 at the end of the chapter.]

3.5   Heat Energy and Specific Heat

Heat energy is one of many forms of energy. Energy may be converted from one form to several other forms. For example, electrical energy can be converted into heat energy in a heater. In this process none of the energy is destroyed. The only change is in the form of the energy. The **law of conservation of energy** states that energy may be converted from one form to another but may be neither created nor destroyed.

In contrast to temperature, which does not depend on the quantity of material, the heat energy contained in a sample of matter depends on the mass of the material. There is twice as much heat energy stored in 50 g of water at 100°C as in 25 g of water at 100°C. It will take twice as much energy to heat 50 g of water from 25 to 100°C as will be required to heat 25 g of water from 25 to 100°C.

Units of Heat Energy

The most common unit of heat energy is the calorie (cal). A **calorie** is the amount of energy required to raise the temperature of 1 of water 1°C measured between 14.5 and 15.5°C. The kilocalorie (kcal), equivalent to 1000 cal, is the unit used in dietary tables, where it is referred to as a Calorie (Cal). Thus the dietetic Calorie is the amount of energy required to raise the temperature of 1 kg of water by 1°.

Each person has caloric requirements that depend on, among other things, body size, metabolic rate, and physical activity. A rough estimate of a minimum caloric intake to maintain a sedentary patient in a hospital bed is 1.0 Calorie per hour per kilogram of body weight ($1.0 \text{ Cal hr}^{-1} \text{ kg}^{-1}$).

The Joule is the unit of energy used in the SI system; 1 cal is equal to 4.184 Joules (J).

Example 3.10

How many Calories are required for a 165 lb patient in a hospital in one day?

Solution   First find the mass equivalent of 165 lb.

$$165 \text{ lb} \times \frac{1 \text{ kg}}{2.2 \text{ lb}} = 75 \text{ kg}$$

Next, the number of Calories required per hour is calculated.

$$75 \text{ kg} \times \frac{1.0 \text{ Cal}}{\text{kg hr}} = 75 \frac{\text{Cal}}{\text{hr}}$$

Finally, the number of Calories required per day is calculated.

$$75 \frac{\text{Cal}}{\text{hr}} \times 24 \frac{\text{hr}}{\text{day}} = 1800 \frac{\text{Cal}}{\text{day}}$$

Example 3.11

How many Joules are required to heat 15 g of water from 14.5 to 15.5°C?

Solution   From the definition of calories we can determine the heat energy required to heat 15 g of water.

$$15 \text{ g} \times 1.0°\text{C} \times 1 \frac{\text{cal}}{\text{g °C}} = 15 \text{ cal}$$

The number of Joules is larger by a factor of 4.184.

$$15 \text{ cal} \times 4.184 \frac{\text{J}}{\text{cal}} = 65.2702 \text{ Joules}$$

The answer is 65 J as required by the two significant figures in 15 cal.

[*Additional examples may be found in 3.31–3.36 at the end of the chapter.*]

Specific Heat

If a calorie of heat energy is added to 1 g of a substance, will the temperature increase by the same amount as in the case of water, that is, 1°C? The answer is no. One calorie of heat energy raises the temperature of 1 g of iron by 9°C, whereas the same amount of heat energy raises the temperature of 1 g of lead by 33°C. Therefore the amount of heat energy required to increase the temperature of a substance by a constant amount such as 1° depends on the substance.

The **specific heat** of a substance is the quantity of heat energy required to change the temperature of 1 g of the substance by 1°C. If you recall the definition of the calorie, you will see that the specific heat of water is exactly 1 cal g^{-1} $°\text{C}^{-1}$. The specific heats of several substances are listed in Table 3.2.

Table 3.2   **Specific Heats (cal g^{-1} °C^{-1}) of Substances at 25°C**

Solids	Specific heat	Liquids	Specific heat	Gases	Specific heat
aluminum	0.215	alcohol	0.587	ammonia	0.502
calcium	0.156	bromine	0.113	argon	0.124
copper	0.092	chloroform	0.231	chlorine	0.114
gold	0.031	ether	0.555	oxygen	0.219
iron	0.106	mercury	0.033	methane	0.523
silver	0.057	octane	0.532	nitrogen	0.249
zinc	0.093	water	1.000		

The specific heat of a material can be determined using the following relationship, which gives the heat lost or gained when a substance is cooled or heated.

calories = g of substance $\times \Delta t \times$ specific heat

The symbol Δt represents the change in temperature that results from adding or removing calories. For example, you can calculate the change in temperature if 12 cal were added to a 75 g sample of silver whose specific heat is 0.057 cal g^{-1} °C^{-1}.

$$12 \text{ cal} = 75 \text{ g} \times \Delta t \times 0.057 \frac{\text{cal}}{\text{g °C}}$$

$$\frac{12 \text{ cal}}{75 \text{ g} \times 0.057 \frac{\text{cal}}{\text{g °C}}} = \Delta t = 2.8°C$$

If the silver was originally at 25.0°C, the temperature after adding 12 calories would be 27.8°C.

A convenient way of experimentally determining specific heats depends on the fact that heat energy flows from a region of high temperature to one of low temperature. If two substances at different temperatures are placed in contact with each other, heat energy will flow until both substances are at the same temperature. The heat energy lost by one substance will be gained by the second substance. Let's consider substance A at some high temperature placed in contact with substance B at some lower temperature. After a period of time the two substances reach the same final temperature. If the specific heat of substance A is known, then the heat energy lost can be calculated based on the change in temperature and the mass of the substance.

(g of A) $\times$ (specific heat of A) $\times$ (Δt of A) = cal from A
    where Δt of A = (t of final A) $-$ (t of original A)

The heat energy gained by B must be equal to the heat energy lost by A. If the mass of the substance and the change in temperature for B is used, then the specific heat of B can be calculated.

(g of B) $\times$ (specific heat of B) $\times$ (Δt of B) = cal to B
    where Δt of B = (t of original B) $-$ (t of final B)

Example 3.12

A 176 g sample of a metal is heated to 100.0°C and placed in 226 g of water at 25°C. The temperature of the water rises to 26.7°C. What is the specific heat of the metal?

Solution First determine the heat energy gained by the water as it is heated from 25.0°C to 26.7°C.

$$\text{heat gained} = 226\,\cancel{g} \times 1.00\,\frac{\text{cal}}{\cancel{g}\,\text{°C}} \times (26.7 - 25.0)\text{°C}$$

$$\text{heat gained} = 38\underline{4}\ \text{cal}$$

The heat energy lost by the metal must also be $38\underline{4}$ cal. Now the specific heat of the metal can be calculated based on the decrease in temperature from 100°C to 26.7°C.

$$38\underline{4}\ \text{cal} = 176\ \text{g} \times \text{specific heat} \times (100.0 - 26.7)\text{°C}$$

$$\frac{38\underline{4}\ \text{cal}}{176\ \text{g} \times 73.3\text{°C}} = \text{specific heat} = 0.029\underline{8}\,\frac{\text{cal}}{\text{g °C}}$$

The temperature change of 1.7°C for the water limits the number of significant figures to be used in the correct answer, 0.030 cal g^{-1} °C^{-1}.

[*Additional examples may be found in 3.37–3.40 at the end of the chapter.*]

3.6 States of Matter

Matter can be described and classified according to its physical form or state. There are three states: gaseous, liquid, and solid. In our experience almost every material is thought of as existing in a single state. Air is a gas, motor oil is a liquid, and iron is a solid. However, most matter can exist in any of the three states under the proper conditions.

Water is one of the few substances that most people have observed in all three states (Figure 3.5). At normal atmospheric pressure, water exists as a solid below 0°C and as a gas above 100°C. Although we use the terms ice and steam to describe the solid and gaseous states of water, to the chemist the substance is still water.

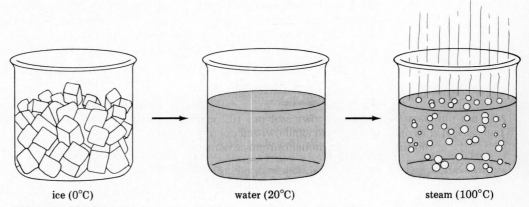

Figure 3.5   ice (0°C)   water (20°C)   steam (100°C)

The states of water

Water can exist in three states: solid, liquid, and gas. When a change of state occurs, the physical appearance changes, but there is no change in chemical composition.

A **gas** has no characteristic shape or volume and completely fills any container in which it is placed. At normal atmospheric pressure and temperature the densities of gases are in the range of 0.0001–0.01 g/mL. When gases are heated, they expand and their densities decrease compared to the surrounding cooler air. Hence hot air rises.

Gases are highly compressible. Under increased pressure, the volume of a gas decreases. This compressibility allows a large volume of oxygen to be stored in small cylinders such as those used in hospitals. The compressed air cylinders used by scuba divers also provide a large volume of air.

Liquids have a constant volume at a specified temperature and pressure, but do not have a characteristic shape. They fill a container from the bottom up. The densities of liquids are greater than those of gases and range from 0.5 to 1.5 g/mL. One notable exception is mercury, a liquid metal, with a density of 13.5 g/mL.

Liquids are only slightly compressible. If hydraulic fluids were compressible, the pressure exerted by your foot on a brake pedal would result only in compressing the brake fluid rather than causing the brakes to exert pressure and stop the car.

Solids are recognized by their definite volume and rigid shape. For all practical purposes, solids are incompressible. For example, the volume of a piece of coal beneath the enormous pressure of the earth is the same as it is on the surface of the earth.

The densities of solids are in the range of 0.5–20 g/cm^3. Metals such as lead, with a density of 11.3 g/cm^3, and gold, with a density of 19.3 g/cm^3, are among the densest common solids.

Most solids expand slightly when heated. For a temperature increase of 1°C, the expansion is usually less than 0.01%. Such a small expansion means that solids such as wood, metals, and plastics can be used for construction.

3.7 Physical and Chemical Properties

Every pure substance can be identified by a set of properties that is characteristic of that substance. Several substances may have some properties in common, but no two substances have a completely common set. These properties are conveniently divided into physical and chemical properties. Both physical and chemical properties are used to classify matter. For example lead, copper, and iron are classified as metals based on their physical and chemical properties.

Physical Properties

Physical properties are characteristics that are determined without altering the chemical composition of the substance. Some examples of physical properties include boiling point, melting point, density, electrical conductivity, odor, and color. Pure water is a colorless, odorless, and tasteless liquid that freezes at 0°C and boils at 100°C. Both the qualitative adjectives and the quantitative numbers describe the physical properties of water. Similarly, chlorine can be described as a yellow-green gas with a suffocating odor and a sharp, sour taste; it boils at −34.5°C and freezes at −101.6°C.

Physical properties are used by chemists to identify substances. For example, aspirin melts at 143°C, whereas sugar melts at 186°C. A chemist who is given a white material known to be either aspirin or sugar can easily determine which it is by measuring the temperature at which it melts.

Matter can change in appearance without changing its chemical composition. For example, the tungsten filament in a light bulb changes appearance when the electricity is turned on. The

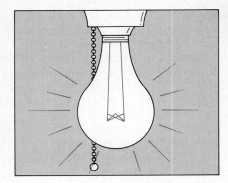

Figure 3.6
A physical change
When a current is passed through a light bulb, there is a change in its physical appearance.
However, no change in composition occurs.

glowing tungsten filament is still chemically the same, and when the light is turned off, the filament reverts to its original appearance (Figure 3.6). The composition of the filament is unchanged. Any conversion of matter that occurs without a change in the composition or structure of the substance is called a **physical change.**

The most common examples of physical changes that you have observed are changes of state. The melting of ice and boiling of water are both examples of physical changes. The melting points and boiling points of some substances are given in Table 3.3. At each of these temperatures, a change of state occurs, which is a physical change.

Although most substances change from solid to liquid and then to a gas as the temperature is increased, there are exceptions. The change of a substance from a solid to a gas without passing through the liquid state is called **sublimation.** Dry ice, which is solid carbon dioxide, sublimes into a gas at −78.5°C.

Table 3.3   **Physical Properties of Selected Substances**

Substance	Melting point (°C)	Boiling point (°C)
ammonia	−77.7	−33.4
bromine	−7.2	58.8
carbon monoxide	−199	−191.5
copper	1083.4	2567
chlorine	−101.0	−34.5
methane	−182.5	−164
salt	801	1413
chloroform	−63.5	61.7
octane	−56.8	125.7

Example 3.13

In what state will chloroform exist at 25°C? (See Table 3.3)

Solution Chloroform melts at −63.5°C and boils at 61.7°C. At 25°C chloroform must be a liquid because that temperature is above the freezing point but is below the boiling point.

[*Additional examples may be found in 3.43 and 3.44 at the end of the chapter.*]

Chemical Properties

Chemical properties describe the characteristics of a substance that govern the kind of changes in composition that the substance can undergo during chemical reactions. For example, water reacts violently with sodium metal to produce hydrogen gas and a caustic substance known as sodium hydroxide. Chlorine reacts with sodium to form sodium chloride, known as table salt. In both examples, the substances formed are of different composition than the initial materials used.

Let's consider a flashbulb (Figure 3.7) and compare it to the light bulb discussed in the previous section. When a small current is passed into a flashbulb, which consists of fine wires of magnesium metal in a gaseous atmosphere of oxygen, a flash results. Is this a physical change similar to that of turning on a light? The answer is no. The bulb will not flash again, and the contents of the bulb have been altered. A white powdery substance known as magnesium oxide has been formed. This substance has very different physical and chemical properties from the original materials. A **chemical change** or reaction is a process in which the composition or structure of one or more substances is altered.

As indicated in Chapter 1, chemists use chemical equations to record the chemical reactions of matter. In the case of the flashbulb, the word equation is

$$\text{magnesium} + \text{oxygen} \longrightarrow \text{magnesium oxide}$$

Recall that in an equation, the substances to the left of the arrow are reactants. The substances to the right of the arrow in the equation are the products. In the flashbulb experiment, both magnesium and oxygen are reactants, whereas magnesium oxide is a product.

Example 3.14

Methane gas burns in the oxygen of air and produces carbon dioxide and water. Is this process a physical or chemical change?

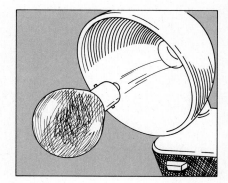

Figure 3.7
A chemical change
When a flashbulb is used, its chemical composition is changed as a result of a reaction between magnesium and oxygen to produce magnesium oxide.

Solution The substances undergo a change in composition. Therefore the process is a chemical change. You can represent this chemical change using an equation in which the methane and oxygen are reactants and the carbon dioxide and water are products.

methane + oxygen $\longrightarrow$ carbon dioxide + water

[*Additional examples may be found in 3.45–3.52 at the end of the chapter.*]

It has been shown experimentally many times that the sum of the masses of the products of a chemical reaction is equal to the sum of the masses of the starting materials. These observations have led to the **law of conservation of mass:** there is no experimentally detectable gain or loss of mass during an ordinary chemical reaction.

The flashbulb provides a good example of the law of conservation of mass since both the reactants and products are sealed in the bulb. If the mass of the bulb is determined both before and after the flash, it can be shown that the chemical reaction occurred without a change in mass (Figure 3.8).

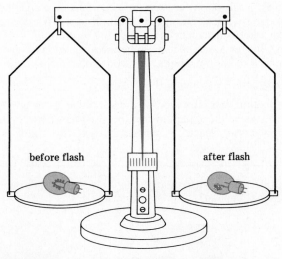

before flash   after flash

Figure 3.8   magnesium + oxygen $\longrightarrow$ magnesium oxide
The law of conservation of mass
The mass of a flashbulb is unchanged after a
chemical reaction has occurred. All of the
products are contained within the flashbulb.

Example 3.15

A 50.0 g sample of lead is heated with sulfur to form lead sulfide. The amount of lead sulfide formed is 57.7 g. How much sulfur reacted with the lead?

Solution The gain in mass must be due to the amount of sulfur that reacted with the lead to give lead sulfide.

$$\text{mass of lead sulfide} - \text{mass of lead} = \text{mass of sulfur}$$
$$57.7 \text{ g} - 50.0 \text{ g} = 7.7 \text{ g}$$

Example 3.16

A 5.00 g sample of a red-colored substance containing mercury is heated. The only visible product is 4.63 g of the grey liquid metal mercury. Explain what occurred.

Solution The decrease in mass of 0.37 g is the result of a chemical reaction. This mass must be accounted for as another product that cannot be seen, that is, a gas.

[Additional examples may be found in 3.53–3.56 at the end of the chapter.]

Chemical reactions release or absorb energy. **Chemical energy** is the energy stored in substances that can be released during a chemical reaction. As the subject is developed in this text, you will see that the chemical energy stored in various substances determines the reactions that they will undergo.

Green plants store chemical energy in complex chemicals produced from carbon dioxide and water. The energy stored is obtained from the radiant energy of light. Energy is released when the plants are metabolized by animals. No energy has been created or destroyed, and the total process is in accord with the law of conservation of energy. The energy has only changed form.

The energy required or produced by chemical reactions may be in the form of electrical, heat, or light energy. A reaction that occurs and releases heat energy is **exothermic,** whereas a reaction in which heat energy is required is **endothermic.**

3.8 Mixtures and Pure Substances

Mixtures

Before matter can be correctly classified it is necessary to recognize that most matter we encounter every day is a complex mixture. Air contains essentially four gases, whereas gasoline is a mixture of at least 20 liquids. Practically all food consists of an incredible number of components. The human body contains hundreds of thousands of substances.

A **mixture** can be separated into pure components by physical methods. Physical methods include any process by which mixtures are separated without changing the identity of the individual components. For example, a pepperoni and cheese pizza is a mixture, and its components can be separated manually.

A mixture of iron filings and sulfur can easily be seen to contain two components. Iron filings are heavy and hard chips of iron. Sulfur is a light fluffy yellow powder. The mixture can be physically separated using a magnet (Figure 3.9). The iron filings are attracted to the magnet and the yellow sulfur remains.

If you put salt and water in a cooking pot, a mixture results, although by looking at the material you could not tell that it is a mixture. The mixture can still be separated by physical methods; that is, you could boil off the water and the salt would remain in the pot. Each of the

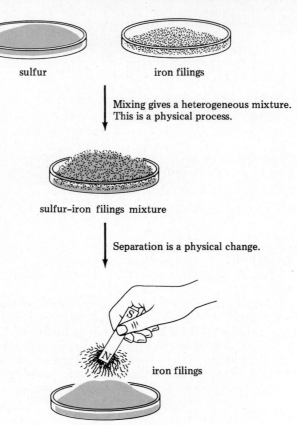

sulfur   iron filings

Mixing gives a heterogeneous mixture.
This is a physical process.

sulfur–iron filings mixture

Separation is a physical change.

iron filings

Figure 3.9   sulfur

Example of physical change

components of the mixture is a pure substance. Neither the salt nor the water can be further separated into simpler materials by physical methods.

Many procedures have been developed by chemists to separate mixtures into their component pure substances. These methods are based on differences in the physical properties of matter. For example, various liquids have different boiling points. The process of **distillation,** a method used to separate liquids, is based on the differences in boiling points of the liquids. When a mixture of liquids is heated, the substance with the lowest boiling point is converted into a gas and ''boils'' from the mixture. The substances with the higher boiling points remain in the liquid.

Mixtures can be classified into two categories, homogeneous and heterogeneous mixtures. A **homogeneous mixture** has the same physical and chemical properties throughout. The salt–water example is a homogeneous mixture. **Heterogeneous mixtures** have physical and chemical properties that are not uniform throughout the sample. A pepperoni pizza is obviously a heterogeneous mixture.

All mixtures can have variable compositions. For example, a salt–water mixture might contain a pinch of salt or a teaspoon of salt. A pepperoni pizza could have a few slices of pepperoni

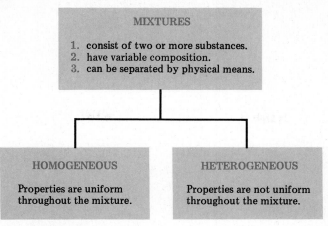

MIXTURES

1. consist of two or more substances.
2. have variable composition.
3. can be separated by physical means.

HOMOGENEOUS

Properties are uniform
throughout the mixture.

HETEROGENEOUS

Properties are not uniform
throughout the mixture.

Figure 3.10

The classification of mixtures

or many slices. A summary of the characteristics of homogeneous and heterogeneous mixtures is given in Figure 3.10.

Homogeneous mixtures are commonly called **solutions.** Solutions of all three states are possible. The air we breathe is a gaseous solution. Salt in water is a liquid solution. An alloy, such as brass, which contains zinc and copper, is a solid solution. We will discuss solutions further in Chapter 14.

Pure Substances

Pure substances have uniform properties, have a definite composition, and cannot be separated into components by physical methods. Using a variety of physical methods, chemists have succeeded in isolating more than 7 million pure substances.

There are relatively few materials in common use that are pure, that is, consist of a single substance. Copper used in electrical wiring is pure; however, steel is a mixture of iron and other substances. Sugar is pure, but table salt has additives such as potassium iodide to help prevent goiter and magnesium chloride to prevent caking. A medicine may consist of a single substance, although mixtures are common for many over-the-counter products.

Each pure substance has its own set of physical and chemical properties. For example, helium is a colorless and odorless gas of low density that does not burn in air and does not react with other elements. Ethyl alcohol is a liquid that mixes with water to form solutions and will burn in air. Table salt (sodium chloride) is a white solid that mixes with water to form solutions and will not burn in air.

There are two classes of pure substances—elements and compounds—with elements forming the smaller class. Of the 7 million pure substances there are only 108 elements. At one time an element was defined as a substance that cannot be made from or decomposed into simpler substances. This definition now has been modified because elements are composed of even simpler components called electrons, protons, and neutrons. These components are discussed in Chapter 4. **Elements** are now defined as substances that cannot be constructed from or decomposed into simpler substances by ordinary chemical processes.

A **compound** is composed of two or more elements in fixed proportion by mass. A compound can be converted into other compounds or elements only by chemical reactions. Why is a compound different from a homogeneous mixture? Why can't a compound be separated into

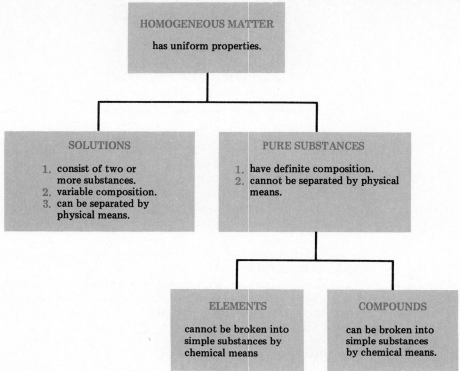

Figure 3.11

The classification of homogeneous matter

elements by physical means? Compounds are composed of elements joined together by forces called bonds. Bonds are discussed in Chapter 10. Bonds cannot be destroyed by physical changes. Only chemical reactions affect bonds. In a mixture, the components are simply mixed together and are not bonded to each other. A summary of the characteristics of elements and compounds is given in Figure 3.11.

3.9 Elements

Like many words in English, the chemical names of the elements find their roots in other languages. The element bismuth originally was named from the German words "*weisse masse*" or white mass. Over time, through contractions and changes in pronunciation, the name changed from wismat to bismat and finally to bismuth.

Some elements have been named after their place of discovery. Germanium, discovered in 1886, and francium, discovered in 1939, indicate nationalistic allegiances of the scientists who discovered them. The elements berkelium, californium, and americium are among the elements synthesized by nuclear reactions by scientists at the University of California at Berkeley.

Another practice in naming elements involves using the name of an individual for honorary

Table 3.4   Symbols of the Elements Derived from Earlier Names

Present name	Symbol	Former name
antimony	Sb	stibium
copper	Cu	cuprum
gold	Au	aurum
iron	Fe	ferrum
lead	Pb	plumbum
mercury	Hg	hydrargyrum
potassium	K	kalium
silver	Ag	argentum
sodium	Na	natrium
tin	Sn	stannum
tungsten	W	wolfram

purposes. A number of elements produced by nuclear reactions are named for famous scientists. These include curium, einsteinium, fermium, lawrencium, and mendelevium.

Symbols have long been used to represent the elements. In the alchemical era the symbols were figures chosen to create a barrier between the alchemist and others and to maintain the secrecy of the early science. Currently used symbols are more logical and are chosen to facilitate communication. A **chemical symbol** is a chemical shorthand or abbreviation of the name of the element. Chemical symbols for most elements are one or two letters. (Some recently discovered elements obtained by nuclear reactions have been assigned symbols containing three letters.) When more than one letter is used, only the first is capitalized.

The symbols of 11 elements are derived from older names and bear no resemblance to the currently used name. These elements and their symbols are listed in Table 3.4. The first letter of the name is used for only 12 of the elements. The majority of elements have symbols consisting of two letters. The symbols of many elements consist of the first two letters of the name or the first and third letter of the name. However, there are also symbols in which some other letter contained in the name is used. The necessity for using other than the first or first two letters in the name of the element is the result of the similarities in the names of the elements. Consider the example of carbon, calcium, cadmium, and californium, whose symbols are C, Ca, Cd, and Cf, respectively. A variety of choices of letters is necessary to provide symbols for these four elements whose first two letters are identical. The symbols of some elements are given in Table 3.5. You should start to learn the symbols for those elements described in this text as they appear. All the known elements and their symbols are listed inside the back cover of this text.

Table 3.5   Symbols of Some Elements

First letter		First and second letters		First and third letters		First letter and some other letter	
H	hydrogen	He	helium	Cl	chlorine	Pd	palladium
C	carbon	Li	lithium	Mg	magnesium	Pt	platinum
N	nitrogen	Ca	calcium	Mn	manganese		
O	oxygen	Al	aluminum	As	arsenic		
F	fluorine	Br	bromine	Cd	cadmium		
P	phosphorus	Ni	nickel	Cr	chromium		
S	sulfur	Si	silicon	Rb	rubidium		

Example 3.17

Examine the chemical symbols of the two metals magnesium and manganese. Determine how the symbols were chosen.

Solution The third letter of the name has been chosen as the lowercase letter of the symbol. Thus magnesium has the symbol Mg, whereas manganese has the symbol Mn. The two symbols are easily confused. For example, examine the second syllable of each name. In magnesium the second syllable starts with n, whereas the second syllable of manganese starts with g. Thus when each element is pronounced, one might feel that the correct letter for the symbol should be the sound that starts the second syllable. This would give you the wrong chemical symbol.

[Additional examples may be found in 3.61–3.72 at the end of the chapter.]

About 90% of the mass of the universe is hydrogen. It is the nuclear fusion of hydrogen that provides the energy of the sun and other stars. Helium, which makes up 9% of the mass of the universe, is the second most abundant element.

If only the surface of the Earth is considered, the abundance of the elements is considerably different. The percent composition by mass listed in Table 3.6 is for the ten-mile-thick shell of the Earth, the atmosphere, and the oceans. Only ten elements make up approximately 99% of this small part of the universe. If the entire Earth is considered including its core, the five most abundant elements are iron, oxygen, silicon, magnesium, and nickel.

The relative abundance of the elements in the human body is also given in Table 3.6. About 25 elements are present, but only ten of these occur in quantities greater than 0.1%. Many

Table 3.6 **Elemental Abundance (weight percent)**

Element	Earth crust, sea, and air	Total earth	Atmosphere	Human body
hydrogen	0.88		0.00005	10.0
oxygen	49.20	29.5	20.9	65.0
carbon	0.08		0.03	18.0
nitrogen	0.03		78.0	3.0
calcium	3.39	1.1		2.0
potassium	2.40			0.2
silicon	25.70	15.2		
magnesium	1.93	12.7		0.04
phosphorus	0.11			1.1
sulfur	0.06	1.9		0.2
aluminum	7.50	1.1		
sodium	2.64			0.1
iron	4.71	34.6		
titanium	0.58			
chlorine	0.19			0.1
argon			0.93	
boron				
nickel	0.02	2.4		
neon			0.0018	

elements present in small quantities are nevertheless important to life. Cobalt occurs in vitamin B_{12}; zinc, manganese, and magnesium are part of some enzymes that are responsible for regulating chemical reactions in living systems.

3.10 Compounds

The composition of a compound is independent of its source. The composition of water, for example, is always 88.81% oxygen and 11.19% hydrogen, whether it is obtained by the purification of seawater or rainwater. The composition can be determined by chemically breaking the compound into its elements. Alternatively, the elements can be recombined in the proper amounts in a chemical reaction to produce water. These facts are summarized in the **law of definite proportions:** When elements combine to form compounds, they do so in definite proportions by mass.

The law of definite proportions can be illustrated with the reaction represented in Figure 3.12. Iron and sulfur can combine to form the compound iron sulfide. If 55.85 g of iron is heated with 32.06 g of sulfur, 87.91 g of iron sulfide results. No iron or sulfur remains. If 90.00 g of iron is heated with 32.06 g of sulfur, then 87.91 g of iron sulfide results, but 34.15 g of iron is unreacted. Iron and sulfur can only combine in the ratio of 55.85 g of iron to 32.06 g of sulfur to form iron sulfide. As further verification of the law of definite proportions, consider the experiment in which 55.85 g of iron is heated with 40.00 g of sulfur. Again 87.91 g of iron sulfide is formed, but in this case 7.94 g of sulfur remains unreacted.

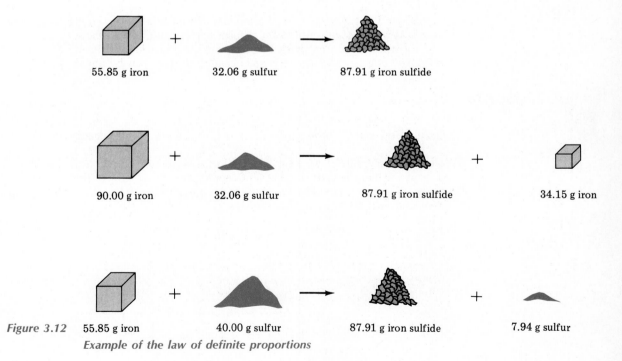

55.85 g iron + 32.06 g sulfur → 87.91 g iron sulfide

90.00 g iron + 32.06 g sulfur → 87.91 g iron sulfide + 34.15 g iron

55.85 g iron + 40.00 g sulfur → 87.91 g iron sulfide + 7.94 g sulfur

Figure 3.12

Example of the law of definite proportions

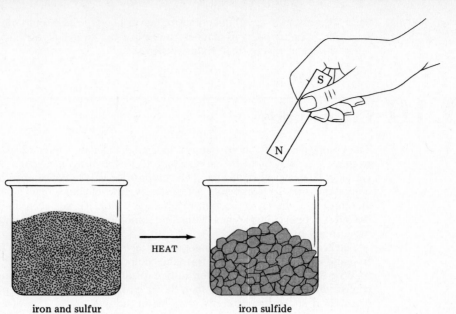

Figure 3.13   iron and sulfur   iron sulfide

Example of chemical change
A chemical reaction of iron and sulfur produces iron sulfide. The iron sulfide
cannot be separated into iron and sulfur by using a magnet.

In Figure 3.9 it was shown that a mixture of iron filings and sulfur can be separated using a magnet. When iron and sulfur react to form a compound, neither the iron nor the sulfur remains as an element. Thus a magnet cannot separate the elements that make up iron sulfide (Figure 3.13). The components of compounds cannot be separated by physical means.

Example 3.18

A 50.0 g sample of lead reacts completely with 7.8 g of sulfur to produce 57.8 g of lead sulfide. How much sulfur is required to react with 25.0 g of lead to produce lead sulfide?

Solution   The ratio of the mass of lead to the mass of sulfur to form lead sulfide must be a constant and must be in the ratio of 50.0 g of lead for every 7.8 g of sulfur. The given quantity is 25.0 g of lead and the desired quantity is grams of sulfur. Thus the following equation uses a factor unit method to solve the problem.

$$25.0 \text{ g lead} \times \frac{7.8 \text{ g sulfur}}{50.0 \text{ g lead}} = 3.9 \text{ g sulfur}$$

The answer is one half that of the 7.8 g of sulfur given in the first part of the problem. This quantity is reasonable because the 25.0 g of lead is one half that of the lead used in the first part of the problem.

[*Additional examples may be found in 3.81–3.84 at the end of the chapter.*]

3.11 Metallic and Nonmetallic Elements

Early in chemical history it was recognized that many elements have similar physical and chemical properties. These properties were used to classify the elements into two groups—metals and nonmetals. A comparison of some of the properties of metals and nonmetals is given in Table 3.7. About three quarters of the elements are metals. With the exception of mercury, metals are solids at room temperature. About half of the nonmetals are gases, such as chlorine, nitrogen, and oxygen. Argon, helium, krypton, neon, radon, and xenon, a group of very unreactive gases known as the noble gases, are also nonmetals. The only nonmetal that is a liquid at room temperature is bromine. A few solid nonmetals, such as phosphorus and sulfur, have low melting points.

Metals as a class have higher densities than nonmetals. In addition, metals are good conductors of heat and electricity. Many metals are ductile (can be drawn into wires) and malleable (can be rolled into sheets). Among the familiar metals are aluminum, chromium, copper, gold, iron, lead, nickel, silver, tin, and zinc. Other, less familiar, metals include calcium, magnesium, potassium, sodium, titanium, tungsten, and uranium.

There are chemical differences between metals and nonmetals that are examined in detail in later chapters. However, at this point, it can be pointed out that most elements will combine with oxygen to form compounds called oxides. In general, the oxides of metals are high-melting solids, whereas the oxides of nonmetals are gases, volatile liquids, or low-melting solids.

There are a few elements whose properties are intermediate between metals and nonmetals. These elements, known as semimetals or metalloids, are antimony, arsenic, boron, germanium, silicon, and tellurium. The physical properties are somewhat closer to those of metals, whereas the chemical properties are somewhat closer to those of nometals. Many of the semimetals are used in the production of semiconductors.

Table 3.7 **Selected Physical Properties of Metals, Nonmetals, and Semimetals**

Metals	Semimetals	Nonmetals
high electrical conductivity	intermediate electrical conductivity	insulators
malleable ductile	brittle	not malleable, not ductile
bright, shiny	intermediate reflectance	dull
high-melting, nonvolatile oxides	volatile, low-melting oxides	volatile, low-melting oxides

Summary

The study of chemistry is based on a classification of matter according to its composition, structure, and reactions. The mass of matter per unit volume is its density. The density of matter is related to its state. The three states are gas, liquid, and solid. Gases have lower densities than liquids or solids. Gases do not have definite volume or shape; a gas expands to fill the available space. Liquids have definite volume, but take the shape of their container. Solids have definite volume and shape.

Temperature determines the direction of heat flow between samples of matter; heat flows from a hot object to a cold object. The three temperature scales are Fahrenheit, Celsius, and Kelvin. The effect of heat energy on the temperature of a substance is given by the specific heat of the substance. The energy change in a physical process or a chemical reaction can be measured in calories or Joules. The dietetic Calorie is a kilocalorie.

Every substance can be identified based on its set of physical and chemical properties. Physical changes do not alter the composition of the substance, whereas chemical changes are associated with changes in composition in a process called a chemical reaction. Chemical reactions occur with no gain or loss of mass. The energy contained in substances is called chemical energy. Chemical reactions may be exothermic or endothermic.

Most matter exists in complex mixtures that may be homogeneous or heterogeneous. Mixtures are of variable composition. Pure substances may be separated from mixtures by physical means. Pure substances have a definite composition; their components cannot be separated by physical means. The two types of pure substances are elements and compounds. Compounds are composed of elements joined by forces called bonds. Compounds have a definite composition.

Elements are classified as metals or nonmetals based on their physical and chemical properties. Metals form the larger class of elements. Some elements known as semimetals or metalloids have intermediate properties between metals and nonmetals.

New Terms

A **calorie** is a unit of energy used in chemistry to describe the change in energy in a chemical reaction or a physical process.

The **Celsius** scale is a temperature scale on which the melting point and boiling point of water are 0°C and 100°C, respectively.

A **chemical change** is a process in which the composition or structure of one or more substances is altered.

Chemical energy is the energy stored in substances that can be released during a chemical reaction.

Chemical properties are characteristics of matter that govern changes in chemical composition during chemical reactions.

A **chemical symbol** is a shorthand or abbreviation of the name of an element.

Compounds are pure substances composed of elements joined together by forces called bonds.

Density is the ratio of mass per unit volume.

Distillation is a physical method used to separate liquids based on their boiling points.

Elements are pure substances that cannot be decomposed into any simpler substance(s) by ordinary chemical reactions.

Endothermic processes require heat energy from the surroundings.

Exothermic processes release heat energy to the surroundings.

The **Fahrenheit** scale is a temperature scale on which the melting point and boiling point of water are 32°F and 212°F, respectively.

A **gas** is a state of matter that has no characteristic shape or volume.

Heterogeneous mixtures have physical and chemical properties that are not uniform throughout.

Homogeneous mixtures have physical and chemical properties that are uniform throughout.

The **Kelvin** scale is a temperature scale on which 0 K is −273.15°C.

The **law of conservation of energy** states that energy may be converted from one form to another but may be neither created nor destroyed.

The **law of definite proportions** states that when elements combine to form compounds, they do so in definite proportions by mass.

A **liquid** is a state of matter that has a definite volume but no characteristic shape.

A **mixture** can be separated into pure substances by physical means.

A **physical change** occurs without a change in the composition of matter.

Physical properties are characteristics of matter that do not involve changes in the chemical composition of substances.

A **pure substance** has uniform properties, has a definite composition and cannot be separated into components by physical means.

A **solid** is a state of matter that has a definite volume and shape.

Solutions are homogeneous mixtures.

Specific gravity is the ratio of the density of a substance to the density of water at the same temperature.

Specific heat is the heat energy required to raise the temperature of 1 gram of a substance by 1°C.

Sublimation is the change of a substance from a solid into a gas without passing through the liquid state.

Temperature is a measure of the degree of "hotness" of a material.

Exercises

Terminology

3.1 Explain in your own words each of the following laws.
(a) law of conservation of mass
(b) law of conservation of energy
(c) law of definite proportions

3.2 Explain the difference between the two terms of each pair.
(a) density and specific gravity
(b) chemical and physical change
(c) mixture and compound
(d) element and compound
(e) mixture and pure substance
(f) reactant and product
(g) temperature and energy

Density

3.3 A cube of gold that is 2.00 cm on a side has a mass of 154.4 g. What is the density of gold?

3.4 A syringe contains 5.0 mL of a solution. The mass of the solution is 5.5 g. What is the density of the solution?

3.5 Osmium is the densest element known. What is its density in grams per cubic centimeter if 2.72 g has a volume of 0.121 mL?

3.6 The mass of 10 cm^3 of uranium is 189.6 g. What is the density of uranium?

3.7 When a 36.7 g sample of iron is placed in a graduated cylinder containing 20.0 mL the water level rises to 24.7 mL. What is the density of iron?

3.8 An empty graduated cylinder weighs 45.8772 g. After 50.0 mL of a liquid is added, the cylinder and its contents weigh 85.7998 g. What is the density of the liquid?

3.9 Pure silver has a density of 10.5 g/cm^3. A 22.8 g sample reputed to be silver displaces 2.0 mL of water. Is the sample pure silver?

3.10 The density of platinum is 22.5 g/cm^3. What is the mass of 5.9 cm^3 of platinum?

3.11 A 25.0 mL sample of bromine (density 3.12 g/mL) is added to a graduated cylinder. What is the mass of the bromine?

3.12 Concentrated hydrochloric acid has a density of 1.19 g/mL. What is the mass of 50.0 mL of hydrochloric acid?

3.13 The density of ether, a volatile liquid used as an anesthetic, is 0.71 g/mL. What volume would 355 g of ether occupy?

3.14 Ethyl alcohol has a density of 0.789 g/mL. What volume of ethyl alcohol must be poured into a beaker to give 19.8 g of ethyl alcohol?

3.15 The densities of zinc, chromium, and iron are 7.89, 7.20, and 7.86 g/cm^3. A metal with a volume of 12.3 mL has a mass of 96.7 g. What is the metal?

3.16 A liquid with a volume of 8.98 mL has a mass of 7.89 g. The densities of the liquids octane, ethanol, and benzene are 0.702, 0.789, and 0.879 g/mL, respectively. What is the liquid?

Specific gravity

3.17 A syringe contains 10.0 mL of a solution. The mass of the solution is 11.05 g. What is the specific gravity of the solution?

3.18 The specific gravity of turpentine is 0.87. What is the mass of 25.0 mL of turpentine?

3.19 The specific gravity of octane is 0.802. You add 15 mL of octane to a vessel. What is the mass of the octane?

3.20 The specific gravity of concentrated hydrochloric acid is 1.19. You need 25 g of concentrated hydrochloric acid. What volume do you use?

Temperature

3.21 The temperature on a warm day in Columbus, Ohio, is 95°F. What is the temperature in degrees Celsius?

3.22 On a cold day in Vermont, the temperature was −31°F. What was the temperature in degrees Celsius?

3.23 A doctor tells an American in Paris that his oral temperature is 38°C. What is his temperature in degrees Fahrenheit?

3.24 A European cookbook indicates that fondant must be cooked to 115°C. To what temperature should you heat the ingredients if you have a Fahrenheit candy thermometer?

3.25 Some surgical instruments are sterilized at 120°C. What is the temperature in Fahrenheit?

3.26 Ethylene glycol, a compound used in antifreeze, freezes at −11.5°C. What is the temperature on the Fahrenheit scale?

3.27 Heat stroke occurs at a body temperature of 41°C. What is the temperature equivalent in degrees Fahrenheit?

3.28 In a hypothermia case, the body temperature dropped to 28.7°C. What is the temperature equivalent in degrees Fahrenheit?

3.29 Silver melts at 1234 K. What is the melting point of silver in Celsius?

3.30 Helium boils at 4 K. What is this temperature in Celsius?

Energy

3.31 A quantity of heat equal to 500 cal is added to 25 g of water at 20°C. What will be the final temperature?

3.32 A water bed contains 1000 L of water. How many kilocalories of heat energy are required to heat the water in the bed from 15°C to body temperature (37.0°C)?

3.33 One ounce of cereal gives 112 Cal of energy on oxidation. How many kilograms of water can be heated from 20 to 30°C by "burning" the cereal?

3.34 A 110 lb woman eats 10 oz of carbohydrates, 2 oz of fat, and 18 oz of protein in one day. The energy released by metabolizing 1 g of either carbohydrate or protein is 4.1 kcal. The energy released by metabolizing 1 g of fat is 9.3 kcal. Calculate the caloric content of the woman's intake in kcal.

3.35 The metabolic energy requirements of a laboratory mouse is about 3.8×10^3 cal per day. How many Calories are required to feed the mouse?

3.36 One cup of zucchini contains about 25 Cal. Explain the meaning of this statement.

Specific Heat

3.37 It takes 0.092 cal to heat 1.0 g of copper by 1°C. If 500 cal are required to heat a sample of copper from 15 to 90°C, what is the mass of the copper?

3.38 It takes 0.12 cal to heat 1.0 g of iron by 1.0°C. How many calories are needed to raise the temperature of 45 g of iron from 14 to 47°C?

3.39 A 139 g sample of tin at 55.8°C is added to 140 mL of water at 23.2°C; the resultant temperature is 24.8°C. What is the specific heat of the tin?

3.40 A 109 g sample of silver at 94.1°C is added to 115 mL of water at 23.2°C; the resultant temperature is 26.8°C. What is the specific heat of the silver?

States of Matter

3.41 Classify each of the following substances as solid, liquid, or gas at room temperature.
(a) methane   (b) iron   (c) mercury
(d) sulfur   (e) brass   (f) salt

3.42 Classify each of the following substances as solid, liquid, or gas at room temperature.
(a) alcohol   (b) oxygen   (c) aspirin
(d) sugar   (e) antifreeze   (f) gasoline

3.43 Nicotine has a melting point and a boiling point of −79°C and 247°C, respectively. In what state will nicotine exist at 25°C?

3.44 Benzene has a melting point and a boiling point of 5.5°C and 80.1°C, respectively. In what state will benzene exist at 0°C?

Chemical and Physical Properties

3.45 Indicate whether the process described is an example of a physical or a chemical change.
(a) chopping of wood   (b) boiling of water
(c) rusting of iron   (d) melting of wax

3.46 Indicate whether the process described is an example of a physical or a chemical change.
(a) digesting food   (b) distilling moonshine
(c) burning wood   (d) melting of iron

3.47 Indicate whether the process described is an example of a physical or a chemical change.
(a) photosynthesis   (b) burning of gasoline
(c) burning toast   (d) boiling alcohol

3.48 Indicate whether the process described is an example of a physical or a chemical change.
(a) lighting a match   (b) grinding beef into hamburger
(c) souring of milk   (d) putting sugar in coffee

3.49 Indicate whether the process described is an example of a physical or a chemical change.
(a) breaking a glass   (b) cooking an egg
(c) spoiling of food   (d) turning on a light

3.50 Which of the following is not a chemical change?
(a) photosynthesis
(b) cooking an egg
(c) boiling alcohol

3.51 What are the physical and chemical properties of this page?

3.52 Describe the physical and chemical properties of iron.

Law of Conservation of Mass

3.53 A 2.317 g sample of silver oxide is heated, and 2.157 g of silver results. Does this information indicate that the law of conservation of mass has been violated?

3.55 Silver is tarnished by sulfur. In a laboratory experiment 0.216 g of silver is heated with excess sulfur that is then burned off to form a gas. The black solid remaining weighs 0.248 g. How much sulfur combined with the silver?

3.54 A 1.06 g sample of pure aluminum reacts slowly in air to form 2.00 g of a white powder. The substance contains only aluminum and oxygen. How much oxygen reacted with the aluminum?

3.56 In a laboratory experiment, 0.243 g of magnesium is burned with oxygen to produce 0.403 g of a white solid. How much oxygen combined with the magnesium?

Mixtures and Pure Substances

3.57 Make a list of ten materials used every day by you. Identify them as mixtures, elements, or compounds.

3.59 How might each of the following mixtures be separated?
(a) iron filings and sand
(b) oil and water
(c) salt and sand

3.58 Some people claim that they avoid foods having "chemicals" in them. Is this really possible?

3.60 Is tap water a pure substance or a mixture? How could you support your answer experimentally?

Elemental Symbols

3.61 What elements do the following symbols represent?
(a) O   (b) N   (c) F   (d) P   (e) H
(f) C   (g) I   (h) U   (i) S

3.63 What elements do the following symbols represent?
(a) Na   (b) K   (c) Ag   (d) Au   (e) Sb
(f) Fe   (g) Pb   (h) Cu   (i) Hg

3.65 Match the following names with their symbols.
(a) silver   (1) Ba
(b) helium   (2) As
(c) zinc   (3) Pb
(d) barium   (4) Ag
(e) uranium   (5) Li
(f) arsenic   (6) U
(g) iron   (7) Cr
(h) lead   (8) He
(i) lithium   (9) Fe
(j) chromium   (10) Zn

3.67 Some elements were named for the planets of our solar system. Identify these elements from the list of names of the elements.

3.69 List ten of the many elements whose elemental symbol consists of the first and second letter of its name.

3.71 Examine the elemental symbols for neodymium, neon, and neptunium. Determine how the symbols for these three elements with the same first two letters were selected.

3.62 What elements do the following symbols represent?
(a) Li   (b) Cl   (c) Ca   (d) Mg   (e) Br
(f) Si   (g) Ni   (h) Zn   (i) Ba

3.64 What elements do the following symbols represent?
(a) Al   (b) Cd   (c) As   (d) Se   (e) Ne
(f) He   (g) Ti   (h) Mn   (i) Ge

3.66 Match the following names with their symbols.
(a) sodium   (1) Ti
(b) potassium   (2) P
(c) calcium   (3) Ni
(d) chlorine   (4) Na
(e) tin   (5) Ca
(f) gold   (6) K
(g) nickel   (7) Au
(h) phosphorus   (8) Cl
(i) sulfur   (9) Sn
(j) titanium   (10) S

3.68 Some elements were named from mythology. Identify these elements from the list of names of the elements.

3.70 List ten of the many elements whose elemental symbol consists of the first and third letter of its name.

3.72 Examine the elemental symbols for thallium, thorium, and thulium. Determine how the symbols for these three elements with the same first two letters were selected.

Elemental Abundance

3.73 What is the most abundant element by mass in the universe?

3.75 What are the two most abundant elements in the entire Earth?

3.74 What are the five most abundant elements in the Earth's crust?

3.76 What are the two most abundant elements in the atmosphere?

Elements and Compounds

3.77 From the following equation classify A, B, and C as elements or compounds. If a classification is impossible, indicate why.

$$A + B \longrightarrow C$$

3.79 From the following equation classify A, B, X, and Y as elements or compounds. If a classification is impossible, indicate why.

$$A + B \longrightarrow Y + Z$$

3.78 From the following equation classify X, Y, and Z as elements or compounds. If a classification is impossible, indicate why.

$$X \longrightarrow Y + Z$$

3.80 From the following equation classify W, X, Y, and Z as elements or compounds. If a classification is impossible, indicate why.

$$X + Y + Z \longrightarrow W$$

Law of Definite Proportions

3.81 A 100 g sample of lead reacts completely with exactly 15.4 g of sulfur. What will occur when 20 g of lead is reacted with 3 g of sulfur?

3.83 A 25.0 g sample of lead will react completely with 3.90 g of oxygen. What will occur when 20.0 g of lead is reacted with 3.12 g of oxygen?

3.82 A 83.7 g sample of iron will react completely with 48.1 g of sulfur. What will occur when 27.9 g of iron is reacted with 16.8 g of sulfur?

3.84 Exactly 7 g of element A reacts completely with 12 g of element B. How many pounds of A are required to react completely with 80 lb of B?

Metals and Nonmetals

3.85 List some physical properties of elements that can be used to classify metals and nonmetals.

3.87 Lutetium is an element whose density is 9.84 g/cm^3. Its melting point and boiling point are 1663 and 3395°C, respectively. Would you classify lutetium as a metal or nonmetal?

3.86 The density of lithium is 0.534 g/cm^3. Is this fact inconsistent with its classification as a metal?

3.88 Fluorine is an element whose melting point and boiling point are −219 and −188°C, respectively. Would you classify fluorine as a metal or nonmetal?

Additional Exercises

3.89 A patient's urine sample has a density of 1.02 g/mL. How many grams of urine are eliminated on a day in which 1250 mL is excreted?

3.91 The density of air is 1.3 g/L. What is the mass of air in a room that is 5 m long, 3 m wide and 2.5 m high?

3.93 The densities of octane and ethyl alcohol are 0.702 and 0.789 g/mL, respectively. You have 10.0 g of each substance. Which sample occupies the larger volume?

3.95 A 45.2 g sample of lead is placed in a graduated cylinder containing 20.0 mL, and the water level rises to 24.0 mL. What is the specific gravity of lead?

3.97 Industrial diamonds can be produced from graphite at high pressure at 2000°C. What is the temperature equivalent in degrees Fahrenheit?

3.99 The surface of Titan, a moon of Saturn, is −180°C. What is this temperature in kelvins?

3.101 How many kilocalories will be given off if 50 g of water is cooled from 90 to 25°C?

3.103 An athlete uses 10 Cal in running the 100 yd dash. How many grams of water could be heated from 0 to 100°C by this amount of energy?

3.90 Ethylene glycol, which is used as an antifreeze, has a density of 1.0088 g/mL. How many grams of ethylene glycol are in 125 mL?

3.92 The density of hydrogen gas at 25°C and normal atmospheric pressure is 0.08 g/L. The Hindenburg dirigible contained 1.9×10^8 L. How many kilograms of hydrogen did the Hindenburg contain?

3.94 The densities of nickel and iron are 8.90 and 7.86 g/cm^3, respectively. You have 100 g of each metal. Which sample occupies the larger volume?

3.96 The specific gravity of alcohol is 0.80. You are told to add 12 g of alcohol to a flask and you do not have a balance. What do you do?

3.98 A scientist reports a study of matter at 13 K. What is the temperature in Celsius and Fahrenheit?

3.100 The boiling point of liquid oxygen is 55 K. What is this temperature in Celsius?

3.102 A brownie has 200 Cal. How many grams of water could be heated from 37 to 39°C by this amount of energy?

3.104 A 150 lb person playing handball uses 600 Cal/hr. How many grams of water could be heated from 0 to 100°C by this amount of energy?

3.105 A 107 g sample of copper at 57.8°C is added to 148 mL of water at 23.5°C; the resultant temperature is 25.6°C. What is the specific heat of the copper?

3.107 Methanol, a poisonous alcohol, has a melting point and a boiling point of −93.9 and 65.0°C, respectively. In what state will methanol exist at 85°C?

3.109 Which of the following is not a physical change?
(a) lighting a match
(b) breaking a glass
(c) melting gold
(d) freezing water

3.111 A 50.0 g sample of lead reacts completely with oxygen to give 57.7 g of lead oxide.
(a) How much oxygen reacted with the lead?
(b) How much oxygen would be required to react with 25.0 g of lead?
(c) What will happen if 50.0 g of lead is reacted with 10.0 g of oxygen?
(d) What will happen if 40.0 g of lead is reacted with 7.7 g of oxygen?

3.113 A 10.000 g sample of silver iodide contains 5.405 g of iodine. How many grams of iodine are present in a 32.00 g sample of silver iodide?

3.106 A 128 g sample of zinc at 51.0°C is added to 125 mL of water at 20.7°C; the resultant temperature is 23.3°C. What is the specific heat of the zinc?

3.108 Isopropyl alcohol, used in water solution as rubbing alcohol, has a melting point and a boiling point of −89.5 and 82.4°C, respectively. In what state will isopropyl alcohol exist at −95°C?

3.110 Which of the following is not a chemical change?
(a) evaporation of dry ice
(b) burning of gasoline
(c) photosynthesis
(d) explosion of TNT

3.112 A 12.0 g sample of zinc reacts with oxygen to give 17.9 g of zinc oxide.
(a) How much oxygen reacted with the zinc?
(b) How much oxygen would be required to react with 6.0 g of zinc?
(c) What will happen if 12.0 g of zinc is reacted with 7.0 g of oxygen?
(d) What will happen if 10.0 g of zinc is reacted with 5.9 g of oxygen?

3.114 A 100 mg sample of vitamin C contains 40.8 mg of carbon. How many grams of carbon are present in a 0.500 g tablet of vitamin C?

4 Atoms, Molecules, and Ions

Learning Objectives

After studying Chapter 4 you should be able to

1. Identify three subatomic particles by name, mass, and charge.
2. Describe the structure of the atom.
3. Express the number of subatomic particles in an atom using elemental symbols.
4. Calculate the atomic weight from elemental abundance and vice versa.
5. Represent molecules by their molecular formulas.
6. Distinguish between ions formed by atoms losing and gaining electrons.
7. Identify common ions by name and symbol.
8. Write formulas for ionic compounds based on the ions present.

4.1 What Is the Smallest Unit of Matter?

We learned in the last chapter that each pure substance has its own set of physical and chemical properties. For example, helium is a colorless and odorless gas of low density that does not burn in air and does not react with other elements. Ethyl alcohol is a liquid that mixes with water to form solutions and will burn in air. Table salt (sodium chloride) is a white solid that mixes with water to form solutions and will not burn in air.

What is responsible for the properties of substances such as the element helium or the compounds ethyl alcohol and sodium chloride? If you take successively smaller samples of matter, how small a piece will still have the described properties?

Let's consider the gas in a child's helium-filled balloon. Is there a point at which the helium sample could no longer be subdivided and still be considered helium? The answer is yes. Eventually the smallest unit of helium is reached. This unit is called the atom.

The contents of a bottle of pure ethyl alcohol divided into smaller and smaller samples is still ethyl alcohol. However, eventually the smallest unit of ethyl alcohol is reached. The unit is a molecule. The ethyl alcohol molecule consists of two carbon atoms, six hydrogen atoms, and one oxygen atom bonded together to form the unit that has the properties of ethyl alcohol. The atoms of elements in a molecule are combined in whole number ratios, never as fractional parts of atoms. It is this ratio of atoms that accounts for the definite proportion by mass of compounds.

A tablet of table salt can be crushed into fine grains. Each grain could be further subdivided. However, the smallest amount of sodium chloride that has the properties of table salt is a pair of ions, the sodium ion and the chloride ion. Ions, which are similar to atoms but are electrically charged, are discussed in Section 4.7. Ionic compounds consist of aggregates of oppositely charged atoms called ions. The ions are attracted to each other in a structure consisting of alternating ions of opposite charge. Thus there is no pair of ions that can be considered to be a molecule. However there is still a simplest ratio of ions as required by the need to balance the charges of the ions and maintain electrical neutrality. This subject is discussed further in Section 4.8.

In all samples of matter there comes a point at which a substance can no longer be subdivided without changing its characteristics. This point is reached when one of three types of very small particles known as atoms, molecules, or ions results. These particles are called submicroscopic because they are very much smaller than the smallest size that can be seen with a microscope. Our current concept of atoms, molecules, and ions is based on the original atomic theory of Dalton which is discussed in Section 4.2.

In this chapter we will also learn that atoms, molecules, and ions consist of even simpler particles. These particles, known as **subatomic particles,** are protons, electrons, and neutrons. The subatomic particles are discussed in Section 4.3 and in greater detail in Chapters 8 and 10.

4.2 Atomic Theory

The atomic theory of matter, proposed by an English schoolteacher, John Dalton, was accepted by many scientists early in the nineteenth century. Dalton suggested the idea of atoms to account for the laws of chemistry, such as the law of definite proportions and the law of conservation of mass. His theory, published in 1803, is called the atomic theory.

1. There are simple substances called elements that are composed of tiny indivisible particles called atoms.
2. Atoms of the same element are identical, but differ from those of any other element. Each type of atom has a definite mass that is different from that of any other atom.
3. Atoms combine only in certain whole number combinations to produce molecules.
4. Compounds are composed of molecules. All molecules of a compound are identical to each other.
5. In chemical reactions atoms are transferred or rearranged to produce different substances. The identity of each atom is unchanged and its mass is unchanged.

Atomic Mass Units

How small is the atom? Hydrogen, the smallest atom, has a mass of 1.673×10^{-24} g. A helium atom is four times heavier and has a mass of 6.646×10^{-24} g. The small mass of an atom is difficult to visualize. In 1.0 g of hydrogen atoms there are 6.0×10^{23} atoms!

$$1.0 \, g \times \frac{1 \text{ atom}}{1.673 \times 10^{-24} \, g} = 6.0 \times 10^{23} \text{ atoms}$$

To give you an idea of how large this number is, let's consider having all the people of the world count the atoms in this 1.0 g sample. If each of the 4 billion people of the world count one atom a second for eight hours a day it would take 14,000,000 years to finish the counting!

Example 4.1 _____

The mass of the helium atom is 6.646×10^{-24} g. How many atoms are there in a 4.0 g sample of helium gas?

Solution The conversion factor needed has the units atoms/g so that the unit of grams can be eliminated and the unit of atoms will be obtained.

$$4.0 \, g \times \frac{1 \text{ atom}}{6.646 \times 10^{-24} \, g} = 6.0 \times 10^{23} \text{ atoms}$$

[_Additional examples may be found in 4.7 and 4.8 at the end of the chapter._]

Note that the number of atoms in the 4.0 g sample of helium in Example 4.1 is the same as that in the 1.0 g sample of hydrogen just discussed. Why is this the case? The sample of helium has four times the mass of the sample of hydrogen. However, each helium atom weighs four times as much as a hydrogen atom. Since the ratio of the masses of the atoms is the same as the ratio of the masses of the samples, the number of atoms must be the same.

Since the masses of atoms are so very small, a special unit is used to describe the mass. The quantity 1.6603×10^{-24} g is called an **atomic mass unit (amu).** Thus, the hydrogen atom has a mass of 1.008 amu.

$$1.673 \times 10^{-24} \, g \times \frac{1 \text{ amu}}{1.6603 \times 10^{-24} \, g} = 1.008 \text{ amu}$$

The advantage in using atomic mass units is that exponential notation is avoided when discussing the relative masses of atoms.

Example 4.2

The mass of the helium atom is 6.646×10^{-24} g. What is the mass of the helium atom in atomic mass units?

Solution In order to convert the quantity given in grams into atomic mass units, a conversion factor with the units amu/g is required.

$$6.646 \times 10^{-24} \, g \times \frac{1 \text{ amu}}{1.6603 \times 10^{-24} \, g} = 4.003 \text{ amu}$$

Note that the mass of helium in atomic mass units is four times that of the hydrogen atom in the same units. This must be the case because the mass of the helium atom is four times that of the hydrogen atom in grams.

[*Additional examples may be found in 4.5 and 4.6 at the end of the chapter.*]

Atomic Radii

A spherical particle is used as a model for the atom. The radii of various atoms range from 0.5×10^{-8} to 2.4×10^{-8} cm. Because of these very small numbers, it is convenient to express the radii of atoms in units that avoid exponential notation. An older unit, the Ångstrom (Å), is commonly used. One Ångstrom unit is equal to 10^{-8} cm. Thus, the radii of atoms range from 0.5 to 2.4 Å (Table 4.1). An alternate unit in the metric system is the nanometer, in which case the range of atomic radii is 0.05 to 0.24 nm.

A distance of 1 Å is difficult to comprehend. The radius of the magnesium atom is 1.60 Å. If magnesium atoms were arranged in a straight line touching each other, it would take 31 million atoms to equal 1 cm. We go into details about the size of atoms in Chapter 10.

Table 4.1 **Atomic Radii of Some Atoms[a]**

Name	Radius (Å)	Relative mass (amu)	Name	Radius (Å)	Relative mass (amu)	Name	Radius (Å)	Relative mass (amu)
helium	0.50	4.00	lithium	1.34	6.94	fluorine	0.71	19.00
neon	0.65	20.18	sodium	1.57	22.99	chlorine	0.99	35.45
argon	0.95	39.95	potassium	1.96	39.10	bromine	1.14	79.90
krypton	1.00	83.80	rubidium	2.16	85.47	iodine	1.33	126.90
xenon	1.30	131.30	cesium	2.35	132.90			

[a] For the elements in each column, the size of the atoms increases as the mass in amu increases. More details of this relationship will be given in Chapter 9.

Table 4.2 **Properties of Subatomic Particles**

Name	Mass (g)	Relative mass (amu)	Charge (coulombs)	Relative charge
proton	1.6726×10^{-24}	1.0073	1.60×10^{-19}	+1
neutron	1.6749×10^{-24}	1.0087	0	0
electron	9.109×10^{-28}	0.0005486	-1.60×10^{-19}	-1

4.3 Subatomic Particles

Dalton's postulate that atoms are indivisible was disproved about a century ago. We now know that atoms can be broken into more fundamental subatomic particles called electrons, protons, and neutrons. The data for these particles are summarized in Table 4.2.

The Electron

The electron was discovered as the result of experiments in electricity. In 1897, the English physicist J. J. Thomson was working with a cathode ray tube that is similar to the modern television tube. He showed that a variety of gaseous atoms could be broken down into charged particles in these tubes. Each gas produced negatively charged particles called electrons. An **electron** has a mass of 9.109×10^{-28} g or 0.0005486 amu. This mass is 1/1837 that of the lightest atom, hydrogen. The mass of an electron usually can be regarded as zero when discussing the mass of the atom.

The charge of an electron is -1.60×10^{-19} coulomb. Such a quantity is cumbersome. However, all charges of subatomic particles and atomic particles are multiples of this quantity. Therefore, the electron charge is usually replaced with a relative charge of -1 without units. The hydrogen atom has one electron, but other atoms have more electrons. The helium atom has two electrons; the carbon atom has six electrons; the oxygen atom has eight electrons.

The Proton

A **proton** has a mass of 1.6725×10^{-24} g, which to four significant figures is equal to 1.008 amu. The charge of a proton is $+1.60 \times 10^{-19}$ coulomb. This quantity is usually replaced by a relative charge of $+1$ without units. Since all atoms contain electrons, they also contain the same number of protons because all atoms are electrically neutral.

The proton is the common unit of positive matter in all atoms and atomic particles. A hydrogen atom has only one proton as well as only one electron. Since the mass of the electron is so small, the majority of the mass of the hydrogen atom is due to the proton.

All atoms contain protons. The number of protons in an atom is characteristic of the element. The helium atom has two protons; the carbon atom has six protons; the oxygen atom has eight protons.

The Neutron

The mass of a helium atom is 4.00 amu. The helium atom contains two electrons and two protons. Based on this number of subatomic particles the helium atom should only weigh 2.00 amu! What accounts for the other 2.00 amu of the helium atom? This "missing" mass of the helium atom is due to neutrons. The English physicist, James Chadwick, showed in 1932 that a subatomic particle called a neutron exists. The **neutron** has a mass of 1.6748×10^{-24} g, or essentially 1 amu, and has no charge. The helium atom has two neutrons. All atoms, with the exception of hydrogen, contain neutrons.

The Nucleus and the Atom

In 1911, Ernest Rutherford, an English scientist, showed that most of the mass of the atom is concentrated in a small dense region called the nucleus. Other experiments led to the conclusion that the **nucleus** is composed of protons and neutrons. Although the radius of the nucleus depends on the number of protons and neutrons, the average diameter of the nucleus is 10^{-13} cm.

Since the radius of an atom is about 10^5 larger than the nucleus and the mass of atom is concentrated in the nucleus, most of the rest of the atom is "empty" space. The electrons move in this region about the nucleus. The closest analogy to the atom is our solar system. Although our solar system is large, it is essentially empty space. The planets move about in this space as they revolve about the center occupied by the sun.

4.4   Atomic Number, Mass Number, and Isotopes

Atoms differ from each other because of the number of subatomic particles that they contain. The number of protons determines the identity of the atom. For example, the hydrogen atom contains one proton, whereas the helium atom contains two protons.

The **atomic number** of an element is equal to the number of protons in the nucleus of an atom of that element. The atomic numbers of hydrogen and helium are 1 and 2, respectively. The atomic numbers of all of the elements are listed in a Table of Elements inside the back cover of this book. Since the number of electrons in an atom must equal the number of protons, the atomic number also indicates the number of electrons in a neutral atom.

The total number of protons plus neutrons in the nucleus of an atom is called the **mass number.** Since helium has two neutrons and two protons in its nucleus, its mass number is 4.

Elemental Symbols

Since atoms of one element differ from those of another element by the number of their subatomic particles, it is convenient to have this information included with the elemental symbol.

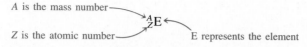

The subscript Z gives the number of protons in the atom. The mass number A is equal to the number of protons and neutrons in the atom. Thus the number of neutrons is equal to $A - Z$.

Example 4.3 _____

How many protons, neutrons, and electrons are contained in an atom represented by $^{14}_{6}C$?

Solution

$$^{14}_{6}C = {}^{A}_{Z}E$$

The elemental symbol represents the element carbon. There are six protons in the element, as indicated by the subscript. Therefore, carbon also contains six electrons. The number of neutrons given by the quantity $A - Z$ is eight.

Example 4.4

How many neutrons are contained in an atom represented by $^{238}_{92}U$?

Solution

$$^{238}_{92}U = {}^{A}_{Z}E$$

The elemental symbol represents the element uranium. There are a total of 238 protons and neutrons in this atom. There are 92 protons in the element, as indicated by the subscript. Therefore, using the formula $A - Z$ gives $238 - 92 = 146$ neutrons.

[*Additional examples may be found in 4.21–4.24 at the end of the chapter.*]

Figure 4.1

An analogy for isotopes
Identical twins may have different masses. Similarly, isotopes of the same element have different masses.

Table 4.3 Naturally Occurring Isotopes of Some Common Elements

Element	Isotope	Isotopic mass	Natural abundance (wt %)
hydrogen	$^{1}_{1}H$	1.0078	99.985
	$^{2}_{1}H$	2.0141	0.015
	$^{3}_{1}H$	3.0160	trace
carbon	$^{12}_{6}C$	12.0000	98.89
	$^{13}_{6}C$	13.0034	1.11
	$^{14}_{6}C$	14.0032	trace
oxygen	$^{16}_{8}O$	15.9949	99.759
	$^{17}_{8}O$	16.9991	0.037
	$^{18}_{8}O$	17.9992	0.204
magnesium	$^{24}_{12}Mg$	23.9850	78.70
	$^{25}_{12}Mg$	24.9858	10.13
	$^{26}_{12}Mg$	25.9826	11.17
chlorine	$^{35}_{17}Cl$	34.9689	75.53
	$^{37}_{17}Cl$	36.9659	24.47

Isotopes

We now know that different atoms of an element may have different masses. These atoms, called isotopes, have the same number of protons but differ in the number of neutrons that they contain. Thus, the **isotopes** of an element have identical atomic numbers but different mass numbers. It is the number of protons in the nucleus (the atomic number) that determines the identity of the element, not the number of neutrons. Of course all the isotopes of a single element have the same number of electrons because they have the same number of protons.

An analogy may be drawn between isotopes and identical twins. Identical twins have the same chromosomal makeup. However, the twins need not always be exactly the same weight (Figure 4.1). They are still twins by the chromosomal criterion regardless of their weights. In a similar manner, in spite of their difference in mass, isotopes of an element are the same element because they contain the same number of protons.

There are two naturally occurring isotopes of chlorine. Chlorine has an atomic number of 17 and contains 17 protons as well as 17 electrons. However, one chlorine isotope has 18 neutrons in the nucleus, whereas the other isotope has 20 neutrons in the nucleus. The mass numbers of these isotopes are 35 and 37, respectively. The two isotopes are represented by the symbols $^{35}_{17}Cl$ and $^{37}_{17}Cl$, respectively. A list of the isotopes of some elements is given in Table 4.3.

Example 4.5

What do the symbols $^{235}_{92}U$ and $^{238}_{92}U$ mean? What relationship exists between the atoms represented by these symbols?

Solution The subscript 92 indicates that the element represented by U (uranium) contains 92 protons. Because the symbols represent atoms that contain the same number of protons but differ in mass number, the atoms are isotopes of each other. The number of neutrons in $^{235}_{92}U$ is $235 - 92 = 143$. In the other isotope $^{238}_{92}U$ the number of neutrons is $238 - 92 = 146$.

Example 4.6

The radioisotope $^{131}_{53}I$ is used to treat cancer of the thyroid gland as well as hyperthyroidism. Describe the atomic composition of this isotope.

Solution

$$^{131}_{53}I = {}^A_Z E$$

There are 53 protons and also 53 electrons as $Z = 53$. The number of neutrons is $A - Z = 131 - 53 = 78$.

[*Additional examples may be found in 4.25–4.30 at the end of the chapter.*]

4.5 Atomic Weights

The majority of elements have at least two naturally occurring isotopes. Therefore the mass of a sample of atoms of any element is the sum of all of the masses of all of the atoms in the sample. How then can we indicate the atomic weight of an element when the element is in fact a mixture of isotopes? The answer is that we use the average of the masses of the isotopes taking into account the relative abundance of the various isotopes. The **atomic weight** of an element is the weighted average of the naturally occurring isotopes of the element expressed in atomic mass units. The atomic weight of each element is listed inside the back cover of this book. The atomic weight scale is based on the assignment of 12.0000 amu to the mass of the most common isotope of carbon, $^{12}_{6}C$, also known as carbon-12.

The calculation of a weighted average is similar to the procedure that may be used in determining your grade in a course. Suppose that your two examinations account for 20% and 30% of your final grade, and the final examination is 50% of your final grade. If you earn 85 on the first examination, 80 on the second examination, and 90 on the final examination, your course grade would be 86. This score is obtained by multiplying each grade by the decimal equivalent of the weighting percentage and then summing.

$$85 \times 0.20 = 17$$
$$80 \times 0.30 = 24$$
$$90 \times 0.50 = \underline{45}$$
$$86$$

Atomic weights can be calculated by multiplying the mass of each isotope by its fractional abundance. For example, boron consists of 19.6% boron of mass 10.013 amu and 80.4% boron of mass 11.009 amu.

$$10.013 \text{ amu} \times 0.196 = 1.96 \text{ amu}$$
$$11.009 \text{ amu} \times 0.804 = \underline{8.85 \text{ amu}}$$
$$10.81 \text{ amu}$$

The atomic weight of boron is 10.81 amu.

Example 4.7

The two principal isotopes of copper are $^{63}_{29}Cu$ and $^{65}_{29}Cu$. Their atomic masses are 62.9298 and 64.9278 amu, respectively. The atomic weight of copper is 63.546 amu. Which isotope is the more abundant?

Solution The weighted average of the isotopes, the atomic weight, will be closer to the mass of the most abundant isotope. The value of 63.546 is closer to 62.9298 than to 64.9278. Thus the most abundant isotope is $^{63}_{29}Cu$.

Example 4.8

The natural abundances of $^{37}_{17}Cl$ and $^{35}_{17}Cl$ are 24.47% and 75.53%, respectively. The atomic masses of the two isotopes are 36.96590 and 34.96885 amu, respectively. What is the atomic weight of chlorine?

Solution The weighted average will be closer to 35 amu than to 37 amu because there is a larger amount of the isotope of lighter atomic mass. The calculation is

$$(0.2447 \times 36.96590 \text{ amu}) + (0.7553 \times 34.96885 \text{ amu}) = 35.457528 \text{ amu}$$

The correct answer to the required four significant figures is 35.46 amu.

Example 4.9

Carbon occurs in nature as a mixture of $^{12}_{6}C$ and $^{13}_{6}C$. (There is a very small amount of $^{14}_{6}C$, which can be neglected in this problem.) The atomic masses are 12.0000 and 13.003 amu, respectively. The atomic weight of carbon is 12.011 amu. What is the percent abundance of $^{12}_{6}C$?

Solution The percent abundance of the two isotopes must total 100%, or in terms of decimal equivalents the total must be 1.00. Assume that the decimal equivalent of the abundance of $^{12}_{6}C$ is D. The decimal equivalent of the abundance of $^{13}_{6}C$ would then be $1 - D$ because the sum of the decimal equivalents must equal 1.00. The expression for the weighted average calculation of the atomic weight is

$$D \times (12.0000) + (1 - D) \times (13.003) = 12.011$$
$$12.0000D + 13.003 - 13.003D = 12.011$$
$$12.0000D - 13.003D = 12.011 - 13.003$$
$$-1.003D = -0.992$$
$$D = (-0.992)/(-1.003) = 0.989$$

Thus the percent abundance of the $^{12}_{6}C$ isotope is 98.9% and the abundance of the other isotope is 1.1%.

[*Additional examples may be found in 4.33–4.36 at the end of the chapter.*]

4.6 Molecules and Molecular Formulas

Elements

Some elements consist of atoms as the simplest unit of the substance. For example, neon, argon, krypton, xenon, and radon are gaseous elements that exist as atoms. These elements are said to be monatomic. However, many elements consist of discrete units containing several atoms. An electrically neutral unit of matter consisting of two or more atoms is a **molecule.** The common gases nitrogen, oxygen, and hydrogen consist of units that contain two atoms; they are diatomic molecules. Thus Dalton's concept that elements exist as single atoms is not entirely correct, since many elements exist as molecules.

Among the elements that are also diatomic are fluorine, chlorine, bromine and iodine. Molecules that contain more than two atoms are polyatomic. Phosphorus consists of molecules of four atoms, whereas sulfur exists in molecules of eight atoms.

Molecular Formulas

A representation of the molecule that indicates the number and kind of each atom contained in the molecule is called the **molecular formula.** The molecular formula gives the number of atoms of each type as a subscript to the right of the elemental symbol. Thus the diatomic molecule chlorine is represented as Cl_2. The elemental forms of phosphorus and sulfur are represented as P_4 and S_8, respectively.

Cl_2 subscript on the right indicates 2 chlorine atoms per molecule

Example 4.10

Oxygen exists in the gaseous state predominately as O_2, but O_3 is found in the upper atmosphere. The latter form of oxygen, known as ozone, is responsible for protecting us from harmful radiation from the sun. What do these two symbols represent?

Solution The subscript to the right of the elemental symbol represents the number of atoms contained in the molecule. Thus the form that exists in the upper atmosphere contains three atoms per molecule, whereas ordinary oxygen contains two atoms per molecule. Ozone is triatomic, whereas ordinary oxygen is diatomic.

[Additional examples may be found in 4.37–4.40 at the end of the chapter.]

Compounds

With the exception of ionic compounds (Section 4.9), the molecule is the smallest possible unit of a compound. Molecules of compounds contain two or more different types of atoms, whereas molecules of elements contain only one type of atom. Molecular formulas of compounds provide information about the number and kind of each element present in the molecule.

Water consists of molecules containing two hydrogen atoms and one oxygen atom. The molecular formula of water is H_2O. The subscript 2 to the right of the symbol for hydrogen indicates the number of hydrogen atoms contained in the molecule. No subscript follows the symbol of oxygen, which means that only one atom of that type is contained in the molecule.

Table 4.4 **Molecular Formulas of Some Organic Compounds**

Molecular formula	Common name
C_3H_8	propane
C_8H_{18}	octane
$C_6H_8O_6$	vitamin C
$C_6H_{12}O_6$	glucose
$C_8H_8O_3$	oil of wintergreen
$C_{20}H_{30}O$	vitamin A
$C_3H_5N_3O_9$	nitroglycerin
$C_{12}H_{12}N_2O_3$	phenobarbital
$C_{13}H_{16}N_2O_2$	novocaine
$C_{16}H_{18}N_2O_4S$	penicillin G
$C_{63}H_{88}N_{14}O_{14}PCo$	vitamin B_{12}

The molecule could be represented by OH_2 equally well. It is a matter of convention that chemists have decided to use H_2O rather than OH_2.

A molecular formula does not give details of the structure of the molecule. Consider for example the molecular formulas of the organic compounds listed in Table 4.4. The possible ways in which the many atoms in these complex molecules can be bonded is very large. Details of the structures of compounds will be given in Chapter 10.

Example 4.11 _____

One of the major sources of nitrogen for fertilizing fields is ammonia. Ammonia is represented by the molecular formula NH_3. What does this molecular formula mean?

Solution The molecular formula indicates that the compound contains nitrogen and hydrogen. Because no subscript appears to the right of N, there is only one atom of nitrogen per molecule. There are three atoms of hydrogen per molecule. The ammonia molecule consists of four atoms—one of nitrogen and three of hydrogen.

Example 4.12 _____

Aspirin, which is acetylsalicylic acid, contains nine atoms of carbon, eight atoms of hydrogen, and four atoms of oxygen. Write the molecular formula.

Solution Although you do not know the convention of the order of the elemental symbols to be used in the molecular formula, let us use C followed by H and finally by O. The molecular formula is then given by $C_9H_8O_4$.

Example 4.13 _____

The insecticide parathion has the molecular formula $C_{10}H_{14}NO_5PS$. How many atoms are there in a molecule of parathion?

Solution The subscripts after each elemental symbol give the number of atoms of that element in the molecule. There are 10 atoms of carbon, 14 atoms of hydrogen, 1 nitrogen atom, 5 atoms of oxygen, 1 phosphorus atom, and 1 sulfur atom. The total number of atoms in the molecule is 32.

[*Additional examples may be found in 4.41–4.46 at the end of the chapter.*]

4.7 Ions

In Section 4.1 it was noted that some matter consists of ions. An **ion** results from the loss or gain of electrons from atoms. If an atom loses one or more electrons, a positively charged ion results. A positive ion is called a **cation.** The sodium atom can lose an electron to form a sodium cation. Note, however that the nucleus with its number of protons and neutrons is unchanged. Therefore, the sodium ion is not another element. If an atom gains one or more electrons, a negatively charged ion results. A negative ion is called an **anion.** The chlorine atom can gain an electron to form a chloride anion.

Symbols of Ions

An ion is written with its relative charge as a superscript to the right of the elemental symbol. The superscript bears a positive or negative sign indicating the charge and an integer indicating how many electrons have been removed from or added to the neutral atom.

When an electron is removed from the sodium atom, the sodium ion results. Equations representing this process are

$$\left(11\,p\right)\,11\right)\,e^- \longrightarrow \left(11\,p\right)\,10\right)\,e^- + e^-$$

$$\text{Na}\begin{pmatrix}11\text{ protons}\\11\text{ electrons}\end{pmatrix} \longrightarrow \text{Na}^+\begin{pmatrix}11\text{ protons}\\10\text{ electrons}\end{pmatrix} + 1\,e^- \text{ (electron)}$$

$$\text{Na} \longrightarrow \text{Na}^+ + e^-$$

The nucleus, represented by the circle, contains 11 protons. The number of neutrons is not indicated because they do not affect the charge of the ion. The electrons are represented as a part of an "orbit" at a distance from the nucleus. The symbol e^- represents the electron with its relative charge of -1. Since the atomic number of sodium is 11 we know that the atom contains 11 protons and 11 electrons. When an electron is removed from the sodium atom, the resultant sodium ion has 11 protons and 10 electrons. The sodium ion has one unit of positive charge and is represented by Na^+.

The calcium atom (atomic number 20) can lose two electrons. The resulting ion has 18 electrons and 20 protons. The calcium ion has two units of positive charge as is represented by Ca^{2+}. The following equations summarize the formation of the calcium ion.

$$\left(20\,p\right)\,20\right)\,e^- \longrightarrow \left(20\,p\right)\,18\right)\,e^- + 2\,e^-$$

$$\text{Ca}\begin{pmatrix}20\text{ protons}\\20\text{ electrons}\end{pmatrix} \longrightarrow \text{Ca}^{2+}\begin{pmatrix}20\text{ protons}\\18\text{ electrons}\end{pmatrix} + 2\,e^- \text{ (electrons)}$$

$$\text{Ca} \longrightarrow \text{Ca}^{2+} + 2\,e^-$$

As an example of an anion, consider the addition of an electron to the chlorine atom (atomic number 17) to yield the chloride ion. The chloride ion contains 17 protons and 18 electrons and is symbolized by Cl^-.

$$1\,e^- + \left(17\,p\right)\,17\,e^- \longrightarrow \left(17\,p\right)\,18\,e^-$$

$$1\,e^- + Cl \begin{pmatrix} 17 \text{ protons} \\ 17 \text{ electrons} \end{pmatrix} \longrightarrow Cl^- \begin{pmatrix} 17 \text{ protons} \\ 18 \text{ electrons} \end{pmatrix}$$

$$e^- + Cl \longrightarrow Cl^-$$

In a similar manner, the oxygen atom (atomic number 8) can accept two electrons to yield the oxide ion O^{2-}. The oxide ion has eight protons and ten electrons.

$$2\,e^- + \left(8\,p\right)\,8\,e^- \longrightarrow \left(8\,p\right)\,10\,e^-$$

$$2\,e^- + O \begin{pmatrix} 8 \text{ protons} \\ 8 \text{ electrons} \end{pmatrix} \longrightarrow O^{2-} \begin{pmatrix} 8 \text{ protons} \\ 10 \text{ electrons} \end{pmatrix}$$

$$2\,e^- + O \longrightarrow O^{2-}$$

Example 4.14

The atomic number of aluminum is 13. Aluminum can lose three electrons to form a cation. How many protons and electrons does the cation have? Write a representation of the ion.

Solution The atomic number is equal to the number of protons in an atom. In addition, the atomic number of an atom is also equal to the number of electrons because the number of protons and electrons are equal in a neutral atom. Aluminum has 13 electrons as well as 13 protons. When aluminum loses three electrons, ten electrons remain. The resulting cation has three units of positive charge and is represented by Al^{3+}.

[*Additional examples may be found in 4.47–4.50 at the end of the chapter.*]

Names of Ions

The anions derived from single atoms of nonmetallic elements are named by adding -ide to the root of the element's name. A list of common anions and their names is given in Table 4.5. You should memorize the ions and their characteristic charges. The reasons why each of these ions is formed will be discussed in Chapter 10.

The cations derived from single atoms of metallic elements are named as the ion of the element. A list of some common cations is given in Table 4.6. Some elements form only one cation, but others can form several cations. In the cases of multiple cations derived from a single element, the charge of the ion is indicated in the name. Two systems of naming the cations are in use. The older system uses the endings -ous and -ic to indicate the lower and higher charged ions, respectively. This system is being replaced by use of Roman numerals within parentheses to indicate the charge. The details of naming ions and compounds will be given in Chapter 11.

Ions that consist of several atoms held together by chemical bonds similar to those involved in molecules are called **polyatomic ions.** These polyatomic ions differ from molecules in that they bear a charge. A polyatomic ion may be positive or negative, but negatively charged polyatomic

Table 4.5 Simple Anions

Anion	Name
F^-	fluoride
Cl^-	chloride
Br^-	bromide
I^-	iodide
O^{2-}	oxide
S^{2-}	sulfide
Se^{2-}	selenide
N^{3-}	nitride
P^{3-}	phosphide

Table 4.6 **Common Simple Cations**

Cation	Name	Cation	Name	Cation	Name
Li^+	lithium ion	Mg^{2+}	magnesium ion	Al^{3+}	aluminum ion
Na^+	sodium ion	Ca^{2+}	calcium ion	Sc^{3+}	scandium ion
K^+	potassium ion	Ba^{2+}	barium ion	Fe^{3+}	ferric ion
Rb^+	rubidium ion	Ni^{2+}	nickel ion		iron(III) ion
Cs^+	cesium ion	Zn^{2+}	zinc ion		
Cu^+	cuprous ion	Fe^{2+}	ferrous ion		
	copper(I) ion		iron(II) ion		
Ag^+	silver ion	Cu^{2+}	cupric ion		
			copper(II) ion		
		Hg_2^{2+}	mercurous ion		
			mercury(I) ion		
		Hg^{2+}	mercuric ion		
			mercury(II) ion		

Table 4.7 **Polyatomic Ions**

Formula	Name
NO_3^-	nitrate
NO_2^-	nitrite
SO_4^{2-}	sulfate
SO_3^{2-}	sulfite
HSO_4^-	bisulfate
HSO_3^-	bisulfite
PO_4^{3-}	phosphate
CO_3^{2-}	carbonate
HCO_3^-	bicarbonate
OH^-	hydroxide
CN^-	cyanide
MnO_4^-	permanganate
ClO_4^-	perchlorate
ClO_3^-	chlorate
ClO_2^-	chlorite
ClO^-	hypochlorite
NH_4^+	ammonium

ions are more common. A list of some polyatomic ions is given in Table 4.7. Remember that the superscript on the right indicates the charge on the ion, whereas a subscript represents the number of atoms of that element occurring in the ion. Thus the nitrate ion, NO_3^-, consists of one atom of nitrogen, three atoms of oxygen, and one electron more than the individual neutral atoms contain.

You should learn the symbols for both simple and polyatomic ions. You should know the elements present in the ion as well as the charge on the ion. This information is needed in order to write correct formulas for ionic compounds (Section 4.8).

The polyatomic ions containing oxygen commonly have suffixes -ate and -ite. Although the suffix does not indicate the number of oxygen atoms in the polyatomic ion, the ion with the -ate suffix contains more oxygen atoms than that with the -ite suffix. Thus, nitrate and nitrite are NO_3^- and NO_2^- respectively, whereas sulfate and sulfite are SO_4^{2-} and SO_3^{2-}, respectively.

There are several cases where two elements form more than two polyatomic ions. Therefore, more endings than -ate and -ite are required. For example, Cl and O actually form four different polyatomic ions. In Table 4.7 are listed perchlorate (ClO_4^-), chlorate (ClO_3^-), chlorite (ClO_2^-), and hypochlorite (ClO^-). The prefix per- is used to mean over. Thus perchlorate has one more oxygen atom than chlorate. The prefix hypo- means under. Thus hypochlorite is related to chlorite by having one less oxygen atom. These prefixes are also used with other similar ions such as hypoiodite for IO^- and perbromate for BrO_4^-.

4.8 Ionic Compounds

As indicated earlier in this chapter, some compounds such as sodium chloride consist of ions. Compounds consisting of ions are called **ionic compounds** and are structurally different from molecular compounds. Ionic compounds consist of oppositely charged ions held together by electrostatic forces of attraction. These forces, called ionic bonds, are discussed in Chapter 10.

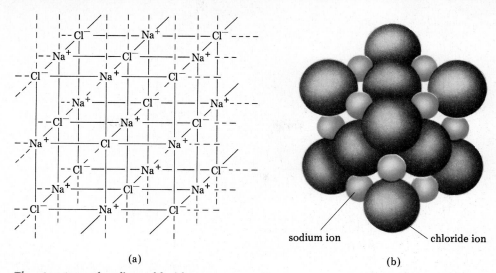

Figure 4.2

(a)   (b)

The structure of sodium chloride
(a) Schematic drawing showing the alternating pattern of sodium and chloride ions. (b) The relative sizes of the ions. Sodium ions are smaller than chloride ions.

Structure of Ionic Compounds

Ionic compounds are crystalline solids at room temperature. These solids consist of ions arranged in repeating three-dimensional patterns. For example, in a crystal of sodium chloride (Figure 4.2) each sodium ion is surrounded by six chloride ions. Each chloride ion is also surrounded by six sodium ions. The resulting arrangement is like three-dimensional tic-tac-toe.

In a crystal of sodium chloride, the number of sodium ions equals the number of chloride ions. As a whole, the crystal is electrically neutral no matter what its size. The entire crystal is held together by electrostatic forces between the cations and anions.

Formulas of Ionic Compounds

Ionic compounds are electrically neutral overall, since the charge of the cations is balanced by the charge of the anions. In table salt, sodium chloride, electrical neutrality is the result of an equal number of Na^+ and Cl^- ions.

Na^+ charge of $+1$
Cl^- charge of -1
           net charge 0

Sodium chloride is represented by NaCl even though there is no sodium chloride molecule. The formula indicates only the 1:1 ratio of the Na^+ and Cl^- ions in the compound.

charge of cation not shown
but understood

charge of anion not shown
but understood

NaCl

subscript of 1 implied
but not shown

subscript of 1 implied
but not shown

Table 4.8 Names of Some Ionic Compounds

Formula	Name	Formula	Name	Formula	Name
$LiClO_4$	lithium perchlorate	$CaCl_2$	calcium chloride	$Mg(CN)_2$	magnesium cyanide
$LiBr$	lithium bromide	CaS	calcium sulfide	$Zn(OH)_2$	zinc hydroxide
$NaNO_3$	sodium nitrate	$CaCO_3$	calcium carbonate	$ZnSO_4$	zinc sulfate
$NaHCO_3$	sodium bicarbonate	$CaSO_4$	calcium sulfate	NH_4Cl	ammonium chloride
KCl	potassium chloride	$Ca_3(PO_4)_2$	calcium phosphate	$(NH_4)_2SO_4$	ammonium sulfate
$KMnO_4$	potassium permanganate	MgO	magnesium oxide	$(NH_4)_3PO_4$	ammonium phosphate
K_3PO_4	potassium phosphate	MgF_2	magnesium fluoride		

In calcium chloride there is one Ca^{2+} ion for every two Cl^- ions, and the chemists' representation of the compound is $CaCl_2$. However, there is no $CaCl_2$ molecule. A crystal of $CaCl_2$ contains Ca^{2+} and Cl^- ions in a $1:2$ ratio.

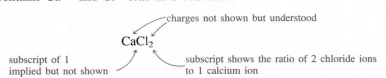

charges not shown but understood

$CaCl_2$

subscript of 1 implied but not shown

subscript shows the ratio of 2 chloride ions to 1 calcium ion

In an ionic compound, the positive charges must balance the negative charges. Since the charge on the anion may not be equal to that on the cation, the number of anions will not always be equal to the number of cations.

$$\text{positive charge} + \text{negative charge} = 0$$

$$\left(\frac{\text{charge}}{\text{cation}} \times \text{relative number of cations}\right) + \left(\frac{\text{charge}}{\text{anion}} \times \text{relative number of anions}\right) = 0$$

For example, in $CaCl_2$, there are two chloride ions for every calcium ion.

$$\left(\frac{\text{charge}}{\text{calcium cation}} \times \begin{array}{c}\text{relative number}\\ \text{of } Ca^{2+}\end{array}\right) + \left(\frac{\text{charge}}{\text{chloride anion}} \times \begin{array}{c}\text{relative number}\\ \text{of } Cl^-\end{array}\right) = 0$$

$$(+2)(1) + (-1)(2) = 0$$

A number of ionic compounds are listed in Table 4.8. You should check the charges of the ions contained in the compounds and show that all of the compounds listed in Table 4.8 are electrically neutral. Note that whenever more than one polyatomic ion is required to form a compound, parentheses are used to enclose the polyatomic ion. The subscript to the right of the parenthesis indicates the number of such polyatomic ions required to maintain electrical neutrality with the ions of opposite charge. Thus $Ca_3(PO_4)_2$ is an ionic compound that consists of three doubly charged calcium ions for every two triply charged phosphate ions.

Example 4.15

Given the fact that aluminum forms a $+3$ ion, write the formula for aluminum oxide.

Solution The compound that consists of Al^{3+} and O^{2-} ions must be electrically neutral. Therefore, the total charge of the cations and the anions in the formula must be balanced. The lowest common multiple of 3 and 2 is 6. Thus, three oxide ions will have a total charge of -6, and two aluminum ions will have a total charge of $+6$. By convention the formula is Al_2O_3 rather than O_3Al_2 (see discussion following on the names of ionic compounds).

[*Additional examples may be found in 4.61 and 4.62 at the end of the chapter.*]

Example 4.16

An oxide of chromium has the formula Cr_2O_3. What is the charge of the chromium ion?

Solution The charge of the oxide ion is -2. For three oxide ions, the total negative charge is -6. The total charge of chromium ions must be $+6$. Thus, the charge of each chromium ion must be $+3$.

Example 4.17

Treatment of teeth with fluoride ions produces fluoroapatite; the formula is $Ca_5(PO_4)_3F$. Show that this material is electrically neutral.

Solution Calcium exists in ionic compounds as Ca^{2+}. The phosphate ion has a -3 charge, and the fluoride ion has a -1 charge. The five Ca^{2+} ions have a total charge of $5(+2) = +10$. The three phosphate ions and one fluoride ion have a total charge of $3(-3) + (-1) = -10$. Thus, the compound is electrically neutral.

Names of Ionic Compounds

Ionic compounds are named for the ions that they contain. The name of the positive ion is given first, then the name of the negative ion. Thus, sodium chloride is the name for NaCl. The order of the ions in the formula is the positive ion followed by the negative ion. In the case of elements that form more than one cation, it is necessary to indicate the charge on the ion by use of the correct name for the ion. For example, $FeCl_2$ and $FeCl_3$ must be distinguished by different names. The names ferrous chloride and ferric chloride tell you that the ions in the two compounds are $+2$ and $+3$ ions, respectively. A newer alternate set of names is iron(II) chloride and iron(III) chloride. A detailed discussion of the names of ionic compounds is given in Chapter 11.

Example 4.18

What is the name for the compound Cu_2O?

Solution The oxide ion has a -2 charge. Therefore the charge on each of the two copper ions must be $+1$. The correct name for this ion is cuprous or copper(I). The compound is then named cuprous oxide or copper(I) oxide.

Example 4.19

What is the name for the compound $Fe_2(SO_4)_3$?

Solution   The sulfate ion has a -2 charge. Therefore the total negative charge of the three sulfate ions is -6. Thus each iron must have a $+3$ charge. The correct name for this ion is ferric or iron(III). The compound is then named ferric sulfate or iron(III) sulfate.

[*Additional examples may be found in 4.63–4.66 at the end of the chapter.*]

Summary

An atom is an electrically neutral particle that is the smallest representative unit of an element. The nucleus of an atom contains neutrons and protons; the proton has a positive charge. Surrounding the nucleus are a number of electrons equal to the number of protons in the nucleus. An electron is negatively charged.

Atoms are described by their atomic number and their mass number. The atomic number is the number of protons in the atom. The mass number is the sum of the number of protons and neutrons. Atoms that have the same number of protons but differ in the number of neutrons are called isotopes of the element. Elemental symbols with a subscript to the left of the symbol representing the atomic number and a superscript to the left of the symbol representing the mass number are used to give information about the atomic composition of the isotope.

The atomic weight of an element is a weighted average of the masses of the isotopes of the element present in nature.

A neutral unit consisting of atoms bonded to each other is called a molecule. Some elements exist as molecules and are represented by the elemental symbol and a subscript to the right to indicate the number of atoms in the molecule. Molecular compounds are represented by molecular formulas that indicate the number and types of atoms present in the molecule. Subscripts to the right of each symbol indicate the number of atoms of that element present.

A loss or gain of electrons by an atom yields an ion. Positive ions are cations; negative ions are anions. Ions held together by electrostatic forces of attraction result in ionic compounds. The relative number of each type of ion is such that the total charge of the cations equals the total charge of the anions. Ionic compounds are named by using the name of the positive ion followed by the name of the negative ion. The formulas for ionic compounds use subscripts to indicate the simplest ratio of ions contained in the substance.

New Terms

An **atomic mass unit** (amu) is equal to 1.6603×10^{-24} g and is used as the basic unit in describing atomic mass.

Anions are negatively charged atomic particles that result from the gain of electrons.

The **atomic number** of an element is equal to the number of protons in the nucleus of an atom of the element.

The **atomic weight** of an element is a weighted average of the masses of the isotopes present in natural abundance.

Cations are positively charged atomic particles that result from the loss of electrons.

The **electron** is a subatomic particle with a mass of 9.109×10^{-28} g and a charge of -1.60×10^{-19} coulomb (represented by a relative charge of -1). One or more electrons are present in all atoms.

Ionic compounds consist of collections of anions and cations in sufficient number to achieve electrical neutrality. The ions are attracted by electrostatic forces.

Isotopes of an element have the same number of protons but differ in the number of neutrons.

The **mass number** of an isotope is equal to the sum of the number of protons and neutrons in the nucleus of the atom.

Molecules are neutral units of atoms that possess the properties of the substance.

A **molecular formula** is a representation of a molecule that indicates the number and type of each atom present in the molecule.

A **neutron** is a neutral subatomic particle that is found in the nucleus of an atom. It has a mass of 1 amu.

The **nucleus** is a region in the center of an atom that contains the protons and neutrons.

A **polyatomic ion** is a group of bonded atoms that bear a positive or negative charge.

A **proton** is a subatomic particle with a mass of 1 amu and a charge of 1.60×10^{-19} coulomb (represented by a relative charge of $+1$).

Subatomic particles are the electrons, protons, and neutrons contained in atoms.

Exercises

Terminology

4.1 Explain the difference between each of the following terms.
   (a) electron and proton
   (b) proton and neutron

4.3 Using your own words, describe the arrangement of atomic particles in an atom.

4.2 Explain the difference between each of the following terms.
   (a) mass number and atomic number
   (b) cation and anion

4.4 Explain how modern atomic theory differs from Dalton's atomic theory.

Properties of the Atom

4.5 Calculate the mass of an atom in atomic mass units for each of the following isotopes.
   (a) an isotope of magnesium whose mass is 3.98×10^{-23} g
   (b) an isotope of oxygen whose mass is 2.99×10^{-23} g
   (c) an isotope of uranium whose mass is 3.90×10^{-22} g

4.7 An atom of argon has a mass of 6.63×10^{-23} g. How many atoms are in a 40.0 g sample of argon?

4.9 The atomic radius of arsenic is 1.21 Å. What is its radius in centimeters?

4.11 The atomic radius of sulfur is 0.104 nm. What is its radius in Angstroms?

4.6 Calculate the mass of an atom in atomic mass units for each of the following isotopes.
   (a) an isotope of beryllium whose mass is 1.16×10^{-23} g
   (b) an isotope of silicon whose mass is 4.65×10^{-23} g
   (c) an isotope of radium whose mass is 3.70×10^{-22} g

4.8 An atom of fluorine has a mass of 3.16×10^{-23} g. How many atoms are in a 19.0 g sample of fluorine?

4.10 The atomic radius of sodium is 1.57×10^{-8} cm. What is its radius in Angstroms?

4.12 The atomic radius of indium is 150 pm. What is its radius in nanometers?

Subatomic Particles

4.13 Why are relative weights of subatomic particles frequently used rather than the absolute masses?

4.15 What relationship, if any, exists between the number of electrons and protons in an atom?

4.17 What are the sizes of the nucleus and the atom?

4.14 Why is the mass of the electron considered to be zero on a relative mass basis?

4.16 What relationship, if any, exists between the number of electrons and neutrons in an atom?

4.18 What are the charges of the electron, proton, and neutron?

Elemental Symbols

4.19 What is the atomic number of an element? How is it represented in a symbol?

4.21 Indicate the number of subatomic particles present in each of the following isotopes.
   (a) $^{16}_{8}\text{O}$ (b) $^{23}_{11}\text{Na}$ (c) $^{27}_{13}\text{Al}$
   (d) $^{32}_{16}\text{S}$ (e) $^{40}_{18}\text{Ar}$ (f) $^{40}_{20}\text{Ca}$

4.23 Write the symbol for each of the following isotopes that contain the indicated number of protons and neutrons.
   (a) fluorine, 9 protons and 10 neutrons
   (b) silicon, 14 protons and 16 neutrons
   (c) silicon, 14 protons and 14 neutrons
   (d) phosphorus, 15 protons and 16 neutrons
   (e) potassium, 19 protons and 20 neutrons
   (f) titanium, 22 protons and 26 neutrons

4.20 What is the mass number of an element? How is it represented in a symbol?

4.22 Indicate the number of subatomic particles present in each of the following isotopes.
   (a) $^{19}_{9}\text{F}$ (b) $^{31}_{15}\text{P}$ (c) $^{39}_{19}\text{K}$
   (d) $^{79}_{34}\text{Se}$ (e) $^{85}_{37}\text{Rb}$ (f) $^{96}_{42}\text{Mo}$

4.24 Write the symbol for each of the following isotopes that contain the indicated number of protons and neutrons.
   (a) cesium, 55 protons and 77 neutrons
   (b) selenium, 34 protons and 45 neutrons
   (c) cadmium, 48 protons and 64 neutrons
   (d) tin, 50 protons and 68 neutrons
   (e) barium, 56 protons and 81 neutrons
   (f) mercury, 80 protons and 119 neutrons

Isotopes

4.25 Explain the significance of the symbols used to write the three isotopes of hydrogen: $^{1}_{1}\text{H}$, $^{2}_{1}\text{H}$, $^{3}_{1}\text{H}$.

4.27 The radioisotope $^{131}_{53}\text{I}$ is used to treat cancer of the thyroid. How does this isotope differ from $^{127}_{53}\text{I}$, which is required for normal functioning of the thyroid gland?

4.26 Explain the significance of the symbols used to write the three isotopes of oxygen: $^{18}_{8}\text{O}$, $^{17}_{8}\text{O}$, $^{16}_{8}\text{O}$.

4.28 The radioisotope $^{226}_{88}\text{Ra}$ was used in cancer therapy. Indicate the number of protons, electrons, and neutrons in this isotope.

4.29 Zinc (atomic number 30) is required in trace amounts for normal growth. The predominant isotope of zinc has a mass number of 64. What is the symbol for this isotope?

4.30 An isotope of chromium whose mass number is 51 is used in compounds to determine the position of the placenta in pregnant women. The atomic number of chromium is 24. What is the symbol of this isotope?

Atomic Weights

4.31 Why aren't the atomic weights in the Table of Atomic Weights all integers?

4.32 Mercury consists of seven naturally occurring isotopes. Explain how these isotopes affect the atomic weight of mercury.

4.33 Neon exists as neon-20, neon-21, and neon-22, where the numbers are the mass numbers of the isotopes. From the listed atomic weight of neon, determine which isotope is present in the largest amount.

4.34 The element bromine exists as the isotopes $^{79}_{35}Br$ and $^{81}_{35}Br$. The atomic weight is 79.9. Approximately how much of each isotope is present in nature?

4.35 Gallium (atomic weight 69.72) has two naturally occurring isotopes, the predominant one being $^{69}_{31}Ga$ with an isotopic weight of 68.9257 and abundance of 60.47%. Calculate the isotopic weight of the other isotope.

4.36 Copper (atomic weight 63.546) has two naturally occurring isotopes, the predominant one being $^{63}_{29}Cu$ with an isotopic weight of 62.9298 and an abundance of 69.09%. Calculate the isotopic weight of the other isotope.

Molecules and Molecular Formulas

4.37 Indicate which of the following elements exist as molecules. How many atoms are contained in each molecule?
(a) He   (b) H   (c) P
(d) Ar   (e) S   (f) N

4.38 Indicate which of the following elements exist as molecules. How many atoms are contained in each molecule?
(a) Cl   (b) Ne   (c) O
(d) F   (e) Kr   (f) I

4.39 Selenium (atomic number 34) resembles sulfur in many of its properties and has the same number of atoms in a molecule as sulfur. What is the molecular formula of selenium?

4.40 Arsenic (atomic number 33) resembles phosphorus in many of its properties and has the same number of atoms in a molecule as phosphorus. What is the molecular formula of arsenic?

4.41 What is meant by each of the following formulas for molecular compounds?
(a) H_2S (b) HCl   (c) N_2H_4
(d) C_2H_2 (e) C_2H_6 (f) H_2O_2

4.42 What is meant by each of the following formulas for molecular compounds?
(a) H_2SO_4 (b) HNO_3 (c) N_2O_4
(d) C_4H_{10} (e) $C_2H_6O_2$ (f) S_2Cl_2

4.43 A molecule of octane contains eight carbon atoms and 18 hydrogen atoms. What is the molecular formula of octane?

4.44 Nicotine has the molecular formula $C_{10}H_{14}N_2$. What does this formula mean to you?

4.45 Vitamin A has the molecular formula $C_{20}H_{30}O$. How many atoms are contained in a vitamin A molecule?

4.46 Vitamin D_1 has the molecular formula $C_{56}H_{88}O_2$. How many atoms are contained in a vitamin D_1 molecule?

Ions

4.47 Selenium can gain two electrons to form an ion. What is the symbol of the ion? How many electrons does this ion contain?

4.48 Phosphorus can gain three electrons to form an ion. What is the symbol of the ion? How many electrons does this ion contain?

4.49 Manganese can lose two electrons to form an ion. What is the symbol of the ion? How many electrons does this ion contain?

4.50 Scandium can lose three electrons to form an ion. What is the symbol of the ion? How many electrons does this ion contain?

4.51 Write proper symbols for each of the following ions.
(a) oxide ion   (b) sulfide ion   (c) iodide ion
(d) bromide ion   (e) nitride ion   (f) fluoride ion

4.52 Write proper symbols for each of the following ions.
(a) sodium ion   (b) calcium ion   (c) magnesium ion
(d) potassium ion (e) lithium ion   (f) zinc ion

4.53 What is the name of each of the following ions?
(a) S^{2-} (b) I^- (c) O^{2-}
(d) F^- (e) N^{3-} (f) Br^-

4.54 What is the name of each of the following ions?
(a) K^+ (b) Ca^{2+} (c) Zn^{2+}
(d) Mg^{2+} (e) Na^+ (f) Li^+

4.55 Write the proper symbol for the following ions.
(a) sulfate ion   (b) phosphate ion
(c) hydroxide ion   (d) ammonium ion
(e) cyanide ion   (f) carbonate ion

4.56 Write the proper symbol for the following ions.
(a) sulfite ion   (b) bicarbonate ion
(c) nitrite ion   (d) permanganate ion
(e) bisulfate ion   (f) bisulfite ion

Ionic Compounds

4.57 Write the correct formula for each of the following compounds.
  (a) lithium fluoride   (b) zinc oxide
  (c) magnesium carbonate   (d) potassium nitrate
  (e) sodium cyanide   (f) aluminum sulfide
  (g) calcium hypochlorite   (h) barium nitrite

4.58 Write the correct formula for each of the following compounds.
  (a) lithium phosphate   (b) zinc sulfate
  (c) magnesium permanganate   (d) sodium bisulfate
  (e) calcium hydroxide   (f) zinc phosphate
  (g) potassium sulfite   (h) silver nitrate

4.59 Write the correct formula for each of the following compounds.
  (a) ferric fluoride   (b) cuprous oxide
  (c) ferrous sulfate   (d) cupric cyanide
  (e) ferric oxide   (g) cupric nitrate

4.60 Write the correct formula for each of the following compounds.
  (a) iron(II) chloride   (b) copper(I) oxide
  (c) lead(II) sulfide   (d) mercury(II) bromide
  (e) ferric sulfide   (f) cupric sulfide
  (g) ferrous sulfide   (h) cuprous sulfide

4.61 Write the correct formulas for compounds containing the following ions
  (a) Fe^{3+} and Cl^- (b) Na^+ and OH^-
  (c) Mg^{2+} and OH^- (d) Cd^{2+} and S^{2-}
  (e) Mn^{2+} and F^- (f) Cs^+ and N^{3-}

4.62 Write the correct formulas for compounds containing the following ions
  (a) Al^{3+} and O^{2-} (b) Na^+ and CO_3^{2-}
  (c) Mg^{2+} and MnO_4^- (d) Cd^{2+} and PO_4^{3-}
  (e) Zn^{2+} and SO_3^{2-} (f) K^+ and S^{2-}

4.63 Name the following compounds.
  (a) $Ca(OH)_2$ (b) $LiClO_4$
  (c) Na_3PO_4 (d) K_2SO_4
  (e) $NaNO_3$ (f) NH_4NO_2
  (g) $MgCl_2$ (h) $AlBr_3$

4.64 Name the following compounds.
  (a) $Zn(CN)_2$ (b) $Cd(ClO_3)_2$
  (c) $Mg_3(PO_4)_2$ (d) $LiNO_3$
  (e) $Ba(HSO_3)_2$ (f) $(NH_4)_2SO_4$
  (g) $Mg(HCO_3)_2$ (h) $ScCl_3$

4.65 Name the following compounds.
  (a) $Ba(OH)_2$ (b) $KClO_2$
  (c) Cs_3PO_4 (d) K_2SO_3
  (e) $NaNO_2$ (f) NH_4NO_3
  (g) $CaBr_2$ (h) $GaCl_3$

4.66 Name the following compounds.
  (a) $ZnCO_3$ (b) $Cd(ClO_2)_2$
  (c) $Ca_3(PO_4)_2$ (d) KNO_3
  (e) $Ba(HSO_4)_2$ (f) $(NH_4)_2SO_3$
  (g) $Sr(HCO_3)_2$ (h) ScI_3

Additional Exercises

4.67 Why is the mass of the hydrogen atom essentially equal to the mass of a proton?

4.68 What are the masses of the electron, proton, and neutron in atomic mass units?

4.69 Calculate the mass in grams of an atom of each of the following isotopes.
  (a) $^{16}_{8}O$ (b) $^{23}_{11}Na$ (c) $^{27}_{13}Al$
  (d) $^{32}_{16}S$ (e) $^{40}_{18}Ar$ (f) $^{40}_{20}Ca$

4.70 Indicate the number of subatomic particles present in each of the following isotopes.
  (a) $^{108}_{47}Ag$ (b) $^{119}_{50}Sn$ (c) $^{133}_{55}Cs$
  (d) $^{197}_{79}Au$ (e) $^{207}_{82}Pb$ (f) $^{243}_{95}Am$

4.71 An atom of yttrium has a mass of 1.48×10^{-22} g. How many atoms are in a 88.9 g sample of yttrium.

4.72 The atomic radius of barium is 198 pm. What is its radius in Angstroms?

4.73 What relationship, if any, exists between the number of protons and neutrons in an atom?

4.74 The $^{14}_{6}C$ isotope of ancient wooden objects is used to determine the age of the object. In what way does this isotope differ from $^{12}_{6}C$?

4.75 There are two naturally occurring isotopes of bromine, $^{79}_{35}Br$ and $^{81}_{35}Br$. In what way do they differ?

4.76 The radioisotope $^{60}_{27}Co$ is used in cancer therapy. Indicate the number of protons, electrons, and neutrons in this isotope.

4.77 Silicon contains isotopes $^{28}_{14}Si$, $^{29}_{14}Si$, and $^{30}_{14}Si$ in 92.2, 4.7, and 3.1% abundance, respectively. The atomic masses are 27.97693, 28.97649, and 29.97376 amu, respectively. Calculate the atomic weight of silicon.

4.78 Magnesium contains isotopes $^{24}_{12}Mg$, $^{25}_{12}Mg$, and $^{26}_{12}Mg$ in 78.7, 10.1, and 11.2% abundance, respectively. The atomic masses are 23.98504, 24.98584, and 25.98259 amu, respectively. Calculate the atomic weight of magnesium.

4.79 A molecule of TNT contains seven carbon atoms, five hydrogen atoms, three nitrogen atoms, and six oxygen atoms. Write the molecular formula of TNT.

4.80 Table sugar has the molecular formula $C_{12}H_{22}O_{11}$. What does this formula mean to you?

4.81 The insecticide malathion has the molecular formula $C_{10}H_{19}O_6PS_2$. How many atoms are contained in a malathion molecule?

4.82 A molecule of nitroglycerin has the molecular formula $C_3H_5N_3O_9$. How many atoms are contained in a nitroglycerin molecule?

4.83 Cerium can form two ions. One results from the loss of three electrons and the other from the loss of four electrons. What are the symbols of the ions? How many electrons does each ion contain?

4.84 Thallium can form two ions. One results from the loss of one electron and the other from the loss of three electrons. What are the symbols of the ions? How many electrons does each ion contain?

4.85 What is the difference in the number of electrons contained in the ferrous and ferric ion?

4.86 What is the difference in the number of electrons contained in the cuprous and cupric ion?

4.87 Write the proper symbol for the following ions.
(a) perchlorate ion   (b) hypochlorite ion
(c) chlorite ion   (d) chlorate ion
(e) hypoiodite ion   (f) bromate ion

4.88 Write the proper symbols for the following ions.
(a) hypobromite ion   (b) iodate ion
(c) iodite ion   (d) perbromate ion
(e) periodate ion   (f) bromite ion

4.89 Explain what each of the following symbols for ions represents.
(a) $WO_4{}^{2-}$ (b) $HPO_4{}^{2-}$ (c) $P_2O_7{}^{4-}$
(d) $VO_4{}^{+}$ (e) $SeO_2{}^{2-}$ (f) $S_2O_8{}^{2-}$
(g) $H_2BO_3{}^{-}$ (h) $AlF_6{}^{3-}$ (i) $P_4O_{12}{}^{4-}$

4.90 Write the correct formulas for compounds containing the following ions.
(a) Sc^{3+} and CN^{-} (b) Ni^{2+} and $HSO_3{}^{-}$
(c) Zn^{2+} and $ClO_3{}^{-}$ (d) Hg^{2+} and $NO_3{}^{-}$
(e) Ag^{+} and $SO_4{}^{2-}$ (f) Sr^{2+} and N^{3-}

4.91 Determine how many atoms of each element are present in each of the following compounds.
(a) $Zn(CN)_2$ (b) $Cd(ClO_3)_2$
(c) $Mg_3(PO_4)_2$ (d) Li_2SO_3
(e) $Ba(HSO_3)_2$ (f) $(NH_4)_2SO_4$
(g) $Mg(HCO_3)_2$ (h) $Sc(HSO_4)_3$

4.92 Antacids are used to neutralize excess stomach acid. Write the formula of the indicated component of the listed antacids.
(a) calcium carbonate found in Tums
(b) sodium bicarbonate found in Alka-Seltzer
(c) magnesium hydroxide found in Milk of Magnesia
(d) aluminum hydroxide found in Maalox

4.93 The antacid in Rolaids is $NaAl(OH)_2CO_3$. Show that this substance is electrically neutral.

4.94 The oxalate ion is $C_2O_4{}^{2-}$. What is the formula for scandium(III) oxalate?

5 Composition of Compounds

Learning Objectives

After studying Chapter 5 you should be able to

1. Calculate the molecular weight of a molecule.
2. Calculate the formula weight of an ionic compound.
3. Calculate the percent composition of a substance given the chemical formula.
4. Relate the mass, the number of moles, and the number of units of matter in a sample.
5. Calculate the empirical formula from percent composition or combining masses.
6. Determine the molecular formula of a compound from the empirical formula and the molecular weight.

5.1 Compounds—Submicroscopic and Macroscopic Views

In discussing matter and its reactions, we consider compounds more frequently than elements because there are many more of them. In the preceding chapter you learned that on a submicroscopic level, compounds are of two types—molecular and ionic. Molecules are composed of atoms that are bonded to each other to form a discrete unit as in the case of the water molecule, which consists of two hydrogen atoms and one oxygen atom. Ionic compounds are composed of cations and anions in a ratio such that electrical neutrality is maintained. For example, in sodium chloride the ratio of Na^+ ions to Cl^- ions is $1:1$, but the ratio of Na^+ ions to O^{2-} ions in sodium oxide is $2:1$. In ionic compounds a collection of cations and anions is held together by the attraction of the positive and negative charges.

As indicated in Chapter 1, composition can be discussed both in terms of the number of items and the mass of those items. Now that we have an understanding of the number of atoms and ions in compounds, it is time to shift our focus to the mass of the components in compounds. How can we deal with the mass of the water molecule or the sodium and chloride ions in sodium chloride when we cannot see or directly weigh these materials? In this chapter you will learn how to express the mass of molecules and ions on a submicroscopic level. The method used is similar to that presented in Chapter 4 to express the mass of atoms in atomic mass units (amu).

It is also necessary to deal with samples of compounds on a macroscopic level—that is, amounts that we can see and weigh. These samples contain large numbers of atoms and ions. In this chapter we will learn how to determine the number of atoms and ions in macroscopic samples of compounds. Thus a connection will be established between the submicroscopic and macroscopic levels of chemistry.

5.2 Molecular and Formula Weights of Compounds

Recall from Chapter 4 that the mass of an atom can be expressed in terms of atomic mass units (amu). Although the atomic weights of various atoms are known with accuracies from 0.01 to 0.0001 amu, the values will usually be rounded to the nearest 0.1 amu in this and following chapters. This procedure provides a sufficient number of significant figures for the calculations in this text. Thus the masses of the hydrogen and carbon atoms are expressed as 1.0 and 12.0 amu, respectively.

Molecular compounds are represented by molecular formulas using the symbols of the constituent atoms and subscripts to indicate how many of each type of atom are present in the molecule. The **molecular weight** of a molecule is the sum of the masses in atomic mass units of the component atoms indicated in the molecular formula. The molecule methane, whose molecular formula is CH_4, has a molecular weight of 16.0 amu to the nearest 0.1 amu.

mass of 1 carbon atom	$= 1 \times 12.0$ amu	$= 12.0$ amu
mass of 4 hydrogen atoms	$= 4 \times\ \ 1.0$ amu	$=\ \ \underline{4.0\text{ amu}}$
mass of 1 methane molecule		$= 16.0$ amu

Example 5.1

What is the molecular weight of vitamin C, $C_6H_8O_6$? Use atomic weights to the nearest 0.1 amu.

Solution Sum the atomic weights of the constituent atoms, taking into account the number of atoms of each kind present in the molecule.

mass of 6 carbon atoms $= 6 \times 12.0$ amu $= 72.0$ amu
mass of 8 hydrogen atoms $= 8 \times 1.0$ amu $= 8.0$ amu
<u>mass of 6 oxygen atoms</u> $= 6 \times 16.0$ amu $= \underline{96.0}$ amu
mass of 1 $C_6H_8O_6$ molecule $= 176.0$ amu

The molecular weight is 176.0 amu.

[*Additional examples may be found in 5.7–5.14 at the end of the chapter.*]

Ionic substances do not exist as molecules, but rather as collections of ions described by a formula unit giving the simplest ratio of ions that are present in the compound. The **formula weight** of an ionic compound is the sum of the atomic weights of the atoms indicated by a formula unit of the substance.

Example 5.2

Limestone is calcium carbonate, $CaCO_3$. What is the formula weight of calcium carbonate? Use atomic weights to the nearest 0.1 amu.

Solution Calcium carbonate is an ionic compound consisting of calcium ions and carbonate ions in a 1:1 ratio. The formula weight is obtained by summing the atomic weights of the atoms.

mass of 1 calcium atom $= 1 \times 40.1$ amu $= 40.1$ amu
mass of 1 carbon atom $= 1 \times 12.0$ amu $= 12.0$ amu
<u>mass of 3 oxygen atoms</u> $= 3 \times 16.0$ amu $= \underline{48.0}$ amu
mass of 1 $CaCO_3$ unit $= 100.1$ amu

The formula weight is 100.1 amu.

Example 5.3

What is the formula weight of zinc phosphate, $Zn_3(PO_4)_2$? Use atomic weights to the nearest 0.1 amu.

Solution Zinc phosphate is an ionic compound consisting of zinc ions and phosphate ions in a 3:2 ratio. The formula weight is obtained by summing the atomic weights of the constitutent atoms. Note that the subscript 2 following the parenthesis after the phosphate ion means that all atoms within the parentheses must be multiplied by 2. There are 2 atoms of phosphorus and 8 atoms of oxygen in the formula unit of this compound.

mass of 3 zinc atoms	$= 3 \times 65.4$ amu	$= 196.2$ amu
mass of 2 phosphorus atoms	$= 2 \times 31.0$ amu	$= 62.0$ amu
mass of 8 oxygen atoms	$= 8 \times 16.0$ amu	$= 128.0$ amu
mass of 1 $Zn_3(PO_4)_2$ unit		$= 386.2$ amu

The formula weight is 386.2 amu.

[*Additional examples may be found in 5.15–5.22 at the end of the chapter.*]

5.3 Percent Composition

The **percent composition** of a given element in a compound is equal to the mass of that element divided by the total mass of all of the elements in the compound with the quotient multiplied by 100. For example, a 44.0 g sample of carbon dioxide contains 12.0 g of carbon and 32.0 g of oxygen. The mass percent of carbon in carbon dioxide is 27.3% and the mass percent of oxygen is 72.7%. Note that three significant figures are required in the answer based on the number of significant figures in the given quantities.

$$\frac{12.0 \text{ g carbon}}{44.0 \text{ g carbon dioxide}} \times 100 = 27.2727 = 27.3\% \text{ carbon}$$

$$\frac{32.0 \text{ g oxygen}}{44.0 \text{ g carbon dioxide}} \times 100 = 72.7272 = 72.7\% \text{ oxygen}$$

Example 5.4

A 250 mg tablet of vitamin C contains 102 mg of carbon, 12 mg of hydrogen, and 136 mg of oxygen. What is the mass percent composition of vitamin C?

Solution   To solve this problem, the mass of each individual element must be divided by the total mass and the quotient multiplied by 100.

$$\% \text{ C} = \frac{102 \text{ mg C}}{250 \text{ mg vitamin C}} \times 100 = 40.8\% \text{ C}$$

$$\% \text{ H} = \frac{12 \text{ mg H}}{250 \text{ mg vitamin C}} \times 100 = 4.8\% \text{ H}$$

$$\% \text{ O} = \frac{136 \text{ mg O}}{250 \text{ mg vitamin C}} \times 100 = 54.4\% \text{ O}$$

[*Additional examples may be found in 5.23 and 5.24 at the end of the chapter.*]

The percent composition of a compound can also be calculated from the molecular formula of a molecular compound or the formula unit of an ionic compound. Consider the molecule ethyl alcohol, C_2H_6O, whose molecular weight is 46.0 amu. Of this weight, 24.0 amu is due to the two atoms of carbon, 6.0 amu is due to the six atoms of hydrogen, and 16.0 amu is due to the single atom of oxygen. The percent composition is calculated as follows.

$$\% \text{ C} = \frac{24.0 \text{ amu C}}{46.0 \text{ amu C}_2\text{H}_6\text{O}} \times 100 = 52.1739 = 52.2\% \text{ C}$$

$$\% \text{ H} = \frac{6.0 \text{ amu H}}{46.0 \text{ amu C}_2\text{H}_6\text{O}} \times 100 = 13.0435 = 13\% \text{ H}$$

$$\% \text{ O} = \frac{16.0 \text{ amu O}}{46.0 \text{ amu C}_2\text{H}_6\text{O}} \times 100 = 34.7826 = 34.8\% \text{ O}$$

Example 5.5

All aluminum products have a coating of aluminum oxide, Al_2O_3, resulting from the reaction of the aluminum with oxygen. What is the percent composition of aluminum oxide? Use atomic weights to the nearest 0.1 amu.

Solution   First determine the formula weight. This process will also give you the weights of each component of the compound.

mass of 2 aluminum atoms	= 2 × 27.0 amu =	54.0 amu
mass of 3 oxygen atoms	= 3 × 16.0 amu =	48.0 amu
mass of 1 Al_2O_3 unit		= 102.0 amu

The percent composition is

$$\% \text{ Al} = \frac{54.0 \text{ amu Al}}{102.0 \text{ amu Al}_2\text{O}_3} \times 100 = 52.9412\% = 52.9\% \text{ Al}$$

$$\% \text{ O} = \frac{32.0 \text{ amu O}}{102.0 \text{ amu Al}_2\text{O}_3} \times 100 = 47.0588\% = 47.1\%$$

[*Additional examples may be found in 5.31 and 5.32 at the end of the chapter.*]

Example 5.6

What is the percent composition of ammonium carbonate, $(NH_4)_2CO_3$, which is used in smelling salts? Use atomic weights to the nearest 0.1 amu.

Solution   Determine the formula weight. Remember that all atoms within parentheses must be multiplied by the subscript after the parenthesis.

mass of 2 nitrogen atoms	= 2 × 14.0 amu	= 28.0 amu
mass of 8 hydrogen atoms	= 8 × 1.0 amu	= 8.0 amu
mass of 1 carbon atom	= 1 × 12.0 amu	= 12.0 amu
mass of 3 oxygen atoms	= 3 × 16.00 amu	= 48.0 amu
mass of 1 $(NH_4)_2CO_3$ unit		= 96.0 amu

The percent composition is

$$\% \ N = \frac{28.0 \ \text{amu N}}{96.0 \ \text{amu (NH}_4)_2\text{CO}_3} \times 100 = 29.1667\% = 29.2\% \ N$$

$$\% \ H = \frac{8.0 \ \text{amu H}}{96.0 \ \text{amu (NH}_4)_2\text{CO}_3} \times 100 = 8.3333\% = 8.3\% \ H$$

$$\% \ C = \frac{12.0 \ \text{amu C}}{96.0 \ \text{amu (NH}_4)_2\text{CO}_3} \times 100 = 12.5\% \ C$$

$$\% \ O = \frac{48.0 \ \text{amu O}}{96.0 \ \text{amu (NH}_4)_2\text{CO}_3} \times 100 = 50.0\% \ O$$

[*Additional examples may be found in 5.33 and 5.34 at the end of the chapter.*]

Example 5.7

Glucose in the blood provides you with chemical energy. What is the percent composition of glucose, $C_6H_{12}O_6$? Use atomic weights to the nearest 0.1 amu.

Solution   The molecular weight is 180.0 amu.

mass of 6 carbon atoms $= 6 \times 12.0 \ \text{amu} = \quad 72.0 \ \text{amu}$
mass of 12 hydrogen atoms $= 12 \times 1.0 \ \text{amu} = \quad 12.0 \ \text{amu}$
mass of 6 oxygen atoms $= 6 \times 16.0 \ \text{amu} = \quad \underline{96.0 \ \text{amu}}$
mass of $C_6H_{12}O_6$ molecule $= 180.0 \ \text{amu}$

$$\% \ C = \frac{72.0 \ \text{amu C}}{180.0 \ \text{amu } C_6H_{12}O_6} \times 100 = 40.0\% \ C$$

$$\% \ H = \frac{12.0 \ \text{amu H}}{180.0 \ \text{amu } C_6H_{12}O_6} \times 100 = 6.6667\% = 6.67\% \ H$$

$$\% \ O = \frac{96.0 \ \text{amu O}}{180.0 \ \text{amu } C_6H_{12}O_6} \times 100 = 53.3333\% = 53.3\% \ O$$

[*Additional examples may be found in 5.27–5.30 at the end of the chapter.*]

5.4 Avogadro's Number and the Mole

Matter can be dealt with on two levels in chemistry. Up to this point we have been concerned with matter at the submicroscopic level. However, it is also necessary to be able to deal with matter that can be seen and weighed, that is, on the macroscopic level. In the chemical laboratory, substances are weighed in terms of grams rather than atomic mass units. Recall that 1 amu is 1.67×10^{-23} g. This value is incredibly small, and no balance can detect such a small quantity. Approximately 10^{16} atoms are required to be detected on the most sensitive balance. Accordingly, chemists have devised a system for comparing quantities of substances containing large collections of atoms, molecules, or ions. Three quantities—Avogadro's number, the mole, and the molar mass—are used to provide the connection between the submicroscopic and macroscopic levels of matter.

1. **Avogadro's number** is the number of atoms contained in exactly 12 g of carbon-12. The number is 6.02×10^{23}.
2. A **mole** of any substance contains Avogadro's number of structural units.
3. The **molar mass** of a substance is the mass in grams of 1 mole of that substance.

Avogadro's Number

Amadeo Avogadro was a nineteenth century Italian scientist who contributed to the development of the concept of atomic weights. In his honor the quantity connecting the submicroscopic and macroscopic levels of matter in chemistry is named after him. Before we discuss Avogadro's number, it may be useful to your understanding of this quantity to first present an analogy with common items. In the analogy, the submicroscopic level is represented by single coins and macroscopic level by a large weight of coins. The quarter and the penny weigh 5.668 and 3.015 g, respectively. The ratio of the mass of a quarter to the mass of a penny is 1.880. Now consider taking 40 pennies to form a "stack" of pennies (Figure 5.1). The mass of pennies is 120.6 g. What would be the mass of a "stack" of 40 quarters? If we count out and weigh a "stack" of 40 quarters, we find that the mass is 227.4 g. However, since we know the ratio of the masses of the individual coins, the ratio of the masses of identical numbers of those coins, such as 40 pennies and 40 quarters, also must be the same (Figure 5.1). We may calculate the mass of 40 quarters using the mass of 40 pennies and a ratio of the masses of the individual coins.

$$120.6 \text{ g pennies} \times \frac{5.668 \text{ g quarter}}{3.015 \text{ g penny}} = 227.4 \text{ g quarters}$$

The idea illustrated by this example using a "stack" of 40 coins is that if the masses of single items are known, then the masses of equal numbers of those items must stand in the same ratio. You should note that this is true regardless of the number of items. There could be 40, 4000, or even 4×10^{12} and the ratio of the masses of quarters to pennies is still 1.880.

Now let's consider a mint that stamps coins. It might have to make a large number of pennies, which weighs 3.015 tons. Now that's a lot of pennies! We can calculate the number of pennies because we know the mass of an individual penny. However, let's not bother to do the calculation. Whatever the number is, we will call it a "mint number." Now let's consider the assignment of producing a "mint number" of quarters. We don't know the value of a "mint number"! Furthermore, even if we did, would we have to count the quarters to be sure that the correct number was produced? The answer is no. The weight of a "mint number" of quarters must be 1.880 times greater than a mint number of pennies, or 5.668 tons. The important point is that you don't have to count the individual items if you want equal numbers of two different items. It is only necessary to weigh out enough of each item so that the ratios of the masses of the samples is the same as the ratio of the masses of the individual items.

Now suppose that we want to weigh out the same number of atoms of two different elements. If we weigh atoms in ratios according to their atomic weights, we will obtain samples containing the same numbers of atoms. The atomic weights of carbon and sulfur are 12.01 and 32.06, respectively. Thus, if we have 12.01 g of carbon and 32.06 g of sulfur, the two samples will contain the same number of atoms (Figure 5.2). This conclusion is correct even though the samples consist of mixtures of isotopes. The atomic weights represent an average of the weights of the individual isotopes. The average sulfur atom, whose atomic weight is 32.06 amu, is 2.669 times heavier than the average carbon atom, whose atomic weight is 12.01 amu. There-

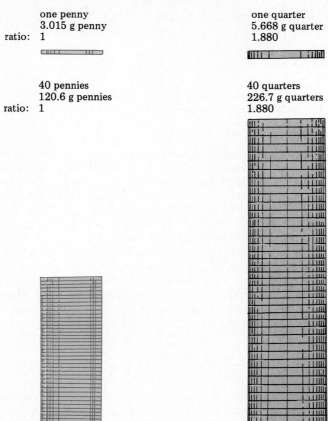

one penny
3.015 g penny
ratio: 1

one quarter
5.668 g quarter
1.880

40 pennies
120.6 g pennies
ratio: 1

40 quarters
226.7 g quarters
1.880

Figure 5.1

An analogy for the mole concept
*The ratio of the mass of a quarter to the mass of a penny is
1.880. The ratio of the mass of a stack of 40 quarters to the
mass of a stack of 40 pennies also is 1.880.*

fore, when we weigh 2.669 times as much sulfur as carbon, we obtain exactly the same number
of atoms.

The number of atoms in exactly 12 g of carbon-12 atoms is called Avogadro's number. How
many atoms of sulfur are in 32.06 g of sulfur? Avogadro's number is the answer because the
mass of sulfur is 2.669 times heavier than the mass of carbon, but the individual sulfur atoms are
2.669 times heavier than the individual carbon atoms. Therefore, the same number of atoms
must be in each sample. We can generalize that there are Avogadro's number of atoms of any
element in a sample having the mass in grams numerically equal to its atomic mass. This
concept is the connecting link between the mass and the number of atoms or molecules in a
sample of matter.

In the analogy using coins, we were able to use the "mint number" without knowing its
value. If you were curious and decided to calculate it, you would have obtained 9.091×10^5. In
the example using atoms, we also do not need to know the value of Avogadro's number at this

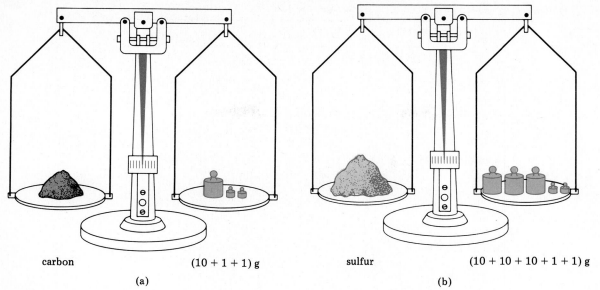

carbon	$(10 + 1 + 1)$ g	sulfur	$(10 + 10 + 10 + 1 + 1)$ g
(a)		(b)	

Figure 5.2
The weight of equal number of atoms
In experiment (a), the sample of carbon weighs 12 g. In experiment (b), the sample of sulfur
weighs 32 g. Both samples contain the same number of atoms because the ratio of the masses
of the individual atoms is the same as the ratio of the masses of the samples.

point. However, its value is 6.022045×10^{23}. In this text, 6.02×10^{23} to three significant figures is more convenient to use.

As previously mentioned, the use of Avogadro's number as a connection between the submicroscopic and the macroscopic levels can be extended to molecules. A methane molecule, CH_4, contains one carbon atom. In 16.0 amu of methane there is 12.0 amu of carbon. Therefore, in 16.0 g of methane there must be 12.0 g of carbon. How many molecules of methane are contained in 16.0 g? The answer is Avogadro's number because there are Avogadro's number of carbon atoms present in the 12.0 g of carbon and one carbon atom is required to produce each molecule of methane.

If we weigh molecules in ratios according to their molecular weights, the samples will contain the same number of molecules. For example, 16.0 g of methane and 44.0 g of carbon dioxide contain the same number of molecules. The molecular weights of methane and carbon dioxide are 16.0 and 44.0 amu, respectively, and therefore any quantities of these two compounds in the ratio of 16.0 to 44.0 must contain the same number of molecules. Since 16.0 g of methane contains Avogadro's number of molecules, it follows that 44.0 g of carbon dioxide also contains Avogadro's number of molecules. We can generalize that there are Avogadro's number of molecules in a sample having the mass in grams numerically equal to its molecular weight.

Although there are no discrete units in ionic compounds that correspond to those represented in formulas such as $CaCO_3$, it is nevertheless convenient to extend the use of Avogadro's number to ionic compounds. Avogadro's number of carbon atoms can be used to produce Avogadro's number of $CaCO_3$ units. The formula weight of $CaCO_3$ is 100.1 amu. From 12.0 g of carbon we can produce 100.1 g of $CaCO_3$, which then must have Avogadro's number of

Table 5.1 Avogadro's Number of Particles and the Mole

Substance	Formula	Number of particles in a mole
argon	Ar	6.02×10^{23} Ar atoms
nitrogen	N_2	6.02×10^{23} N_2 molecules $2 \times (6.02 \times 10^{23})$ N atoms
carbon dioxide	CO_2	6.02×10^{23} CO_2 molecules 6.02×10^{23} C atoms $2 \times (6.02 \times 10^{23})$ O atoms
magnesium oxide	MgO	6.02×10^{23} MgO formula units 6.02×10^{23} Mg^{2+} ions 6.02×10^{23} O^{2-} ions
zinc fluoride	ZnF_2	6.02×10^{23} ZnF_2 formula units 6.02×10^{23} Zn^{2+} ions $2 \times (6.02 \times 10^{23})$ F^- ions
lithium carbonate	Li_2CO_3	6.02×10^{23} Li_2CO_3 formula units $2 \times (6.02 \times 10^{23})$ Na^+ ions 6.02×10^{23} CO_3^{2-} ions

$CaCO_3$ units. We can generalize that there are Avogadro's number of formula units of an ionic compound in a sample having the mass in grams numerically equal to its formula weight. A summary of how Avogadro's number relates atoms, molecules, and ionic compounds is given in Table 5.1.

The Mole and Molar Mass

You are familar with several terms relating the number of items in a sample. A dozen is 12, a gross is 144, and a ream is 500. In chemistry the quantity 6.02×10^{23}, Avogadro's number, is the number of items in a mole. Thus a mole of any substance has the same number of structural units as a mole of any other substance. A mole of carbon contains the same number of atoms as a mole of sulfur. A mole of ethyl alcohol has the same number of molecules as a mole of glucose.

The mass of a mole of a substance is the molar mass and has the units grams per mole (g/mole). What is the molar mass of carbon-12? Avogadro's number of carbon atoms is contained in exactly 12 g of carbon atoms and Avogadro's number of atoms is a mole. Thus, the molar mass of carbon is 12 g/mole. It follows that the molar mass of any substance is numerically equal to the mass of the substance expressed on the atomic weight scale.

A mole of helium by definition contains 6.02×10^{23} helium atoms. A mole of argon by definition has 6.02×10^{23} argon atoms. Since the atomic weights of helium and argon are 4.0 and 40.0 amu, respectively, a mole of helium weighs 4.0 g, whereas a mole of argon weighs 40.0 g. The molar masses of helium and argon are 4.0 and 40.0 g/mole, respectively.

For compounds consisting of molecules, a mole contains 6.02×10^{23} molecules. A mole of carbon dioxide (CO_2) contains 6.02×10^{23} carbon dioxide molecules. A mole of glucose contains 6.02×10^{23} glucose ($C_6H_{12}O_6$) molecules. The weight of a mole, of course, depends on the substance. The molar mass of each of these substances is numerically equal to its individual molecular weight.

The subscripts in the formulas of molecular compounds represent the number of atoms of

Table 5.2   Some Relationships Between Moles of Substances

Substance	Formula	Mass of 1 mole	Number of moles of components
helium	He	4.0 g	1 mole of He atoms
oxygen	O_2	32.0 g	1 mole of O_2 molecules 2 moles of O atoms
water	H_2O	18.0 g	1 mole of H_2O molecules 2 moles of H atoms 1 mole O atoms
sodium chloride	NaCl	58.4 g	1 moles of NaCl formula units 1 mole Na^+ ions 1 mole Cl^- ions
calcium chloride	$CaCl_2$	111.0 g	1 mole of $CaCl_2$ formula units 1 mole of Ca^{2+} ions 2 moles of Cl^- ions
sodium sulfate	Na_2SO_4	142.0	1 mole of Na_2SO_4 formula units 2 moles of Na^+ ions 1 mole of SO_4^{2-} ions

each element present in a molecule of the substance. Therefore, the subscripts also indicate the number of moles of the atoms of the elements present in a mole of molecules. In one mole of carbon dioxide there are 1 mole of carbon atoms and 2 moles of oxygen atoms. One mole of carbon dioxide contains 6.02×10^{23} atoms of carbon, and $2 \times (6.02 \times 10^{23})$ atoms of oxygen.

A mole of an ionic compound contains 6.02×10^{23} formula units. Thus, 1 mole of $CaCO_3$ contains 6.02×10^{23} calcium ions and 6.02×10^{23} carbonate ions, which account for the 6.02×10^{23} formula units of $CaCO_3$. In 1 mole of sodium carbonate, Na_2CO_3, there are $2 \times (6.02 \times 10^{23})$ sodium ions and 6.02×10^{23} carbonate ions. A summary of the number of moles of atoms, molecules, and ions present in a mole of various substances is given in Table 5.2.

5.5   Calculations Using Moles

The definition of a mole in terms of Avogadro's number gives us two conversion factors that are reciprocals of each other.

$$\frac{6.02 \times 10^{23} \text{ entities}}{1 \text{ mole entities}} \quad \text{and} \quad \frac{1 \text{ mole entities}}{6.02 \times 10^{23} \text{ entities}}$$

We can use these factors to determine the number of entities on the submicroscopic scale and relate them to our counting unit of the macroscopic scale, the mole. If we are considering molecules, then it follows that

$$\text{number of molecules} \times \frac{1 \text{ mole}}{6.02 \times 10^{23} \text{ molecules}} = \text{number of moles}$$

$$\text{number of moles} \times \frac{6.02 \times 10^{23} \text{ molecules}}{1 \text{ mole}} = \text{number of molecules}$$

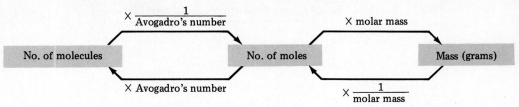

Figure 5.3

Relationships among molecules, moles, and mass

The molar mass of a substance is the mass in grams of 1 mole of that substance.

$$\text{molar mass} = \frac{\text{grams}}{\text{mole}}$$

We can use the molar mass or its reciprocal to relate the number of moles and the number of grams of a substance. In the case of determining the mass of a given number of moles we use

number of moles × molar mass = mass

$$\text{moles} \times \frac{\text{grams}}{\text{mole}} = \text{grams}$$

To determine the number of moles in a given mass we use

number of grams × $\dfrac{1}{\text{molar mass}}$ = moles

$$\text{grams} \times \frac{\text{moles}}{\text{grams}} = \text{moles}$$

In summary the quantities Avogadro's number, the mole, and the molar mass are related as shown in Figure 5.3.

Example 5.8

How many moles of gold atoms are present in a 30.00 kg ingot of gold?

Solution The atomic weight of gold is 197.0 amu and the molar mass of gold is 197.0 g. The number of moles in the ingot is calculated as follows.

$$3.000 \times 10^4 \text{ g Au} \times \frac{1 \text{ mole Au atoms}}{197.0 \text{ g Au}} = 152.2843 \text{ moles Au atoms}$$

The correct answer to the four required significant figures is 152.3 moles Au atoms.

[*Additional examples may be found in 5.41 and 5.42 at the end of the chapter.*]

Example 5.9

An average adult exhales 1.0×10^3 g of carbon dioxide per day. How many moles of carbon dioxide are exhaled?

Solution The molecular weight of carbon dioxide, CO_2, is 44.0 amu. Therefore, the number of moles in 1.0×10^3 g is calculated by multiplying by the reciprocal of the molar mass.

$$1.0 \times 10^3 \text{ g } CO_2 \times \frac{1 \text{ mole } CO_2}{44.0 \text{ g } CO_2} = 22.7273 \text{ moles } CO_2$$

The correct answer to the two required significant figures is 23 moles CO_2.

[*Additional examples may be found in 5.43–5.46 at the end of the chapter.*]

Example 5.10

Ferric sulfate, $Fe_2(SO_4)_3$, can be applied to soil that is too alkaline. How many moles are contained in 50.0 lb (22.7 kg) of this substance?

Solution First determine the formula weight of the substance.

mass of 2 iron atoms	$= 2 \times 55.8$ amu	$= 111.6$ amu
mass of 3 sulfur atoms	$= 3 \times 32.1$ amu	$= 96.3$ amu
mass of 12 oxygen atoms	$= 12 \times 16.0$ amu	$= 192.0$ amu
formula weight of $Fe_2(SO_4)_3$		$= 399.9$ amu

Thus, the mass of one mole is 399.9 g. The number of moles in the sample is

$$2.27 \times 10^4 \text{ g } Fe_2(SO_4)_3 \times \frac{1 \text{ mole } Fe_2(SO_4)_3}{399.9 \text{ g } Fe_2(SO_4)_3} = 56.7642 \text{ moles } Fe_2(SO_4)_3$$

The correct answer to the three required significant figures is 56.8 moles.

[*Additional examples may be found in 5.47–5.52 at the end of the chapter.*]

Example 5.11

During a severe pollution period in a city, a chemist determines the amount of lead in a cubic meter of air to be 1.5×10^{-8} moles. How many grams were in each cubic meter of air?

Solution The atomic weight of lead is 207.2 amu. Therefore, 1 mole of lead atoms weighs 207.2 g. In order to convert the given quantity in moles into the desired quantity in grams, the conversion factor must be the molar mass in units of g/mole.

$$1.5 \times 10^{-8} \text{ moles Pb} \times \frac{207.2 \text{ g Pb}}{1 \text{ mole Pb}} = 3.108 \times 10^{-6} \text{ g Pb}$$

The correct answer to two significant figures is 3.1×10^{-6} g.

[*Additional examples may be found in 5.57 and 5.58 at the end of the chapter.*]

Example 5.12

Barium sulfate is given as a slurry in water before x-rays are taken of the intestinal tract. How many grams are in 0.600 moles of $BaSO_4$? The formula weight of barium sulfate is 233.4 amu.

Solution One mole of barium sulfate weighs 233.4 g. The factor required to solve the problem is the molar mass that converts the given quantity in moles into the desired quantity in grams.

$$0.600 \text{ mole BaSO}_4 \times \frac{233.4 \text{ g BaSO}_4}{1 \text{ mole BaSO}_4} = 140.04 \text{ g BaSO}_4$$

The correct answer to three significant figures is 1.40×10^2 g.

[*Additional examples may be found in 5.61 and 5.62 at the end of the chapter.*]

Example 5.13

Calculate the mass of 0.00125 mole of vitamin C, $C_6H_8O_6$.

Solution In Example 5.1 the molecular weight of vitamin C was established to be 176.0 amu. Therefore, 1 mole has a mass of 176.0 g, and the mass of 0.00125 mole can be calculated as follows.

$$0.00125 \text{ mole } C_6H_8O_6 \times \frac{176.0 \text{ g } C_6H_8O_6}{1 \text{ mole } C_6H_8O_6} = 0.220 \text{ g } C_6H_8O_6$$

[*Additional examples may be found in 5.59 and 5.60 at the end of the chapter.*]

Example 5.14

A cigarette contains 2.50×10^{-5} g of nickel. What number of atoms does this represent?

Solution The atomic weight of nickel is 58.7 amu. Therefore, 1 mole of nickel atoms weighs 58.7 g. First determine the number of moles of nickel in the cigarette.

$$2.50 \times 10^{-5} \text{ g Ni} \times \frac{1 \text{ mole Ni}}{58.7 \text{ g Ni}} = 4.25\underline{9} \times 10^{-7} \text{ mole Ni}$$

Note that an extra place value is retained to be used in the next step. The answer will be properly rounded to three significant figures after the last step.

The number of atoms of nickel in 1 mole is 6.02×10^{23}. Multiplying the calculated number of moles by a factor having the units atoms/mole will give the desired answer.

$$4.25\underline{9} \times 10^{-7} \text{ moles Ni} \times \frac{6.02 \times 10^{23} \text{ atoms Ni}}{1 \text{ mole Ni}} = 2.5639 \times 10^{17} \text{ atoms Ni}$$

The correct answer to three significant figures is 2.56×10^{17} atoms.

[*Additional examples may be found in 5.67 and 5.68 at the end of the chapter.*]

Example 5.15

How many molecules of vitamin C are in a 0.500 g tablet of vitamin C ($C_6H_8O_6$)?

Solution In order to determine the number of molecules, it is first necessary to determine the number of moles present in the tablet, which has a mass of 0.500 g.

$$0.500 \text{ g } C_6H_8O_6 \times \frac{1 \text{ mole } C_6H_8O_6}{176.0 \text{ g } C_6H_8O_6} = 0.002841 \text{ mole } C_6H_8O_6$$

The number of molecules of $C_6H_8O_6$ in a mole is 6.02×10^{23}. In the tablet, which contains 2.841×10^{-3} mole, the number of molecules is

$$2.841 \times 10^{-3} \text{ mole } C_6H_8O_6 \times \frac{6.02 \times 10^{23} \text{ molecules } C_6H_8O_6}{1 \text{ mole } C_6H_8O_6}$$
$$= 1.7103 \times 10^{21} \text{ molecules } C_6H_8O_6$$

The correct answer to three significant figures is 1.71×10^{21} molecules.

[*Additional examples may be found in 5.69 and 5.70 at the end of the chapter.*]

Example 5.16

Lithium carbonate, Li_2CO_3, is used to treat manic depression. How many lithium ions are in a 0.300 g dose of Li_2CO_3? The formula weight of lithium carbonate is 74.9 amu.

Solution In 1 mole of Li_2CO_3 there are 2 moles of lithium ions. Therefore, first determine the number of moles of Li_2CO_3 in 0.300 g and then multiply by 2 to obtain the number of moles of lithium ions.

$$0.300 \text{ g } Li_2CO_3 \times \frac{1 \text{ mole } Li_2CO_3}{73.9 \text{ g } Li_2CO_3} = 0.004060 \text{ mole } Li_2CO_3$$

$$0.004060 \text{ mole } Li_2CO_3 \times \frac{2 \text{ mole } Li^+ \text{ ions}}{1 \text{ mole } Li_2CO_3} = 0.008120 \text{ mole } Li^+ \text{ ions}$$

The number of lithium ions in 1 mole is 6.02×10^{23}. Therefore the number of lithium ions in the sample is

$$8.120 \times 10^{-3} \text{ mole } Li^+ \times \frac{6.02 \times 10^{23} \text{ } Li^+}{\text{mole}} = 4.8883 \times 10^{21} \text{ } Li^+ \text{ ions}$$

The correct answer to three significant figures is 4.89×10^{21} ions.

[*Additional examples may be found in 5.71 and 5.72 at the end of the chapter.*]

5.6 Empirical Formula

You have learned how to calculate the percent composition of the elements in a compound from its molecular formula or formula unit. However, how are these formulas determined by chemists? In other words, how do chemists know the number of atoms or ions contained in a compound? It might appear that one could then do the reverse operation of the procedure we have just studied, that is, determine the molecular formula from the percent composition.

Table 5.3   Empirical Formulas and Molecular Formulas

Compound	Molecular formula	Empirical formula
acetylene	C_2H_2	CH
benzene	C_6H_6	CH
ethylene	C_2H_4	CH_2
cyclopropane	C_3H_6	CH_2
cyclohexane	C_6H_{12}	CH_2
formaldehyde	CH_2O	CH_2O
acetic acid	$C_2H_4O_2$	CH_2O
glucose	$C_6H_{12}O_6$	CH_2O

However, you cannot. Consider the percent compositions of C_2H_4 and C_4H_8. Both compounds are 85.7% carbon and 14.3% hydrogen. Would you have expected this result? Remember that percent is a part of a whole. Both compounds have the same ratio of carbon atoms to hydrogen atoms—1:2. Therefore their percent compositions must be identical. If two compounds with different molecular formulas have the same percent composition, how can a given percent composition lead to a unique molecular formula? It cannot be done. The only result of converting percent composition to a formula is to give a simplest ratio of atoms, in this case CH_2. The process used to determine the simplest ratio of atoms in a compound is given in Section 5.7.

The **empirical formula** of a compound, or simplest formula, gives only the smallest whole number ratio of the atoms present in a compound. This contrasts with the molecular formula, which gives the total number of atoms of each type of element that are present in one molecule or formula unit of the compound. In the case of water, the empirical formula and the molecular formula are the same. The formula H_2O gives the simplest ratio of two hydrogen atoms for every oxygen atom in the molecule. Thus H_2O is an empirical formula. However, each molecule of water contains two hydrogen atoms and one oxygen atom; therefore the formula H_2O is also the molecular formula. Hydrogen peroxide has the molecular formula H_2O_2. The molecular formula does not correspond to the simplest formula as the ratio of the atoms is one hydrogen atom to one oxygen atom. The empirical formula is HO.

A molecular formula is always a multiple of the empirical formula (Table 5.3), but the factor may be unity, as in the case of water. The empirical formula of the gas used in welding torches, acetylene, is CH. This empirical formula indicates that there is one carbon atom for every hydrogen atom in the molecule. However, this does not necessarily mean that there is only one carbon atom and one hydrogen atom in one molecule of acetylene. The acetylene molecule actually contains two carbon atoms and two hydrogen atoms, and the molecular formula is C_2H_2. In this case the subscripts of the empirical formula are multiplied by 2 to obtain the molecular formula. Benzene, an important major product of the chemical industry, is a liquid and has an empirical formula CH. Again, as in the case of acetylene, the empirical formula only indicates that there is a 1:1 ratio of carbon and hydrogen atoms. The molecular formula of benzene is C_6H_6. One molecule of benzene contains exactly six atoms of carbon and six atoms of hydrogen. The subscripts of the empirical formula of benzene must be multiplied by 6 to obtain the molecular formula.

Example 5.17

What are the empirical formulas and molecular formulas of ammonia and hydrazine? Ammonia contains one nitrogen atom and three hydrogen atoms in a molecule; hydrazine contains two atoms of nitrogen and four atoms of hydrogen in a molecule.

Solution The empirical formula of ammonia is NH_3 because the ratio of nitrogen atoms to hydrogen atoms is $1:3$. The molecular formula for ammonia is also NH_3 since each molecule contains only one atom of nitrogen and three atoms of hydrogen. In the case of hydrazine, the empirical formula and the molecular formula are not the same. The empirical formula is NH_2, as the ratio of nitrogen atoms to hydrogen atoms is $1:2$. However, the molecular formula is N_2H_4; there are two nitrogen atoms and four hydrogen atoms in a molecule.

Example 5.18

The molecular formula of vitamin C is $C_6H_8O_6$. What is the empirical formula of vitamin C?

Solution The empirical formula must give only the smallest ratio of atoms present in the substance. The subscripts 6, 8, and 6 are not the smallest ratio; they are all divisible by 2 to give 3, 4, and 3. Therefore the empirical formula is $C_3H_4O_3$.

[*Additional examples may be found in 5.75 and 5.76 at the end of the chapter.*]

5.7 Determining the Empirical Formula

If the composition of a compound is determined by chemical analysis, the mass percent of the elements in the compound is available. Using the mass percent composition, four simple arithmetical steps are necessary to calculate the empirical formula.

1. Express the composition of the compound in terms of the mass of each constituent element.
2. Convert the mass of each element into its equivalent in moles.
3. Determine the mole ratio for each element by dividing by the smallest number obtained in step 2.
4. If the mole ratios are not all integers, multiply by the simplest factor required to convert all mole ratios to integers.

If grams of each element are given for a compound, then these quantities may be used directly to determine the empirical formula. If percent composition is given, convert the percents to grams for a selected sample. For example, vitamin C contains 40.9% carbon, 4.6% hydrogen, and 54.5% oxygen. It is convenient to use a 100.0 g sample for the calculation, in which case there is 40.9 g of carbon, 4.6 g of hydrogen, and 54.5 g of oxygen in the sample.

Now, convert the mass of the element into its equivalent in moles. This is done by multiplying by the reciprocal of the molar mass in units of moles/g. The molar masses of carbon, hydrogen, and oxygen expressed to the nearest 0.1 g are 12.0, 1.0, and 16.0 g, respectively.

$$40.9 \text{ g C} \times \frac{1 \text{ mole C}}{12.0 \text{ g C}} = 3.41 \text{ moles C}$$

$$4.6 \text{ g H} \times \frac{1 \text{ mole H}}{1.0 \text{ g H}} = 4.6 \text{ moles H}$$

$$54.5 \text{ g O} \times \frac{1 \text{ mole O}}{16.0 \text{ g O}} = 3.41 \text{ moles O}$$

These operations yield the number of moles of each element contained in the 100.0 g sample.

In the third step, the mole ratios that give the relative number of atoms in a molecule or formula unit are obtained by dividing the number of moles of each element by the smallest of these numbers. This process may give you the subscripts of the empirical formula, providing all of the numbers are close to integers within the accuracy of the calculation. In the example considered, the relative number of atoms are not all integers.

$$\text{mole ratio for C} = \frac{3.41}{3.41} = 1.0$$

$$\text{mole ratio for H} = \frac{4.6}{3.41} = 1.3$$

$$\text{mole ratio for O} = \frac{3.41}{3.41} = 1.0$$

The ratio of the subscripts in a formula must be the same as the mole ratios determined in this calculation. However, the formula $C_1H_{1.3}O_1$ is not acceptable because not all subscripts are integers. Multiplication by 3 yields $C_3H_{3.9}O_3$ or within the accuracy of the data, the empirical formula is $C_3H_4O_3$.

Example 5.19

Vinegar contains acetic acid, a compound containing carbon, hydrogen, and oxygen. The percent composition is 40.0% carbon, 6.7% hydrogen, and 53.3% oxygen. What is the empirical formula of acetic acid?

Solution Using a reference 100.0 g sample of acetic acid, there is 40.0 g of carbon, 6.7 g of hydrogen and 53.3 g of oxygen. Convert each of these quantities into moles by using a factor relating moles and mass.

$$40.0 \, \text{g C} \times \frac{1 \, \text{mole C}}{12.0 \, \text{g C}} = 3.33 \, \text{moles C}$$

$$6.7 \, \text{g H} \times \frac{1 \, \text{mole H}}{1.0 \, \text{g H}} = 6.7 \, \text{moles H}$$

$$53.3 \, \text{g O} \times \frac{1 \, \text{mole O}}{16.0 \, \text{g O}} = 3.33 \, \text{moles O}$$

Division of each number of moles by 3.33 yield the numbers 1, 2, and 1. The empirical formula of acetic acid is CH_2O.

Example 5.20

A 2.732 g sample of a compound contains 1.468 g of iron and 1.264 g of sulfur. What is the empirical formula of the compound?

Solution Calculate the relative number of moles of each element in the given mass of each element.

$$1.468 \text{ g Fe} \times \frac{1 \text{ mole Fe}}{55.85 \text{ g Fe}} = 0.0262846 = 0.02628 \text{ moles Fe}$$

$$1.264 \text{ g S} \times \frac{1 \text{ mole S}}{32.06 \text{ g S}} = 0.0394261 = 0.03943 \text{ moles S}$$

Division of each number of moles by 0.02628 yields the numbers 1.00 and 1.50. Since integers do not result, we must multiply by a factor that results in integers for all elements. Multiplication by 2 gives 2 and 3, respectively, for iron and sulfur. The empirical formula is Fe_2S_3.

[*Additional examples may be found in 5.77–5.84 at the end of the chapter.*]

5.8 Determining the Molecular Formula

The molecular formula of a substance may be determined from the empirical formula if the molecular weight or formula weight is known. Methods of determining molecular weights of gases are given in Chapter 12. The molecular weights of liquids and solids can be determined in solution (Chapter 14). Since the molecular formula is always a simple multiple of the empirical formula, the molecular weight is a simple multiple of the weight of the empirical formula unit. For example, the empirical formula of water is H_2O and the molecular formula must be $H_{2n}O_n$, where n may be 1, 2, 3, and so on. Thus the molecular formula could be H_2O, H_4O_2, H_6O_3, and so on. These various possibilities have molecular weights 18 amu, 36 amu, and 54 amu, respectively. The experimental molecular weight of water is 18 amu and the molecular formula must be H_2O.

Acetylene and benzene both have an empirical formula of CH. The molecular formulas may be any of the possibilities given by C_nH_n. In order to obtain the molecular formula, the molecular weight of each substance is needed. The molecular weights of acetylene and benzene ae 26 and 78 amu, respectively. Therefore, the molecular formulas are C_2H_2 and C_6H_6, respectively.

Example 5.21

The empirical formula of nicotine is C_5H_7N. The molecular weight is 162 amu. What is the molecular formula?

Solution If the molecular formula were the same as the empirical formula, the molecular weight would be 5 × 12 amu + 7 × 1 amu + 14 amu = 81 amu. Therefore, the molecular formula must be twice the empirical formula, $C_{10}H_{14}N_2$.

Example 5.22

A gaseous compound contains 80.0% carbon and 20.0% hydrogen. The molecular weight is 30.0 amu. Determine its molecular formula.

Solution Using a reference 100.0 g sample, there is 80.0 g of carbon and 20.0 g of hydrogen.

Convert each of these quantities into moles.

$$80.0 \, \cancel{g \, C} \times \frac{1 \text{ mole C}}{12.0 \, \cancel{g \, C}} = 6.6667 = 6.67 \text{ moles C}$$

$$20.0 \, \cancel{g \, H} \times \frac{1 \text{ mole H}}{1.0 \, \cancel{g \, H}} = 20.0 \text{ moles H}$$

Division of each number of moles by 6.67 yields the numbers 1 and 3. The empirical formula is CH_3. If the molecular formula and the empirical formula were identical, the molecular weight would be 15.0 amu.

$$12.0 \text{ amu} + 3 \times 1.0 \text{ amu} = 15.0 \text{ amu}$$

Since the molecular weight is 30.0 amu, the molecular formula must be twice the empirical formula or C_2H_6.

[*Additional examples may be found in 5.85–5.88 at the end of the chapter.*]

Summary

The molecular weight of a molecule is the sum of the masses in atomic mass units of its component atoms. The formular weight of an ionic compound is the sum of the atomic weights of the atoms indicated by the formula unit of the substance.

The concept of a mole occupies a central position in making chemistry a quantitative science. The mole is a count of the number of units of a substance present in matter. These units may be atoms, molecules, or ions. The number of units in a mole of a substance is called Avogadro's number.

In order to relate the number of units of matter to mass as measured with balances, the individual mass of each unit of matter is needed. The atomic weight is used to measure the mass of atoms, the molecular weight to measure the mass of molecules, and the formula weight to measure the mass of ionic compounds.

The composition of a compound is given by a molecular formula or formula unit in terms of the number of constituent atoms or ions. Using the formula of a compound and the atomic weights of the elements, the percent composition of the elements is calculated.

The empirical formula or simplest ratio of atoms in a compound may be calculated from the percent composition. The molecular formula is a multiple of the empirical formula.

New Terms

Avogadro's number is equal to the number of structural units of a substance in a mole.

The **empirical formula** gives the simplest whole number ratio of the atoms present in a compound.

The **formula weight** is the sum of the atomic weights of the atoms indicated by the formula unit of an ionic compound.

The **molar mass** of a substance is the mass in grams of 1 mole of that substance.

A **mole** is a quantity in grams equal to the atomic weight in the case of atoms, molecular weight in the case of molecules, and formula weight in the case of ionic compounds. A mole contains Avogadro's number of structural units of matter.

The **molecular weight** of a molecule is the sum of the atomic weights of the component atoms of the molecule.

The **percent composition** of an element in a compound is equal to the mass of that element present in the compound divided by the total mass of all elements in the compound with the quotient multiplied by 100.

Exercises

Terminology

5.1 Explain the differences and similarities in the terms molecular weight and formula weight.

5.2 What relationship exists between the number of moles of a substance and the number of structural units of that substance?

5.3 What information is given by the molecular formula of a compound?

5.4 What information is given by the empirical formula of a compound?

Molecular Weights

5.5 Why isn't the molecular weight of carbon dioxide, CO_2, equal to the sum of the atomic weights of carbon and oxygen, that is $12 + 16$?

5.6 A friend in your chemistry class tells you that the molecular weight of SO_3 is equal to $32 + 16 = 48$. What do you tell him?

5.7 Calculate the molecular weight of each of the following substances.

(a) CO   (b) CO_2 (c) SO_2
(d) SO_3 (e) NO   (f) NO_2
(g) SiO_2 (h) SeO_2 (i) OF_2

5.8 Calculate the molecular weight of each of the following substances.

(a) CH_4 (b) SiH_4 (c) GeH_4
(d) NH_3 (e) PH_3 (f) AsH_3
(g) H_2O (h) H_2S (i) H_2Se

5.9 Calculate the molecular weight of each of the following substances.

(a) CCl_4 (b) SiF_4 (c) $GeBr_4$
(d) PBr_3 (e) PCl_5 (f) NF_3
(g) OCl_2 (h) BrF_3 (i) SF_2

5.10 Calculate the molecular weight of each of the following substances.

(a) SF_6 (b) PF_5 (c) BrF_5
(d) SeF_4 (e) XeF_2 (f) XeF_4
(g) ClF_5 (h) IF_7 (i) IF_3

5.11 Calculate the molecular weight of each of the following substances.

(a) H_3BO_3 (b) H_3PO_3 (c) HNO_3
(d) H_5IO_6 (e) H_6TeO_6 (f) $HAsO_2$
(g) H_2SO_3 (h) $H_5P_3O_{10}$ (i) $H_2S_2O_7$

5.12 Calculate the molecular weight of each of the following substances.

(a) $H_3B_3O_3$ (b) H_3PO_4 (c) HNO_2
(d) H_3IO_6 (e) H_3TeO_3 (f) $H_2S_2O_6$
(g) H_2SO_4 (h) $H_4P_2O_6$ (i) $H_2S_2O_4$

5.13 Calculate the molecular weight of each of the following explosives.

(a) $C_3H_5N_3O_9$, nitroglycerin
(b) $C_7H_5N_3O_6$, TNT

5.14 Calculate the molecular weight of each of the following vitamins.

(a) $C_6H_8O_6$, vitamin C
(b) $C_{20}H_{30}O$, vitamin A
(c) $C_{63}H_{88}N_{14}O_{14}PCo$, vitamin B_{12}
(d) $C_{12}H_{17}ClN_4OS$, vitamin B_1 (thiamine)

Formula Weights

5.15 Calculate the formula weight of each of the following compounds.

(a) NaCl   (b) KBr   (c) LiF
(d) MgS   (e) CaSe   (f) SrO
(g) AlN   (h) GaP   (i) InAs

5.16 Calculate the formula weight of each of the following compounds.

(a) Na_2O (b) K_2S (c) Li_2Se
(d) MgF_2 (e) $CaCl_2$ (f) $SrBr_2$
(g) SiO_2 (h) $GaCl_3$ (i) SnO_2

5.17 Calculate the formula weight of each of the following compounds.

(a) GeO_2 (b) ReO_2 (c) OsO_4
(d) WO_3 (e) MnO_2 (f) UO_2
(g) CrO_3 (h) PoO_3 (i) PbO_2

5.18 Calculate the formula weight of each of the following compounds.

(a) Sb_2O_3 (b) Bi_2O_4 (c) Rh_2O_3
(d) Sb_2O_5 (e) Nb_2O_3 (f) V_2O_5

5.19 Calculate the formula weight of each of the following compounds.

(a) Na_2SO_4 (b) K_2CO_3 (c) Li_2SO_3
(d) $Ca(NO_3)_2$ (e) $Mg(ClO_4)_2$ (f) $Ba(HCO_3)_2$
(g) $Al_2(SO_4)_3$ (h) $RbClO$ (i) $KHSO_3$

5.20 Calculate the formula weight of each of the following compounds.

(a) Na_3PO_4 (b) $KMnO_4$ (c) Li_2SO_4
(d) $Ba(OH)_2$ (e) $Ca(CN)_2$ (f) $Mg(HSO_3)_2$
(g) $Fe_2(SO_4)_3$ (h) $Cu(ClO_3)_2$ (i) Ag_2CO_3

5.21 Calculate the formula weight of each of the following compounds.

(a) sodium carbonate   (b) lithium sulfate
(c) potassium phosphate   (d) iron(II) sulfate
(e) copper(II) nitrate   (f) potassium permanganate
(g) magnesium bicarbonate   (h) magnesium hydroxide

5.22 Calculate the formula weight of each of the following compounds.

(a) sodium chlorite   (b) lithium bisulfite
(c) potassium cyanide   (d) iron(III) nitrate
(e) copper(II) phosphate   (f) potassium perchlorate
(g) calcium bisulfite   (h) barium nitrite

Percent Composition

5.23 What is the percent composition for compounds formed by combinations of the elements in the indicated quantities.
(a) 2.50 g of a metal and 2.00 g of oxygen
(b) 1.40 g of an element and 1.60 g of oxygen
(c) 35.0 g of a metal and 20.0 g of sulfur

5.24 What is the percent composition for compounds formed by combinations of the elements in the indicated quantities.
(a) 2.62 g of a metal and 2.38 g of chlorine
(b) 1.31 g of a metal and 2.02 g of chlorine
(c) 4.49 g of a metal and 10.07 g of sulfur

5.25 Which of each of the following pairs of compounds contains the larger percentage of the indicated element.
(a) chlorine in ferric chloride and ferrous chloride
(b) oxygen in sodium oxide and potassium oxide
(c) sulfur in potassium sulfide and calcium sulfide

5.26 Which of each of the following pairs of compounds contains the larger percentage of the indicated element.
(a) chlorine in $CdCl_2$ and $ZnCl_2$
(b) iron in $Fe_2(SO_4)_3$ and $FeSO_4$
(c) oxygen in FeO and Fe_2O_3

5.27 Calculate the percent composition of each of the following substances.
(a) CO (b) CO_2 (c) SO_2
(d) SO_3 (e) NO (f) NO_2
(g) P_4O_6 (h) P_4O_{10} (i) Cl_2O_7
(j) I_2O_5 (k) XeO_3 (l) XeO_4

5.28 Calculate the percent composition of each of the following substances.
(a) CH_4 (b) SiH_4 (c) GeH_4
(d) NH_3 (e) PH_3 (f) AsH_3
(g) C_2H_6 (h) C_2H_4 (i) C_2H_2
(j) N_2H_2 (k) B_2H_6 (l) P_2H_4

5.29 Calculate the percent composition of each of the following substances.
(a) CCl_4 (b) SiF_4 (c) $GeBr_4$
(d) PBr_3 (e) PCl_5 (f) NF_3
(g) SF_6 (h) PF_5 (i) BrF_5
(j) SeF_4 (k) XeF_2 (l) XeF_4

5.30 Calculate the percent composition of each of the following substances.
(a) H_3BO_3 (b) H_3PO_3 (c) HNO_3
(d) H_5IO_6 (e) H_6TeO_6 (f) $HAsO_2$
(g) $H_3B_3O_3$ (h) H_3PO_4 (i) HNO_2
(j) H_3IO_6 (k) H_3TeO_3 (l) $H_2S_2O_6$

5.31 Calculate the percent composition of each of the following substances.
(a) $NaCl$ (b) KBr (c) LiF
(d) MgS (e) $CaSe$ (f) SrO
(g) Na_2O (h) K_2S (i) Li_2Se
(j) SiO_2 (k) $GaCl_3$ (l) SnO_2

5.32 Calculate the percent composition of each of the following substances.
(a) GeO_2 (b) ReO_2 (c) OsO_4
(d) WO_3 (e) MnO_2 (f) UO_2
(g) $TiCl_4$ (h) $ScBr_3$ (i) $SnCl_4$
(j) $CrCl_3$ (k) InF_3 (l) $PdCl_2$

5.33 Calculate the percent composition of each of the following substances.
(a) Na_2SO_4 (b) K_2CO_3 (c) Li_2SO_3
(d) $Ca(NO_3)_2$ (e) $Mg(ClO_4)_2$ (f) $Ba(HCO_3)_2$
(g) Na_3PO_4 (h) $KMnO_4$ (i) Li_2SO_4
(j) $Fe_2(SO_4)_3$ (k) $Cu(ClO_3)_2$ (l) Ag_2CO_3

5.34 Calculate the percent composition of each of the following substances.
(a) Ag_3PO_4 (b) $CsClO_4$ (c) $La_2(SO_4)_3$
(d) $Ni(OH)_2$ (e) $Cd(CN)_2$ (f) $Hg(HSO_3)_2$
(g) $ZnSO_4$ (h) $Co(ClO_2)_2$ (i) $Tl_2(SO_4)_3$
(j) $CaCO_3$ (k) $FeSO_4$ (l) $CuCO_3$

5.35 Calculate the percent composition of each of the following substances.
(a) sodium carbonate   (b) lithium sulfate
(c) potassium phosphate   (d) iron(II) sulfate
(e) copper(II) nitrate   (f) potassium permanganate
(g) potassium cyanide   (h) iron(III) nitrate

5.36 Calculate the percent composition of each of the following substances.
(a) sodium oxide   (b) lithium sulfide
(c) potassium iodide   (d) iron(II) chloride
(e) copper(II) fluoride   (f) zinc oxide
(g) magnesium nitride   (h) silver fluoride

The Mole

5.37 What relationship exists between the mole and Avogadro's number?

5.38 How do you calculate the mass of 1 mole of a substance?

5.39 How do you calculate the number of moles contained in a given mass of a substance?

5.40 Since the mole is so central to the study of chemistry, why hasn't a balance been made that determines the number of moles in a sample rather than the number of grams?

5.41 Calculate the number of moles of atoms in each of the following samples.

5.42 Calculate the number of moles of atoms in each of the following samples.

(a) 46.0 g of sodium (b) 16.0 g of sulfur
(c) 2.0 g of mercury (d) 24.3 g of magnesium
(e) 0.31 g of phosphorus (f) 1.04 g of chromium

5.43 Calculate the number of moles of molecules in each of the following samples.
(a) 180 g of H_2O (b) 0.44 g of CO_2
(c) 92 g of NO_2 (d) 3.00 g of NO
(e) 0.18 g of H_2S (f) 85 g of NH_3

5.45 Calculate the number of moles of molecules in each of the following samples.
(a) 38.4 g CCl_4 (b) 0.104 g of SiF_4
(c) 15.7 g $GeBr_4$ (d) 4.51 g PBr_3
(e) 13.9 g PCl_5 (f) 2.84 g NF_3

5.47 Calculate the number of moles of formula units in each of the following samples.
(a) 0.584 g NaCl (b) 5.95 g KBr
(c) 25.9 g LiF (d) 28.2 g MgS
(e) 2.38 g CaSe (f) 0.518 g SrO

5.49 Calculate the number of moles of formula units in each of the following samples.
(a) 0.532 g GeO_2 (b) 0.546 g ReO_2
(c) 1.27 g OsO_4 (d) 4.64 g WO_3
(e) 0.435 g MnO_2 (f) 67.5 g UO_2

5.51 Calculate the number of moles of formula units of each of the following samples.
(a) 21.2 g sodium carbonate
(b) 22.0 g lithium sulfate
(c) 21.2 g potassium phosphate
(d) 25.5 g iron(II) sulfate
(e) 3.75 g copper(II) nitrate
(f) 3.95 g potassium permanganate

5.53 Which of the following contains the larger number of moles?
(a) 2 g of helium or 10 g of argon
(b) 8 g of CH_4 or 33 g of CO_2
(c) 10 g of CaO or 10 g of CaS

5.55 Which of the following contains the largest number of moles of nitrogen atoms?
(a) 0.1 mole of N_2O_5
(b) 0.2 mole of N_2O_3
(c) 0.3 mole of NO_2
(d) 0.25 mole of N_2O_4
(e) 0.3 mole of N_2O

5.57 Calculate the mass of the following samples.
(a) 1.0 mole of sodium (b) 0.10 mole of lithium
(c) 0.40 mole of mercury (d) 0.25 mole of magnesium
(e) 0.35 mole of arsenic (f) 0.25 mole of chromium

5.59 Calculate the mass of each of the following samples.
(a) 1.0 mole of H_2O (b) 0.40 mole of CO_2
(c) 0.50 mole of NO_2 (d) 3.00 moles of NO
(e) 0.20 mole of H_2S (f) 0.75 mole of NH_3

(a) 78.2 g of potassium (b) 0.60 g of carbon
(c) 1.97 g of gold (d) 59.3 g of tin
(e) 1.4 g of silicon (f) 0.10 g of calcium

5.44 Calculate the number of moles of molecules in each of the following samples.
(a) 7.2 g C_3O_2 (b) 0.76 g N_2O_3
(c) 0.124 g N_2O_5 (d) 2.2 g P_4O_6
(e) 0.284 g P_4O_{10} (f) 1.83 g Cl_2O_7

5.46 Calculate the number of moles of molecules in each of the following samples.
(a) 1.78 g P_4S_{10} (b) 1.42 g P_2I_4
(c) 0.538 g Si_2Cl_6 (d) 2.30 g S_4N_4
(e) 21.9 g Mo_2Cl_{10} (f) 13.5 g S_2Cl_2

5.48 Calculate the number of moles of formula units in each of the following samples.
(a) 12.4 g Na_2O (b) 0.274 g K_2S
(c) 0.186 g Li_2Se (d) 0.623 g MgF_2
(e) 27.7 g $CaCl_2$ (f) 0.491 g $SrBr_2$

5.50 Calculate the number of moles of formula units in each of the following samples.
(a) 5.68 g Na_2SO_4 (b) 3.46 g K_2CO_3
(c) 18.8 g Li_2SO_3 (d) 0.827 g $Ca(NO_3)_2$
(e) 11.2 g $Mg(ClO_4)_2$ (f) 5.19 g $Ba(HCO_3)_2$

5.52 Calculate the number of moles of formula units of each of the following samples.
(a) 18.1 g sodium chlorite
(b) 17.6 g lithium bisulfite
(c) 6.51 g potassium cyanide
(d) 4.84 g iron(III) nitrate
(e) 15.2 g copper(II) phosphate
(f) 3.46 g potassium perchlorate

5.54 Which of the following contains the larger number of moles?
(a) 10 g of FeO or 10 g of Fe_2O_3
(b) 15 g of NaCl or 15 g of KCl
(c) 24 g of SO_2 or 24 g of SO_3

5.56 Which of the following contains the largest number of moles of oxygen atoms?
(a) 0.1 mole of N_2O_5
(b) 0.2 mole of N_2O_3
(c) 0.3 mole of NO_2
(d) 0.25 mole of N_2O_4
(e) 0.3 mole of N_2O

5.58 Calculate the mass of the following samples.
(a) 2.0 moles of potassium (b) 0.60 mole of carbon
(c) 0.0010 mole of gold (d) 5.00 moles of tin
(e) 1.5 moles of silicon (f) 0.10 mole of calcium

5.60 Calculate the mass of the following sampoles.
(a) 2.0 moles C_3O_2 (b) 0.25 mole of N_2O_3
(c) 0.10 mole of N_2O_5 (d) 2.50 moles of P_4O_6
(e) 0.200 mole of P_4O_{10} (f) 0.150 mole of Cl_2O_7

5.61 Calculate the mass of each of the following samples.
  (a) 0.500 mole NaCl   (b) 0.020 mole of KBr
  (c) 2.00 moles of LiF   (d) 1.5 moles of MgS
  (e) 0.40 mole of CaSe   (f) 0.100 mole of SrO

5.63 Diammonium phosphate, $(NH_4)_2HPO_4$, is a fertilizer widely used to supply both nitrogen and phosphorus. How many moles of compound are present in a 100 lb sack (47.5 kg) of this fertilizer?

5.65 The minimum daily requirement of the amino acid lysine $(C_6H_{14}N_2O_2)$ is 0.80 g. How many moles is this?

5.62 Calculate the mass each of the following samples.
  (a) 1.50 moles of Na_2O (b) 0.200 mole of K_2S
  (c) 0.40 mole of Li_2Se (d) 0.600 mole of MgF_2
  (e) 0.020 mole of $CaCl_2$ (f) 0.500 mole of $SrBr_2$

5.64 Ascorbic acid (vitamin C) has the molecular formula $C_6H_8O_6$. The recommended daily allowance of ascorbic acid in the diet of a human is 45 mg. How many moles of ascorbic acid is this?

5.66 The herbicide Treflan $(C_{13}H_{16}N_3O_4F_3)$ is applied at the rate of 454 g (1 lb) per acre to control weeds. What is the application in moles per acre?

Avogadro's Number

5.67 Calculate the number of atoms present in each of the following samples.
  (a) 20.0 g of mercury   (b) 120 g of carbon
  (c) 53.2 g of palladium   (d) 0.635 g of copper
  (e) 0.195 g of potassium   (f) 2.16 g of silver

5.69 Calculate the number of molecules present in each of the following samples.
  (a) 2.8 g of CO   (b) 90 g of H_2O
  (c) 32 g of O_2 (d) 0.40 g of CH_4
  (e) 0.85 g of NH_3 (f) 3.2 g of SO_2

5.71 Calculate the number of formula units present in each of the following samples.
  (a) 1.24 g Na_2O (b) 27.6 g of K_2S
  (c) 6.23 g MgF_2 (d) 55.5 g $CaCl_2$
  (e) 12.0 g SiO_2 (f) 17.6 g $GaCl_3$

5.73 Calculate the number of indicated atoms present in each of the following.
  (a) oxygen atoms in 0.1 mole of SO_3
  (b) carbon atoms in 0.02 mole of C_3O_2
  (c) oxygen atoms in 0.2 mole of N_2O_5
  (d) phosphorus atoms in 0.25 mole of P_4O_6
  (e) boron atoms in 0.2 mole of B_5H_{11}

5.68 Calculate the number of atoms in each of the following samples.
  (a) 46.0 g of sodium   (b) 16.0 g of sulfur
  (c) 2.0 g of mercury   (d) 24.3 g of magnesium
  (e) 0.31 g of phosphorus   (f) 1.04 g of chromium

5.70 Calculate the number of molecules present in each of the following samples.
  (a) 36.0 g C_3O_2 (b) 19.0 g N_2O_3
  (c) 22.0 g P_4O_6 (d) 14.2 g P_4O_{10}
  (e) 36.6 g Cl_2O_7 (f) 66.8 g I_2O_5

5.72 Calculate the number of formula units present in each of the following samples.
  (a) 1.64 g Na_3PO_4 (b) 7.90 g $KMnO_4$
  (c) 11.0 g Li_2SO_4 (d) 8.57 g $Ba(OH)_2$
  (e) 18.4 g $Ca(CN)_2$ (f) 3.73 g $Mg(HSO_3)_2$

5.74 Calculate the number of indicated atoms present in each of the following.
  (a) chlorine atoms in 2.0 moles of CCl_4
  (b) fluorine atoms in 0.020 mole of SF_6
  (c) fluorine atoms in 0.030 mole of IF_7
  (d) sulfur atoms in 4.0 moles of P_4S_{10}
  (e) chlorine atoms in 0.0010 mole of Mo_2Cl_{10}

Empirical Formula

5.75 The molecular formula of ethylene glycol, used in antifreeze, is $C_2H_6O_2$. What is the empirical formula?

5.77 A dry cleaning fluid is composed of 14.5% C and 85.5% Cl. What is the empirical formula of the compound?

5.79 The light-emitting diode used in some electronic calculators contains 69.2% Ga and 30.8% P. What is the empirical formula of the compound?

5.81 Given the following percent compositions, determine the empirical formula.
  (a) 47.26% Cu and 52.74% Cl
  (b) 72.25% Mg and 27.75% N
  (c) 72.25% Fe and 27.64% O
  (d) 58.93% Na and 41.07% S

5.76 Hydroquinone, a compound used in developing photographs, has the molecular formula $C_6H_6O_2$. What is the empirical formula?

5.78 Iron carbide, present in cast iron, has the weight percentage 93.31% Fe and 6.69% C. What is the empirical formula of the iron carbide?

5.80 A sample of a Freon propellant in an aerosol can is analyzed and is found to contain 0.423 g C, 2.50 g Cl and 1.34 g F. What is the empirical formula of this substance?

5.82 Given the following percent compositions, determine the empirical formula of each compound.
  (a) 69.02% g Na and 30.98% P
  (b) 70.92% K and 29.08% S
  (c) 56.01% V and 43.99% O
  (d) 52.92% Al and 47.08% O

5.83 Given the following percent compositions, determine the empirical formula of each compound.
(a) 44.3% Cu and 55.7% Br
(b) 62.6% Sn and 37.4% Cl
(c) 25.9% Fe and 74.1% Br
(d) 51.9% Cr and 48.1% S

5.84 Given the following percent compositions, determine the empirical formula of each compound.
(a) 49.4% K, 20.3% S, and 30.3% O
(b) 62.5% Pb, 8.5% N, and 29.0% O
(c) 44.9% K, 18.4% S, and 36.7% O
(d) 38.9% Ba, 29.4% Cr, and 31.7% O

Molecular Formulas

5.85 Adipic acid is used in the manufacture of nylon. The composition of this acid is 49.3% C, 6.9% H, and 43.8% O by mass. The molecular weight is 146 amu. What is the molecular formula?

5.86 The stimulant caffeine has a molecular weight of 194.2 amu and the percent composition by mass is 49.5% C, 5.2% H, 28.8% N, and 16.5% O. What is the molecular formula of caffeine?

5.87 Estrone, a female sex hormone, is composed of 80.0% C, 8.2% H, and 11.8% O. The molecular weight is 270 amu. What is the molecular formula of estrone?

5.88 A sodium metaphosphate, used in detergents, is 22.5% Na, 30.4% P and 47.1% O. The formula weight is 612 amu. What is the formula of the compound?

Additional Exercises

5.89 Explain how the molecular weight of a compound is calculated from the molecular formula.

5.90 Explain how the formula weight of an ionic compound is calculated.

5.91 Calculate the molecular weight of each of the following substances.
(a) C_3O_2 (b) N_2O_3 (c) N_2O_5
(d) P_4O_6 (e) P_4O_{10} (f) Cl_2O_7
(g) I_2O_5 (h) XeO_3 (i) XeO_4

5.92 Calculate the molecular weight of each of the following substances.
(a) C_2H_6 (b) C_2H_4 (c) C_2H_2
(d) N_2H_2 (e) B_2H_6 (f) P_2H_4
(g) Si_2H_6 (h) P_3H_5 (i) B_5H_{11}

5.93 Calculate the molecular weight of each of the following substances.
(a) P_4S_{10} (b) P_2I_4 (c) Si_2Cl_6
(d) S_4N_4 (e) Mo_2Cl_{10} (f) S_2Cl_2

5.94 Calculate the molecular weight of each of the following substances.
(a) $N_3S_3Cl_3$ (b) $S_4N_4F_4$ (c) $P_3N_3Cl_6$
(d) ClO_2F (e) BrO_2F (f) IO_3F
(g) XeO_2F_2 (h) SO_2F_2 (i) COF_2

5.95 Calculate the formula weight of each of the following compounds.
(a) $LaCl_3$ (b) UCl_4 (c) $FeCl_3$
(d) $TiCl_4$ (e) $ScBr_3$ (f) $SnCl_4$
(g) $CrCl_3$ (h) InF_3 (i) $PdCl_2$

5.96 Calculate the formula weight of each of the following compounds.
(a) Ag_3PO_4 (b) $CsClO_4$ (c) $La_2(SO_4)_3$
(d) $Ni(OH)_2$ (e) $Cd(CN)_2$ (f) $Hg(HSO_3)_2$
(g) $ZnSO_4$ (h) $Co(ClO_2)_2$ (i) $Tl_2(SO_4)_3$

5.97 Calculate the formula weight of each of the following ores.
(a) TiO_2, rutile
(b) Na_3AlF_6, cryolite
(c) $FeCr_2O_4$, chromite
(d) $CaWO_4$, scheelite
(e) $Be_3Al_2Si_6O_{18}$, beryl

5.98 Calculate the formula weight of each of the following compounds.
(a) sodium oxide (b) lithium sulfide
(c) potassium iodide (d) iron(II) chloride
(e) copper(II) fluoride (f) zinc oxide
(g) magnesium nitride (h) silver fluoride

5.99 What is the percent composition for compounds formed by combinations of the elements in the indicated quantities.
(a) 1.11 g of copper and 1.39 g of bromine
(b) 5.70 g of copper and 14.3 g of bromine
(c) 8.15 g of zinc and 2.00 g of oxygen

5.100 Which of each of the following pairs of compounds contains the larger percentage of the indicated element.
(a) oxygen in SO_2 and SO_3
(b) nitrogen in N_2H_4 and NH_3
(c) oxygen in N_2O_3 and N_2O_5

5.101 Calculate the percent composition of each of the following substances.
(a) $C_3H_5N_3O_9$, nitroglycerin
(b) $C_7H_5N_3O_6$, TNT
(c) $C_6H_8O_6$, vitamin C
(d) $C_{20}H_{30}O$, vitamin A
(e) $C_{63}H_{88}N_{14}O_{14}PCo$, vitamin B_{12}
(f) $C_{12}H_{17}ClN_4OS$, vitamin B_1 (thiamine)

5.102 Calculate the percent composition of each of the following minerals.
(a) Na_3AlF_6, cryolite
(b) $CaWO_4$, scheelite
(c) $Be_3Al_2Si_6O_{18}$, beryl
(d) TiO_2, rutile
(e) $FeCr_2O_4$, chromite

5.103 Calculate the number of moles of molecules in each of the following samples.
(a) 4.0 g CH_4 (b) 6.42 g SiH_4
(c) 3.83 g GeH_4 (d) 0.425 g H_2O_2
(e) 11.3 g PH_3 (f) 4.05 g H_2Se
(g) 0.060 g N_2H_2 (h) 3.36 g B_2H_6

5.104 Calculate the number of moles of molecules in each of the following samples.
(a) 15.4 g H_3BO_3 (b) 0.410 g H_3PO_3
(c) 2.10 g HNO_3 (d) 1.64 g H_2SO_3
(e) 3.92 g H_3PO_4 (f) 0.235 g HNO_2
(g) 4.58 g H_3IO_6 (h) 1.96 g H_2SO_4

5.105 Calculate the number of moles of molecules in each of the following samples.
(a) 1.76 g $C_6H_8O_6$, vitamin C
(b) 1.14 g $C_{20}H_{30}O$, vitamin A
(c) 5.64 g $C_{63}H_{88}N_{14}O_{14}PCo$, vitamin B_{12}
(d) 11.4 g $C_{12}H_{17}ClN_4OS$, vitamin B_1 (thiamine)

5.106 Calculate the number of moles of formula units in each of the following samples.
(a) 58.3 g Sb_2O_3 (b) 12.0 g Bi_2O_4
(c) 0.508 g Rh_2O_3 (d) 3.24 g Sb_2O_5
(e) 0.585 g Nb_2O_3 (f) 3.65 g V_2O_5

5.107 Calculate the number of moles of formula units of each of the following samples.
(a) 6.56 g Na_3PO_4 (b) 3.16 g $KMnO_4$
(c) 4.40 g Li_2SO_4 (d) 6.85 g $Ba(OH)_2$
(e) 1.84 g $Ca(CN)_2$ (f) 3.75 g $Mg(HSO_3)_2$

5.108 Calculate the number of moles of formula units of each of the following samples.
(a) 3.10 g sodium oxide
(b) 9.19 g lithium sulfide
(c) 41.5 g potassium iodide
(d) 31.7 g iron(II) chloride
(e) 5.08 g copper(II) fluoride
(f) 20.3 g zinc oxide

5.109 Which of the following contains the largest number of moles of hydrogen atoms?
(a) 0.2 mole of CH_4
(b) 0.1 mole of C_2H_6
(c) 0.3 mole of C_2H_4
(d) 0.2 mole of C_3H_8
(e) 0.1 mole of C_6H_6

5.110 Calculate the mass of each of the following.
(a) 0.10 mole of helium atoms
(b) 0.20 mole of argon atoms
(c) 2.5 moles of neon atoms
(d) 5.0 moles of sulfur atoms
(e) 0.10 mole of oxygen atoms
(f) 0.40 mole of chlorine atoms

5.111 Calculate the mass of each of the following samples.
(a) 0.15 mole of CH_4 (b) 0.25 mole of SiH_4
(c) 3.00 moles of GeH_4 (d) 0.40 mole of H_2O_2
(e) 1.5 moles of PH_3 (f) 0.50 mole of H_2Se
(g) 0.060 mole of N_2H_2 (h) 3.50 moles of B_2H_6

5.112 Calculate the mass of each of the following.
(a) 0.10 mole of $C_9H_8O_4$ (aspirin)
(b) 3.0 moles of $C_{14}H_9Cl_5$ (DDT)
(c) 0.001 mole of $C_{20}H_{30}O$ (vitamin A)
(d) 0.002 mole of $C_{16}H_{18}N_2O_4S$ (Penicillin G)

5.113 Calculate the mass of each of the following samples.
(a) 0.500 mole of GeO_2 (b) 0.250 mole of ReO_2
(c) 0.0100 mole of OsO_4 (d) 0.0250 mole of WO_3
(e) 0.0400 mole of MnO_2 (f) 0.000150 mole of UO_2

5.114 Calculate the number of atoms in each of the following samples.
(a) 78.2 g of potassium   (b) 0.60 g of carbon
(c) 1.97 g of gold   (d) 59.3 g of tin
(e) 1.4 g of silicon   (f) 0.10 g of calcium

5.115 Calculate the number of indicated atoms present in each of the following.
(a) carbon atoms in 0.0010 mole of $C_{20}H_{30}O$, vitamin A
(b) hydrogen atoms in 0.0200 mole of $C_{63}H_{88}N_{14}O_{14}PCo$, vitamin B_{12}
(c) nitrogen atoms in 0.01 mole of $C_{12}H_{17}ClN_4OS$, vitamin B_1

5.116 Calculate the number of indicated ions present in each of the following.
(a) nitrate ions in 2.0 moles of $Ca(NO_3)_2$
(b) perchlorate ions in 0.01 mole of $Mg(ClO_4)_2$
(c) sulfate ions in 0.3 mole of $Al_2(SO_4)_3$
(d) carbonate ions in 2.5 moles of Ag_2CO_3
(e) bisulfite ions in 4.0 moles of $Hg(HSO_3)_2$

5.117 A compound consists of 68.2% silver and 37.2% tellurium. What is the empirical formula?

5.118 Magnesium pyrophosphate contains 21.9% Mg, 27.8% P and 50.3% O. What is its empirical formula?

5.119 Given the following percent compositions, determine the empirical formula of each compound.
(a) 85.7% C and 14.3% H
(b) 80.0% C and 20.0% H
(c) 92.3% C and 7.7% H

5.120 Given the following percent compositions, determine the empirical formula of each compound.
(a) 38.7% C, 9.67% H, and 51.6% O
(b) 40.0% C, 6.66% H, and 53.3% O
(c) 54.6% C, 9.09% H, and 36.3% O

5.121 Vitamin D_1 has the molecular formula $C_{56}H_{88}O_2$. What is the empirical formula?

5.122 Fructose, a sugar found in honey, is 40.0% C, 6.7% H and 53.3% O. It has a molecular weight of 180 amu. What is the molecular formula of fructose?

6 Chemical Equations and Reactions

Learning Objectives

After studying Chapter 6 you should be able to

1. Balance chemical equations.

2. Recognize five types of chemical reactions.

3. Complete and balance equations for five types of chemical reactions

6.1 Chemical Equations

The law of conservation of mass is a statement of the fact that in ordinary chemical reactions matter is neither created nor destroyed. At the macroscopic level, this means that the total mass of the products is equal to the total mass of the reactants used in the reaction. At the submicroscopic level, the atomic theory accounts for these observations. In chemical reactions atoms do not disappear nor appear. All atoms arranged in structures in the reactants are redistributed into new combinations in the products.

Chemical equations are used to symbolize chemical reactions and indicate the number of atoms involved and how they are redistributed. All chemical equations must be balanced to reflect the law of conservation of mass, which means that the number of atoms of each kind must be the same in the products as in the reactants.

In Chapter 3, a word equation was used to express the chemical reaction of magnesium and oxygen in a flashbulb to yield magnesium oxide. Now we can use our knowledge of chemical formulas to make these equations more detailed. By replacing the words with the correct chemical formula for each reactant and product, we have a better idea of what is happening at the submicroscopic level. All balanced equations must meet the following requirements:

1. Correct formulas must be used to represent the substances involved.
2. The reactants must be placed on the left side and the products on the right side of the equation.
3. Atoms of any element, either in its elemental form or in a compound, that appear on one side of the equation must also appear on the other side of the equation and in the same amount.

The balanced equation for the flashbulb reaction is

$$2\,Mg + O_2 \longrightarrow 2\,MgO$$

The numbers appearing before each substance are coefficients and indicate the number of units of that substance involved relative to all other substances. Thus, two atoms of magnesium react with one molecule of oxygen to yield two formula units of magnesium oxide. There are two atoms of magnesium in both the reactants and products. There are also two atoms of oxygen in both the reactants and products.

In the first part of this chapter you will learn how to write equations and determine the coefficients of each substance in an equation. In the second part of this chapter you will see how balanced chemical equations are classified according to the type of chemical reaction. Five classes of chemical reactions will be considered.

6.2 Symbols in Equations

The Chemicals

Chemical formulas are symbolic representations of substances; chemical equations are a collection of symbols used to represent chemical reactions. The substances on the left of the equation are reactants, regardless of whether they are elements or compounds. The substances on the right of the equation are products and may be either elements or compounds. The generally accepted scheme for writing chemical reactions is

reactant (1) + reactant (2) $\longrightarrow$ product(1) + product (2)

In this example only two reactants and products are given; however, there may be any number of reactants and products.

The Coefficients

A balanced chemical equation contains numbers placed to the left of the chemical formulas. These numbers, called **coefficients,** indicate the number of units of that substance relative to all other reactants or products of the reaction. As indicated in Section 6.1, the coefficients in the reaction of magnesium with oxygen represent a submicroscopic accounting of the number of atoms involved.

Symbols of States

The physical states of the reactants and products may be denoted by writing (s), (l), or (g) to the right of the symbol for solid, liquid, and gas, respectively. Water, for example, might be produced as a liquid in one reaction but as a gas in another reaction, and the use of (l) or (g) gives the reader this information. Many reactions occur in water solution. The symbol (aq), meaning aqueous solution, is placed to the right of the chemical symbol for such reactants and products. This information is important. For example, there is a significant difference in reactivity of NaCl(aq) as compared to NaCl(s). The reaction in a flashbulb with the states indicated would be

$$2 \, Mg(s) + O_2(g) \longrightarrow 2 \, MgO(s)$$

The states of the substances need not always be indicated. A simpler although older convention uses arrows to indicate the gaseous or solid products in a reaction. An arrow pointing upward ($\uparrow$) denotes the release of a gas, as when hydrogen forms from the reaction of a metal with an acid.

$$Zn + 2 \, HCl \longrightarrow ZnCl_2 + H_2\uparrow$$

A solid product precipitating out of a solution is shown by an arrow pointing downward ($\downarrow$), as in the formation of silver chloride.

$$AgNO_3 + NaCl \longrightarrow AgCl\downarrow + NaNO_3$$

Connecting Symbols

On each side of the equation a plus sign is used to separate each of the reactants and each of the products. It is read as "plus" or "and." Several symbols are used to separate the reactants and products. The single arrow $\rightarrow$ or an equal sign is read as "yields" or "produces." A double arrow $\rightleftharpoons$ indicates that the reaction may proceed from the left to right as written or in the reverse direction, that is from right to left. (Forward and reverse reactions are discussed in Chapter 15.) A listing of these symbols and others used in chemical equations is given in Table 6.1.

Symbols Above the Arrow

The symbol Δ placed above the arrow represents the fact that the reaction is run at a high temperature. For example, the conversion of limestone ($CaCO_3$) into lime (CaO) requires heat.

$$CaCO_3(s) \xrightarrow{\Delta} CaO(s) + CO_2(g)$$

Table 6.1   Symbols Used in Chemical Equations

$\rightarrow$	yields or produces
$\rightleftharpoons$	reaction proceeds in both directions
(s)	solid; written immediately after the substance
(l)	liquid; written immediately after the substance
(g)	gas; written immediately after the substance
(aq)	substance is dissolved in water
+	plus
$\uparrow$	a gas; written immediately after a gaseous product
$\downarrow$	a solid deposited from solution; written immediately after the substance
Δ	heat

The exact temperature required may be written over the arrow instead of the symbol Δ, as in the equation for the industrial preparation of methyl alcohol.

$$2\,H_2(g) + CO(g) \xrightarrow{\;350°C\;} CH_3OH(g)$$

Occasionally, a symbol for an element or a compound also appears above the arrow. This symbol represents a catalyst, which causes an increase in the rate of the reaction. At the end of the reaction, the catalyst can be recovered because, unlike the reactants, a catalyst is not altered in the reaction. For example, the production of methyl alcohol requires a mixture of zinc and chromium oxides as the catalyst. We discuss catalysts further in Chapter 15.

$$2\,H_2(g) + CO(g) \xrightarrow[\;ZnO/Cr_2O_3\;]{\;350°C\;} CH_3OH(g)$$

6.3   Balancing Equations

Now let's consider how we obtain a balanced chemical equation. Although there is no hard and fast set of rules that can be used to balance equations, some helpful general guidelines are given in Table 6.2. With practice using these guidelines, you will be able to balance equations given in this text.

How do we balance the equation for the conversion of magnesium and oxygen into magnesium oxide? We start by properly representing the chemical constitution of the reactants and products. Since oxygen exists as a diatomic molecule, we must represent that structure as O_2 in the chemical equation. Magnesium oxide consists of Mg^{2+} and O^{2-} ions in equal numbers and is represented as MgO (Figure 6.1). We then write

$$Mg + O_2 \longrightarrow MgO$$

This equation is not balanced because there are two oxygen atoms in the O_2 molecule on the left but only one in the MgO on the right. The equation must not be balanced by changing the formula of magnesium oxide to MgO_2 since that formula does not correctly represent the product obtained in the reaction. However, the coefficient 2 can be placed in front of MgO to indicate that two units of MgO are formed (Figure 6.1).

$$Mg + O_2 \longrightarrow 2\,MgO$$

Table 6.2   **Guidelines to Balance Chemical Equations**

1. Write the correct formulas for the reactants and products on the opposite sides of the yield arrow. Once the correct formula is written, do not alter it or any of its subscripts during the balancing process.

2. Disregarding hydrogen, oxygen, and polyatomic ions, find the substance containing the largest number of atoms of a single element. Balance the number of this element by placing the proper coefficient in front of the substance containing this element on the other side of the equation. A coefficient placed in front of a formula multiplies every atom contained in that formula.

3. Proceed to balance the atoms of other elements by the same process. Check to see if in balancing one element, others have become unbalanced. Readjust the coefficients on both sides of the equation to achieve the necessary balance.

4. Balance polyatomic ions on each side of the equation as a single unit.

5. Balance hydrogen and oxygen not previously considered in polyatomic ions.

6. Check all coefficients to insure that they are the lowest possible whole numbers. If the coefficients are fractions, multiply by a number to make the fraction a whole number. All other coefficients must be multiplied by the same quantity to maintain a balanced equation. If all the coefficients are divisible by a common factor, do so to achieve the lowest possible whole number.

7. Check the entire equation to ensure that all atoms are balanced.

When a coefficient of 2 is placed in front of MgO, it means that there are two magnesium and two oxygen atoms. Now the oxygen atoms on the left and right sides of the equation are equal in number, but there are two magnesium atoms on the right side and only one on the left side. This situation can be corrected by placing the coefficient 2 in front of Mg on the left side of the equation (Figure 6.1).

$$2\,Mg + O_2 \longrightarrow 2\,MgO$$

The equation is now balanced. Note that the following are also balanced equations.

$$Mg + \tfrac{1}{2}O_2 \longrightarrow MgO$$
$$4\,Mg + 2\,O_2 \longrightarrow 4\,MgO$$
$$6\,Mg + 3\,O_2 \longrightarrow 6\,MgO$$

These equations, however, are either a fraction or a multiple of the accepted balanced equation in which the coefficients are in the smallest whole number ratio possible. Only the balanced equation with the smallest whole number coefficients should be written.

Now let's consider a somewhat more complicated reaction. Aluminum hydroxide, $Al(OH)_3$, reacts with sulfuric acid, H_2SO_4, to yield aluminum sulfate and water according to the following unbalanced equation.

$$Al(OH)_3 + H_2SO_4 \longrightarrow Al_2(SO_4)_3 + H_2O$$

How can we balance this equation? Using guideline 2 from Table 6.2 we ignore hydrogen, oxygen, and the polyatomic ions hydroxide and sulfate. Therefore we first balance the two

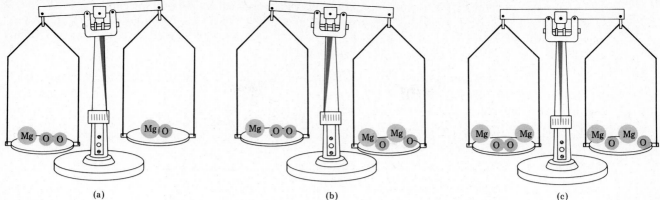

(a)	(b)	(c)

Figure 6.1

Balancing an equation
In (a) the number of
atoms of oxygen in the
reactants exceeds the
oxygen atoms in the
product. In (b) the
number of atoms of
magnesium in the
product exceeds the
magnesium atoms in
the reactant. In (c) the
numbers of atoms in
the product and
reactants are equal.

atoms of aluminum in $Al_2(SO_4)_3$ on the right side of the equation by placing a coefficient 2 in front of $Al(OH)_3$ on the left side of the equation.

$$2\ Al(OH)_3 + H_2SO_4 \longrightarrow Al_2(SO_4)_3 + H_2O$$

Since only hydrogen, oxygen, and the sulfate ions remain unbalanced, the sulfate ion is balanced next. The three sulfate units in the $Al_2(SO_4)_3$ are balanced by placing the coefficient 3 in front of H_2SO_4.

$$2\ Al(OH)_3 + 3\ H_2SO_4 \longrightarrow Al_2(SO_4)_3 + H_2O$$

Now only the oxygen not incorporated in the sulfate ion and the hydrogen remain unbalanced. There are six hydrogen atoms contained in the two units of $Al(OH)_3$ and six hydrogen atoms in the three units of H_2SO_4, for a total of 12 that must be balanced by a coefficient in front of H_2O. The coefficient needed is 6 because there are two hydrogen atoms in an H_2O molecule.

$$2\ Al(OH)_3 + 3\ H_2SO_4 \longrightarrow Al_2(SO_4)_3 + 6\ H_2O$$

The six oxygen atoms in two units of $Al(OH)_3$ are also brought into balance by the six units of H_2O. There is a total of 18 oxygen atoms incorporated in the reactants as well as the products. Therefore, the equation is balanced.

Finally let's consider a reaction in which it is necessary to multiply by a factor to eliminate fractional coefficients. The hydrocarbon butane, C_4H_{10}, burns in oxygen to yield carbon dioxide and water.

$$C_4H_{10} + O_2 \longrightarrow CO_2 + H_2O$$

The molecular formula of butane has four carbon atoms and ten hydrogen atoms. Using guideline 2 from Table 6.2, we start our balancing process with carbon. The coefficient 4 is placed in front of CO_2.

$$C_4H_{10} + O_2 \longrightarrow 4\ CO_2 + H_2O$$

Next, the hydrogen is balanced since there are no other elements besides oxygen that need to be balanced. Hydrogen is chosen over oxygen because it appears in only one compound on each side of the equation. The coefficient 5 placed in front of H_2O balances the ten hydrogen atoms of butane.

$$C_4H_{10} + O_2 \longrightarrow 4\ CO_2 + 5\ H_2O$$

Finally we balance the oxygen atoms. The oxygen atoms in the products now total 13, eight in the 4 CO_2 and five in the 5 H_2O. In order to balance the oxygen atoms in the O_2 molecule, a coefficient of $6\frac{1}{2}$ or $\frac{13}{2}$ is necessary. In $6\frac{1}{2}$ oxygen molecules there are 13 oxygen atoms.

$$C_4H_{10} + \tfrac{13}{2}\, O_2 \longrightarrow 4\, CO_2 + 5\, H_2O$$

Finally, the fractional coefficient must be eliminated to obtain the proper balanced equation by multiplying all coefficients by 2.

$$2\, C_4H_{10} + 13\, O_2 \longrightarrow 8\, CO_2 + 10\, H_2O$$

Example 6.1

Balance the equation for the combustion of propane, C_3H_8.

$$C_3H_8 + O_2 \longrightarrow CO_2 + H_2O$$

Solution   There are three atoms of carbon in one molecule of propane. Therefore, three molecules of CO_2 must be produced. The coefficient 3 is placed in front of the formula CO_2.

$$C_3H_8 + O_2 \longrightarrow 3\, CO_2 + H_2O$$

There are eight atoms of hydrogen contained in one molecule of C_3H_8. Therefore, four molecules of H_2O must be produced. Placing the coefficient 4 in front of the formula H_2O yields

$$C_3H_8 + O_2 \longrightarrow 3\, CO_2 + 4\, H_2O$$

The equation is not balanced because there are ten atoms of oxygen represented on the right of the arrow and only two on the left. In order to balance the equation, the coefficient 5 must be placed in front of O_2.

$$C_3H_8 + 5\, O_2 \longrightarrow 3\, CO_2 + 4\, H_2O$$

[*Additional examples may be found in 6.13 and 6.14 at the end of this chapter.*]

Example 6.2

Balance the equation for the reaction of magnesium hydroxide and phosphoric acid.

$$Mg(OH)_2 + H_3PO_4 \longrightarrow Mg_3(PO_4)_2 + H_2O$$

Solution   Disregarding the OH^- and the polyatomic ion PO_4^{3-}, the compound with the greatest number of atoms is $Mg_3(PO_4)_2$, which contains 3 magnesium atoms. Balancing the $Mg(OH)_2$ with the coefficient 3 produces

$$3\, Mg(OH)_2 + H_3PO_4 \longrightarrow Mg_3(PO_4)_2 + H_2O$$

Since only the phosphate ion, hydrogen, and oxygen remain unbalanced, the phosphate ion is balanced next. There are two phosphate ions in the one unit of $Mg_3(PO_4)_2$, whose coefficient has been established. The coefficient 2 is placed before H_3PO_4 to achieve a phosphate balance.

$$3\, Mg(OH)_2 + 2\, H_3PO_4 \longrightarrow Mg_3(PO_4)_2 + H_2O$$

Finally, either the hydrogen or oxygen may be balanced. There are 12 hydrogens in 3 Mg(OH)$_2$ and 2 H$_3$PO$_4$ combined. The coefficient 6 placed in front of H$_2$O achieves the balance of hydrogen.

$$3 \text{ Mg(OH)}_2 + 2 \text{ H}_3\text{PO}_4 \longrightarrow \text{Mg}_3(\text{PO}_4)_2 + 6 \text{ H}_2\text{O}$$

There are a total of 14 oxygen atoms in both the reactants and products, thus the reaction is balanced.

[*Additional examples may be found in 6.15 and 6.16 at the end of the chapter.*]

Example 6.3

Balance the following equation.

$$\text{Se} + \text{BrF}_5 \longrightarrow \text{SeF}_6 + \text{BrF}_3$$

Solution The element present in the largest amount is fluorine in BrF$_5$. However, the fluorine appears in two products and cannot be balanced directly based on the fluorine present in the reactant. Therefore, either selenium or bromine must be balanced first. The element of choice is bromine, because once bromine is brought into balance it will also fix the amount of fluorine available from the reactant BrF$_5$. Some of the fluorine will be fixed by the coefficient of BrF$_3$, and the remaining fluorine can be accounted for with a proper coefficient for SeF$_6$.

$$\text{Se} + 1 \text{ BrF}_5 \longrightarrow \text{SeF}_6 + 1 \text{ BrF}_3$$

Three of the five fluorine atoms of BrF$_5$ are accounted for in BrF$_3$. The remaining two fluorine atoms must be balanced with a coefficient for SeF$_6$ which must be $\frac{1}{3}$.

$$\text{Se} + 1 \text{ BrF}_5 \longrightarrow \tfrac{1}{3} \text{SeF}_6 + 1 \text{ BrF}_3$$

Now, the amount of selenium is fixed by the coefficient established for SeF$_6$.

$$\tfrac{1}{3} \text{Se} + 1 \text{ BrF}_5 \longrightarrow \tfrac{1}{3} \text{SeF}_6 + 1 \text{ BrF}_3$$

Multiplication of all quantities by 3 results in integers in the final balanced equation.

$$\text{Se} + 3 \text{ BrF}_5 \longrightarrow \text{SeF}_6 + 3 \text{ BrF}_3$$

[*Additional examples may be found in 6.3–6.12 at the end of the chapter.*]

6.4 Types of Chemical Reactions

The number of known and potential chemical reactions involving the 108 elements and approximately 7 million compounds is astronomically large. In order to develop a systematic understanding of chemical reactions, it is useful to classify them according to the type of process that occurs. The placement of a reaction into a particular category is based on certain defined similarities in common with other reactions. Recognizing a reaction as one of a class allows us to predict the products of the reaction.

At this point in your study of chemistry, we will consider only five simple types of reactions.

1. Combination reactions.
2. Decomposition reactions.
3. Single replacement reactions.
4. Double replacement reactions.
5. Neutralization reactions.

A large number of reactions are included in these five categories. If you continue to study chemistry, it will become apparent that there are other ways to classify reactions. However, this classification scheme is sufficient for the kinds of reactions we study in this text.

6.5 Combination Reactions

A **combination reaction** involves the direct union of two or more substances to produce one new substance. The two or more combining materials may be either elements or compounds, or a combination of the two. Of course, the product must be a compound. A general equation that summarizes combination reactions can be written, using X and Y to symbolize the reactants. A combination reaction is illustrated in Figure 6.2.

$$X + Y \longrightarrow XY$$

One common example of a combination reaction involves the reaction of metals with the nonmetal oxygen to form metal oxides which are most commonly ionic compounds. For example, lithium (a reactive metal) combines with oxygen to yield lithium oxide.

$$4\,Li(s) + O_2(g) \xrightarrow{\Delta} 2\,Li_2O(s)$$

Freshly prepared aluminum reacts with oxygen to give aluminum oxide, which forms a hard coating over the metal.

$$4\,Al(s) + 3\,O_2(g) \longrightarrow 2\,Al_2O_3(s)$$

Most nonmetals combine with oxygen to produce molecular compounds. One example is the reaction in which oxygen combines with sulfur to yield sulfur dioxide.

$$S(s) + O_2(g) \xrightarrow{\Delta} SO_2(g)$$

Sulfur dioxide is one of the seriously detrimental pollutants in the air. It results from the burning of any sulfur-containing fuels. With proper experimental conditions, sulfur also can combine with oxygen to yield sulfur trioxide.

$$2\,S(s) + 3\,O_2(g) \longrightarrow 2\,SO_3(g)$$

The nonmetal carbon also can combine with oxygen to yield two possible compounds. In a limited amount of oxygen, carbon monoxide is formed. If sufficient oxygen is present, the product is carbon dioxide.

$$2\,C(s) + O_2(g) \longrightarrow 2\,CO(g)$$
$$C(s) + O_2(g) \longrightarrow CO_2(g)$$

Oxygen does not react with nitrogen at ordinary temperatures. However, at temperatures above 2000°C in the internal combustion engine, these two elements do combine to yield nitrogen oxide (nitric oxide).

Combination

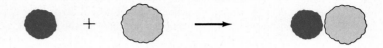

Decomposition

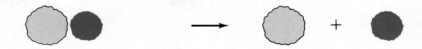

Single replacement

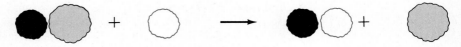

Double replacement

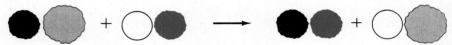

Figure 6.2

General types of chemical reactions
The differently shaded circles in each of the general representations of chemical reactions represent either single atoms or groups of atoms.

$$N_2(g) + O_2(g) \longrightarrow 2\,NO(g)$$

Hydrogen can combine directly with elements. One reaction of major industrial importance is the Haber process, in which ammonia is formed by the direct combination of hydrogen and nitrogen. Ammonia is used to produce fertilizer and to form a variety of nitrogen compounds.

$$N_2(g) + 3\,H_2(g) \longrightarrow 2\,NH_3(g)$$

Compounds can also undergo combination reactions. Metal oxides can react with water to produce metal hydroxides, which are called **bases.** Since a base is produced, the oxide from which it is derived is called a **basic oxide.** Bases are discussed in Chapter 16.

$$\underset{\text{basic oxide}}{H_2O(l) + Na_2O(s)} \longrightarrow \underset{\text{base}}{2\,NaOH(aq)}$$

$$\underset{\text{basic oxide}}{H_2O(l) + MgO(s)} \longrightarrow \underset{\text{base}}{Mg(OH)_2(aq)}$$

$$\underset{\text{basic oxide}}{3\,H_2O(l) + Al_2O_3(s)} \longrightarrow \underset{\text{base}}{2\,Al(OH)_3(s)}$$

Oxides of nonmetals are called **acidic oxides** because they may combine with water to yield acids. Some acids are compounds of hydrogen combined with a simple anion or polyatomic anion. These compounds are also discussed in Chapter 16.

$$N_2O_5(g) + H_2O(l) \longrightarrow 2\,HNO_3(aq)$$
acidic oxide   nitric acid

$$P_4O_{10}(s) + 6\,H_2O(l) \longrightarrow 4\,H_3PO_4(aq)$$
acidic oxide   phosphoric acid

$$SO_3(g) + H_2O(l) \longrightarrow H_2SO_4(aq)$$
acidic oxide   sulfuric acid

6.6 Decomposition Reactions

In a **decomposition reaction,** a single substance undergoes a breakdown into two or more simple substances. The reactant is a compound, but the products may be either elements or compounds. Usually heat is required to cause decomposition of a compound. A general representation of the decomposition reaction is given below and is illustrated in Figure 6.2.

$$XY \longrightarrow X + Y$$

The decomposition of mercury(II) oxide to yield mercury and oxygen exemplifies this type of reaction. This reaction was studied by Lavoisier, whose quantitative findings on this reaction helped establish the law of conservation of mass.

$$2\,HgO(s) \xrightarrow{\Delta} 2\,Hg(l) + O_2(g)$$

A common process used in the undergraduate laboratory to produce oxygen involves the decomposition of potassium chlorate. In this reaction manganese dioxide is used as a catalyst.

$$2\,KClO_3(s) \xrightarrow[\Delta]{MnO_2} 2\,KCl(s) + 3\,O_2(g)$$

Some carbonates will decompose to yield oxides and carbon dioxide when heated. For example, calcium carbonate decomposes in the following way.

$$CaCO_3(s) \xrightarrow{\Delta} CaO(s) + CO_2(g)$$

Calcium carbonate is one of the materials that is deposited in pipes and in equipment when hard water is used. It is the major constituent of coral skeletons and is also produced by the reaction of atmospheric carbon dioxide with calcium ions contained in the oceans.

Some bicarbonates decompose to yield water, carbon dioxide, and a carbonate salt.

$$2\,NaHCO_3(s) \xrightarrow{\Delta} Na_2CO_3(s) + H_2O(g) + CO_2(g)$$

Sodium bicarbonate may be used to put out fires because the carbon dioxide formed by its decomposition prevents oxygen from feeding the fire.

Molten nitrates may decompose in a variety of ways, as illustrated by the breakdown of sodium nitrate or ammonium nitrate. The decomposition of nitrates may occur with explosive violence.

$$2\,NaNO_3(s) \xrightarrow{\Delta} 2\,NaNO_2(s) + O_2(g)$$

$$NH_4NO_3(s) \xrightarrow{\Delta} N_2O(g) + 2\,H_2O(g)$$

TNT is another nitrogen-containing compound that decomposes at a very rapid rate, resulting in an explosion.

$$2\,\underset{\text{TNT}}{C_7H_5N_3O_6(s)} \longrightarrow 3\,N_2(g) + 5\,H_2O(g) + 7\,CO(g) + 7\,C(s)$$

6.7 Single Replacement Reactions

In a **single replacement** reaction, one element substitutes for or replaces another element in a compound (Figure 6.2). Thus, an element and a compound produce another element and a new compound. The elements involved may be metals or nonmetals. For example, the metal A may replace the metal ion B. In some cases, B may be a hydrogen ion of an acid.

$$A + BY \longrightarrow B + AY$$

The direction of substitution of one metal by another is given by the activity series (Table 6.3). We discuss the activity series more in Chapter 17. Because of the way the activity series is ordered, any metal can substitute for the ion of a metal below it in the series. Thus, zinc can replace the copper in copper(II) sulfate because zinc is above copper in the activity series. A strip of zinc immersed in a solution of copper(II) sulfate disappears, and free copper metal comes out of the solution.

$$Zn(s) + CuSO_4(aq) \longrightarrow ZnSO_4(aq) + Cu(s)$$

If a metal is in contact with an ion of a metal above it in the activity series, then no reaction occurs. Thus, as indicated by the activity series, silver will not react with Na_2SO_4.

$$Ag + Na_2SO_4 \longrightarrow \text{no reaction}$$

Hydrogen is included in the activity series although it is not a metal. Any metal above hydrogen in the series will react with an acid to form a metal salt and hydrogen gas.

$$Zn(s) + H_2SO_4(aq) \longrightarrow ZnSO_4(aq) + H_2(g)$$
$$Sn(s) + 2\,HCl(aq) \longrightarrow SnCl_2(aq) + H_2(g)$$

A metal below hydrogen in the activity series will not react with an acid.

$$Au + HCl \longrightarrow \text{no reaction}$$

There are numerous industrial processes in which metals are produced by reacting a compound of that metal with a more active metal. Usually, these reactions occur in the molten or solid state at high temperatures.

$$Cr_2O_3 + 2\,Al \xrightarrow{\Delta} 2\,Cr + Al_2O_3$$

$$Sb_2S_3 + 3\,Fe \xrightarrow{\Delta} 2\,Sb + 3\,FeS$$

Table 6.3 Activity Series of Metals

Symbol	Activity
Li	
K	
Ba	
Ca	
Na	
Mg	
Al	increasing activity ↑
Zn	
Fe	
Cd	
Ni	
Sn	
H	
Cu	
Hg	
Ag	
Au	

Single replacement reactions involving nonmetals can be represented as follows, where X and Y represent nonmetals.

$$X + BY \longrightarrow BX + Y$$

A reactivity series has been formulated for substitution reactions of a nonmetal for a nonmetallic ion. The activity order is F > Cl > Br > I. Fluorine will replace chloride, bromide, or iodide ions; chlorine will replace bromide or iodide; bromine will replace only iodide. For example, chlorine gas can be used to obtain bromine from the bromide ion contained in sea water.

$$Cl_2(g) + 2\,NaBr(aq) \longrightarrow 2\,NaCl(aq) + Br_2(aq)$$

Many metals are produced by a third type of substitution reaction. In this reaction the non-metal combined with the metal is removed by reaction with another nonmetal. Thus, germanium and tungsten are obtained from their oxides by reacting with hydrogen.

$$GeO_2(s) + 2\,H_2(g) \xrightarrow{\Delta} Ge(s) + 2\,H_2O(g)$$

$$WO_3(s) + 3\,H_2(g) \xrightarrow{\Delta} W(s) + 3\,H_2O(g)$$

A few metal sulfides react with oxygen to yield the free metal and sulfur dioxide.

$$HgS(s) + O_2(g) \xrightarrow{\Delta} Hg(l) + SO_2(g)$$

$$Cu_2S(s) + O_2(g) \xrightarrow{\Delta} 2\,Cu(s) + SO_2(g)$$

Carbon can be used to convert some oxides into the free elements and carbon monoxide.

$$SnO_2(s) + 2\,C(s) \xrightarrow{\Delta} Sn(l) + 2\,CO(g)$$

$$As_4O_6(s) + 6\,C(s) \xrightarrow{\Delta} As_4(g) + 6\,CO(g)$$

6.8 Double Replacement Reactions

In **double replacement reactions,** two compounds exchange atoms or groups of atoms (such as polyatomic ions) to produce two different compounds (Figure 6.2). A general representation of a double replacement reaction is

$$AX + BY \longrightarrow AY + BX$$

where any of the letters may be single atoms or polyatomic ions containing several atoms. In this type of reaction one of the products is usually either a solid that precipitates from solution or a gas.

The formation of an insoluble material that precipitates from solution is the result of the exchange of partners so that the least soluble compound is produced. In order to predict when a double replacement reaction will occur, it is necessary to learn the solubilities of ionic compounds. A list of generalizations of solubilities of compounds in water at room temperature is given in Table 6.4.

Table 6.4   **General Rules for Solubilities of Common Ionic Compounds**

1. All ionic compounds of the lithium (Li^+), sodium, (Na^+), potassium, (K^+), rubidium (Rb^+), and cesium (Cs^+) ions and of the ammonium ion (NH_4^+) are soluble.

2. All nitrates (NO_3^-), chlorates (ClO_3^-), and perchlorates (ClO_4^-) are soluble.

3. The chlorides (Cl^-), bromides (Br^-), and iodides (I^-) of most metals are soluble. The principal exceptions are those of lead (Pb^{2+}), silver (Ag^+), and mercury(I) (Hg_2^{2+}).

4. All sulfates (SO_4^{2-}) are soluble except those of strontium (Sr^{2+}), barium (Ba^{2+}), lead (Pb^{2+}), and mercury(I) (Hg_2^{2+}).

5. All carbonates (CO_3^{2-}), chromates (CrO_4^{2-}), and phosphates (PO_4^{3-}) are insoluble except those of lithium, sodium, potassium, rubidium, cesium, and ammonium.

6. The hydroxides (OH^-) of lithium, sodium, potassium, rubidium, and cesium ions are soluble. The hydroxides of calcium, strontium, and barium are moderately soluble. The rest of the hydroxides are insoluble.

7. The sulfides (S^{2-}) of all metals are insoluble except those of lithium, sodium, potassium, rubidium, cesium, and ammonium.

Consider the reaction of silver nitrate and sodium chloride in aqueous solution. Since silver chloride is insoluble, a double replacement reaction will occur to yield a precipitate. Sodium nitrate remains in solution.

$$AgNO_3(aq) + NaCl(aq) \longrightarrow AgCl(s) + NaNO_3(aq)$$

No apparent reaction will occur with calcium nitrate and sodium chloride because both of the potential products, calcium chloride and sodium nitrate, are soluble in water.

$$Ca(NO_3)_2(aq) + NaCl(aq) \longrightarrow \text{no reaction}$$

Acids or bases may provide the ions for double replacement reactions and result in the formation of a precipitate. In the case of barium chloride and sulfuric acid, the precipitate formed is barium sulfate. In the case of nickel(II) nitrate and potassium hydroxide the precipitate formed is nickel(II) hydroxide.

$$BaCl_2(aq) + H_2SO_4(aq) \longrightarrow BaSO_4(s) + 2\,HCl(aq)$$
$$Ni(NO_3)_2(aq) + 2\,KOH(aq) \longrightarrow 2\,KNO_3(aq) + Ni(OH)_2(s)$$

A gas is formed in some double replacement reactions. When a metal carbonate or bicarbonate reacts with acid, the double replacement reaction should yield carbonic acid (H_2CO_3); however, carbonic acid is unstable and decomposes to yield carbon dioxide and water. The carbon dioxide is a gas.

$$MgCO_3(s) + H_2SO_4(aq) \longrightarrow MgSO_4(aq) + H_2O(l) + CO_2(g)$$

Magnesium carbonate is one of the ingredients in several antacids. Antacids react with the hydrochloric acid (HCl) in the stomach to produce carbon dioxide.

6.9 Neutralization Reactions

Neutralization reactions occur whenever an acid or an acidic oxide reacts with a base or a basic oxide. The various combinations of reactants possible for a neutralization reaction are

1. Acid and base.
2. Basic oxide and acid.
3. Base and acidic oxide.
4. Acidic oxide and basic oxide.

In most neutralization reactions, one of the products is water and the other is often an ionic compound, which is also called a salt. The following general equation shows a neutralization reaction using HX and BOH to represent an acid and a base, respectively.

$$HX + BOH \longrightarrow BX + H_2O$$

You may notice that this general equation resembles that of a double replacement reaction. In a neutralization reaction, however, one of the products is specifically water. In addition one of the components of one reactant is a hydrogen atom and one of the components of the second reactant is a hydroxide ion. In Chapter 16 we establish that acids behave chemically as donors of protons, that is, the hydrogen ion (H^+). The acids that contain this hydrogen ion are written with the hydrogen atom first followed by a simple or polyatomic ion. Examples of such acids include hydrochloric acid (HCl), perchloric acid $(HClO_4)$, nitric acid (HNO_3), sulfuric acid (H_2SO_4), and a phosphoric acid (H_3PO_4). Examples of neutralization reactions of an acid and a base are given below.

$$\underset{\text{acid}}{HNO_3(aq)} + \underset{\text{base}}{KOH(aq)} \longrightarrow \underset{\text{salt}}{KNO_3(aq)} + \underset{\text{water}}{H_2O(l)}$$

$$\underset{\text{acid}}{2\,HCl(aq)} + \underset{\text{base}}{Ca(OH)_2(aq)} \longrightarrow \underset{\text{salt}}{CaCl_2(aq)} + \underset{\text{water}}{2\,H_2O(l)}$$

In each balanced equation, the number of hydrogen ions provided by the acid equals the number of hydroxide ions provided by the base.

The reaction of a basic oxide and an acid is also a neutralization reaction. Recall from Section 6.5 that metal oxides are basic oxides. In the reaction of calcium oxide with perchloric acid, the calcium ion and the hydrogen ion are interchanged. Again the products of the reaction are a salt and water.

$$\underset{\text{basic oxide}}{CaO} + \underset{\text{acid}}{2\,HClO_4} \longrightarrow \underset{\text{salt}}{Ca(ClO_4)_2} + \underset{\text{water}}{H_2O}$$

The reaction of an acidic oxide and a base is another example of a neutralization reaction. Recall from Section 6.5 that oxides of nonmetals are acidic oxides. In the reaction of carbon dioxide, an acidic oxide, and lithium hydroxide, the reaction produces a salt and water.

$$\underset{\text{base}}{2\,LiOH} + \underset{\text{acidic oxide}}{CO_2} \longrightarrow \underset{\text{salt}}{Li_2CO_3} + \underset{\text{water}}{H_2O}$$

In some cases, the acidic oxide combines directly with the base and no water is formed.

$$\underset{\text{base}}{NaOH} + \underset{\text{acidic oxide}}{SO_2} \longrightarrow \underset{\text{salt}}{NaHSO_3}$$

Acidic oxides and basic oxides also can combine directly. In such reactions, no water is formed.

$$CaO \quad + \quad SO_3 \quad \longrightarrow \quad CaSO_4$$

basic oxide   acidic oxide   salt

Summary

A chemical equation is used to represent how substances known as reactants undergo transformation into substances called the products of a reaction. Chemical symbols are used to represent the reactants and products. The coefficients of a balanced chemical equation provide information about the quantities of reactants and products involved on a submicroscopic scale. Additional information that may be conveyed in a chemical equation includes the state of the reactants and products, the conditions necessary to carry out the reaction, and the requirement of catalysts.

Guidelines for balancing equations are useful to establish the coefficients of a chemical equation. The first focus is on the molecule containing the largest number of atoms of a single element, excluding hydrogen, oxygen, and polyatomic ions from considera-

tion. The process is repeated for other elements. Polyatomic ions are next balanced followed by hydrogen and oxygen. The final coefficients must not consist of fractions and must be the lowest possible whole numbers.

Chemical reactions can be categorized according to the type of process involved. Simple reactions can be labeled as combination, decomposition, single replacement, double replacement, and neutralization reactions.

Metal oxides, known as basic oxides, react with water to produce bases. Oxides of nonmetals, known as acidic oxides, react with water to produce acids. Neutralization reactions may involve an acid and a base, a basic oxide and an acid, a base and an acidic oxide, or an acidic oxide and a basic oxide.

New Terms

Acids are compounds of hydrogen that are donors of hydrogen ions.

Acidic oxides are oxides of nonmetals that combine with water to form acids.

One type of **base** is a metal hydroxide.

Basic oxides are oxides of metals that combine with water to form bases.

Coefficients indicate the number of units of a substance relative to other reactants and products in a chemical equation.

Combination reactions involve the direct union of two or more substances to produce one new substance.

In a **decomposition reaction,** a single substance undergoes a breakdown into two or more simpler substances.

In a **double replacement reaction,** two compounds exchange atoms or groups of atoms to produce two different compounds.

In a **neutralization reaction,** an acid or acidic oxide reacts with a base or a basic oxide.

A **salt** is another term for an ionic compound.

In a **single replacement reaction,** one element substitutes for or replaces another element in a compound.

Exercises

Terminology

6.1 Explain the difference between a combination reaction and a decomposition reaction.

6.2 Explain the difference between a single replacement reaction and a double replacement reaction.

Balancing Equations

6.3 Balance each of the following equations.
   (a) $Ba + O_2 \longrightarrow BaO$
   (b) $Fe + Cl_2 \longrightarrow FeCl_3$
   (c) $H_2 + I_2 \longrightarrow HI$
   (d) $HgO \longrightarrow Hg + O_2$

6.4 Balance each of the following equations.
   (a) $O_3 \longrightarrow O_2$
   (b) $N_2O_5 \longrightarrow NO_2 + O_2$
   (c) $SrCO_3 \longrightarrow SrO + CO_2$
   (d) $Fe + O_2 \longrightarrow Fe_2O_3$

6.5 Balance each of the following equations.
(a) $PbCl_2 + Na_2S \longrightarrow PbS + NaCl$
(b) $Mg + Fe_3O_4 \longrightarrow MgO + Fe$
(c) $Au_2O_3 \longrightarrow Au + O_2$
(d) $Cs_2O + H_2O \longrightarrow CsOH$

6.7 Balance each of the following equations.
(a) $Li + Se \longrightarrow Li_2Se$
(b) $H_2O + O_2 \longrightarrow H_2O_2$
(c) $Ca_3P_2 + H_2O \longrightarrow PH_3 + Ca(OH)_2$
(d) $P_4 + Cl_2 \longrightarrow PCl_5$

6.9 Balance each of the following equations.
(a) $Ag_2O \longrightarrow Ag + O_2$
(b) $N_2O \longrightarrow N_2 + O_2$
(c) $Al + Cl_2 \longrightarrow AlCl_3$
(d) $CaO + SiO_2 \longrightarrow CaSiO_3$

6.11 Balance each of the following equations.
(a) $SiO_2 + HF \longrightarrow H_2SiF_6 + H_2O$
(b) $Li_3N + H_2O \longrightarrow LiOH + NH_3$
(c) $Al_4C_3 + HCl \longrightarrow AlCl_3 + CH_4$
(d) $H_2O_2 + N_2H_4 \longrightarrow N_2 + H_2O$

6.13 Balance each of the following equations.
(a) $C_2H_6 + O_2 \longrightarrow CO_2 + H_2O$
(b) $C_5H_{12} + O_2 \longrightarrow CO_2 + H_2O$
(c) $C_5H_{10} + O_2 \longrightarrow CO_2 + H_2O$
(d) $C_5H_8 + O_2 \longrightarrow CO_2 + H_2O$

6.15 Balance each of the following equations.
(a) $H_2SO_4 + Na_2CO_3 \longrightarrow Na_2SO_4 + H_2O + CO_2$
(b) $H_3PO_4 + KOH \longrightarrow K_3PO_4 + H_2O$
(c) $FeCl_2 + NaOH \longrightarrow Fe(OH)_2 + NaCl$
(d) $Fe(OH)_3 + H_2SO_4 \longrightarrow Fe_2(SO_4)_3 + H_2O$

6.6 Balance each of the following equations.
(a) $SO_2 + O_2 \longrightarrow SO_3$
(b) $Al_2O_3 + C \longrightarrow Al + CO_2$
(c) $I_2O_5 + CO \longrightarrow CO_2 + I_2$
(d) $Na + P_4 \longrightarrow Na_3P$

6.8 Balance each of the following equations.
(a) $CH_4 + Cl_2 \longrightarrow CCl_4 + HCl$
(b) $CO + H_2 \longrightarrow CH_4 + H_2O$
(c) $PCl_5 + H_2O \longrightarrow H_3PO_4 + HCl$
(d) $Br_2 + CaI_2 \longrightarrow CaBr_2 + I_2$

6.10 Balance each of the following equations.
(a) $TlI_3 + Na \longrightarrow NaI + Tl$
(b) $CaC_2 + H_2O \longrightarrow Ca(OH)_2 + C_2H_2$
(c) $KClO_3 \longrightarrow KCl + O_2$
(d) $Sb_2S_3 + O_2 \longrightarrow Sb + SO_2$

6.12 Balance each of the following equations.
(a) $PCl_3 + H_2O \longrightarrow H_3PO_3 + HCl$
(b) $Zn + HCl \longrightarrow ZnCl_2 + H_2$
(c) $SiCl_4 + Mg \longrightarrow MgCl_2 + Si$
(d) $Mg + N_2 \longrightarrow Mg_3N_2$

6.14 Balance each of the following equations.
(a) $C_2H_6O + O_2 \longrightarrow CO_2 + H_2O$
(b) $C_5H_{12}O + O_2 \longrightarrow CO_2 + H_2O$
(c) $C_5H_{10}O + O_2 \longrightarrow CO_2 + H_2O$
(d) $C_5H_8O + O_2 \longrightarrow CO_2 + H_2O$

6.16 Balance each of the following equations.
(a) $AgNO_3 + CsCl \longrightarrow CsNO_3 + AgCl$
(b) $H_3PO_4 + Ca(OH)_2 \longrightarrow Ca_3(PO_4)_2 + H_2O$
(c) $NiCO_3 + HCl \longrightarrow NiCl_2 + CO_2 + H_2O$
(d) $(NH_4)_2SO_4 + BaCl_2 \longrightarrow BaSO_4 + NH_4Cl$

Combination Reactions

6.17 Write the expected product of each of the following unbalanced equations for combination reactions and balance the equation.
(a) $Ba + O_2 \longrightarrow$
(b) $H_2 + I_2 \longrightarrow$
(c) $Li + S \longrightarrow$
(d) $Al + Cl_2 \longrightarrow$

6.19 Write the expected product of each of the following unbalanced equations for combination reactions and balance the equation.
(a) $Na_2O + H_2O \longrightarrow$
(b) $CaO + H_2O \longrightarrow$
(c) $BaO + CO_2 \longrightarrow$
(d) $MgO + CO_2 \longrightarrow$

6.18 Write the expected product of each of the following unbalanced equations for combination reactions and balance the equation.
(a) $Ca + Cl_2 \longrightarrow$
(b) $Mg + N_2 \longrightarrow$
(c) $Li + N_2 \longrightarrow$
(d) $K + F_2 \longrightarrow$

6.20 Write the expected product of each of the following unbalanced equations for combination reactions and balance the equation.
(a) $SO_2 + H_2O \longrightarrow$
(b) $Li_2O + H_2O \longrightarrow$
(c) $BaO + SO_3 \longrightarrow$
(d) $CaO + CO_2 \longrightarrow$

Decomposition Reactions

6.21 Write the expected products of each of the following unbalanced equations for decomposition reactions and balance the equation.
(a) $HgO \longrightarrow$
(b) $SrCO_3 \longrightarrow$
(c) $Au_2O_3 \longrightarrow$
(d) $Ag_2O \longrightarrow$

6.22 Write the expected products of each of the following unbalanced equations for decomposition reactions and balance the equation.
(a) $N_2O \longrightarrow$
(b) $KClO_3 \longrightarrow$
(c) $MgCO_3 \longrightarrow$
(d) $H_2O \rightarrow$

Single Replacement Reactions

6.23 Write the expected products of each of the following equations for single replacement reactions and balance the equation.
(a) $KBr + Cl_2 \longrightarrow$
(b) $SnO_2 + H_2 \longrightarrow$
(c) $Mg + Fe_3O_4 \longrightarrow$
(d) $Al_2O_3 + C \longrightarrow$

6.25 Write the expected products of each of the following equations for single replacement reactions and balance the equation.
(a) $Zn + HCl \longrightarrow$
(b) $SiCl_4 + Mg \longrightarrow$
(c) $Br_2 + NaI \longrightarrow$
(d) $ZnO + C \longrightarrow$

6.24 Write the expected products of each of the following equations for single replacement reactions and balance the equation.
(a) $Sb_4O_6 + C \longrightarrow$
(b) $Br_2 + CaI_2 \longrightarrow$
(c) $TlI_3 + Na \longrightarrow$
(d) $Sb_2S_3 + O_2 \longrightarrow$

6.26 Write the expected products of each of the following equations for single replacement reactions and balance the equation.
(a) $Fe + CuCl_2 \longrightarrow$
(b) $NiCl_2 + Zn \longrightarrow$
(c) $F_2 + NaI \longrightarrow$
(d) $Al + HCl \longrightarrow$

Double Replacement Reactions

6.27 Write the expected products of each of the following equations for double replacement reactions and balance the equation.
(a) $Pb(NO_3)_2 + HCl \longrightarrow$
(b) $FeCl_2 + NaOH \longrightarrow$
(c) $Cd(NO_3)_2 + H_2S \longrightarrow$
(d) $AgNO_3 + CsCl \longrightarrow$

6.29 Write the expected products of each of the following equations for double replacement reactions and balance the equation.
(a) $MnSO_4 + (NH_4)_2S \longrightarrow$
(b) $CaCO_3 + H_2SO_4 \longrightarrow$
(c) $CdSO_4 + NaOH \longrightarrow$
(d) $ZnCO_3 + H_3PO_4 \longrightarrow$

6.28 Write the expected products of each of the following equations for double replacement reactions and balance the equation.
(a) $Pb(NO_3)_2 + Na_2S \longrightarrow$
(b) $NiCO_3 + HCl \longrightarrow$
(c) $(NH_4)_2SO_4 + BaCl_2 \longrightarrow$
(d) $Bi(NO_3)_3 + NaOH \longrightarrow$

6.30 Write the expected products of each of the following equations for double replacement reactions and balance the equation.
(a) $Bi(NO_3)_3 + H_2S \longrightarrow$
(b) $Ba(NO_3)_2 + KOH \longrightarrow$
(c) $Pb(NO_3)_2 + K_2CrO_4 \longrightarrow$
(d) $AgNO_3 + Na_2S \longrightarrow$

Neutralization Reactions

6.31 Write the expected products of each of the following equations for neutralization reactions and balance the equation.
(a) $H_3PO_4 + KOH \longrightarrow$
(b) $Fe(OH)_3 + H_2SO_4 \longrightarrow$
(c) $H_3PO_4 + Ca(OH)_2 \longrightarrow$
(d) $Zn(OH)_2 + HNO_3 \longrightarrow$

6.32 Write the expected products of each of the following equations for neutralization reactions and balance the equation.
(a) $Fe(OH)_3 + H_3PO_4 \longrightarrow$
(b) $Zn(OH)_2 + H_2SO_4 \longrightarrow$
(c) $Al(OH)_3 + HNO_3 \longrightarrow$
(d) $KOH + H_2SO_4 \longrightarrow$

Additional Exercises

6.33 Balance each of the following equations.
(a) $U + F_2 \longrightarrow UF_6$
(b) $P + Cl_2 \longrightarrow PCl_5$
(c) $KCl + Br_2 \longrightarrow KBr + Cl_2$
(d) $SnO_2 + H_2 \longrightarrow Sn + H_2O$

6.35 Balance each of the following equations.
(a) $N_2O \longrightarrow N_2 + O_2$
(b) $MnO_2 + HCl \longrightarrow MnCl_2 + Cl_2 + H_2O$
(c) $Br_2 + NaI \longrightarrow NaBr + I_2$
(d) $ZnO + CO \longrightarrow Zn + CO_2$

6.34 Balance each of the following equations.
(a) $Ag + H_2S + O_2 \longrightarrow Ag_2S + H_2O$
(b) $NH_3 + O_2 \longrightarrow NO + H_2O$
(c) $TiO_2 + Cl_2 + C \longrightarrow TiCl_4 + CO$
(d) $Sb_4O_6 + C \longrightarrow Sb + CO$

6.36 Balance each of the following equations.
(a) $PbS + C + O_2 \longrightarrow Pb + CO_2 + SO_2$
(b) $Li + N_2 \longrightarrow Li_3N$
(c) $Cu + Al_2O_3 \longrightarrow CuO + Al$
(d) $ZnS + O_2 \longrightarrow ZnO + SO_2$

6.37 Balance each of the following equations.
  (a) $C_2H_6O_2 + O_2 \longrightarrow CO_2 + H_2O$
  (b) $C_5H_{12}O_2 + O_2 \longrightarrow CO_2 + H_2O$
  (c) $C_5H_{10}O_2 + O_2 \longrightarrow CO_2 + H_2O$
  (d) $C_5H_8O_2 + O_2 \longrightarrow CO_2 + H_2O$

6.39 Write the expected products of each of the following unbalanced equations for decomposition reactions and balance the equation.
  (a) $Na_2CO_3 \longrightarrow$
  (b) $PbCO_3 \longrightarrow$
  (c) $NaHCO_3 \longrightarrow$
  (d) $Ba(HCO_3)_2 \longrightarrow$

6.38 Write the expected product of each of the following unbalanced equations for combination reactions and balance the equation.
  (a) $SO_3 + H_2O \longrightarrow$
  (b) $Al + O_2 \longrightarrow$
  (c) $Mg + S \longrightarrow$
  (d) $Al + N_2 \longrightarrow$

6.40 Write the expected products of each of the following equations for single replacement reactions and balance the equation.
  (a) $Al + SnCl_2 \longrightarrow$
  (b) $Cu(NO_3)_2 + Cd \longrightarrow$
  (c) $Hg(NO_3)_2 + Al \longrightarrow$
  (d) $Zn + Pb(NO_3)_2 \longrightarrow$

7 Stoichiometry

Learning Objectives

After studying Chapter 7 you should be able to

1. Determine from an equation the mole ratio necessary for stoichiometric calculations.
2. Perform mole–mole stoichiometric calculations.
3. Perform mole–mass stoichiometric calculations.
4. Perform mass–mass stoichiometric calculations.
5. Determine the limiting reagent in a reaction.
6. Calculate the theoretical yield and the percent yield for a reaction.

7.1 Stoichiometry and the Balanced Equation

You learned in the preceding chapter that a balanced chemical equation provides information about chemical change on a submicroscopic level. The information is given by the chemical formulas and their coefficients. In Chapter 5, you learned the terms Avogadro's number, mole, and molar mass and how to use these quantities to develop the relationships that connect the submicroscopic and macroscopic levels of chemistry.

In the laboratory our concern is with the macroscopic aspects of chemical change. How much of each reactant must be weighed out, and how much product can be obtained in a chemical reaction? In this chapter you will learn how to determine these quantities from the balanced chemical equation. The mathematical calculation of the quantities of reactants and products in a reaction is called **stoichiometry.** If we know the ratio of the numbers of atoms and molecules involved in a chemical reaction on a submicroscopic level as given by the coefficients in the chemical equation, then the same ratio governs the number of moles of those substances on a macroscopic level. The stoichiometric quantities of reactants and products involved in the reaction can be calculated on a mole or mass basis.

In this chapter we will also learn how to calculate the amount of product formed when the amounts of reactants are not present in stoichiometric quantities. The limiting reagent, the reactant present in the smallest amount based on the stoichiometry of the reaction, controls the amount of product that can be formed.

Finally, we will learn to distinguish between the amount of product that in theory can be prepared and the actual amount that is formed or isolated. The former quantity is the theoretical yield and the latter quantity is the actual yield.

7.2 Methods for Solving Stoichiometry Problems

Although there are a variety of methods that could be used to solve stoichiometry problems, the mole method is recommended. The mole method, which consists of three steps, is rooted in the factor unit method. The actual number of steps involved in working problems by the mole method depends on the units in which the given substance is measured and the units required of the desired substance. The possible steps are

1. The conversion of the given quantities into their equivalent mole quantities.
2. The establishment of the stoichiometric relationship between the given substance and the desired substance from the balanced equation, and the determination of the moles of the desired substance.
3. The conversion of the moles of the desired substance into the required units. The final answer must be expressed to the proper number of significant figures based on the significant figures of the units of the given substance.

The three steps of the mole method are illustrated in Figure 7.1

The central step, step 2, of the mole method, depends on relating the number of moles of one substance to the number of moles of another substance. In order to do this, a mole ratio is used. The **mole ratio** is a ratio between the number of moles of two substances involved in a chemical

Figure 7.1 An outline of the mole method of solving stoichiometric problems

reaction. Thus, for the reaction of hydrogen and nitrogen to give ammonia,

$$N_2(g) + 3 H_2(g) \longrightarrow 2 NH_3(g)$$

there are six mole ratios that may be written. All of these ratios contain the coefficients of the corresponding substances in the equation.

$$\frac{1 \text{ mole } N_2}{3 \text{ moles } H_2} \quad \text{or} \quad \frac{3 \text{ moles } H_2}{1 \text{ mole } N_2}$$

$$\frac{1 \text{ mole } N_2}{2 \text{ moles } NH_3} \quad \text{or} \quad \frac{2 \text{ moles } NH_3}{1 \text{ mole } N_2}$$

$$\frac{3 \text{ moles } H_2}{2 \text{ moles } NH_3} \quad \text{or} \quad \frac{2 \text{ moles } NH_3}{3 \text{ moles } H_2}$$

There are only three mole ratios that relate different pairs of substances for this reaction. The remaining three are reciprocal ratios of the same quantities. Each of the mole ratios applies only to the reaction under consideration. Thus, for a reaction of nitrogen and hydrogen to produce hydrazine, the mole ratios would be different.

$$N_2(g) + 2 H_2(g) \longrightarrow N_2H_4(g)$$

To use the mole method for stoichiometry problems, it is necessary to convert from the known number of moles of the given substance to the number of moles of the desired substance. These quantities can be related by the proper mole ratio. Therefore we may write

moles of desired substance = mole ratio × moles of given substance

The mole ratio then must be

$$\text{mole ratio} = \frac{\text{moles of desired substance}}{\text{moles of given substance}}$$

Because of the relationship established between the macroscopic and submicroscopic levels of chemical change it follows that the mole ratio is numerically equal to the ratio of the coefficients in the balanced equation. Thus we may write

$$\text{mole ratio} = \frac{\text{coefficient of desired substance in balanced equation}}{\text{coefficient of given substance in balanced equation}}$$

Note that the mole ratio is an exact number. Thus, the number of significant figures in an answer depends only on the number of significant figures in the given substance used in the calculation.

A summary of the mole method is given in Table 7.1.

Table 7.1   A Summary of the Mole Method

1. From the mass of the reactant or product given, convert to the mole equivalent by multiplying by the reciprocal of the molar mass, mole/g.

2. Examine the balanced equation and calculate the moles of the desired unknown that can be obtained from the moles of the substances calculated in step 1 by multiplying by the mole ratio.

3. From the moles of the desired substance calculated by step 2, determine the equivalent quantity in terms of mass by multiplying by the molar mass, g/mole.

Example 7.1

Consider the following equation and write two mole ratios relating aluminum and oxygen.

$$4 \, Al(s) + 3 \, O_2(g) \longrightarrow 2 \, Al_2O_3(s)$$

Solution   The coefficients for aluminum and oxygen are 4 and 3, respectively. These coefficients give the ratio of the number of moles of each substance required for the reaction. The mole ratios are

$$F_1 = \frac{4 \text{ moles Al}}{3 \text{ moles O}_2} \qquad F_2 = \frac{3 \text{ moles O}_2}{4 \text{ moles Al}}$$

The ratio F_1 can be used to convert a given number of moles of O_2 into the number of moles of Al required for the reaction. The ratio F_2 can be used to convert a given number of moles of Al into the number of moles of O_2 required for the reaction.

[*Additional examples are given in 7.3–7.6 at the end of the chapter.*]

7.3   Mole–Mole Stoichiometry Problems

In this first type of stoichiometric calculation, you will find that the steps and math involved are easy. However, the method is essential to solve the more complex problems described in the next two sections. Therefore, make sure that you fully understand how to do the problems given in this section before trying the problems in the next section.

We wish to calculate the number of moles of one substance that will react with or be produced from a given number of moles of a second substance. On a macroscopic level a balanced equation establishes the mole relationship between all reactants and products. The coefficient in a balanced equation gives the number of moles of each substance with respect to the number of moles of every other reactant or product in the equation. Thus, in the reaction of nitrogen and hydrogen to produce ammonia, we may write

$$3 \, H_2 \quad + \quad N_2 \quad \longrightarrow \quad 2 \, NH_3$$
$$3 \text{ moles } H_2 + 1 \text{ mole } N_2 \longrightarrow 2 \text{ moles } NH_3$$

Example 7.2

How many moles of ammonia can be produced from 9 moles of hydrogen according to the following equation?

$$3 H_2 + N_2 \longrightarrow 2 NH_3$$

Solution The coefficients of the equation indicate that 3 moles of hydrogen will yield 2 moles of ammonia. The mole ratio necessary for the solution of the problem must cancel the units of moles of hydrogen and leave moles of ammonia.

$$9 \text{ moles } H_2 \times \text{mole ratio} = ? \text{ moles } NH_3$$

$$9 \text{ moles } H_2 \times \frac{2 \text{ moles } NH_3}{3 \text{ moles } H_2} = 6 \text{ moles } NH_3$$

Example 7.3

How many moles of potassium chlorate are required to produce 0.15 mole of oxygen according to the following equation?

$$2 KClO_3(s) \longrightarrow 2 KCl(s) + 3 O_2(g)$$

Solution The necessary mole ratio needed to convert the given number of moles of O_2 into the required number of moles of $KClO_3$ is

$$\frac{2 \text{ moles } KClO_3}{3 \text{ moles } O_2}$$

$$0.15 \text{ mole } O_2 \times \frac{2 \text{ moles } KClO_3}{3 \text{ moles } O_2} = 0.10 \text{ mole } KClO_3$$

Note that the answer is expressed to two significant figures consistent with the two significant figures in the given quantity of O_2. The mole ratio is an exact quantity.

Example 7.4

Determine the number of moles of oxygen required to burn 12 moles of ethane (C_2H_6).

$$2 C_2H_6(g) + 7 O_2(g) \longrightarrow 4 CO_2(g) + 6 H_2O(g)$$

Solution The equation provides the information that 7 moles of oxygen are required to burn 2 moles of ethane. The ratio of the moles of the two substances in question is first established.

$$\text{mole ratio} = \frac{7 \text{ moles } O_2}{2 \text{ moles } C_2H_6}$$

This ratio, when multiplied by the given quantity, 12 moles of ethane, will provide the answer in moles of oxygen.

$$12 \text{ moles } C_2H_6 \times \frac{7 \text{ moles } O_2}{2 \text{ moles } C_2H_6} = 42 \text{ moles } O_2$$

[*Additional examples are given in 7.7–7.16 at the end of the chapter.*]

7.4 Mole–Mass Stoichiometry Problems

In this type of stoichiometry problem, either the quantity of the known substance is given in mass units, or the desired unknown substance must be found in mass units. The remaining quantity is in mole units. If the known substance is given in mole units, then you only need to use steps 2 and 3 of the mole method to obtain the quantity of desired substance in mass units. If the known substance is given in mass units, then you must use only steps 1 and 2 to obtain the desired substance in mole units.

Example 7.5

What mass of antimony can be obtained from 0.15 mole of Sb_2S_3 in the following process?

$$Sb_2S_3(s) + 3\,O_2(g) \longrightarrow 2\,Sb(s) + 3\,SO_2(g)$$

Solution First determine the relationship between Sb_2S_3 and Sb on a mole basis using a mole ratio. This process is step 2 in the mole method.

$$0.15 \text{ mole } Sb_2S_3 \times \frac{2 \text{ moles } Sb}{1 \text{ mole } Sb_2S_3} = 0.30 \text{ mole } Sb$$

Then convert the moles of Sb into grams of Sb by step 3 of the mole method.

$$0.30 \text{ mole } Sb \times \frac{121.75 \text{ g } Sb}{1 \text{ mole } Sb} = 36.525 \text{ g } Sb$$

The answer to the correct number of significant figures is 37 g of Sb.

Example 7.6

How many moles of oxygen are required to metabolically convert 90.0 g of glucose ($C_6H_{12}O_6$) into carbon dioxide and water?

$$C_6H_{12}O_6 + 6\,O_2 \longrightarrow 6\,CO_2 + 6\,H_2O$$

Solution First the number of moles of glucose is calculated as required by step 1 of the mole method. The molecular weight of $C_6H_{12}O_6$ is 180 amu. Thus, 1 mole of glucose has a mass of 180 g.

$$90.0 \text{ g } C_6H_{12}O_6 \times \frac{1 \text{ mole } C_6H_{12}O_6}{180 \text{ g } C_6H_{12}O_6} = 0.500 \text{ mole } C_6H_{12}O_6$$

Now, the moles of oxygen required are calculated from the mole ratio obtained from the coefficients of the balanced equation. This is step 2 of the mole method.

$$0.500 \text{ mole } C_6H_{12}O_6 \times \frac{6 \text{ moles } O_2}{1 \text{ mole } C_6H_{12}O_6} = 3.00 \text{ moles } O_2$$

Example 7.7 _____

How many grams of ethyl alcohol (C_2H_5OH) can be produced from the fermentation of 1.00 moles of cane sugar ($C_{12}H_{22}O_{11}$)?

$$C_{12}H_{22}O_{11} + H_2O \longrightarrow 4\,C_2H_5OH + 4\,CO_2$$

Solution Since the number of moles of cane sugar is given, only steps 2 and 3 of the mole method must be used. First, the moles of the desired C_2H_5OH are calculated by using the mole ratio. Then, the moles of C_2H_5OH are converted into the desired quantity, grams of C_2H_5OH. Both steps can be combined in one mathematical operation.

$$1.00 \text{ mole } C_{12}H_{22}O_{11} \times \underbrace{\frac{4 \text{ moles } C_2H_5OH}{1 \text{ mole } C_{12}H_{22}O_{11}}}_{(\text{step 2})} \times \underbrace{\frac{46.0 \text{ g } C_2H_5OH}{1 \text{ mole } C_2H_5OH}}_{(\text{step 3})} = 184 \text{ g } C_2H_5OH$$

[*Additional examples are given in 7.17–7.24 at the end of the chapter.*]

7.5 Mass–Mass Stoichiometry Problems

Now you should be quite adept at solving stoichiometry problems using moles of the reactants and products. In mole–mole stoichiometry calculations you used step 2 of the mole method. In mole–mass stoichiometry calculations you used either steps 1 and 2 or steps 2 and 3. The technique of solving mass–mass stoichiometry problems involves all three steps of the mole method.

Example 7.8 _____

How many grams of oxygen are required to burn completely 57.0 g of octane?

$$2\,C_8H_{18} + 25\,O_2 \longrightarrow 16\,CO_2 + 18\,H_2O$$

Solution In step 1 the number of moles of octane in the 57.0 g is

$$57.0 \text{ g } C_8H_{18} \times \frac{1 \text{ mole } C_8H_{18}}{114.0 \text{ g } C_8H_{18}} = 0.500 \text{ mole } C_8H_{18}$$

In step 2 the number of moles of oxygen required to burn 0.500 mole of C_8H_{18} is

$$0.500 \text{ mole } C_8H_{18} \times \frac{25 \text{ moles } O_2}{2 \text{ moles } C_8H_{18}} = 6.25 \text{ moles } O_2$$

Finally, in step 3 the mass of oxygen required is

$$6.25 \text{ moles } O_2 \times \frac{32.0 \text{ g } O_2}{1 \text{ mole } O_2} = 2.00 \times 10^2 \text{ g } O_2$$

Example 7.9 _____

How many grams of carbon dioxide are required by a plant to produce 1.80 g of glucose by photosynthesis?

$$6\ CO_2 + 6\ H_2O \longrightarrow C_6H_{12}O_6 + 6\ CO_2$$

Solution All three steps necessary to solve this problem may be combined in one mathematical expression. The corresponding steps of the mole ratio method are indicated in parentheses.

$$1.80\ \text{g } C_6H_{12}O_6 \times \underbrace{\frac{1\ \text{mole } C_6H_{12}O_6}{180.0\ \text{g } C_6H_{12}O_6}}_{\text{(step 1)}} \times \underbrace{\frac{6\ \text{moles } CO_2}{1\ \text{mole } C_6H_{12}O_6}}_{\text{(step 2)}} \times \underbrace{\frac{44.0\ \text{g } CO_2}{1\ \text{mole } CO_2}}_{\text{(step 3)}} = 2.64\ \text{g } CO_2$$

[_Additional examples are given in 7.25–7.34 at the end of the chapter._]

7.6 The Limiting Reagent

Given a certain amount of one material and a balanced equation for its reaction with other substances, it is possible to calculate the exact quantities of each sustance required for the reaction to occur and the amount of each product produced. However, what would happen if instead of exact stoichiometric amounts, excessive amounts of one or more of the reactants were used? How does this affect the amount of the product formed? When this situation exists, the reactant that is not present in excess is called the **limiting reagent.** The amount of product derived from the reaction can be no more than the quantity calculated from the limiting reagent. For any mixture of reactants, the balanced equation for the reaction may be used to determine whether one reactant is present in limiting amounts. The limiting reagent is that material that has the smallest mole-to-coefficient ratio. This ratio is obtained by dividing the moles of the reactant by the coefficient of that reactant in the balanced equation.

The concept of a limiting reagent may be better understood by making an analogy to the effectiveness of an airline in terms of pilots and airplanes. A plane cannot be of any use unless pilots are available to fly it. Similarly, a group of trained pilots is of little use unless they have planes to fly. If two pilots are required per plane, we may write

$$2\ \text{pilots} + 1\ \text{plane} \longrightarrow 1\ \text{available aircraft}$$

If an airline has 100 planes and 300 pilots, you might know intuitively that the airline is limited by the number of planes. You can verify this fact by dividing these quantities by the appropriate coefficients of the equation.

$$\frac{100\ \text{planes}}{1\ \text{plane}} = 100 \qquad \frac{300\ \text{pilots}}{2\ \text{pilots}} = 150$$

The lowest ratio is for the planes, proving them to be the limiting quantity. You can confirm that the planes are the limiting quantity by calculating the number of pilots required for the 100 planes.

$$100\ \text{planes} \times \frac{2\ \text{pilots}}{\text{plane}} = 200\ \text{pilots}$$

There is an excess of pilots available. An alternate way of checking that the planes are the limiting quantity is to make the incorrect assumption that the pilots are the limiting quantity. For the stated number of pilots, 150 planes are required.

$$300 \text{ pilots} \times \frac{1 \text{ plane}}{2 \text{ pilots}} = 150 \text{ planes}$$

There is an insufficient number of planes; the number of pilots is not the limiting quantity.

For the reaction of nitrogen and hydrogen to give ammonia, the balanced equation is

$$N_2 + 3 H_2 \longrightarrow 2 NH_3$$

If 25 moles of nitrogen and 45 moles of hydrogen are available for the reaction, you can determine which reactant is limiting by looking at the ratio of the moles available to the coefficients in the equation. The ratio is lowest for hydrogen; therefore hydrogen is the limiting reagent.

$$\frac{25 \text{ moles } N_2}{1 \text{ mole } N_2} = 25 \qquad \frac{45 \text{ moles } H_2}{3 \text{ moles } H_2} = 15$$

You can confirm that hydrogen is the limiting reagent by calculating the amount of nitrogen required to react with the 45 moles of hydrogen.

$$45 \text{ moles } H_2 \times \frac{1 \text{ mole } N_2}{3 \text{ moles } H_2} = 15 \text{ moles } N_2$$

There are 25 moles of N_2 available and the given 45 moles of H_2 requires only 15 moles of N_2. The N_2 is in excess; therefore H_2 is the limiting reagent.

An alternate way of checking that hydrogen is the limiting reagent is to make the assumption that nitrogen is the limiting reagent. The number of moles of hydrogen required to react with 25 moles of N_2 can be calculated.

$$25 \text{ moles } N_2 \times \frac{3 \text{ moles } H_2}{1 \text{ mole } N_2} = 75 \text{ moles } H_2$$

The required amount of hydrogen is not available. Thus the assumption that nitrogen is the limiting reagent is incorrect.

Example 7.10

Methyl alcohol (CH_3OH) is prepared industrially from carbon monoxide and hydrogen at 375°C and a pressure of 3000 pounds per square inch (psi). How many moles of gaseous methyl alcohol can be produced from 2.0 kg of hydrogen gas and 7.0 kg of carbon monoxide?

$$CO(g) + 2 H_2(g) \longrightarrow CH_3OH(g)$$

Solution From the quantities given in grams it might appear that hydrogen is the limiting reagent. However, it is necessary to determine the limiting reagent on a mole basis and to take into account the coefficients in the equation.

$$2.0 \text{ kg } H_2 \times \frac{1000 \text{ g}}{1 \text{ kg}} \times \frac{1 \text{ mole } H_2}{2.0 \text{ g } H_2} = 1.0 \times 10^3 \text{ moles } H_2$$

$$7.0 \text{ kg } CO \times \frac{1000 \text{ g}}{1 \text{ kg}} \times \frac{1 \text{ mole } CO}{28.0 \text{ g } CO} = 2.5 \times 10^2 \text{ moles } CO$$

Now determine the ratio of the moles available to the coefficients of the equation for each reactant.

$$\frac{1.00 \times 10^3 \text{ moles } H_2}{2 \text{ moles } H_2} = 5.0 \times 10^2$$

$$\frac{2.5 \times 10^2 \text{ moles } CO}{1 \text{ mole } CO} = 2.5 \times 10^2$$

Thus, CO is the limiting reagent. We can confirm this by calculating the amount of hydrogen required to react completely with the carbon monoxide.

$$2.5 \times 10^2 \text{ moles CO} \times \frac{2 \text{ moles } H_2}{1 \text{ mole CO}} = 5.0 \times 10^2 \text{ moles } H_2$$

There is more than enough hydrogen available. Therefore we have confirmed that carbon monoxide is the limiting reagent.

Now the number of moles of methyl alcohol that can be produced is calculated using the number of moles of the limiting reagent.

$$2.5 \times 10^2 \text{ moles CO} \times \frac{1 \text{ mole } CH_3OH}{1 \text{ mole CO}} = 2.5 \times 10^2 \text{ moles } CH_3OH$$

[*Additional examples are given in 7.35–7.42 at the end of the chapter.*]

7.7 Theoretical and Percent Yields

In the preceding sections, you actually learned how to calculate the theoretical yield for a variety of reactions. The **theoretical yield** is the maximum amount of product that could be obtained from the complete reaction of the limiting reagent. However, some of the product may be lost in attempts to isolate and purify it. Furthermore, side reactions may occur to convert the reactants into other products. For example, when carbon and oxygen react to form carbon dioxide, some carbon monoxide also is formed.

$$C + O_2 \longrightarrow CO_2$$
$$2\,C + O_2 \longrightarrow 2\,CO$$

If complete combustion to give carbon dioxide is the desired reaction, then the formation of carbon monoxide is the side reaction. To the extent that some of the carbon is converted to undesired material, the theoretical yield of carbon dioxide will not be obtained.

The amount of the desired product actually obtained from a chemical process is called the **actual yield**. Actual yields can be lowered by the competition of side reactions or by losses that occur when the desired material is separated from any excess reactants or the other products.

The **percent yield** in a chemical reaction is the actual yield divided by the theoretical yield multiplied by 100.

$$\% \text{ yield} = \frac{\text{actual yield}}{\text{theoretical yield}} \times 100$$

Example 7.11

A 0.100 mole sample of $KClO_3$ is decomposed and 4.20 g of O_2 is collected by a student. What is the percent yield of oxygen?

$$2 \, KClO_3 \longrightarrow 2 \, KCl + 3 \, O_2$$

Solution Since the known is given in moles, step 1 of the mole method is unnecessary. First we calculate the relationship between the moles of known and unknown as required by step 2.

$$0.100 \, \text{mole } KClO_3 \times \frac{3 \text{ mole } O_2}{2 \text{ moles } KClO_3} = 0.150 \text{ mole } O_2$$

Now the 0.150 mole of O_2 is converted into the desired gram units as required in step 3.

$$0.150 \, \text{mole } O_2 \times \frac{32.0 \text{ g } O_2}{1 \text{ mole } O_2} = 4.80 \text{ g } O_2$$

The quantity obtained by the student is less than the theoretical yield.

$$\% \text{ yield} = \frac{4.20 \text{ g } O_2 \text{ obtained}}{4.80 \text{ g } O_2 \text{ theoretical}} \times 100 = 87.5\% \text{ yield}$$

[*Additional examples are given in 7.43–7.46 at the end of the chapter.*]

Summary

Stoichiometry is the mathematical relationship between the quantities of reactants and products of a reaction as indicated by a chemical equation. A wide variety of calculations may be done, but all are based on the mole method and use a mole ratio obtained from the coefficients in the balanced equation. The mole ratio is a conversion factor used to interrelate the stoichiometric amounts of the substances involved in a chemical reaction. The three types of calculations described are mole–mole, mass–mole and mass–mass.

It is necessary to identify the single reactant that may control the amount of product that can be formed. The reactant that is present in the smallest mole to coefficient ratio is called the limiting reagent. Stoichiometric calculations must be done based on the limiting reagent.

The theoretical yield is that amount of product that could be obtained from a complete reaction using all of the limiting reagent. The actual yield, the amount of product actually obtained, is usually less than the theoretical yield because of losses in isolating the product or the occurence of side reactions. The percent yield is the actual yield divided by the theoretical yield with the result multiplied by 100.

New Terms

The **actual yield** is the amount of the product actually obtained from a chemical process.

The **limiting reagent** is that material that has the smallest mole to coefficient ratio.

The **mole ratio** is a ratio of the coefficients of two substances given by a balanced equation.

The **percent yield** is the quotient of the actual yield and the theoretical yield with the result multiplied by 100.

Stoichiometry is the numerical relationship between the quantities of reactants and products in a chemical reaction.

The **theoretical yield** is the amount of product that would be obtained from the complete reaction of the limiting reagent in a reaction.

Exercises

Terminology

7.1 How does the actual yield differ from the theoretical yield?

7.2 What is a limiting reagent in a chemical reaction?

The Mole Ratio

7.3 Balance the following equation and write the mole ratio necessary to convert the number of moles of MnO_2 available into the number of moles of Al required for the reaction.

$$MnO_2 + Al \longrightarrow Al_2O_3 + Mn$$

7.4 Balance the following equation and write the mole ratio necessary to convert the number of moles of Na_2O available into the number of moles of P_4O_{10} required for the reaction.

$$Na_2O + P_4O_{10} \longrightarrow Na_3PO_4$$

7.5 Balance the following equation and write the mole ratio necessary to convert the number of moles of $BaCl_2$ available into the number of moles of AgCl produced.

$$BaCl_2 + AgNO_3 \longrightarrow AgCl + Ba(NO_3)_2$$

7.6 Balance the following equation and write the mole ratio necessary to convert the number of moles of $C_4H_{10}O_2$ available into the number of moles of O_2 required for the reaction.

$$C_4H_{10}O_2 + O_2 \longrightarrow CO_2 + H_2O$$

Mole–Mole Stoichiometry

7.7 How many moles of oxygen are required to react completely with 5.0 moles of sulfur dioxide to yield sulfur trioxide?

$$2 SO_2 + O_2 \longrightarrow 2 SO_3$$

7.8 How many moles of carbon dioxide are produced in the combustion of 0.5 mole of ethane?

$$2 C_2H_6 + 7 O_2 \longrightarrow 4 CO_2 + 6 H_2O$$

7.9 How many moles of oxygen are required to completely convert 2 moles of iron(II) sulfide to iron(III) oxide?

$$4 FeS + 7 O_2 \longrightarrow 2 Fe_2O_3 + 4 SO_2$$

7.10 How many moles of nitrogen dioxide (NO_2) are produced when 3 moles of oxygen are produced in the decomposition of nitric acid by light?

$$4 HNO_3 \longrightarrow 4 NO_2 + 2 H_2O + O_2$$

7.11 Calculate the number of moles of nitrogen dioxide (NO_2) produced from the complete reaction of 2.3 moles of nitrogen oxide (NO).

$$2 NO + O_2 \longrightarrow 2 NO_2$$

7.12 How many moles of ammonia may be produced from 1.2 moles of nitrogen, assuming that sufficient hydrogen is available.

$$N_2 + 3 H_2 \longrightarrow 2 NH_3$$

7.13 How many moles of sulfur dioxide would be produced in the reaction of iron(II) sulfide with oxygen if 0.35 mole of O_2 is consumed?

$$4 FeS + 7 O_2 \longrightarrow 2 Fe_2O_3 + 4 SO_2$$

7.14 How many moles of oxygen are required to completely burn 0.40 mole of ethane?

$$2 C_2H_6 + 7 O_2 \longrightarrow 4 CO_2 + 6 H_2O$$

7.15 How many moles of hydrogen are required to react completely with 12 moles of carbon monoxide to form methyl alcohol (CH_3OH)?

$$CO + 2 H_2 \longrightarrow CH_3OH$$

7.16 Carefully controlled decomposition of ammonium nitrate can yield dinitrogen oxide (N_2O). How many moles of dinitrogen oxide could be produced from the decomposition of 0.10 mole of ammonium nitrate?

$$NH_4NO_3 \longrightarrow N_2O + 2 H_2O$$

Mole–Mass Stoichiometry

7.17 What mass of carbon tetrachloride (CCl_4) can be formed from the reaction of 0.25 mole of methane (CH_4) with chlorine.

$$CH_4 + 4 Cl_2 \longrightarrow CCl_4 + 4 HCl$$

7.18 What mass of oxygen will be produced from the decomposition of 0.025 mole of potassium chlorate?

$$2 KClO_3 \longrightarrow 2 KCl + 3 O_2$$

7.19 How many moles of potassium chlorate can be produced from the reaction of 71.0 g of chlorine?

$$3 Cl_2 + 6 KOH \longrightarrow 5 KCl + KClO_3 + 3 H_2O$$

7.20 How many grams of ethane can be burned by using 0.70 mole of oxygen?

$$2 C_2H_6 + 7 O_2 \longrightarrow 4 CO_2 + 6 H_2O$$

7.21 How many moles of hydrogen are required to convert 159 g of copper(II) oxide into copper?

$$CuO + H_2 \longrightarrow Cu + H_2O$$

7.23 How many grams of zinc are required to produce 0.20 mole of hydrogen gas from the following chemical reaction?

$$Zn + 2\,HCl \longrightarrow ZnCl_2 + H_2$$

7.22 How many moles of sulfur dioxide can be produced by burning 16 g of sulfur?

$$S + O_2 \longrightarrow SO_2$$

7.24 Hydrogen and oxygen are produced by the electrolysis of water. How many moles of hydrogen gas can be produced from 1.80 g of water?

$$2\,H_2O \longrightarrow 2\,H_2 + O_2$$

Mass–Mass Stoichiometry

7.25 How many grams of copper can be produced from the reaction of 159 g of copper(II) oxide with hydrogen?

$$CuO + H_2 \longrightarrow Cu + H_2O$$

7.27 How many grams of calcium carbide (CaC_2) are required to produce 5.2 g of acetylene (C_2H_2) by the following reaction?

$$CaC_2 + 2\,H_2O \longrightarrow Ca(OH)_2 + C_2H_2$$

7.29 How many grams of copper can be produced by roasting 1590 g of copper(I) sulfide?

$$Cu_2S + O_2 \longrightarrow 2\,Cu + SO_2$$

7.31 White phosphorus (P_4) is used in military incendiary devices because it ignites spontaneously in air. How many grams of phosphorus (P_4) will react with 25.0 g of oxygen (O_2)?

$$P_4 + 5\,O_2 \longrightarrow P_4O_{10}$$

7.33 Silicon carbide, used as an abrasive on sandpaper, is prepared as indicated in the following equation. How much SiO_2 (in grams) is required to produce 738 g of silicon carbide.

$$SiO_2 + 3\,C \longrightarrow SiC + 2\,CO$$

7.26 How many grams of carbon are required to produce 6.54 g of zinc by the following reaction?

$$ZnO + C \longrightarrow Zn + CO$$

7.28 How many grams of carbon tetrachloride (CCl_4) can be produced from 4.0 g of methane (CH_4) by the following reaction?

$$CH_4 + 4\,Cl_2 \longrightarrow CCl_4 + 4\,HCl$$

7.30 Solutions of sodium hypochlorite ($NaClO$) are sold as laundry bleach. How many grams of chlorine (Cl_2) are needed to react with 60.0 g of sodium hydroxide ($NaOH$)?

$$2\,NaOH + Cl_2 \longrightarrow NaCl + NaClO + H_2O$$

7.32 Freshly exposed aluminum surfaces react with oxygen to form a tough oxide coating that protects the metal from further corrosion. How many grams of oxygen are required to react with 8.09 g of aluminum?

$$4\,Al + 3\,O_2 \longrightarrow 2\,Al_2O_3$$

7.34 Silver tarnishes in the presence of hydrogen sulfide according to the given balanced equation. What mass of "tarnish" is produced from 0.136 g of H_2S?

$$4\,Ag + 2\,H_2S + O_2 \longrightarrow 2\,Ag_2S + 2\,H_2O$$

Limiting Reagent

7.35 How many moles of hydrogen gas can be produced from the reaction of 3.27 g of zinc and 3.65 g of HCl?

$$Zn + 2\,HCl \longrightarrow ZnCl_2 + H_2$$

7.37 How many moles of carbon dioxide can be produced from the reaction of 1.6 g of oxygen and 1.1 g of propane (C_3H_8)?

$$C_3H_8 + 5\,O_2 \longrightarrow 3\,CO_2 + 4\,H_2O$$

7.39 How many grams of silver chloride can be produced from 16.99 g of silver nitrate and 2.92 g of sodium chloride?

$$AgNO_3 + NaCl \longrightarrow AgCl + NaNO_3$$

7.41 Mercury and sulfur react to form mercuric sulfide (HgS). A 3.00 g sample of mercury is reacted with 1.00 g of sulfur. What mass of what substance remains unreacted?

$$Hg + S \longrightarrow HgS$$

7.36 How many grams of water will be produced from the reaction of 1.12 moles of hydrogen and 1.12 moles of oxygen?

$$2\,H_2 + O_2 \longrightarrow 2\,H_2O$$

7.38 How many grams of ammonia can be produced from the reaction of 4.0 g of hydrogen and 14 g of nitrogen?

$$N_2 + 3\,H_2 \longrightarrow 2\,NH_3$$

7.40 How many grams of cane sugar ($C_{12}H_{22}O_{11}$) can be produced from 8.8 g of carbon dioxide and 3.6 g of water?

$$12\,CO_2 + 11\,H_2O \longrightarrow C_{12}H_{22}O_{11} + 12\,O_2$$

7.42 What is the limiting reagent and what mass of P_4S_{10} is produced when 3.10 g of P_4 reacts with 8.50 g of S_8?

$$4\,P_4 + 5\,S_8 \longrightarrow 4\,P_4S_{10}$$

Percent Yield

7.43 A student reacts 3.27 g of Zn with excess HCl and obtains 6.47 g of $ZnCl_2$. What is the percent yield?

$$Zn + 2\,HCl \longrightarrow ZnCl_2 + H_2$$

7.45 A student reacts 2.23 g of Fe with excess HCl and obtains 4.76 g of $FeCl_2$. What is the percent yield?

$$Fe + 2\,HCl \longrightarrow FeCl_2 + H_2$$

7.44 A student reacts 4.16 g of $BaCl_2$ with excess Na_2SO_4 and obtains 4.06 g of $BaSO_4$. What is the percent yield?

$$BaCl_2 + Na_2SO_4 \longrightarrow BaSO_4 + 2\,NaCl$$

7.46 A student reacts 4.25 g of $AgNO_3$ with excess NaCl and obtains 3.58 g of AgCl. What is the percent yield?

$$AgNO_3 + NaCl \longrightarrow AgCl + NaNO_3$$

Additional Exercises

7.47 Balance the following equation and write the mole ratio necessary to convert the number of moles of Bi_2S_3 available into the number of moles of SO_2 produced.

$$Bi_2S_3 + O_2 \longrightarrow Bi_2O_3 + SO_2$$

7.49 Ammonium nitrate (NH_4NO_3) may decompose with explosive violence to produce N_2 gas. How many moles of NH_4NO_3 are required to produce 2.4 moles of N_2?

$$2\,NH_4NO_3 \longrightarrow 2\,N_2 + O_2 + 4\,H_2O$$

7.51 Nitrogen dioxide (NO_2) reacts with water to form HNO_3. Calculate the number of moles of HNO_3 that can be formed from 0.06 mole of NO_2.

$$3\,NO_2 + H_2O \longrightarrow 2\,HNO_3 + NO$$

7.53 What mass of $CaCO_3$ is produced simultaneously with the formation of 0.15 mole of CO_2 from the decomposition of $Ca(HCO_3)_2$?

$$Ca(HCO_3)_2 \longrightarrow CaCO_3 + CO_2 + H_2O$$

7.55 How many moles of CO_2 are produced simultaneously with 112.2 g of CaO in the thermal decomposition of limestone ($CaCO_3$)?

$$CaCO_3 \longrightarrow CaO + CO_2$$

7.57 Pure silver metal can be formed by heating Ag_2CO_3. How many grams of Ag_2CO_3 must be decomposed to produce 12.7 g of Ag?

$$2\,Ag_2CO_3 \longrightarrow 4\,Ag + 2\,CO_2 + O_2$$

7.59 Zinc (Zn) and S react to form ZnS, a substance used in phosphors that coat the inner surface of TV tubes. How many grams of ZnS can be formed when 12.0 g of Zn reacts with 6.50 g of S?

$$Zn + S \longrightarrow ZnS$$

7.61 Zinc metal can be obtained by the following reaction. What is the maximum amount of Zn that can be obtained from 75.0 g of Zn and 50.0 g of CO?

$$ZnO + CO \longrightarrow Zn + CO_2$$

7.48 Balance the following equation and write the mole ratio necessary to convert the number of moles of Ca_3P_2 available into the number of moles of PH_3 produced.

$$Ca_3P_2 + H_2O \longrightarrow Ca(OH)_2 + PH_3$$

7.50 Nitrogen oxide (NO), which is produced during lightning flashes in the atmosphere, reacts rapidly with O_2 to yield NO_2. How many moles of NO_2 result from the reaction of 8 moles of NO?

$$2\,NO + O_2 \longrightarrow 2\,NO_2$$

7.52 Sulfur dioxide (SO_2) reacts with water to form H_2SO_3. How many moles of SO_2 are required to form 0.05 mole of H_2SO_3?

$$SO_2 + H_2O \longrightarrow H_2SO_3$$

7.54 How many grams of Fe can be produced in a blast furnace from Fe_2O_3 from a supply of 25 moles of CO?

$$Fe_2O_3 + 3\,CO \longrightarrow 2\,Fe + 3\,CO_2$$

7.56 Thin films of silicon (Si), used in fabrication of electronic components, may be prepared by the decomposition of silane (SiH_4). What mass (in grams) of SiH_4 is required to prepare 0.306 g of Si?

$$SiH_4 \longrightarrow Si + 2\,H_2$$

7.58 Tungsten (W) metal, used to make incandescent bulb filaments, is produced by the following reaction. How many grams of W can be obtained from 3.67 g of WO_3?

$$WO_3 + 3\,H_2 \longrightarrow W + 3\,H_2O$$

7.60 Silver tarnishes in the presence of H_2S according to the following equation. How much Ag_2S is produced from a mixture of 0.950 g Ag, 0.140 g H_2S, and 0.0800 g O_2?

$$4\,Ag + 2\,H_2S + O_2 \longrightarrow 2\,Ag_2S + 2\,H_2O$$

7.62 How many grams of AgBr can be produced from 18.0 g of $AgNO_3$ and 3.50 g of NaBr?

$$AgNO_3 + NaBr \longrightarrow AgBr + NaNO_3$$

8 Electronic Structure of the Atom

Learning Objectives

After studying Chapter 8 you should be able to

1. Describe the shells or energy levels of an atom using either integer values or alphabetic symbols.

2. List the subshells within a given shell.

3. Give the number of orbitals within a given subshell.

4. Write the electron distribution within a subshell taking into account the orbitals and electron spin.

5. Write the electron configuration for an element given the atomic number.

6. Represent the valence electrons of atoms by electron-dot symbols.

8.1 The Bohr Atom and the Modern Atom

In Chapter 4 a model of the atom was presented that consisted of protons and neutrons, located in a small region at the center of the atom called the nucleus, and electrons, located in a larger volume of space around the nucleus. This model pictures an atom as mostly empty space; the diameter of the nucleus is about 10^{-5} that of the atom.

For a variety of reasons it is necessary to have a more detailed picture of the position of the electrons in the atom. This knowledge is required to help us understand why and how compounds are formed (Chapter 10). The details of molecular structure in turn help us understand why various types of matter exist as gases, liquids, and solids. In addition, the knowledge of electronic structure in atoms and molecules enables us to understand why chemical reactions occur. For all of these reasons, it is important that we discuss the arrangement of the electrons about the nucleus.

Information about arrangement of electrons in the atom was first obtained from an analysis of the line spectra of the elements. To understand what we mean by line spectra, let's consider the spectrum obtained by passing white light from the sun or from an incandescent light bulb through a prism (Figure 8.1). The result is a continuous "rainbow" of all colors from red to violet. Now consider that when high voltages are applied to elements in the gaseous state colored light results. You have observed this phenomenon in neon lights. When this type of light is passed through a prism, a series of narrow colored lines rather than a "rainbow" of colors is observed (Figure 8.1). The number of lines and their colors are characteristic for each element.

Niels Bohr, a Danish physicist, suggested that the light given off by the element hydrogen was due to the position and distance of the electron from the nucleus. Bohr theorized that there are a number of possible orbits for the electron at various distances from the nucleus. The electron could be in any of these orbits but not between orbits. When an atom absorbs energy, an electron "jumps" from a low-energy orbit near the nucleus to a higher energy orbit farther

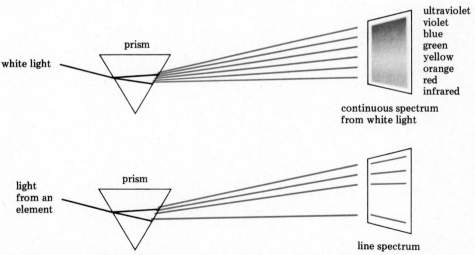

Figure 8.1

A continuous spectrum of sunlight and a line spectrum of an element
In sunlight all of the colors form a continuous spectrum. In the light emitted from excited atoms of an element, only a few lines of characteristic color are observed.

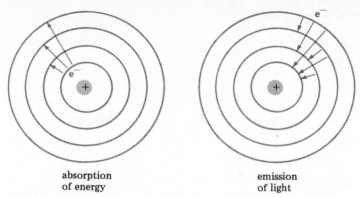

Figure 8.2

absorption
of energy

emission
of light

The Bohr atom
When energy is added to the atom, the electrons absorb energy
and move to higher energy orbits. When the electrons return to a
lower orbit, the energy is released as light energy. The color of
the light is characteristic of the difference in energy between the
energy levels.

away (Figure 8.2) resulting in an ''excited atom.'' When the electron ''falls'' back to its lower orbit, energy is released in the form of light. Only certain colors of light are possible, because only certain amounts of energy can be released. This energy is equal to the energy difference between the orbits.

Bohr's theory of the atom was consistent with the data at that time; however, his work dealt with only the simpler atoms, such as hydrogen, that contain few electrons. As a result of more evidence on more elements, the Bohr concept has been modified. It is no longer thought that electrons exist in planet-like orbits about the nucleus. Rather, electrons occupy regions of space about the nucleus called orbitals. Although the words orbit and orbital resemble each other, there is a difference in the meaning of the two terms. An orbit is a defined path through space, whereas an orbital is a volume of space. The electron is somewhere in the volume of space but is not confined to a specific pathway. There are many orbitals in an atom. They differ in their energy, shape, and location with respect to the nucleus.

The modern picture of the electronic arrangement in the atom is based on a complex theoretical approach called quantum mechanics, which is beyond the learning objectives of this book. However, the conclusions of the theory are fairly simple and will be summarized here. You can use this material to understand both the periodic law (Chapter 9) and why elements form compounds (Chapter 10).

8.2 Shells

The various electrons in the orbital of an atom do not have the same energy; thus, their locations within the atom differ. In general, the higher the energy of the electron, the farther away from the nucleus it will be. The electrons in the orbitals of an atom are grouped into certain principal energy levels or **shells.** The electrons of a shell have a certain defined average energy. The shells are designated by integers $n = 1, 2, 3, 4$, and so on. These shells have been also desig-

Table 8.1	**Energy Levels or Shells**		
Energy level	Integer designation, n	Letter designation
first	1	K
second	2	L
third	3	M
fourth	4	N
fifth	5	O
sixth	6	P
seventh	7	Q

nated K, L, M, N, O, and so on. However, the integers are more commonly used today (Table 8.1). The energy of the electrons in the energy levels increases in the order $1 < 2 < 3 < 4$, and so on. Therefore, the distance between the shell of electrons and the nucleus increases in the same order.

There are limitations on the number of electrons that can exist in a specific energy level. The maximum number of electrons in a shell is given by $2n^2$, where n is the number of the energy level. For example, the number of electrons that may exist in energy level 1 or the K shell is $2 \times 1^2 = 2$. For the second energy level, the number of electrons is $2 \times 2^2 = 8$. The number of electrons in each of the principal energy levels is listed in Table 8.2.

Example 8.1

What is the maximum number of electrons that may be in the N shell?

Solution   The N shell is an alternate way of indicating the fourth energy level when n = 4. Using the formula $2n^2$, we obtain $2 \times 4^2 = 32$ electrons.

[*Additional examples may be found in 8.9 and 8.10 at the end of the chapter.*]

Table 8.2	**Electrons in Energy Levels**	
Energy level	Integer designation, n	Maximum number of electrons in level
first	1	$2 \times 1^2 = 2$
second	2	$2 \times 2^2 = 8$
third	3	$2 \times 3^2 = 18$
fourth	4	$2 \times 4^2 = 32$
fifth	5	$2 \times 5^2 = 50$[a]
sixth	6	$2 \times 6^2 = 72$[a]
seventh	7	$2 \times 7^2 = 98$[a]

[a]The theoretical values for these three energy levels have not been observed because no elements have the necessary number of electrons to completely fill the shells.

8.3 Subshells

Within the energy levels there are energy sublevels that are called subshells. Each subshell is characterized by the shape of the space in which the electrons are located. The higher energy levels contain more electrons and also have more subshells. Of the known atoms, electrons exist in four types of **subshells,** which are labeled by the lowercase letters s, p, d, and f. Other subshells may have electrons when energy is added to an atom to produce an "excited atom." These are labeled g, h, i, and so on; however, we will not consider these in any detail in this text.

The order of the energy of the subshells is $s < p < d < f$. The relationships between the shells and subshells are shown in Table 8.3. The number of subshells is equal to the shell number. There are restrictions on the number of electrons that each of the subshells can hold. The maximum number of electrons in each of the subshells is as follows: 2 in the s subshell, 6 in the p subshell, 10 in the d subshell, and 14 in the f subshell. The total sum of the number of electrons possible within the subshells is equal to the total number of electrons possible within the shell.

The shells and subshells in an atom are represented by symbols such as $2s$, $4d$, $5p$, and so on. The number represents the energy level or shell, and the letter represents the subshell within that energy level.

Table 8.3   **Shells and Subshells**

Shell, n	Subshells[a]	Number of electrons	Total $2n^2$
1	s	2	2
2	s, p	2 + 6	8
3	s, p, d	2 + 6 + 10	18
4	s, p, d, f	2 + 6 + 10 + 14	32
5	s, p, d, f, g	2 + 6 + 10 + 14 + 18	50
6	s, p, d, f, g, h	2 + 6 + 10 + 14 + 18 + 22	72
7	s, p, d, f, g, h, i	2 + 6 + 10 + 14 + 18 + 22 + 26	98

[a]The g, h, and i subshells do not contain electrons in known elements in their normal state.

Example 8.2

What is the maximum number of electrons that can occupy the $5p$ subshell?

Solution   The indication of the fifth energy level is not required to solve this problem. All p subshells, regardless of the energy level may contain no more than 6 electrons.

Example 8.3

Show that the maximum number of electrons in the fourth energy level is equal to the sum of the maximum number of electrons in each of the subshells of that level.

Solution   The total number of electrons in the fourth energy level is given by the formula $2n^2$ or $2 \times 4^2 = 32$ electrons. The fourth energy level has s, p, d, and f subshells. By adding the maximum number of electrons in each subshell we have $2 + 6 + 10 + 14 = 32$ electrons.

[*Additional examples may be found in 8.19–8.22 at the end of the chapter.*]

8.4   Orbitals

As was previously stated, according to quantum mechanical theory each subshell consists of orbitals. An **orbital** is a region of space in which an electron is most likely to be found. The characteristics of orbitals are as follows.

1. Each type of orbital has a characteristic shape.
2. All orbitals within a subshell are of the same energy.
3. The orbital volume and average distance from the nucleus increase as the value of n increases.
4. Each orbital can contain a maximum of two electrons.
5. Electrons move rapidly and "occupy" the entire orbital volume.

It is useful to picture an orbital as a cloud of electrons. Clouds in the sky look quite big but do not contain much matter. Since the electron has a small mass, the average distribution of matter within the orbital is very low or cloud-like.

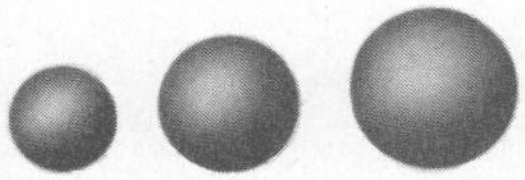

The s orbitals are spherical, as shown in Figure 8.3. The $2s$ orbital has a larger volume than the $1s$ orbital, and the $3s$ orbital is larger than the $2s$ orbital. However, each s orbital can still only contain two electrons.

There are three p orbitals within each p subshell, and each orbital can be occupied by two electrons. The three p orbitals are identical in shape and are located at right angles to one another (Figure 8.4). Note that each orbital consists of two teardrop shapes. The electrons within a p orbital can occupy any of the area shown as a p orbital; they are not confined one to each of the two teardrop shapes. The orbital describes a volume of space within which two electrons may exist. There are five d orbitals in one d subshell and seven f orbitals in one f subshell. The much more complex shapes of d and f orbitals will not be described in this text.

1s   2s   3s

Figure 8.3

The shape of s orbitals
These spherical volumes represent regions in which the electrons of s orbitals may be located.

Example 8.4

Based on the number of electrons that can occupy an orbital and the number of orbitals in a subshell, calculate the maximum number of electrons that can occupy the $4f$ subshell.

Solution   The number of electrons in a subshell is related to the number of orbitals in that subshell. Each orbital regardless of type can only hold two electrons. In an f subshell, regardless of the shell in which it is located, there are 7 orbitals. Thus the number of electrons is twice the number of orbitals, or $2 \times 7 = 14$ electrons.

[*Additional examples may be found in 8.27–8.30 at the end of the chapter.*]

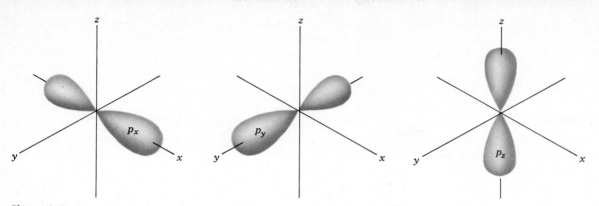

Figure 8.4
The shape of p orbitals
The three orbitals of the p subshell are arranged perpendicular to each other. Each orbital may contain two electrons.

8.5 Electron Spin

Electrons within an orbital repel each other because their charges are identical. However, the electrons have another property, called **electron spin,** which causes some attraction between two electrons in the same orbital. The spin occurs about an axis, much as the earth rotates about its north–south axis.

Electrons spin or rotate in two possible directions (referred to as clockwise and counterclockwise). These spins are often represented by arrows. The arrows do not refer to a specific direction of spin. However, an arrow pointing up ($\uparrow$) means that the spin of an electron is the opposite of an electron represented by an arrow pointing down ($\downarrow$).

Two electrons occupying an orbital always have opposite spins. A spinning, charged particle creates a magnetic field. Thus, two electrons spinning in opposite directions in an orbital produce opposite magnetic fields. You know that the north pole of one magnet attracts the south pole of another magnet. Thus, electrons with opposite spins magnetically attract each other. Electrons with opposite spins within an orbital are said to be **paired.** A single electron in an orbital or several electrons with the same spin distributed in different orbitals are said to be unpaired.

Electrons have the same electrical charge regardless of their spin; as a consequence they tend to repel each other and "spread out" among the orbitals within the subshell. Thus, three electrons within the p subshell will be distributed one to each of the three orbitals but, in addition, the electrons have the same spin. Horizontal lines may be used to represent the three orbitals of the p subshell.

$$\underline{\uparrow} \ \underline{\uparrow} \ \underline{\uparrow} \qquad\qquad \underline{\uparrow\downarrow} \ \underline{\uparrow} \ \underline{}$$

preferred "spread-out" arrangement a hypothetical arrangement
of electrons in orbitals   of electrons

The rational for this "spread out" arrangement, called Hund's rule, is the electrical repulsion between electrons that causes them to avoid each other. **Hund's rule** states that electrons will not enter an orbital containing another electron if an empty orbital of the same energy is available. By having electrons in different orbitals, electrical repulsion is minimized. Also, remember that two electrons can be in the same orbital only if they have opposite spins.

Example 8.5

There are seven electrons present in the d subshell of an atom. Is the following arrangement of the electrons acceptable?

$$\underline{\uparrow\downarrow} \ \underline{\uparrow\downarrow} \ \underline{\uparrow\downarrow} \ \underline{\uparrow} \ \underline{}$$

Solution   The arrangement is not acceptable. The maximum number of orbitals has not been occupied by electrons. Up to a total of five electrons in a d subshell can be distributed one to each orbital. Then each of the other two electrons must be paired with an electron already in an orbital. The accepted arrangement is

$$\underline{\uparrow\downarrow} \ \underline{\uparrow\downarrow} \ \underline{\uparrow} \ \underline{\uparrow} \ \underline{\uparrow}$$

[*Additional examples may be found in 8.41 and 8.42 at the end of the chapter.*]

8.6   Electron Configuration

The picture of the atom developed in the previous section describes where electrons may be located or "reside" in an atom. In a similar way, you could describe your residence within a city by giving the apartment building, the floor on which you live in the building, and your apartment number. In the case of the electron, the energy level is the apartment building, the subshell is the floor, and the orbital is the apartment. In the orbital "apartment" only two occupants are allowed and they must have opposite spins. Now let's consider the electrons within an atom and how they are distributed.

The arrangement of the electrons within an atom is called the **electron configuration.** The rules for predicting the electron configuration are quite simple. First, electrons are located in the lowest energy subshell available. This results in the most stable atom. Second, electrons must first "spread out" among the orbitals within the subshell, until all orbitals have one electron. Third, no more than two electrons may exist in the same orbital, and then they must have opposite spins.

The order of increasing energies of subshells is $1s$, $2s$, $2p$, $3s$, $3p$, $4s$, $3d$, $4p$, $5s$, $4d$, $5p$, $6s$, $4f$, $5d$, $6p$, $7s$, $5f$, $6d$ (Figure 8.5). Examination of this order reveals that not all of the subshells within a given shell will be filled before the next shell starts to fill. For example, the $4s$ subshell intervenes between the $3p$ and the $3d$ subshells. The ordering of shells and subshells occurs in a complicated manner.

A method of memorizing the order of energies of the subshells is given in Figure 8.6. Start from the $1s$ subshell, and follow the solid line from the start of the arrow line to the head of the arrow. Then proceed via the dotted line to the start of the next solid line arrow. This procedure reproduces the order of the energies previously given.

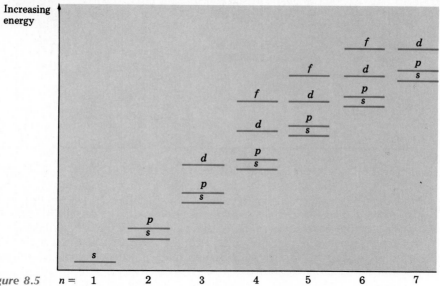

Figure 8.5

Energies of subshells in an atom

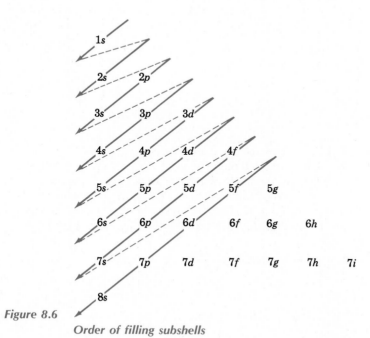

Figure 8.6

Order of filling subshells

We now have the information needed to predict electron configurations. A convenient way of showing energy levels, subshells, and the number of electrons that they contain is as follows.

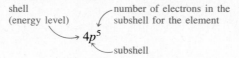

8.7 Electron Configurations of the First 20 Elements

Let's predict some electron configurations. The first 20 elements can be discussed easily. The lightest element, hydrogen, contains one electron in the $1s$ subshell, which is the lowest energy subshell available. The electronic configuration of hydrogen is written $1s^1$. The superscript denotes the number of electrons in the subshell. Helium has two electrons, both occupying the $1s$ subshell. The electronic configuration of helium is $1s^2$, and the subshell is filled. The electrons in helium are paired.

Lithium has three electrons. Two electrons fill the $1s$ subshell; the third is in the subshell with the next highest energy, which is the $2s$. The lithium electronic configuration is $1s^22s^1$. The electron in the $2s$ subshell is unpaired. Beryllium has the electronic configuration $1s^22s^2$. The fourth electron of beryllium completes the $2s$ subshell, which has a capacity of two electrons. The two electrons in the $2s$ subshell are paired.

In boron, the $2p$ subshell, the subshell with the next lowest energy, is occupied by a single electron. The electron configuration of boron is $1s^22s^22p^1$. A p subshell can hold a maximum of six electrons, and this subshell is filled by additional electrons added for the elements carbon, nitrogen, oxygen, fluorine, and neon. The electron configurations for these elements are

B	$1s^22s^22p^1$
C	$1s^22s^22p^2$
N	$1s^22s^22p^3$
O	$1s^22s^22p^4$
F	$1s^22s^22p^5$
Ne	$1s^22s^22p^6$

With neon, the second energy level is complete. This element is a member of a group of elements known as the noble or rare gases (Chapter 9).

The next eight elements, sodium through argon, have electrons in the third energy level. For sodium, the first ten electrons are distributed the same as for the noble gas, neon. The eleventh electron is located in the $3s$ subshell. The electron configuration of sodium can be given by either of two designations: The first shows the location of all of the 11 electrons; the second representation is a shorthand where [Ne] represents the electron configuration corresponding to neon.

Na　$1s^22s^22p^63s^1$　　or　　$[Ne]3s^1$

The magnesium atom contains one more electron than sodium. This electron is located in the $3s$ subshell as well and is paired with the other electron in the subshell.

Mg　$1s^22s^22p^63s^2$　　or　　$[Ne]3s^2$

For the elements, aluminum, silicon, phosphorus, sulfur, chlorine, and argon, the additional electrons are in the $3p$ subshell.

Al	$1s^22s^22p^63s^23p^1$	or	$[Ne]3s^23p^1$
Si	$1s^22s^22p^63s^23p^2$	or	$[Ne]3s^23p^2$
P	$1s^22s^22p^63s^23p^3$	or	$[Ne]3s^23p^3$
S	$1s^22s^22p^63s^23p^4$	or	$[Ne]3s^23p^4$
Cl	$1s^22s^22p^63s^23p^5$	or	$[Ne]3s^23p^5$
Ar	$1s^22s^22p^63s^23p^6$	or	$[Ne]3s^23p^6$

With the element argon the $3s$ and $3p$ subshells are completed in a manner similar to neon, which had completed the $2s$ and $2p$ subshells. Argon is also a member of the group of noble gases. Thus, chemists use [Ar] as a shorthand to indicate the electron configuration of the 18 electrons in argon and for any other element that also has 18 electrons distributed in subshells in a manner identical to argon.

Potassium and calcium, which have one and two more electrons than argon respectively, might be expected to have electrons in the $3d$ subshell. The third energy level can contain up to 18 electrons, and only eight electrons are present in the third energy level of argon. However, the order of filling subshells as given earlier indicates that electrons will be located in the $4s$ subshell before the $3d$ subshell is filled. Accordingly, the electron configurations of potassium and calcium are

K	$1s^22s^22p^63s^23p^64s^1$	or	$[Ar]4s^1$
Ca	$1s^22s^22p^63s^23p^64s^2$	or	$[Ar]4s^2$

Note the use of the alternative [Ar] to simplify the representation of the first 18 electrons present in potassium and calcium.

8.8 Electron Configurations for the Elements Beyond Calcium

The first 20 elements do not have electrons in the d and f subshells. As indicated earlier, the order of the energies of these subshells is complicated. Nevertheless, by using this order, you can write the electron configuration of any element. A simpler way of writing electron configurations based on the periodic table is given in Chapter 10. A list of the electron configurations of the elements is given in Table 8.4. The electron configurations of the noble gases krypton, xenon, and radon are used as shorthand to write the electron configuration of other elements in the table.

Kr	$1s^22s^22p^63s^23p^64s^23d^{10}4p^6$	or	[Kr]
Xe	$1s^22s^22p^63s^23p^64s^23d^{10}4p^65s^24d^{10}5p^6$	or	[Xe]
Rn	$1s^22s^22p^63s^23p^64s^23d^{10}4p^65s^24d^{10}5p^66s^24f^{14}5d^{10}6p^6$	or	[Rn]

For each element, the s and p subshells of the highest energy level have 2 and 6 electrons, respectively.

A close examination of Table 8.4 reveals some ''discrepancies'' from the rules as you learned them. However, these are few in number and involve considerations that are not important for an understanding of chemistry at the level of this text.

Table 8.4   Electron Configuration of the Elements

Element	Atomic number	Electron structure	Element	Atomic number	Electron structure
H	1	$1s^1$	Xe	54	$[Kr]5s^24d^{10}5p^6$
He	2	$1s^2$	Cs	55	$[Xe]6s^1$
Li	3	$1s^22s^1$	Ba	56	$[Xe]6s^2$
Be	4	$1s^22s^2$	La	57	$[Xe]6s^25d^1$
B	5	$1s^22s^22p^1$	Ce	58	$[Xe]6s^24f^15d^1$
C	6	$1s^22s^22p^2$	Pr	59	$[Xe]6s^24f^3$
N	7	$1s^22s^22p^3$	Nd	60	$[Xe]6s^24f^4$
O	8	$1s^22s^22p^4$	Pm	61	$[Xe]6s^24f^5$
F	9	$1s^22s^22p^5$	Sm	62	$[Xe]6s^24f^6$
Ne	10	$1s^22s^22p^6$	Eu	63	$[Xe]6s^24f^7$
Na	11	$[Ne]3s^1$	Gd	64	$[Xe]6s^24f^75d^1$
Mg	12	$[Ne]3s^2$	Tb	65	$[Xe]6s^24f^9$
Al	13	$[Ne]3s^23p^1$	Dy	66	$[Xe]6s^24f^{10}$
Si	14	$[Ne]3s^23p^2$	Ho	67	$[Xe]6s^24f^{11}$
P	15	$[Ne]3s^23p^3$	Er	68	$[Xe]6s^24f^{12}$
S	16	$[Ne]3s^23p^4$	Tm	69	$[Xe]6s^24f^{13}$
Cl	17	$[Ne]3s^23p^5$	Yb	70	$[Xe]6s^24f^{14}$
Ar	18	$[Ne]3s^23p^6$	Lu	71	$[Xe]6s^24f^{14}5d^1$
K	19	$[Ar]4s^1$	Hf	72	$[Xe]6s^24f^{14}5d^2$
Ca	20	$[Ar]4s^2$	Ta	73	$[Xe]6s^24f^{14}5d^3$
Sc	21	$[Ar]4s^23d^1$	W	74	$[Xe]6s^24f^{14}5d^4$
Ti	22	$[Ar]4s^23d^2$	Re	75	$[Xe]6s^24f^{14}5d^5$
V	23	$[Ar]4s^23d^3$	Os	76	$[Xe]6s^24f^{14}5d^6$
Cr	24	$[Ar]4s^13d^5$	Ir	77	$[Xe]6s^24f^{14}5d^7$
Mn	25	$[Ar]4s^23d^5$	Pt	78	$[Xe]6s^14f^{14}5d^9$
Fe	26	$[Ar]4s^23d^6$	Au	79	$[Xe]6s^14f^{14}5d^{10}$
Co	27	$[Ar]4s^23d^7$	Hg	80	$[Xe]6s^24f^{14}5d^{10}$
Ni	28	$[Ar]4s^23d^8$	Tl	81	$[Xe]6s^24f^{14}5d^{10}6p^1$
Cu	29	$[Ar]4s^13d^{10}$	Pb	82	$[Xe]6s^24f^{14}5d^{10}6p^2$
Zn	30	$[Ar]4s^23d^{10}$	Bi	83	$[Xe]6s^24f^{14}5d^{10}6p^3$
Ga	31	$[Ar]4s^23d^{10}4p^1$	Po	84	$[Xe]6s^24f^{14}5d^{10}6p^4$
Ge	32	$[Ar]4s^23d^{10}4p^2$	At	85	$[Xe]6s^24f^{14}5d^{10}6p^5$
As	33	$[Ar]4s^23d^{10}4p^3$	Rn	86	$[Xe]6s^24f^{14}5d^{10}6p^6$
Se	34	$[Ar]4s^23d^{10}4p^4$	Fr	87	$[Rn]7s^1$
Br	35	$[Ar]4s^23d^{10}4p^5$	Ra	88	$[Rn]7s^2$
Kr	36	$[Ar]4s^23d^{10}4p^6$	Ac	89	$[Rn]7s^26d^1$
Rb	37	$[Kr]5s^1$	Th	90	$[Rn]7s^26d^2$
Sr	38	$[Kr]5s^2$	Pa	91	$[Rn]7s^25f^26d^1$
Y	39	$[Kr]5s^24d^1$	U	92	$[Rn]7s^25f^36d^1$
Zr	40	$[Kr]5s^24d^2$	Np	93	$[Rn]7s^25f^46d^1$
Nb	41	$[Kr]5s^14d^4$	Pu	94	$[Rn]7s^25f^6$
Mo	42	$[Kr]5s^14d^5$	Am	95	$[Rn]7s^25f^7$
Tc	43	$[Kr]5s^24d^5$	Cm	96	$[Rn]7s^25f^76d^1$
Ru	44	$[Kr]5s^14d^7$	Bk	97	$[Rn]7s^25f^9$
Rh	45	$[Kr]5s^14d^8$	Cf	98	$[Rn]7s^25f^{10}$
Pd	46	$[Kr]4d^{10}$	Es	99	$[Rn]7s^25f^{11}$
Ag	47	$[Kr]5s^14d^{10}$	Fm	100	$[Rn]7s^25f^{12}$
Cd	48	$[Kr]5s^24d^{10}$	Md	101	$[Rn]7s^25f^{13}$
In	49	$[Kr]5s^24d^{10}5p^1$	No	102	$[Rn]7s^25f^{14}$
Sn	50	$[Kr]5s^24d^{10}5p^2$	Lr	103	$[Rn]7s^25f^{14}6d^1$
Sb	51	$[Kr]5s^24d^{10}5p^3$	Rf	104	$[Rn]7s^25f^{14}6d^2$
Te	52	$[Kr]5s^24d^{10}5p^4$	Ha	105	$[Rn]7s^25f^{14}6d^3$
I	53	$[Kr]5s^24d^{10}5p^5$	—	106	$[Rn]7s^25f^{14}6d^4$

Example 8.6

An isotope of strontium is radioactive and is produced in nuclear explosions. Write the electron configuration of $^{90}_{38}Sr$.

Solution The subscript number, 38, equals the number of protons or electrons in the atom. The electrons are located in the lowest energy subshells available to them. The ordering of subshells is

$$1s < 2s < 2p < 3s < 3p < 4s < 3d < 4p < 5s$$

Electrons up to a total of 38 are then placed within each subshell to obtain

$$1s^2 2s^2 2p^6 3s^2 3p^6 4s^2 3d^{10} 4p^6 5s^2$$

The first 36 electrons are arranged according to the electron configuration of krypton. Thus a shorthand representation is

$$[Kr]5s^2$$

Example 8.7

What is the arrangement of electrons in $^{60}_{27}Co$? This isotope, called cobalt-60, is used in cancer therapy.

Solution There are 27 electrons in this isotope. Write down the order of energies of the subshells, filling each in order.

$$1s^2 2s^2 2p^6 3s^2 3p^6 4s^2$$

The above arrangement accounts for 20 electrons. The remaining seven electrons are placed in an incomplete $3d$ subshell to give

$$1s^2 2s^2 2p^6 3s^2 3p^6 4s^2 3d^7$$

The first 18 electrons are arranged in the electron configuration of argon. Thus a shorthand representation is

$$[Ar]4s^2 3d^7$$

[*Additional examples are given in 8.35–8.38 at the end of the chapter.*]

8.9 Electron Configuration and Electron Pairing

Now let's consider a more detailed account of the electrons in an atom. Consider the elements boron, carbon, nitrogen, oxygen, fluorine, and neon whose electron configurations involve electrons in the $2p$ subshell. In this series the $2p$ subshell contains from one to six electrons. According to Hund's rule, the electrons will be located as shown in the orbital diagrams of Table 8.5. The lines in the orbital diagrams represent the orbitals, and the electron spin is represented with arrows. The order of filling the $2p$ subshell according to Hund's rule tells us

Table 8.5   **Illustration of Hund's Rule and Electron Configuration**

Element	Atomic number	1s	2s	2p	Electron configuration
H	1	↑			$1s^1$
He	2	↑↓			$1s^2$
Li	3	↑↓	↑		$1s^2 2s^1$
Be	4	↑↓	↑↓		$1s^2 2s^2$
B	5	↑↓	↑↓	↑	$1s^2 2s^2 2p^1$
C	6	↑↓	↑↓	↑ ↑ __	$1s^2 2s^2 2p^2$
N	7	↑↓	↑↓	↑ ↑ ↑	$1s^2 2s^2 2p^3$
O	8	↑↓	↑↓	↑↓ ↑ ↑	$1s^2 2s^2 2p^4$
F	9	↑↓	↑↓	↑↓ ↑↓ ↑	$1s^2 2s^2 2p^5$
Ne	10	↑↓	↑↓	↑↓ ↑↓ ↑↓	$1s^2 2s^2 2p^6$

the number of pairs of electrons and the number of unpaired electrons. Nitrogen, for example, has three unpaired electrons. The number and type of bonds formed by elements are based on such considerations (Chapter 10).

The number of unpaired electrons can be determined by remembering that the electrons are distributed one to an orbital up to the number of orbitals available. Then any additional electrons must be paired. Thus for the d subshell, the first five electrons will occupy different orbitals. For the f subshell, the first seven electrons will occupy different orbitals.

Example 8.8

Consider the electron configuration of cobalt given in Example 8.7. Describe the electrons located in the $3d$ subshell.

Solution   There are seven electrons in a subshell that may contain up to ten electrons. There are five orbitals in the subshell. In order to "spread out" the electrons according to Hund's rule, one electron must first be located in each orbital. The remaining two electrons must be paired with electrons already present in the orbitals. The final representation of the d subshell of cobalt is

↑↓ ↑↓ ↑ ↑ ↑

[*Additional examples are given in 8.41–8.48 at the end of the chapter.*]

8.10 Electron-Dot Symbols

The highest energy level of an atom containing electrons in s and p subshells is called the **valence energy level** or **valence shell.** The eight electrons that can occupy these subshells are known as valence electrons. Most of the chemistry of the elements to be discussed in this text is

Table 8.6   Electron-Dot Symbols of Selected Elements

Element	Electron configuration	Electron-dot symbols
^7_3Li	$1s^22s^1$	Li· or L̇i or ·Li or Li
^9_4Be	$1s^22s^2$	Be: or B̈e or Be or :Be
$^{11}_5B$	$1s^22s^22p^1$	:B or B̈· or :B or B·
$^{12}_6C$	$1s^22s^22p^2$	:C̈ or C̈· or :C· , etc.
$^{14}_7N$	$1s^22s^22p^3$	·N̈· or ·N: or ·N· or :N·
$^{16}_8O$	$1s^22s^22p^4$	·Ö: or ·Ö· or :Ö· , etc.
$^{19}_9F$	$1s^22s^22p^5$	:F̈· or :F̈: or ·F̈: or :F:

based on the fact that there is a maximum of eight valence electrons. All of the other electrons, together with the nucleus are collectively called the **core**. This division is useful because the valence electrons form chemical bonds (Chapter 10). Because of the importance of valence electrons we display them in electron-dot symbols (Table 8.6).

Electron-dot symbols are written according to the following rules.

1. The symbol of the element represents the core.
2. Dots representing valence electrons are placed on either side of, above, or below the symbol.
3. All positions about the symbol are equivalent.
4. A maximum of two electrons per side, top, or bottom of the symbol is allowed. Thus, only eight valence electrons can be placed about the symbol.
5. Valence electrons are indicated by a pair of dots for the pairs of electrons and a single dot for an unpaired electron in the valence shell.
6. Hydrogen and helium are exceptions to the maximum of eight electrons. They have a maximum of two electrons.

These rules are used to obtain symbols from which we can derive models to explain chemical observations. Several examples in Table 8.6 illustrate the use of electron dot symbols.

Summary

Our model for the electronic structure of the atom was devised initially from the explanation for line spectra obtained from the elements. The Bohr model of the atom explains the nature of the light emitted from electronically excited hydrogen atoms. Bohr theorized that there are a number of possible orbits for the electron at various distances from the nucleus. The electron could be in any of these orbits but not between orbits. When an atom absorbs energy, an electron "jumps" from a low-energy orbit to a higher energy orbit farther away, resulting in an "excited atom." When the electron "falls" back to its lower orbit, energy is released in the form of light.

The modern model of the electronic structure of atoms is more complex. Electrons within an atom have certain energies but their positions are described in terms of shells, subshells, orbitals, and electron spin.

The maximum number of electrons that can occupy a shell is given by $2n^2$, where n is the number of the shell. Each shell contains one or more subshells designated by s, p, d, and f. The number of subshells in a particular shell is equal to the value of n. Each subshell can hold a specific maximum number of electrons. These values are 2, 6, 10. and 14 for the s, p, d, and f subshells, respectively. Each subshell consists of one or more orbitals. For the s, p, d, and f subshells there are 1, 3, 5, and 7 orbitals, respectively. No more than two electrons may occupy any orbital, and the two electrons must have opposite spins.

Electron configurations for atoms may be written by using simple rules. First, electrons are located in the lowest energy subshell until the subshell is filled. Second, within a subshell, the electrons are distributed in different orbitals until each orbital has one elec-tron. When two electrons are located in the orbital, the electrons have opposite spins.

New Terms

The **core** used in electron-dot symbols represents the nucleus and all electrons other than those in the valence shell.

The **electron configuration** of an atom is a description of the arrangement of the electrons in the atom by shells, subshells, and orbitals.

Electron spin is a property of the electron and may be either clockwise or counterclockwise.

The **electron-dot symbol** gives the number of valence electrons as dots located around the elemental symbol.

Hund's rule states that electrons tend to avoid the same orbital so that electrons will locate singly in orbitals of equal energy before pairing occurs.

An **orbital** describes the region in space in an atom where no more than two electrons may be found.

Paired electrons are two electrons of opposite spin in the same orbital.

The **shell** of an atom is a description of those electrons with characteristic energies. The shells are designated by an integer n.

The **subshells** located within a shell are characterized by a shape according to type. The subshells are labeled s, p, d, and f.

Valence shell electrons are electrons of the s and p subshell in the highest occupied energy level.

Exercises

Terminology

8.1 Distinguish between a shell and a subshell.

8.2 Distinguish between a subshell and an orbital.

Electrons and Symbols

8.3 For each of the following, indicate how many electrons are contained in one atom.
(a) $^{16}_{8}O$ (b) $^{36}_{18}Ar$ (c) $^{9}_{4}Be$
(d) $^{12}_{6}C$ (e) $^{31}_{15}P$ (f) $^{7}_{3}Li$

8.4 For each of the following, indicate how many electrons are contained in one atom.
(a) $^{14}_{7}N$ (b) $^{28}_{14}Si$ (c) $^{11}_{5}B$
(d) $^{19}_{9}F$ (e) $^{32}_{16}S$ (f) $^{4}_{2}He$

8.5 For each of the following, indicate how many electrons are contained in one atom.
(a) $^{39}_{19}K$ (b) $^{84}_{35}Br$ (c) $^{52}_{24}Cr$
(d) $^{64}_{29}Cu$ (e) $^{72}_{32}Ge$ (f) $^{84}_{36}Kr$

8.6 For each of the following, indicate how many electrons are contained in one atom.
(a) $^{131}_{54}Xe$ (b) $^{85}_{37}Rb$ (c) $^{128}_{52}Te$
(d) $^{89}_{39}Y$ (e) $^{119}_{50}Sn$ (f) $^{96}_{42}Mo$

Energy Levels

8.7 Write the integer value n for the following energy levels.
(a) L   (b) K   (c) N
(d) M   (e) O   (f) P

8.8 Write the letter equivalent for the following energy levels.
(a) first   (b) third   (c) sixth
(d) second   (e) fifth   (f) fourth

8.9 List the number maximum of electrons that may be located in each of the four lowest energy levels.

8.10 Why aren't more than seven energy levels discussed in this chapter or in other chemistry texts?

8.11 List the number of electrons contained in each of the following by principal energy levels.
(a) $^{14}_{7}N$ (b) $^{28}_{14}Si$ (c) $^{11}_{5}B$
(d) $^{19}_{9}F$ (e) $^{32}_{16}S$ (f) $^{4}_{2}He$

8.12 List the number of electrons contained in each of the following by principal energy levels.
(a) $^{16}_{8}O$ (b) $^{36}_{18}Ar$ (c) $^{9}_{4}Be$
(d) $^{12}_{6}C$ (e) $^{31}_{15}P$ (f) $^{7}_{3}Li$

8.13 List the number of electrons contained in each of the following by principal energy levels.
(a) $^{39}_{19}K$ (b) $^{24}_{12}Mg$ (c) $^{27}_{13}Al$
(d) $^{40}_{20}Ca$ (e) $^{72}_{32}Ge$ (f) $^{70}_{31}Ga$

8.14 List the number of electrons contained in each of the following by principal energy levels.
(a) $^{23}_{11}Na$ (b) $^{35}_{17}Cl$ (c) $^{13}_{6}C$
(d) $^{20}_{10}Ne$ (e) $^{37}_{18}Ar$ (f) $^{2}_{1}H$

8.15 How many electrons are contained in the valence shell of each of the following?
(a) $^{14}_{7}N$ (b) $^{28}_{14}Si$ (c) $^{11}_{5}B$
(d) $^{19}_{9}F$ (e) $^{32}_{16}S$ (f) $^{4}_{2}He$

8.17 How many electrons are contained in the valence shell of each of the following?
(a) $^{39}_{19}K$ (b) $^{24}_{12}Mg$ (c) $^{27}_{13}Al$
(d) $^{40}_{20}Ca$ (e) $^{72}_{32}Ge$ (f) $^{70}_{31}Ga$

8.16 How many electrons are contained in the valence shell of each of the following?
(a) $^{16}_{8}O$ (b) $^{36}_{18}Ar$ (c) $^{9}_{4}Be$
(d) $^{12}_{6}C$ (e) $^{31}_{15}P$ (f) $^{7}_{3}Li$

8.18 How many electrons are contained in the valence shell of each of the following?
(a) $^{23}_{11}Na$ (b) $^{35}_{17}Cl$ (c) $^{13}_{6}C$
(d) $^{20}_{10}Ne$ (e) $^{36}_{18}Ar$ (f) $^{2}_{1}H$

Subshells

8.19 How many subshells are there in each of the following?
(a) the second shell
(b) the fourth energy level
(c) the third energy level
(d) the fifth energy level

8.21 What is the total number of subshells associated with the $n = 4$ level?

8.23 What subshells are occupied by electrons in the valence level?
(a) $^{16}_{8}O$ (b) $^{36}_{18}Ar$ (c) $^{9}_{4}Be$
(d) $^{12}_{6}C$ (e) $^{31}_{15}P$ (f) $^{7}_{3}Li$

8.25 What subshells are occupied by electrons in the valence level?
(a) $^{39}_{19}K$ (b) $^{24}_{12}Mg$ (c) $^{27}_{13}Al$
(d) $^{40}_{20}Ca$ (e) $^{72}_{32}Ge$ (f) $^{70}_{31}Ga$

8.20 How many subshells are there in each of the following?
(a) the K shell
(b) the M shell
(c) the N shell
(d) the L shell

8.22 What is the total number of subshells associated with the $n = 6$ level?

8.24 What subshells are occupied by electrons in the valence level?
(a) $^{14}_{7}N$ (b) $^{28}_{14}Si$ (c) $^{11}_{5}B$
(d) $^{19}_{9}F$ (e) $^{32}_{16}S$ (f) $^{4}_{2}He$

8.26 What subshells are occupied by electrons in the valence level?
(a) $^{23}_{11}Na$ (b) $^{35}_{17}Cl$ (c) $^{13}_{6}C$
(d) $^{20}_{10}Ne$ (e) $^{85}_{37}Rb$ (f) $^{2}_{1}H$

Orbitals

8.27 How many orbitals are in each of the following subshells?
(a) $3s$ (b) $2p$ (c) $4d$
(d) $6s$ (e) $3p$ (f) $5d$

8.29 How many electrons may be in each of the following orbitals?
(a) the $3s$ orbital
(b) one of the orbitals in the $3p$ subshell
(c) one of the orbitals in the $4d$ subshell
(d) one of the orbitals in the $5f$ subshell

8.31 Which in the following list of atomic orbital designations is impossible?
(a) $7s$ (b) $1p$ (c) $5d$
(d) $2d$ (e) $4f$ (f) $6p$

8.33 Which in the following list of atomic orbital designations are possible?
(a) $4p$ (b) $2d$ (c) $3s$
(d) $1f$ (e) $3g$ (f) $5g$

8.28 How many orbitals are in each of the following subshells?
(a) $5s$ (b) $4f$ (c) $2s$
(d) $3d$ (e) $4p$ (f) $4s$

8.30 How many electrons may be in each of the following orbitals?
(a) the $6s$ orbital
(b) one of the orbitals in the $6p$ subshell
(c) one of the orbitals in the $5d$ subshell
(d) one of the orbitals in the $4f$ subshell

8.32 Which in the following list of atomic orbital designations is impossible?
(a) $3f$ (b) $7p$ (c) $6g$
(d) $5f$ (e) $4d$ (f) $3d$

8.34 Which in the following list of atomic orbital designations are possible?
(a) $6p$ (b) $4g$ (c) $3f$
(d) $8s$ (e) $2d$ (f) $6f$

Electron Configuration

8.35 Write the electronic arrangement of each of the following.
(a) $^{16}_{8}O$ (b) $^{36}_{18}Ar$ (c) $^{9}_{4}Be$
(d) $^{12}_{6}C$ (e) $^{31}_{15}P$ (f) $^{7}_{3}Li$

8.37 Write the electronic arrangement of each of the following.
(a) $^{14}_{7}N$ (b) $^{28}_{14}Si$ (c) $^{11}_{5}B$
(d) $^{19}_{9}F$ (e) $^{32}_{16}S$ (f) $^{4}_{2}He$

8.36 Write the electronic arrangement of each of the following.
(a) $^{39}_{19}K$ (b) $^{24}_{12}Mg$ (c) $^{27}_{13}Al$
(d) $^{40}_{20}Ca$ (e) $^{72}_{32}Ge$ (f) $^{70}_{31}Ga$

8.38 Write the electronic arrangement of each of the following.
(a) $^{23}_{11}Na$ (b) $^{35}_{17}Cl$ (c) $^{13}_{6}C$
(d) $^{20}_{10}Ne$ (e) $^{85}_{37}Rb$ (f) $^{2}_{1}H$

8.39 Which of the following electron configurations are correct for an atom in its normal state?

(a) $1s^1 2s^1$ (b) $1s^2 2s^2 2p^3$
(c) $[\text{Ne}]3s^2 3p^3 4s^1$ (d) $[\text{Ne}]3s^2 3p^6 4f^4$
(e) $1s^2 2s^2 2p^4 3s^2$

8.40 Which of the following represents an excited state of an atom?

(a) $1s^2 2s^1$ (b) $[\text{Ne}]3s^2 3p^6 4s^2 3d^1$
(c) $[\text{Ne}]3s^2 3p^6 4s^2 3d^8$ (d) $[\text{Ne}]3s^2 3p^6 4s^1 3d^2$
(e) $1s^2 2s^2 2p^6 3s^1$

Hund's Rule

8.41 On the basis of the indicated number of electrons in each subshell, determine the number of unpaired electrons.

(a) $3d^6$ (b) $3p^5$ (c) $4s^1$
(d) $2p^3$ (e) $4d^7$ (f) $5f^{10}$

8.42 On the basis of the indicated number of electrons in each subshell, determine the number of unpaired electrons.

(a) $4p^5$ (b) $5d^7$ (c) $5s^2$
(d) $3p^4$ (e) $4d^8$ (f) $4f^9$

8.43 On the basis of the electronic configuration, predict the number of unpaired electrons in each of the following elements.

(a) $^{14}_{7}\text{N}$ (b) $^{28}_{14}\text{Si}$ (c) $^{11}_{5}\text{B}$
(d) $^{19}_{9}\text{F}$ (e) $^{32}_{16}\text{S}$ (f) $^{4}_{2}\text{He}$

8.44 On the basis of the electronic configuration, predict the number of unpaired electrons in each of the following elements.

(a) $^{16}_{8}\text{O}$ (b) $^{36}_{18}\text{Ar}$ (c) $^{9}_{4}\text{Be}$
(d) $^{12}_{6}\text{C}$ (e) $^{31}_{15}\text{P}$ (f) $^{7}_{3}\text{Li}$

8.45 On the basis of the electronic configuration, predict the number of unpaired electrons in each of the following elements.

(a) $^{39}_{19}\text{K}$ (b) $^{24}_{12}\text{Mg}$ (c) $^{27}_{13}\text{Al}$
(d) $^{40}_{20}\text{Ca}$ (e) $^{72}_{32}\text{Ge}$ (f) $^{70}_{31}\text{Ga}$

8.46 On the basis of the electronic configuration, predict the number of unpaired electrons in each of the following elements.

(a) $^{23}_{11}\text{Na}$ (b) $^{35}_{17}\text{Cl}$ (c) $^{13}_{6}\text{C}$
(d) $^{20}_{10}\text{Ne}$ (e) $^{37}_{18}\text{Ar}$ (f) $^{2}_{1}\text{H}$

8.47 On the basis of the electronic configuration, predict the number of unpaired electrons in each of the following elements.

(a) $^{45}_{21}\text{Sc}$ (b) $^{55}_{25}\text{Mn}$ (c) $^{58}_{28}\text{Ni}$
(d) $^{80}_{35}\text{Br}$ (e) $^{79}_{34}\text{Se}$ (f) $^{66}_{30}\text{Zn}$

8.48 Which of the following has the largest number of unpaired electrons?

(a) $^{39}_{19}\text{K}$ (b) $^{24}_{12}\text{Mg}$ (c) $^{27}_{13}\text{Al}$
(d) $^{31}_{15}\text{P}$ (e) $^{35}_{17}\text{Cl}$ (f) $^{32}_{16}\text{S}$

Electron-Dot Symbols

8.49 Depict each of the following by electron-dot symbols.

(a) $^{16}_{8}\text{O}$ (b) $^{36}_{18}\text{Ar}$ (c) $^{9}_{4}\text{Be}$
(d) $^{12}_{6}\text{C}$ (e) $^{31}_{15}\text{P}$ (f) $^{7}_{3}\text{Li}$

8.50 Depict each of the following by electron-dot symbols.

(a) $^{14}_{7}\text{N}$ (b) $^{28}_{14}\text{Si}$ (c) $^{11}_{5}\text{B}$
(d) $^{19}_{9}\text{F}$ (e) $^{32}_{16}\text{S}$ (f) $^{4}_{2}\text{He}$

8.51 Depict each of the following by electron-dot symbols.

(a) $^{39}_{19}\text{K}$ (b) $^{84}_{35}\text{Br}$ (c) $^{24}_{12}\text{Mg}$
(d) $^{40}_{20}\text{Ca}$ (e) $^{72}_{32}\text{Ge}$ (f) $^{84}_{36}\text{Kr}$

8.52 Depict each of the following by electron-dot symbols.

(a) $^{131}_{54}\text{Xe}$ (b) $^{85}_{37}\text{Rb}$ (c) $^{128}_{52}\text{Te}$
(d) $^{127}_{52}\text{Te}$ (e) $^{119}_{50}\text{Sn}$ (f) $^{70}_{31}\text{Ga}$

Additional Exercises

8.53 For each of the following, indicate how many electrons are contained in one atom.

(a) $^{20}_{10}\text{Ne}$ (b) $^{36}_{18}\text{Ar}$ (c) $^{2}_{1}\text{H}$
(d) $^{23}_{11}\text{Na}$ (e) $^{35}_{17}\text{Cl}$ (f) $^{13}_{6}\text{C}$

8.54 For each of the following, indicate how many electrons are contained in one atom.

(a) $^{45}_{21}\text{Sc}$ (b) $^{75}_{33}\text{As}$ (c) $^{59}_{27}\text{Co}$
(d) $^{48}_{22}\text{Ti}$ (e) $^{55}_{26}\text{Fe}$ (f) $^{106}_{46}\text{Pd}$

8.55 Explain what would be observed if an excited helium atom with an electron in the fourth energy level were converted back to ordinary helium.

8.56 Is it necessary to list the mass number in order to determine the electronic arrangement of an atom? Explain.

8.57 List the number of electrons contained in each of the following by principal energy levels.

(a) $^{45}_{21}\text{Sc}$ (b) $^{55}_{25}\text{Mn}$ (c) $^{58}_{28}\text{Ni}$
(d) $^{80}_{35}\text{Br}$ (e) $^{79}_{34}\text{Se}$ (f) $^{66}_{30}\text{Zn}$

8.58 How many electrons are contained in the valence shell of each of the following?

(a) $^{79}_{34}\text{Se}$ (b) $^{75}_{33}\text{As}$ (c) $^{84}_{36}\text{Kr}$
(d) $^{85}_{37}\text{Rb}$ (e) $^{127}_{53}\text{I}$ (f) $^{114}_{49}\text{In}$

8.59 Can any shell contain any type of subshell?

8.60 How many orbitals are in the g subshell?

8.61 What subshells are occupied by electrons in the valence level?

(a) $^{79}_{34}\text{Se}$ (b) $^{75}_{33}\text{As}$ (c) $^{84}_{36}\text{Kr}$
(d) $^{87}_{38}\text{Sr}$ (e) $^{114}_{49}\text{In}$ (f) $^{131}_{53}\text{I}$

8.62 How many orbitals are in each of the following subshells?

(a) $3s$ (b) $6p$ (c) $7s$
(d) $5p$ (e) $5f$ (f) $6d$

8.63 What is the total number of orbitals in each of the following energy levels?

(a) second   (b) fourth   (c) third

8.64 What is the total number of subshells in each of the following energy levels?

(a) first   (b) fifth   (c) sixth

8.65 Write the electronic arrangement of each of the following.

(a) $^{79}_{34}$Se   (b) $^{75}_{33}$As   (c) $^{84}_{36}$Kr

(d) $^{87}_{38}$Sr   (e) $^{114}_{49}$In   (f) $^{131}_{53}$I

8.66 Write the electronic arrangement of each of the following.

(a) $^{66}_{30}$Zn   (b) $^{48}_{22}$Ti   (c) $^{59}_{28}$Ni

(d) $^{51}_{23}$V   (e) $^{55}_{26}$Fe   (f) $^{59}_{27}$Co

8.67 Which of the following has the smallest number of unpaired electrons?

(a) $^{45}_{21}$Sc   (b) $^{55}_{26}$Fe   (c) $^{69}_{30}$Zn

(d) $^{75}_{33}$As   (e) $^{79}_{34}$Se   (f) $^{40}_{20}$Ca

8.68 Which of the following has the largest number of unpaired electrons?

(a) $^{79}_{35}$Br   (b) $^{75}_{33}$As   (c) $^{55}_{25}$Mn

(d) $^{65}_{35}$Zn   (e) $^{70}_{31}$Ga   (f) $^{118}_{50}$Sn

8.69 Depict each of the following by electron-dot symbols.

(a) $^{20}_{10}$Ne   (b) $^{36}_{18}$Ar   (c) $^{2}_{1}$H

(d) $^{23}_{11}$Na   (e) $^{35}_{17}$Cl   (f) $^{13}_{6}$C

8.70 Depict each of the following by electron-dot symbols.

(a) $^{126}_{53}$I   (b) $^{87}_{38}$Sr   (c) $^{114}_{49}$In

(d) $^{88}_{38}$Sr   (e) $^{133}_{55}$Cs   (f) $^{122}_{51}$Sb

9

The Periodic Table

Learning Objectives

After studying Chapter 9 you should be able to

1. State the periodic law for the elements in its modern form.

2. Distinguish between periods and groups in the periodic table.

3. Indicate the relationship of the electron configuration of the elements to their positions in the periodic table.

4. Locate the metals, nonmetals, and metalloids within the periodic table.

5. Identify trends in atomic radii, ionization energies, and electronegativities within the periodic table.

9.1 Are the Elements Related?

Elements are the smallest class of pure substances, but they react to form millions of other pure substances called compounds. Therefore to understand chemistry it is reasonable to focus first on the elements. Some elements are similar to each other but very different from other elements. Sodium and potassium are both reactive metals and have similar properties. Chlorine and bromine have similar chemical properties, but these nonmetals are distinctly different from sodium and potassium.

Why do some elements have similar properties? Is it possible to classify, arrange, or categorize the elements to better understand chemistry? Raising such questions is important to the development of knowledge. Chemists, like other scientists, search for an underlying natural order that may become apparent when a large number of observations are categorized according to some common feature. It then may be possible to discover a reason for the underlying order, which in turn may lead to a theory.

There is no simple procedure that can be used to select which particular facts can be arranged into a classification scheme that will reveal a natural order. In the case of the elements, there is no way to know which of the many physical and chemical properties, if any, are related to some fundamental feature of nature. Therefore, trial and error are part of the process of searching for a classification scheme that will lead to a hypothesis and ultimately to a theory.

In this chapter we will learn how several individuals sought to arrange the elements to emphasize similarities in their properties. Eventually in the mid-nineteenth century a table was prepared in which elements with similar properties were placed next to each other. A modern version of this table is called the **periodic table.**

Once a periodic table was prepared, it was a natural step for chemists to wonder what is responsible for the underlying order implied by the table. In this chapter you will learn that electron configuration described in the previous chapter can be used to explain the similarities and differences in the properties of the elements as revealed by the periodic table. Correlations between the properties of the elements and electron configuration will then be discussed.

9.2 Early Classifications of Elements

The earliest classification of elements involved two large groups called metals and nonmetals (Section 3.11). In terms of physical properties, metals have higher densities than nonmetals. Chemically, although there are exceptions, metals form high-melting oxides that are basic, whereas nonmetals form oxides that are gases, volatile liquids, or low melting solids that are acidic. This type of classification, however, is far too broad to help anyone understand chemical reactions. For example, sodium is a fairly soft, light, and reactive metal, whereas tungsten is a hard, heavy, and unreactive metal.

Dobereiner's Triads

In 1829, the German chemist Johann Dobereiner tried to classify the elements into smaller and simpler subgroups. He observed that elements with similar physical and chemical properties fall into groups of three. He called these related groups of three elements **triads.** One such triad includes the elements chlorine, bromine, and iodine. These elements form compounds of the same general formula such as NaCl, NaBr, and NaI. Similarly, the triad of elements calcium,

Table 9.1   Dobereiner's Triads

Element	Atomic weight	Averaged atomic weight	Density (g/mL)	Melting point (°C)
chlorine	35.5		1.6	−101
bromine	79.9	81.2	3.1	−7
iodine	126.9		4.9	113.5
sulfur	32.1		2.1	95.5
selenium	79.0	79.8	· 4.8	217
tellurium	127.6		6.2	452
calcium	40.1		1.6	842
strontium	87.6	88.7	2.5	769
barium	137.3		3.5	725

strontium, and barium form the compounds $CaCl_2$, $SrCl_2$, and $BaCl_2$. In each of these triads, the atomic weight of the intermediate element is approximately the average of the atomic weights of the other two elements (Table 9.1). Similarly, the density of the intermediate element is approximately the average of the densities of the other two elements.

There are other triads of elements with similar chemical and physical properties. However, groups of only three elements are not significant because there was a smaller number of known elements to classify when Dobereiner made his observations. Currently we can identify groups containing as many as six elements that have similar chemical and physical properties.

Newlands' Law of Octaves

It was not until 1866, when the English chemist John Newlands examined the then known elements, that an order of great significance was detected. Newlands arranged the elements in increasing order of their atomic weights, and he noted that chemically similar elements occur at regular intervals in the list (Figure 9.1). Every eighth element has similar properties. Lithium, the second in the list, is comparable to sodium, the ninth in the list. Likewise, magnesium and calcium, which also have similar properties, are tenth and seventeenth in the list. The similarity of this repetition of the properties of the elements to the octave of the musical scale prompted Newlands to postulate a law of octaves. The **law of octaves** states that chemically similar elements reoccur in octaves when arranged in order of increasing atomic weight.

There are some deficiencies in Newlands' octaves because several elements do not fit well into this scheme. For example chromium, which is placed in the same position as aluminum in a subsequent octave, has substantially different chemical properties than aluminum. The metals

Figure 9.1

	no.		no.		no.		no.		no.		no.		no.
H	1	Li	2	Be	3	B	4	C	5	N	6	O	7
F	8	Na	9	Mg	10	Al	11	Si	12	P	13	S	14
Cl	15	K	16	Ca	17	Cr	18	Ti	19	Mn	20	Fe	21

Newlands' octaves
The elements are arranged in order of increasing atomic weights as they were known at that time. The numbers are the order in a list.

manganese and iron do not resemble the nonmetals phosphorus and sulfur, which occur in the preceeding octave. Unfortunately, when Newlands listed the elements by increasing atomic weight, he did not consider two factors. First, he did not allow for the possibility that not all of the elements might have been discovered; for example, gallium, which resembles aluminum, and germanium, which resembles silicon, had not been discovered at the time that Newlands postulated his classification scheme. Second, he did not question whether all of the atomic weights were correct; for example, the atomic weight of titanium was incorrect. Nevertheless, Newland's observation that there is a regularity of the properties of elements arranged in order of increasing atomic weight was an important contribution to chemistry.

9.3 Mendeleev and the Periodic Law

In 1869, the Russian chemist Dimitri Mendeleev succeeded in arranging the known elements in a systematic table. Like Newlands, Mendeleev found that when the elements are placed in the order of increasing atomic weight, elements with similar properties occur at regular or periodic intervals.

Just what does the term periodic mean? Think of a long fence along a field. The fence is supported by posts placed at intervals. If you walk along the fence you would periodically find a fence post. The posts resemble each other but are not exactly identical. Another analogy for the periodic concept is the order of the seasons. Every year we have the season of summer. No two summers are exactly alike, but similar seasons periodically occur each year separated by other seasons.

We can see the periodic pattern discovered by Mendeleev in Figure 9.2. The table of elements, given in order of increasing atomic weight, is based on the chemical properties of the

Row	Group I	Group II	Group III	Group IV	Group V	Group VI	Group VII	Group VIII
1	H = 1							
2	Li = 7	Be = 9.4	B = 11	C = 12	N = 14	O = 16	F = 19	
3	Na = 23	Mg = 24	Al = 27.3	Si = 28	P = 31	S = 32	Cl = 35.5	
4	K = 39	Ca = 40	? = 44	Ti = 48	V = 51	Cr = 52	Mn = 55	Fe = 56, Co = 59, Ni = 59
5	Cu = 63	Zn = 63	? = 68	? = 72	As = 75	Se = 78	Br= 80	
6	Rb = 85	Sr = 87	Y = 88	Zr = 90	Nb = 94	Mo = 96	? = 100	Ru = 104, Rh = 104, Pd = 106
7	Ag = 108	Cd = 112	In = 113	Sn = 118	Sb = 122	Te = 125	I = 127	
8	Cs = 133	Ba = 137	Dy = 138	Ce = 140				
9								
10			Er = 178	La = 180	Ta = 182	W = 184		Os = 195, Ir = 197, Pt = 198
11	Au = 199	Hg = 200	Tl = 204	Pb = 207	Bi = 208			
12				Th = 231	U = 240			

Figure 9.2

Mendeleev's periodic table

elements. For example, the elements lithium, sodium, and potassium of atomic weights 6.9, 23.0, and 39.1 amu, respectively, are separated by many intervening elements, but their properties are similar. All of these elements are reactive metals. Each of these metals reacts with water to produce a metal hydroxide and hydrogen gas.

$$2 \, Li(s) + 2 \, H_2O(l) \longrightarrow 2 \, LiOH(aq) + H_2(g)$$

$$2 \, Na(s) + 2 \, H_2O(l) \longrightarrow 2 \, NaOH(aq) + H_2(g)$$

$$2 \, K(s) + 2 \, H_2O(l) \longrightarrow 2 \, KOH(aq) + H_2(g)$$

Because Mendeleev focused on the chemical properties of the elements in the periodic table, he concluded that certain atomic weights were incorrect. For example, he found that the atomic weight of chromium is actually 52.0 amu, not 43.4 amu as thought at that time. Although there was a place in the table for chromium between calcium and titanium, based on the incorrect value for its atomic weight, the properties of chromium did not fit with this placement. This incorrect value was responsible for its improper placement in Newlands' octaves. Mendeleev relied on chemical and physical properties to guide him to the proper placement of the elements in the table.

The absence of elements with certain physical and chemical properties also indicated that not all elements had been discovered. Mendeleev left many gaps in his table where there were no known elements whose properties would fit. In Mendeleev's time the next known element with a higher atomic weight than calcium was titanium. If titanium were placed after calcium, it would fall under aluminum in Group III of the table. However, titanium forms compounds more similar to the elements of Group IV. Accordingly, Mendeleev placed titanium in Group IV and left a space for an undiscovered element in Group III, which turned out to be scandium.

The classification of the elements attracted considerable attention. An order of nature had been unveiled, and the several empty spaces in the table led to the search for new elements. The location of the spaces allowed Mendeleev and other chemists to make predictions of the chemical and physical properties of the unknown elements. These predictions guided the search for new elements. Mendeleev suggested that elements similar to aluminum and silicon should exist. Gallium (similar to aluminum) and germanium (similar to silicon) were discovered in 1871 and 1886, respectively.

Example 9.1

The melting points of sodium and rubidium are 97.8 and 39.0°C, respectively. Predict the melting point of potassium.

Solution Potassium is between sodium and rubidium in the Group I column in the Mendeleev table. Therefore, its melting point should be between that of sodium and that of rubidium. The actual melting point is 63.5°C, a value that is approximately the average of the other two melting points.

[*Additional examples are given in 9.3 and 9.4 at the end of the chapter.*]

The predictions of chemical and physical properties by Mendeleev were based on the properties of the elements surrounding the space in his table. He predicted the properties of "ekasilicon" (Table 9.2), which are close to those eventually found for germanium. Eka (Greek) means first beyond or first after.

Table 9.2   **Prediction of the Properties of an Unknown Element**

	Ekasilicon	Germanium
atomic weight	72	72.32
specific gravity	5.5	5.47
color	dark grey	greyish white
formula of oxide	EsO_2	GeO_2
specific gravity of oxide	4.7	4.70
formula of chloride	$EsCl_4$	$GeCl_4$
specific gravity of chloride	1.9	1.887
boiling point of chloride	below 100°C	83°C

At the time that Mendeleev made his statement of the periodic behavior of the elements, the scientific world did not know about electrons, protons, and neutrons. Some 30 years after Mendeleev's observations, it became evident that the properties of elements are really related to their atomic number. Of course, as the atomic numbers of elements increase, so do the atomic weights. However, since atomic weight depends on the number of neutrons and protons and on the relative abundance of the isotopes, a few pairs of elements are in a reversed order when arranged by atomic weight compared to an arrangement by atomic number. For example, argon is before potassium by atomic number, but the atomic weight of argon is larger than that of potassium. Only by arranging the elements by atomic number does a completely logical classification result. The modern **periodic law** is stated: The properties of elements are periodic functions of their atomic numbers.

9.4   The Modern Periodic Table

Mendeleev presented his classification of the elements in a periodic table that contained only 63 elements. A blank space was left in the table if there was no known element with the proper anticipated properties and atomic weight based on the periodic concept. By 1950 all the blank spots were filled. New elements have since been made by nuclear reactions, but they all have high atomic numbers and have been added to the end of the periodic table.

The modern periodic table is shown in Figure 9.3. Both the atomic number and the atomic weight are given in the rectangular box containing the symbol of the element. The elements are arranged by increasing atomic number in horizontal rows so that elements with similar properties fall into a column.

Periods of the Periodic Table

The horizontal rows in the periodic table are called **periods.** The seven periods are numbered sequentially from 1 through 7 from the top to the bottom of the table. Note that the periods of the table contain different numbers of elements. The first period contains only two elements, hydrogen and helium. The second and third periods have eight elements each, whereas the fourth and fifth periods have eighteen elements each. This means that there are ten elements in the fourth and fifth periods that have no counterparts in the two earlier periods. The table is split to accommodate these ten elements. Beyond the fifth period, there are even more elements.

Figure 9.3
The modern periodic table

Period	IA	IIA	IIIB	IVB	VB	VIB	VIIB	VIII			IB	IIB	IIIA	IVA	VA	VIA	VIIA	0
1	1 H 1.0079																	2 He 4.003
2	3 Li 6.941	4 Be 9.012											5 B 10.81	6 C 12.011	7 N 14.007	8 O 15.999	9 F 18.998	10 Ne 20.179
3	11 Na 22.990	12 Mg 24.305											13 Al 26.982	14 Si 28.086	15 P 30.974	16 S 32.06	17 Cl 35.453	18 Ar 39.948
4	19 K 39.098	20 Ca 40.08	21 Sc 44.956	22 Ti 47.88	23 V 50.942	24 Cr 51.996	25 Mn 54.938	26 Fe 55.847	27 Co 58.933	28 Ni 58.69	29 Cu 63.546	30 Zn 65.38	31 Ga 69.72	32 Ge 72.59	33 As 74.922	34 Se 78.96	35 Br 79.904	36 Kr 83.80
5	37 Rb 85.4678	38 Sr 87.62	39 Y 88.906	40 Zr 91.22	41 Nb 92.906	42 Mo 95.94	43 Tc (98)	44 Ru 101.07	45 Rh 102.906	46 Pd 106.42	47 Ag 107.868	48 Cd 112.41	49 In 114.82	50 Sn 118.69	51 Sb 121.75	52 Te 127.60	53 I 126.904	54 Xe 131.29
6	55 Cs 132.905	56 Ba 137.3	57 * La 138.906	72 Hf 178.49	73 Ta 180.948	74 W 183.85	75 Re 186.207	76 Os 190.2	77 Ir 192.22	78 Pt 195.08	79 Au 196.966	80 Hg 200.59	81 Tl 204.383	82 Pb 207.2	83 Bi 208.980	84 Po (209)	85 At (210)	86 Rn (222)
7	87 Fr (223)	88 Ra 226.025	89 ** Ac 227.028	104 Unq (261)	105 Unp (262)	106 Unh (263)	107 Uns (262)		109 Une (266)									

transition elements

lanthanides *	58 Ce 140.12	59 Pr 140.9077	60 Nd 144.24	61 Pm (145)	62 Sm 150.36	63 Eu 151.96	64 Gd 157.25	65 Tb 158.925	66 Dy 162.50	67 Ho 164.930	68 Er 167.26	69 Tm 168.934	70 Yb 173.04	71 Lu 174.967
actinides **	90 Th 232.0381	91 Pa 231.0359	92 U 238.02	93 Np 237.0482	94 Pu (244)	95 Am (243)	96 Cm (247)	97 Bk (247)	98 Cf (251)	99 Es (252)	100 Fm (257)	101 Md (258)	102 No (259)	103 Lr (260)

Elements 58–71 of the sixth period and elements 90–103 of the seventh period have no counterparts in earlier periods. They are placed outside the main table rather than splitting it again.

Groups of the Periodic Table

Each vertical column in the periodic table is called a **group.** A group is sometimes called a **family.** For example, fluorine, chlorine, bromine, iodine, and astatine are in Group VIIA and form a family of elements called the **halogens.** Other families are also given names. For example, elements in Groups IA and IIA are called **alkali metals** and **alkaline earth metals,** respectively.

Group 0 at the right side of the periodic table contains the noble gases which are also called rare or inert gases. These elements were unknown at the time Mendeleev proposed his table. The atomic numbers of the noble gases place them after the elements of the halogen family.

Elements in groups designated by Roman numerals I through VII in combination with the letter A and in Group 0 (or VIIIA) are called **representative elements.** Elements in Groups IB through VIIB and the three columns designated VIII are called **transition metals.** The two rows outside the periodic table contain **inner transition metals.** Elements 58 through 71 in the first row are called **lanthanides,** and elements 90 through 103 in the second row are **actinides.**

Each element in the table belongs to a group and a period. Thus, fluorine is in Group VIIA of the second period, whereas chlorine, which is located below fluorine, is in Group VIIA of the third period. The element oxygen, immediately to the left of fluorine, is in Group VIA of the second period.

Example 9.2 _____

Locate gallium (atomic number 31) in the modern periodic table. What are the period number and group number of the element? What elements should resemble gallium?

Solution Gallium is located in Period 4 and in Group IIIA in the periodic table. The elements aluminum and indium are above and below gallium, respectively. Thus, these two elements should have properties similar to gallium.

_____

[*Additional examples are given in 9.11–9.20 at the end of the chapter.*]

9.5 Electron Configuration and the Periodic Table

When the elements are arranged in the order of increasing atomic number, a periodic relationship of their properties is observed. How are the properties related to the atomic number? The atomic number of an atom is equal to the number of electrons in the atom, and these electrons are systematically located in specific subshells. Thus there is a relationship between the periodic table and the electron configuration of atoms.

The relationship between the energy levels of atoms and the periodic table is shown in Figure 9.4. In the first period, there are only two elements, hydrogen and helium; these have one and two electrons, respectively, in the 1s subshell. The elements lithium (3) and beryllium (4) have electrons in the 2s subshell and are separated from the other six elements in the period. Elements boron (5) through neon (10) involve the addition of six successive electrons to the 2p subshell. Notice that the two elements adding electrons only in the 2s subshell are on the left of the period, whereas the elements having electrons in the 2p subshell are on the right of the period. The period number and the energy level number are both labeled as 2.

The eleventh electron of sodium and the eleventh and twelfth electrons of magnesium occupy

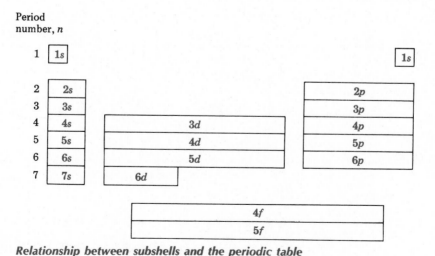

Figure 9.4

Relationship between subshells and the periodic table

the 3s subshell. These elements are located on the left of the period. Again, the six elements on the right of the periodic table involve the filling of a p subshell, which in these elements is the 3p subshell. Furthermore, the period number and the energy level number are both labeled as 3.

Elements 19 and 20 are in the same region in which s subshells were filled in earlier periods. In this case, the 4s subshell is involved for the highest energy electrons. Now, for the first time, there are elements in the central region of the periodic table. The electronic configurations of the ten elements scandium (21) through zinc (30) have electrons in the 3d subshell. In the case of d orbitals, the energy level number 3 is one less than the period number, 4. Then, elements 31–36 appear in the same region where p subshells have been filled previously. In this case, it is the 4p subshell, and these elements are in the fourth period.

The remainder of the periodic table follows in an orderly fashion in the case of s and p subshells. The energy level numbers of d subshells being occupied are always one less than the period number. Outside the general body of the periodic table are located two groups of 14 elements that represent the filling of the 4f and 5f subshells. The energy level numbers of f subshells being occupied are two less than the period number.

Example 9.3

Using the periodic table, determine the electron configuration of silicon.

Solution The electron configuration of silicon (atomic number 14) can be written by moving progressively through the regions of the periodic table until silicon is reached. The first period involves electrons achieving a $1s^2$ configuration. The second period involves two regions corresponding to $2s^2$ and $2p^6$. In the third period, the $3s^2$ region is passed and silicon is the second element in the $3p$ region corresponding to $3p^2$. The entire electron configuration is $1s^2 2s^2 2p^6 3s^2 3p^2$.

Example 9.4

Using the periodic table determine the electron configuration of zirconium (40).

Solution In order to reach zirconium by reading through the periodic table, the $1s$, $2s$, $2p$, $3s$, $3p$, $4s$, $3d$, $4p$, and $5s$ regions must be passed. Then one has to proceed by two elements into the $4d$ region. The entire electron configuration is $1s^2 2s^2 2p^6 3s^2 3p^6 4s^2 3d^{10} 4p^6 5s^2 4d^2$.

[*Additional examples are given in 9.37–9.40 at the end of the chapter.*]

Example 9.5

Using the periodic table, determine the number of unpaired electrons present in the 3d subshell of cobalt (27).

Solution Cobalt is located in the central region of the periodic table corresponding to the filling of a d subshell. Cobalt is in the fourth period and the central region corresponds to the 3d subshell. Cobalt is in the seventh position of the 3d region and has a $3d^7$ electron configuration.

When seven electrons are properly distributed in the *d* subshell, there are four paired and three unpaired electrons.

$$\text{⇅} \quad \text{⇅} \quad \text{↑} \quad \text{↑} \quad \text{↑}$$

[*Additional examples are given in 9.45 and 9.46 at the end of the chapter.*]

9.6 General Information Derived from Groups

The periodic table provides information about the chemical and physical properties of the elements. In this section, some of this information is outlined. In Chapter 10, the chemical properties of the elements and the types of bonding in compounds are discussed.

Physical Properties

For representative elements, the metallic properties increase with increasing atomic number within a group, and the nonmetallic properties decrease. Metallic properties within periods increase toward the left, and nonmetallic properties increase toward the right. Thus, within a group, the dividing line between metals and nonmetals occurs at higher atomic numbers for groups on the right than groups on the left. The stairstep line shown in Figure 9.5 separates the metals and nonmetals. Elements to the right of the line are nonmetals and elements to the left of the line are metals. The semimetals lie immediately to either side of the line. For example, in Group VA the most metallic element is bismuth; antimony and arsenic have borderline properties and are semimetals; phosphorus and nitrogen are nonmetals.

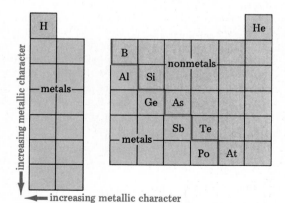

Figure 9.5

Metals, nonmetals, and metalloids of the representative elements.

The elements on each side of the stairstep line are metalloids.

Table 9.3   Formulas of Compounds of Related Elements

Element	Formula of chloride	Formula of oxide	Formula of nitrate	Formula of sulfate
lithium	LiCl	Li_2O	$LiNO_3$	Li_2SO_4
sodium	NaCl	Na_2O	$NaNO_3$	Na_2SO_4
potassium	KCl	K_2O	KNO_3	K_2SO_4
rubidium	RbCl	Rb_2O	$RbNO_3$	Rb_2SO_4
cesium	CsCl	Cs_2O	$CsNO_3$	Cs_2SO_4

Chemical Properties

The number of valence electrons of the representative elements is given directly by the Roman numeral of the group. Since the period number is the same as the energy level being filled, one knows the valence electron configuration at a glance. Thus, all Group IA elements, called the alkali metals, have an s^1 outer shell configuration. For four of the elements the configurations are

Li $1s^2 2s^1$
Na $1s^2 2s^2 2p^6 3s^1$
K $1s^2 2s^2 2p^6 3s^2 3p^6 4s^1$
Rb $1s^2 2s^2 2p^6 3s^2 3p^6 4s^2 3d^{10} 4p^6 5s^1$

Elements within groups have similar chemical properties due to the similarity of electronic configurations. Thus, all alkali metals can lose one electron from the s orbital to form singly charged cations. Therefore, similar compounds such as LiCl, NaCl, KCl, RbCl, and CsCl are formed. The periodic table can be used to predict the formula of compounds of an element in a family based on the other elements in the family. A list of compounds formed from the elements of Group 1A is given in Table 9.3

Predictions of the formulas of compounds have some limitations. Not every element in a group must behave identically. Consider for example the form in which the elements are found. Oxygen exists as O_2 but sulfur exists as S_8. Nitrogen exist as N_2 but phosphorus exists as P_4.

For the halogens, which are nonmetals, the valence electron configurations are $ns^2 np^5$. These elements tend to gain one electron to form singly charged anions. Therefore, similar compounds such as NaF, NaCl, NaBr, and NaI are formed. The halogens also form polyatomic ions in salts such as $NaClO_3$. Therefore, we can predict that $NaBrO_3$ and $NaIO_3$ might also be possible compounds. These compounds actually do exist. One could also predict that $NaFO_3$ might be a possible compound. However, this compound is unknown. Further details of the structure of compounds and the properties of the elements is given in Chapter 10.

9.7   Atomic Radius

The size of an atom reflects the number of electrons and their energies. The overall shape of the atom is considered to be spherical, and the size of an atom is expressed as the radius of a sphere. The radii of atoms are given in Angstrom units (1 Å $= 10^{-8}$ cm) in Figure 9.6 for the representative elements.

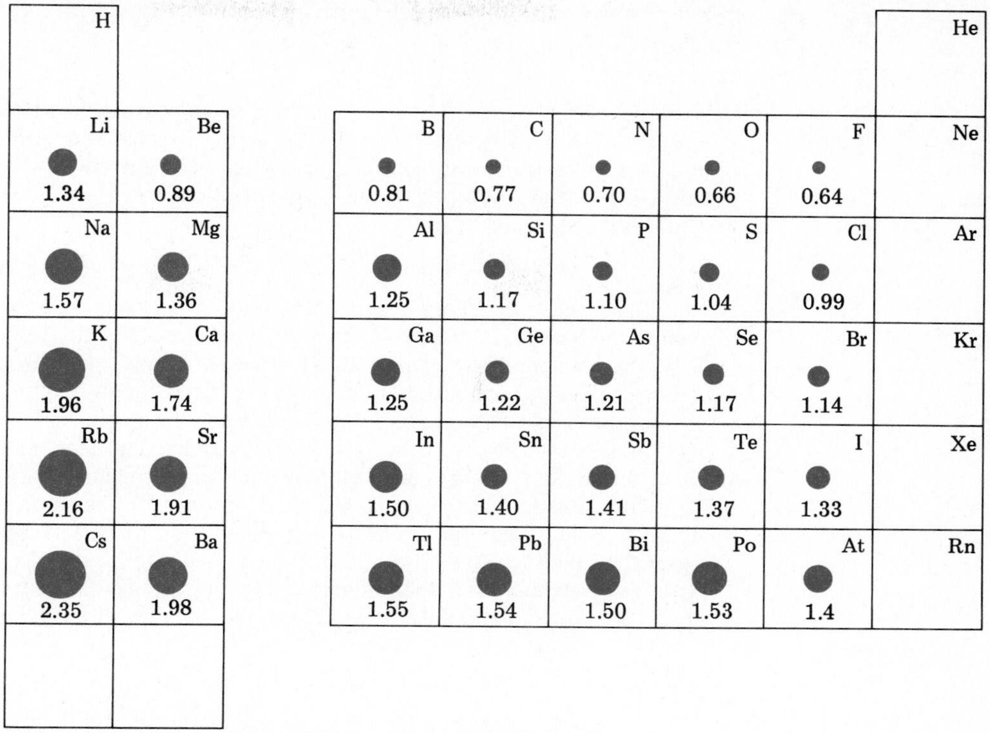

Figure 9.6

H							He
Li 1.34	Be 0.89	B 0.81	C 0.77	N 0.70	O 0.66	F 0.64	Ne
Na 1.57	Mg 1.36	Al 1.25	Si 1.17	P 1.10	S 1.04	Cl 0.99	Ar
K 1.96	Ca 1.74	Ga 1.25	Ge 1.22	As 1.21	Se 1.17	Br 1.14	Kr
Rb 2.16	Sr 1.91	In 1.50	Sn 1.40	Sb 1.41	Te 1.37	I 1.33	Xe
Cs 2.35	Ba 1.98	Tl 1.55	Pb 1.54	Bi 1.50	Po 1.53	At 1.4	Rn

Periodic trends in radii of the representative elements

The atomic radii tend to decrease from left to right in a period of the periodic table. Within a period, the nuclear charge increases in the same direction, and the electrons are located within the same energy level. This increase in nuclear charge tends to draw the electrons toward the nucleus and results in a smaller volume.

From top to bottom in a family of the periodic table, the atomic radii increase. Each successive member has one additional energy level containing electrons. Because the size of an orbital increases with the number of the energy level, the size of the atom tends to increase.

Example 9.6

Using the periodic table, determine the relative sizes of the germanium and bromine atoms.

Solution   The two elements are in period four of the periodic table. The trend within a period is a decrease in atomic radii in going from left to right. Bromine is in Group VIIA, which is to the right of germanium in Group IVA. Thus, bromine has a smaller atomic radius.

[*Additional examples are given in 9.49–9.52 at the end of the chapter.*]

9.8 Ionization Energy

If sufficient energy is added to an atom of a gas, an electron may be removed to form a positive ion. The removal of an electron from an atom is called **ionization**. The amount of energy required to remove the electron is called the **ionization energy**. Ionization energies are expressed in electron volts per atom. The ionization energy of gaseous lithium, Li(g), is 5.4 electron volts (eV) per atom.

$$Li(g) \longrightarrow Li^+(g) + 1\ e^-$$

Because the ionization energy is a measure of the energy required to remove an electron from an element, it indicates the energy of the electron in the element. Higher energy electrons require less additional energy (lower ionization energy) to remove them from the element; conversely, low-energy electrons are closely bound to the atom and require more energy for ionization. The ionization energies of the representative elements are given in Figure 9.7.

The ionization energies of the representative elements tend to increase from left to right within a period. This trend is due to the increase in the nuclear charge that causes an increase in the attraction for the electron. The ionization energy for the fluorine atom is 17.4 eV/atom, a value much higher than that for lithium at the start of the period. Metals have lower ionization energies than nonmetals.

Within a group there is a trend of decreasing ionization energies for elements of higher atomic number. Thus the ionization energy of cesium is 3.9 eV/atom, a value less than the 5.4 eV/atom of lithium.

Figure 9.7

increasing ionization energy ➡

increasing ionization energy ⬆

H 13.6								He 24.6
Li 5.4	Be 9.3		B 8.3	C 11.3	N 14.5	O 13.6	F 17.4	Ne 21.6
Na 5.1	Mg 7.6		Al 6.0	Si 8.2	P 11.0	S 10.4	Cl 13.0	Ar 15.8
K 4.3	Ca 6.1		Ga 6.0	Ge 8.1	As 9.8	Se 9.8	Br 11.8	Kr 14.0
Rb 4.2	Sr 5.7		In 5.8	Sn 7.3	Sb 8.6	Te 9.0	I 10.5	Xe 12.1
Cs 3.9	Ba 5.2		Tl 6.1	Pb 7.4	Bi 7.3	Po 8.4	At	Rn 10.7
Fr	Ra 5.3							

Periodic trends in ionization energies of the representative elements

Example 9.7

Using the periodic table, determine which element, magnesium or barium, has the larger ionization energy. Which element has the greater tendency to lose electrons?

Solution The two elements are in the same group of the periodic table. The trend within a group is a decrease in ionization energy in going from top to bottom. Magnesium, which is in the third period, has a larger ionization energy than barium, which is in the sixth period. Barium, which has the lower ionization energy, has the greater tendency to lose electrons.

[Additional examples are given in 9.53–9.58 at the end of the chapter.]

9.9 Electronegativity

Electronegativity is a measure of the ability of an atom to compete for or attract electrons from another atom in a bond. The electronegativity values of the representative elements are given in Figure 9.8. They are dimensionless numbers and range from about 1 to 4. The larger the electronegativity value, the greater the tendency to attract electrons.

Within a period, the electronegativities increase from left to right. Within a group, the electronegativities decrease with increasing atomic weight. Thus nonmetals are more electronegative than metals. Metals are often referred to as being electropositive; that is, they have only a small tendency to attract electrons and, in fact, will tend to lose electrons to electronegative elements.

The electronegativities of atoms determine the types of chemical bonds they can form. If the electronegativities of the atoms are very different, electron transfer will occur and an ionic bond will result. As the electronegativities become more similar, electrons are shared in covalent bonds. Bonding is discussed in the next chapter.

Figure 9.8

increasing electronegativity ➡

increasing electronegativity

H 2.1						
Li 1.0	Be 1.5	B 2.0	C 2.5	N 3.0	O 3.5	F 4.0
Na 0.9	Mg 1.2	Al 1.5	Si 1.8	P 2.1	S 2.5	Cl 3.0
K 0.8	Ca 1.0	Ga 1.6	Ge 1.8	As 2.0	Se 2.4	Br 2.8
Rb 0.8	Sr 1.0	In 1.7	Sn 1.8	Sb 1.9	Te 2.1	I 2.5
Cs 0.7	Ba 0.9	Tl 1.8	Pb 1.8	Bi 1.9	Po 2.0	At 2.2
Fr 0.7	Ra 0.9					

Periodic trends in electronegativities of the representative elements

Example 9.8

Using the periodic table, decide which element, tellurium (Te) or bromine (Br), would be expected to have the larger electronegativity.

Solution   Bromine is to the right and above tellurium in the periodic table. Electronegativities tend to increase from left to right within a period and to decrease from top to bottom within a group. Thus, iodine, which is to the right of tellurium, will have a larger electronegativity. Bromine, which is above iodine, will have a still larger electronegativity. Thus, bromine has a larger electronegativity than tellurium.

[*Additional examples are given in 9.59 and 9.60 at the end of the chapter.*]

Summary

The periodic law is historically based on the periodic table that provides an ordering of the elements according to their chemical and physical properties. Mendeleev's periodic table was used to predict the existence of new elements and to correct some atomic weights. The modern periodic table has been used to establish and describe relationships between the elements based on the period and group placement of the elements. The classification of elements as metals, metalloids, and nonmetals is shown by the position of the elements in the periodic table.

The experimental basis of the periodic table prepared by Mendeleev is now understood in terms of a description of the electron configuration of atoms. The groups of the periodic table consist of elements with similar electron configurations. Based on the electron configuration of the elements, it is possible to understand trends in the atomic radii, ionization energy, and electronegativity of the elements.

New Terms

The **actinides** are elements 58 through 71, located in a row outside the main periodic table.

The **atomic radius** of an atom is given in Ångstrom units (Å) ($1 \text{ Å} = 10^{-8}$ cm).

Alkali metals are the elements of Group IA.

Alkaline earth metals are the elements of Group IIA.

Electronegativity is a measure of the electron-attracting power of an atom relative to other atoms.

A **family** is a set of chemically related elements located in a group of the periodic table.

A **group** is a vertical column of elements in the periodic table that have similar properties.

The **halogens** are the elements of Group VIIA.

The **inner transition elements** are elements in two rows outside the main periodic table.

Ionization is the removal of an electron from an atom.

The **ionization energy** of an atom is the energy required to remove the highest energy electron and form an ion.

The **lanthanides** are elements 90 through 103, located in a row outside the main periodic table.

The **law of octaves** states that chemically similar elements occur in octaves when arranged in order of increasing atomic weight.

A **period** is a horizontal row in the periodic table.

The **periodic law** describes the periodic recurrence of properties of elements when considered in order of increasing atomic number.

The **periodic table** is an arrangement of elements with elements of similar properties grouped together.

Representative elements are elements in which the s and p subshells of the highest energy level are being filled.

The **transition elements** are elements in which the d or f subshells are being filled.

Exercises

Terminology

9.1   Distinguish between the terms period and group.

9.2   Distinguish between the terms representative element and transition element.

Early Classification of Elements

9.3   Consider selenium as a member of a triad. What two other elements have atomic weights and properties such that their averages would yield the atomic weight and properties of selenium?

9.4   Consider gallium as a member of a triad. What two other elements have atomic weights and properties such that their averages would yield the atomic weight and properties of gallium?

9.5 What relationship exists between Dobereiner's triads and the modern periodic table?

9.6 Why did Dobereiner only discover triad relationships when there are more elements within a family of the modern periodic table?

Mendeleev's Periodic Law

9.7 Cite an example of periodic behavior other than the two examples cited in this chapter.

9.8 How did Mendeleev's periodic law differ from the modern periodic law?

9.9 Mendeleev left a space in his table after molybdenum. What element was missing?

9.10 Examine the periodic table and locate three sets of elements whose positions appear to be reversed based on their atomic weights.

Groups and Periods

9.11 Using the periodic table, indicate the period and group number of each of the following elements. The atomic numbers are given in parentheses.
(a) carbon (6)   (b) tellurium (52)
(c) strontium (38)   (d) phosphorus (15)
(e) silicon (14)   (f) iodine (53)

9.12 Using the periodic table, indicate the period and group number of each of the following elements. The atomic numbers are given in parentheses.
(a) oxygen (8)   (b) indium (49)
(c) calcium (20)   (d) aluminum (13)
(e) chlorine (17)   (f) germanium (32)

9.13 Using the periodic table, indicate the period and group number of each of the following elements. The atomic numbers are given in parentheses.
(a) fluorine (9)   (b) cesium (55)
(c) magnesium (12)   (d) arsenic (33)
(e) argon (18)   (f) lead (82)

9.14 Using the periodic table, indicate the period and group number of each of the following elements. The atomic numbers are given in parentheses.
(a) vanadium (23)   (b) silver (47)
(c) tungsten (74)   (d) nickel (28)
(e) zirconium (40)   (f) gold (79)

9.15 Identify the element located in the indicated period and group.
(a) 3, VA   (b) 4, IIIA   (c) 5, IVA
(d) 6, IA   (e) 7, IIA   (f) 2, VIA

9.16 Identify the element located in the indicated period and group.
(a) 3, VIIA   (b) 4, IA   (c) 6, VA
(d) 5, IIA   (e) 7, IA   (f) 2, IVA

9.17 Identify the element located in the indicated period and group.
(a) 3, IIA   (b) 4, IVA   (c) 5, VIIA
(d) 6, IIA   (e) 2, IIIA   (f) 2, VIIA

9.18 Identify the element located in the indicated period and group.
(a) 6, VB   (b) 4, IIIB   (c) 5, IVB
(d) 6, IB   (e) 7, IIIB   (f) 4, VIB

9.19 Indicate the location of each of the following elements by period number and group designation.
(a) $_7$N   (b) $_{20}$Ca   (c) $_{48}$Cd
(d) $_{35}$Br   (e) $_3$Li   (f) $_{13}$Al

9.20 Indicate the location of each of the following elements by period number and group designation.
(a) $_{50}$Sn   (b) $_{16}$S   (c) $_{54}$Xe
(d) $_{26}$Fe   (e) $_{24}$Cr   (f) $_{29}$Cu

Families of Elements

9.21 Consider the following elements and select those that are representative elements.
(a) Ca   (b) Ru   (c) Se
(d) Ag   (e) Cr   (f) Cl

9.22 Consider the following elements and select those that are representative elements.
(a) Ni   (b) As   (c) I
(d) Au   (e) Sr   (f) Ir

9.23 Consider the following elements and select those that are transition elements.
(a) N   (b) Re   (c) P
(d) Ag   (e) V   (f) Cl

9.24 Consider the following elements and select those that are transition elements.
(a) Ni   (b) Ge   (c) Se
(d) Au   (e) Mg   (f) Mo

9.25 Consider the following elements and select those that are halogens.
(a) Ca   (b) Ru   (c) Se
(d) F   (e) Cr   (f) Cl

9.26 Consider the following elements and select those that are alkali metals.
(a) Mg   (b) As   (c) Al
(d) Cs   (e) Sr   (f) Ba

Metals, Nonmetals, and Metalloids

9.27 Consider the following elements and select those that are metals.
   (a) Ca   (b) Rb   (c) Se
   (d) Ag   (e) Cl   (f) Ni

9.28 Consider the following elements and select those that are nonmetals.
   (a) Zn   (b) C   (c) S
   (d) Pd   (e) Mn   (f) Br

9.29 Consider the following elements and select those that are nonmetals.
   (a) Na   (b) Cr   (c) Te
   (d) Hg   (e) Sc   (f) F

9.30 Consider the following elements and select those that are metalloids.
   (a) Zn   (b) C   (c) S
   (d) Ge   (e) Te   (f) Sr

9.31 Using the periodic table, classify each of the following as a metal or a nonmetal.
   (a) carbon   (b) calcium   (c) mercury
   (d) gallium   (e) iodine   (f) phosphorus

9.32 Using the periodic table, classify each of the following as a metal or a nonmetal.
   (a) strontium   (b) cesium   (c) selenium
   (d) vanadium   (e) tin   (f) sulfur

9.33 Using the periodic table, indicate which member of each pair is the more metallic.
   (a) magnesium and silicon   (b) germanium and bromine
   (c) sulfur and selenium   (d) silicon and tin

9.34 Using the periodic table, indicate which member of each pair is the more metallic.
   (a) phosphorus and antimony   (b) oxygen and tellurium
   (c) indium and tin   (d) sodium and cesium

9.35 According to the periodic table, which element should have the most metallic character?

9.36 Explain why the metalloids are found in a midregion of the periodic table.

Electron Configuration

9.37 In what period and group is an element with each of the following electron configurations located?
   (a) $[Kr]5s^24d^2$ (b) $[Ar]4s^23d^{10}4p^2$
   (c) $[Xe]6s^24f^{14}5d^4$ (d) $[Ne]3s^23p^3$

9.38 In what period and group is an element with each of the following electron configurations located?
   (a) $[Kr]5s^24d^{10}$ (b) $[Ar]4s^23d^{10}4p^5$
   (c) $[Xe]6s^24f^{14}5d^{10}$ (d) $[Ne]3s^23p^4$

9.39 Identify the group that has the following general electron configuration.
   (a) ns^2np^3 (b) $ns^2(n-1)d^1$
   (c) ns^2np^5 (d) ns^2np^2

9.40 Identify the group that has the following general electron configuration.
   (a) $ns^2(n-1)d^3$ (b) ns^2np^4
   (c) ns^2np^1 (d) $ns^2(n-1)d^2$

9.41 Indicate the number of valence electrons for each of the following elements.
   (a) silicon   (b) selenium   (c) phosphorus
   (d) krypton   (e) francium   (f) arsenic

9.42 Indicate the number of valence electrons for each of the following elements.
   (a) oxygen   (b) iodine   (c) aluminum
   (d) calcium   (e) bromine   (f) germanium

9.43 How many electrons are in each of the following?
   (a) the p subshell of a Group IVA element?
   (b) the p subshell of a Group VA element?
   (c) the s subshell of a Group IA element?
   (d) the d subshell of a Group IVB element?

9.44 How many electrons are in each of the following?
   (a) the p subshell of a Group IIIA element?
   (b) the d subshell of a Group IIIB element?
   (c) the p subshell of a Group VIA element?
   (d) the p subshell of a Group VIIA element?

Unpaired Electrons

9.45 Using the periodic table, determine the number of unpaired electrons in each of the following.
   (a) Ca   (b) Rb   (c) Se
   (d) Cd   (e) Cl   (f) Ni

9.46 Using the periodic table, determine the number of unpaired electrons in each of the following.
   (a) Na   (b) As   (c) I
   (d) Sc   (e) Sr   (f) O

Prediction of Physical Properties

9.47 In each of the following series, estimate the missing value for the indicated property of the element.

(a) Element	Atomic radius (Å)
F	0.64
Cl	?
Br	1.14

9.48 In each of the following series, estimate the missing value for the property of the indicated compound.

(a) Compound	Boiling point (°C)
$SiCl_4$	58
$GeCl_4$	83
$SnCl_4$	?

(b)

Element	Density (g/cm^3)
Ca	1.54
Sr	2.60
Ba	?

(c)

Element	Melting point (°C)
Na	?
K	63
Rb	39

(b)

Compound	Melting point (°C)
H_2S	−86
H_2Se	?
H_2Te	−49

(c)

Compound	Density (g/cm^3)
B_2S_3	1.6
Al_2S_3	2.4
Ga_2S_3	?

Atomic Radii

9.49 Explain why the sizes of atoms do not simply increase with increasing atomic weight.

9.51 Indicate which member of each of the following pairs of elements has the larger radius.
(a) F or Cl   (b) Si or S   (c) Li or K
(d) C or F   (e) P or Sb   (f) Ge or Pb

9.50 Which atom has the largest atomic radius?

9.52 Indicate which member of each of the following pairs of elements has the larger radius.
(a) S or F   (b) Sr or Cs   (c) As or In
(d) C or Ga   (e) Ge or Tl   (f) Na or Mg

Prediction of Ionization Energy

9.53 Which element in the periodic table has the highest ionization energy?

9.55 Indicate which member of each of the following pairs of elements has the higher ionization energy.
(a) O or F   (b) Li or K   (c) Cl or Br
(d) S or Se   (e) Ge or Pb   (f) Na or Li

9.57 Which Group IIA element has the lowest ionization energy?

9.54 Which element in the periodic table has the lowest ionization energy?

9.56 Indicate which member of each of the following pairs of elements has the higher ionization energy.
(a) S or F   (b) Sr or Cs   (c) As or In
(d) C or Ga   (e) Ge or Tl   (f) Na or Mg

9.58 Which Group VIIA has the highest ionization energy?

Prediction of Electronegativity

9.59 Indicate which member of each of the following pairs of elements has the higher electronegativity.
(a) O or F   (b) S or Se   (c) Li or K
(d) C or N   (e) Cl or Br   (f) P or Sb

9.60 Indicate which member of each of the following pairs of elements has the higher electronegativity.
(a) S or F   (b) Sr or Cs   (c) As or In
(d) C or Ga   (e) Ge or Tl   (f) Na or Mg

Additional Exercises

9.61 If the inert gases had been discovered prior to Newlands' work, would he have postulated a law of octaves?

9.63 At the time that Mendeleev developed the periodic table, the atomic weight of indium was thought to be 77. Where would this element have been placed if chemical properties were disregarded?

9.65 The order of the periodic table is possible because of certain facts about the elements. What facts are these?

9.62 It could be said that Newlands' law of octaves would have been more successful if the transition elements were not considered. Why?

9.64 Examine the position of nickel and cobalt in the modern periodic table and their atomic weights. Is the placement of the two elements a violation of the periodic law?

9.66 Using the periodic table, indicate the period and group number of each of the following elements. The atomic numbers are given in parentheses.
(a) iron (26)   (b) palladium (46)
(c) mercury (80)   (d) copper (29)
(e) cadmium (48)   (f) osmium (76)

9.67 Identify the element located in the indicated period and group.
  (a) 4, VIIB   (b) 4, IB   (c) 6, VB
  (d) 5, IIIB   (e) 6, IB   (f) 5, VB

9.68 Consider the following elements and select those that are representative elements.
  (a) Zn   (b) C   (c) S
  (d) Pd   (e) Mn   (f) Br

9.69 Consider the following elements and select those that are representative elements.
  (a) O   (b) B   (c) K
  (d) Pt   (e) Zr   (f) Tl

9.70 Consider the following elements and select those that are alkaline earth metals.
  (a) Zn   (b) Ca   (c) Se
  (d) Na   (e) Sc   (f) Ba

9.71 Consider the following elements and select those that are metals.
  (a) Ni   (b) As   (c) I
  (d) Au   (e) Sr   (f) O

9.72 Consider the following elements and select those that are metalloids.
  (a) Nb   (b) As   (c) I
  (d) Sb   (e) Mg   (f) Cu

9.73 Information about the properties of the transition elements is not given in this chapter. Based on their placement, what types of physical and chemical properties are expected?

9.74 In what period and group do the elements with the following electron configurations belong?
  (a) $1s^2 2s^2 2p^3$ (b) $1s^2 2s^2 2p^6 3s^1$
  (c) $1s^2 2s^2 2p^6 3s^2 3p^5$ (d) $1s^2 2s^2 2p^6 3s^2 3p^6 4s^2 3d^5$

9.75 Identify the group that has the following general electron configuration.
  (a) $ns^2(n-1)d^{10}$ (b) $ns^2 np^6$
  (c) $ns^2(n-1)d^5$ (d) $ns^2(n-1)d^8$

9.76 Using the periodic table, determine the number of unpaired electrons in each of the following.
  (a) Zn   (b) C   (c) S
  (d) Se   (e) Ga   (f) Br

9.77 Using the periodic table, determine the number of unpaired electrons in each of the following.
  (a) Na   (b) Sb   (c) Te
  (d) Hg   (e) Pb   (f) Ba

9.78 The element francium is radioactive. It is estimated that about 1 oz of francium may exist naturally. In spite of its rarity, how can one predict its properties?

9.79 How could a chemist predict the properties of an element, atomic number 112, that might be made in the future?

9.80 Predict what the radius of the oxide ion would be compared to that of the oxygen atom.

9.81 Predict what the radius of the magnesium ion would be compared to that of the magnesium atom.

9.82 How is the value of the ionization energy of metals related to their tendency for form cations?

9.83 Why don't nonmetals form cations when combined with other elements?

9.84 How is the electronegativity of nonmetals related to their tendency to form anions?

10 | Chemical Bonds

Learning Objectives

After studying Chapter 10 you should be able to

1. Use electron configurations to explain why atoms combine.

2. Describe the electron transfers necessary between atoms to form ionic bonds.

3. Describe the formation of covalent and polar covalent bonds between atoms.

4. Draw Lewis structures to depict compounds, using electronegativities to predict which bonds will be ionic, covalent, and polar covalent.

5. Use the VSEPR theory to explain the shapes of molecules.

6. Use molecular shapes and electronegativities to predict whether molecules will be polar or nonpolar.

10.1 Compounds and Bonds

Most substances around us and within organisms do not consist of isolated atoms but of ions in ionic compounds or of atoms in combination with one another in molecules. The ionic compound most familiar to you is sodium chloride (NaCl), a white crystalline solid that occurs in enormous deposits within the earth. Sodium chloride was deposited in various sites as primordial seas dried up when the surface of the earth changed. Sodium chloride is the most abundant compound dissolved in today's oceans and is recovered to obtain table salt or to be converted into substances such as sodium hydroxide or chlorine. Sodium chloride is also present in blood and body tissues. The amount of salt in your body is important in maintaining the amount of water in the body and affects blood pressure.

There are thousands of ionic compounds, but molecular compounds number in the millions. The molecules of some substances consist of very few atoms, as in the case of water (H_2O), a triatomic molecule with a molecular weight of 18 amu, and carbon dioxide (CO_2), a triatomic molecule with a molecular weight of 44 amu. Water, as you are well aware, is the wonderfully refreshing substance in which we swim on a hot summer day. In our body tissues, it is responsible for maintaining our temperature within the narrow range required for life. Carbon dioxide, which we exhale, is the end product of the metabolism of our food. Plants use carbon dioxide from the atmosphere as they grow and produce oxygen gas in the process known as photosynthesis.

The molecules of most substances in living organisms are very much larger. For example, vitamin B_{12} has a molecular weight of 1.35×10^3 amu, and hemoglobin has a molecular weight of 6.45×10^4 amu. DNA, the substance that contains the genetic code, has a molecular weight of 2×10^{12} amu in humans.

How are the ions or atoms bonded to each other? The answer is that the electrons of atoms and the forces that result from changes in electronic structure are responsible for bond formation. A clue to what occurs when bonds are formed is provided by classifying compounds based on physical properties.

Some substances such as sodium chloride are brittle and have a high melting point. These substances, known as ionic compounds, have rigid crystalline structures, but other substances have characteristics that are quite different from those of ionic compounds. Carbon dioxide, water, ethyl alcohol, aspirin, and a large number of medicinal compounds are examples. These compounds, which consist of molecules, tend to be gases, liquids, or low-melting solids.

The physical characteristics of the two classes of chemical compounds suggest that bonds may be of two types. Thus, the simplest approach is to describe two bonding classes: ionic bonds and covalent bonds. In this chapter we will learn how the electrons of atoms are involved in the ionic bonding of ionic compounds and covalent bonding of simple molecules. All chemical bonds result from a change in the electronic structure of two or more atoms as they associate with each other. Thus, the number and type of bond formed depend on the electron configuration of the atoms involved in the bond. A good understanding of Chapter 8 is very important to studying bonds in compounds.

10.2 Lewis Concept of Bonding

It is interesting that the modern concept of bonding came from a consideration of elements that do not form compounds! In 1916, G. N. Lewis of the University of California asked why the noble gases of Group 0 did not form compounds. He suggested that there is something unique about the electronic configuration of these atoms that prevents them from combining with other elements. All of the noble gases have filled valence shells. Helium, has an electron configuration of $1s^2$, and the s and p subshells of the highest energy level of the other noble gases contain a total of eight electrons. Lewis suggested that atoms of other elements might have a tendency to combine to obtain an electron configuration of eight electrons in the s and p subshells. In this way the highest energy subshell of the atoms would be made similar to those of the noble gases.

The bonding theory of Lewis is based on the following hypotheses.

1. Only the valence electrons are involved in bonding.
2. In forming an ionic compound, electrons are transferred from one atom to another to form anions and cations, which have an eight-electron valence shell.
3. The ionic bond consists of an attraction between oppositely charged ions.
4. A mutual sharing of electrons between some atoms results in molecules with covalent bonds.
5. The sharing of electrons occurs to provide atoms with a noble gas electron configuration, which is an eight-electron valence shell.

These hypotheses are summarized in the **Lewis octet rule:** Most atoms tend to combine and form bonds by transferring or sharing electrons until each atom is surrounded by eight valence electrons.

10.3 Ionic Bonds

The bond between ions is called an **ionic** or **electrovalent bond.** Compounds containing ionic bonds are called **ionic compounds.** Ionic bonds are formed between two or more atoms by the transfer of one or more electrons between the atoms. This electron transfer produces anions and cations, and there is an attraction between oppositely charged ions.

Let's examine the bonding in sodium chloride. The sodium atom, which has 11 protons and 11 electrons, has a single valence electron in the $3s$ subshell. The chlorine atom, which has 17 protons and 17 electrons, has seven valence electrons represented by $3s^2 3p^5$ (Figure 10.1). In forming an ionic bond, the sodium atom, which has a low ionization energy, loses its valence electron to chlorine. As a result, the sodium ion obtains the same electron configuration as neon ($1s^2 2s^2 2p^6$) and develops a $+1$ charge because there are 11 protons in the nucleus but only 10 electrons about the nucleus.

Na atom

charge of 11 protons	$= +11$
charge of 11 electrons	$= -11$
net charge of Na atom $=$	0

Na$^+$ ion

charge of 11 protons	$= +11$
charge of 10 electrons	$= -10$
net charge of Na$^+$ ion $=$	$+1$

The chlorine atom, which has a high electronegativity, gains an electron and is converted into a chloride ion that has the same electron configuration as argon ($1s^2 2s^2 2p^6 3s^2 3p^6$). The chloride ion has a -1 charge because there are 17 protons in the nucleus but there are 18 electrons about the nucleus.

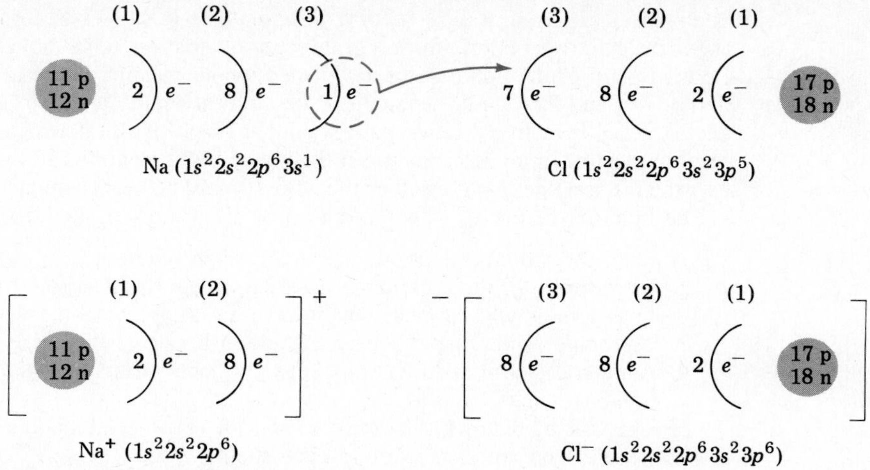

Figure 10.1

Na$^+$ $(1s^2 2s^2 2p^6)$ Cl$^-$ $(1s^2 2s^2 2p^6 3s^2 3p^6)$

The ionic bond in sodium chloride
The transfer of an electron between sodium and chlorine atoms produces ions. The electron configurations of the atoms and ions are given in parentheses.

Cl atom		Cl$^-$ ion	
charge of 17 protons	= +17	charge of 17 protons	= +17
charge of 17 electrons	= −17	charge of 18 electrons	= −18
net charge of Cl atom	= 0	net charge of Cl$^-$ ion	= −1

A shorthand way of showing the formation of sodium chloride from the sodium and chlorine atoms uses Lewis structures. **Lewis structures** represent only the valence electrons; the electron pairs are shown as a pair of dots.

$$\text{Na} \cdot \; + \; :\!\overset{..}{\underset{..}{\text{Cl}}}\!\cdot \longrightarrow \text{Na}^+ + \;:\!\overset{..}{\underset{..}{\text{Cl}}}\!:^-$$

The compound sodium chloride contains an equal number of sodium and chloride ions and is electrically neutral.

Sodium combines with oxygen to form the ionic compound sodium oxide (Na$_2$O). In this compound, each of two sodium atoms loses its $3s^1$ valence electron and is converted into a sodium ion with a neon electron configuration and a +1 charge. Oxygen, which has a $1s^2 2s^2 2p^4$ electron configuration, requires two electrons to produce the neon electron configuration. Thus each of the two sodium atoms transfers its single valence electron to one oxygen atom. The oxide ion produced has two more electrons than protons and therefore has a −2 charge.

O atom		O^{2-} ion	
charge of 8 protons	= +8	charge of 8 protons	= +8
charge of 8 electrons	= −8	charge of 10 electrons	= −10
net charge of O atom	= 0	net charge of O^{2-} ion	= −2

The formation of sodium oxide from the sodium and oxygen atoms is represented by electron-dot symbols as

196

$$2\,Na\cdot \; + \; :\overset{\cdot\cdot}{O}\cdot \; \longrightarrow \; 2\,Na^+ \; + \; :\overset{\cdot\cdot}{\underset{\cdot\cdot}{O}}:^{2-}$$

The compound sodium oxide contains sodium ions and oxide ions in a 2:1 ratio.

In all ionic compounds, the electrons lost by one or more atoms to form a Lewis octet in the cation are accepted by the appropriate number of atoms to result in a Lewis octet in the anion. Ionic compounds are typically the result of combinations of metallic elements, located on the left side of the periodic table, with nonmetals located mostly on the upper right side of the periodic table. The metals tend to lose electrons, whereas the nonmetals tend to gain electrons.

Example 10.1

Describe the ionic bond formed between calcium and chlorine in calcium chloride ($CaCl_2$).

Solution The electron configurations of the valence shell electrons of calcium and chlorine are $4s^2$ and $3s^23p^5$, respectively. Therefore calcium can lose two electrons to produce the Ca^{2+} cation and achieve the stable electron configuration of argon. Chlorine can gain one electron to achieve the stable electron configuration of argon by forming the Cl^- anion.

$$Ca(4s^2) \longrightarrow Ca^{2+}(3s^23p^6) + 2\,e^-$$
$$1\,e^- + Cl(3s^23p^5) \longrightarrow Cl^-(3s^23p^6)$$

In order to balance the electronic requirements of each type of atom, each of the two chlorine atoms accepts one electron from a calcium atom to form $CaCl_2$. In terms of electron-dot symbols, the process is

$$Ca: \; + \; 2\,:\overset{\cdot\cdot}{Cl}\cdot \; \longrightarrow \; Ca^{2+} \; + \; 2\,:\overset{\cdot\cdot}{\underset{\cdot\cdot}{Cl}}:^-$$

Example 10.2

Describe the ionic bond formed between magnesium and oxygen in magnesium oxide (MgO).

Solution The electron configuration of the valence shell of magnesium, a Group IIA element, is $3s^2$. The valence shell electrons of oxygen are $2s^22p^4$. Magnesium can lose two electrons and form Mg^{2+}, which has a neon-like electron configuration. Oxygen can gain two electrons to form O^{2-}, which also has a neon-like electron configuration.

$$Mg(3s^2) \longrightarrow Mg^{2+}(2s^22p^6) + 2\,e^-$$
$$2\,e^- + O(2s^22p^4) \longrightarrow O^{2-}(2s^22p^6)$$

The +2 charge of the magnesium ion electrically balances the −2 charge of the oxide ion, and the resulting compound is MgO. In terms of electron-dot symbols, the process is

$$Mg: \; + \; :\overset{\cdot\cdot}{O}\cdot \; \longrightarrow \; Mg^{2+} \; + \; :\overset{\cdot\cdot}{\underset{\cdot\cdot}{O}}:^{2-}$$

[*Additional examples may be found in 10.19–10.23 at the end of the chapter.*]

10.4 Ions and the Structure of Ionic Compounds

The **ionic radius** is the radius of a spherical ion as it is found in ionic compounds. The radii of ions differ from the radii of atoms. Some examples for Groups IA, IIA, VIA, and VIIA are shown in Figure 10.2. For negative ions, there are one or more extra electrons without a balancing positive charge in the nucleus, so the anion has its electrons more spread out in space than the atom does. For example, the chlorine atom has a radius of 0.99 Å, but the chloride ion, which has one more electron than the chlorine atom—18 electrons attracted by only 17 protons—has a radius of 1.81 Å. For positive ions there is a decrease in the radius compared to the atom. There are fewer electrons about the nucleus than there are protons in the nucleus; as a result the electrons are attracted closer to the nucleus. In the case of the sodium ion, which has 10 electrons attracted by 11 protons in the nucleus, the radius is only 0.97 Å compared with the 1.57 Å radius of the sodium atom.

In ionic compounds, such as sodium chloride, there are no molecules. The bond that exists does not involve two specific ions. A single sodium ion is not attracted exclusively to a single chloride ion. As shown in Figure 10.3, each positive sodium ion is surrounded by negative chloride ions, and each chloride ion is surrounded by sodium ions. A sodium chloride crystal actually consists of large numbers of sodium and chloride ions in which the ratio of cations to anions is 1:1.

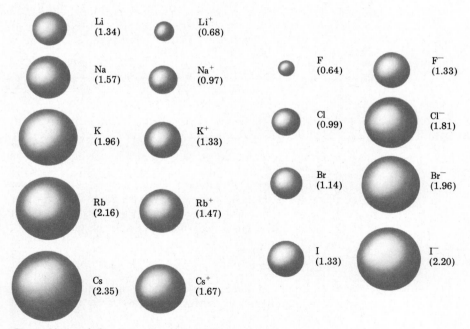

Figure 10.2

Comparison of the sizes of atoms and ions
The atoms and ions are represented as spheres within which the electrons of the atom or ion are located. The radii are given in Å.

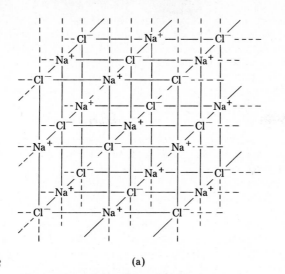

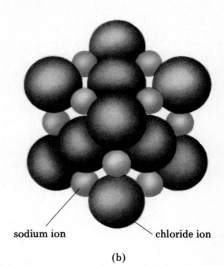

sodium ion   chloride ion

Figure 10.3   (a)   (b)

The structure of sodium chloride
(a) Schematic drawing showing the alternating pattern of the sodium and chloride ions.
(b) The relative sizes of the ions. Sodium ions are smaller than chloride ions.

Example 10.3

The atomic radii of zinc and nitrogen atoms are 1.33 and 0.74 Å, respectively. Compare the size of the Zn^{2+} and N^{3-} ions to those of the neutral atoms.

Solution A loss of two electrons for zinc causes the remaining electrons to be more strongly attracted to the protons in the nucleus. Thus the zinc ion should have an atomic radius less than 1.33 Å. The actual radius is 0.74 Å. The N^{3-} ion has three more electrons than the nitrogen atom. Because there are no balancing protons in the nucleus, the ionic radius is expected to be larger than that of the nitrogen atom. The actual radius is 1.71 Å.

[*Additional examples may be found in 10.25–10.28 at the end of the chapter.*]

10.5 Covalent Bonds

A **covalent bond** exists when an electron pair is shared between two atoms. This type of bond occurs when the difference between the electronegativities of the atoms is too small for an electron transfer, which would form ions, to occur. In the covalent bond of H_2 or F_2, the electron pair is shared equally between the two identical atoms.

 Covalent bonds in molecules hold atoms together in distinct molecular units. Each atom in a molecule ''belongs'' to the other atoms in the molecule. Recall that ionic bonds hold together ions in large aggregates, but ions of the opposite charge do not ''belong'' to each other. There is no molecule of sodium chloride or of any other ionic compound.

The hydrogen molecule forms from two hydrogen atoms, each with one electron in a $1s$ orbital. Since each hydrogen atom needs one additional electron to complete the first energy level and attain the electron configuration of the noble gas, helium, the two identical atoms join so that their two electrons are shared equally. Both atoms acquire a helium-like electron configuration as long as they are bonded together by sharing two electrons. The bonding of two hydrogen atoms to form a hydrogen molecule is pictured below in a Lewis structure.

$$\text{H} \cdot \; + \; \cdot \text{H} \longrightarrow \text{H} \cdot\cdot \text{H}$$

The electrons that are shared between atoms are called **bonding electrons.** The bonding electrons are the "glue" that holds the atoms together. They are located in space between the two positive nuclei of the atoms.

The fluorine molecule is pictured as two fluorine atoms joined by a shared pair of electrons. Each fluorine atom, which has seven valence electrons, requires one electron to fill the second energy level and form a Lewis octet or neon-like electron configuration. Because electron transfer cannot occur between the two identical fluorine atoms, an electron pair, formed from one electron from each fluorine atom, is shared equally. The two shared electrons in the fluorine molecule are bonding electrons. Each fluorine atom has three other pairs of electrons in its valence shell. Pairs of electrons in the valence shell associated with an atom but not involved in bonding are called **nonbonding** electrons, or **lone pair** electrons, or **unshared** electron pairs.

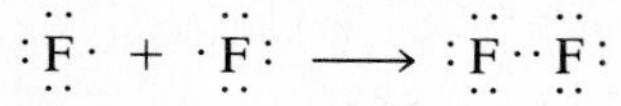

$$: \overset{\cdot\cdot}{\underset{\cdot\cdot}{\text{F}}} \cdot \; + \; \cdot \overset{\cdot\cdot}{\underset{\cdot\cdot}{\text{F}}} : \; \longrightarrow \; : \overset{\cdot\cdot}{\underset{\cdot\cdot}{\text{F}}} \cdot\cdot \overset{\cdot\cdot}{\underset{\cdot\cdot}{\text{F}}} :$$

In representing covalently bound molecules with Lewis structures, the pair of electrons shared between atoms can be represented by a dash. The hydrogen and fluorine molecules are more easily written by this convention.

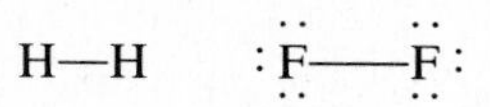

$$\text{H—H} \qquad : \overset{\cdot\cdot}{\underset{\cdot\cdot}{\text{F}}} \!\!—\!\! \overset{\cdot\cdot}{\underset{\cdot\cdot}{\text{F}}} :$$

More than one pair of electrons can be shared between pairs of atoms. If four electrons (two pairs) or six electrons (three pairs) are shared, the bonds are called **double** and **triple bonds,** respectively. The nitrogen molecule consists of two nitrogen atoms bound together by six shared electrons, a triple bond. Each nitrogen atom contributes three electrons to the triple bond.

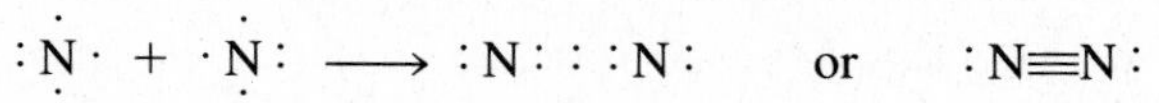

$$: \overset{\cdot}{\text{N}} \cdot \; + \; \cdot \overset{\cdot}{\text{N}} : \; \longrightarrow \; : \text{N} ::: \text{N} : \qquad \text{or} \qquad : \text{N} \equiv \text{N} :$$

The carbon atom can form single, double, or triple bonds with other carbon atoms as well as with some other elements. The carbon atom has four electrons in its valence shell, which can be shared to form four covalent bonds. In ethane, ethylene, and acetylene the carbon atoms are bonded by single, double, and triple covalent bonds, respectively. The remaining valence electrons of carbon make single bonds with hydrogen atoms.

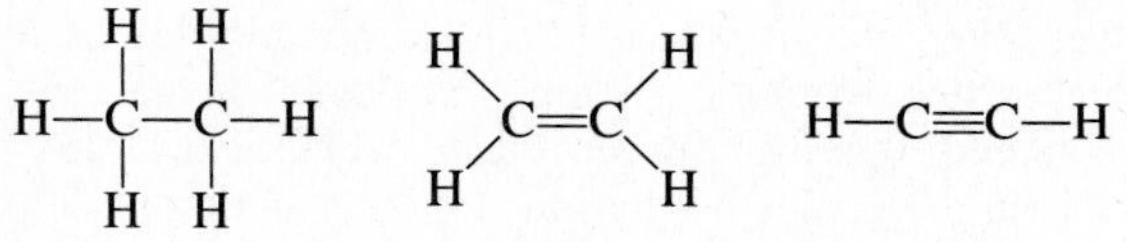

ethane
single bond
sharing 2 e⁻

ethylene
double bond
sharing 4 e⁻

acetylene
triple bond
sharing 6 e⁻

Example 10.4

What type of bond exists in Cl_2? Describe the electrons involved in bond formation. How many unshared electron pairs are there for each atom?

Solution The chlorine atom has a $3s^2 3p^5$ valence shell electron configuration. Each chlorine atom in a molecule needs only one electron to achieve a noble gas configuration. However, since both chlorine atoms are identical, each atom shares one electron to form a single covalent bond.

$$:\overset{..}{\underset{..}{Cl}}\cdot \ + \ \cdot \overset{..}{\underset{..}{Cl}}: \ \longrightarrow \ :\overset{..}{\underset{..}{Cl}}-\overset{..}{\underset{..}{Cl}}:$$

The remaining six electrons in the valence shell of each chlorine atom are present as three unshared electron pairs.

[Additional examples may be found in 10.29 and 10.30 at the end of the chapter.]

10.6 Polar Covalent Bonds

In Sections 10.3 and 10.5, two extreme types of chemical bonds were discussed. In this section the polar covalent bond, which is intermediate between the ionic and the covalent bond, is examined. A covalent bond in which the electrons are shared unequally in a bond between two different atoms is called a **polar covalent bond.** The hydrogen chloride molecule consists of two atoms, each of which needs one electron to form a noble gas electron configuration. Chlorine has a larger electronegativity than hydrogen, but the chlorine atom's attraction for electrons is not strong enough to remove an electron from hydrogen to form an ionic bond. Consequently, the bonding electrons in hydrogen chloride are shared unequally in a polar covalent bond. The molecule is represented by the conventional Lewis structure, even though the shared electron pair is associated to a larger extent with chlorine.

$$^{\delta +}H-\overset{..}{\underset{..}{Cl}}:{}^{\delta -}$$

As a result of the unequal sharing of the bonding electrons, the chlorine end of the molecule acquires a partial negative charge, and the hydrogen end acquires a partial positive charge. The symbol δ (the Greek lowercase letter delta) is used to denote the fractional charge located at a site within a molecule. The hydrogen chloride (HCl) molecule is said to possess a **dipole** (two poles), which is a pair of opposite charges of equal magnitude at a distance from each other.

It should be understood that only a displacement of electrons toward the more electronegative chlorine atom occurs in hydrogen chloride. There is no transfer of electrons between atoms as in sodium chloride. Thus there are no ions in hydrogen chloride, and the entire molecule is electrically neutral even though there are partial charges on each atom (Figure 10.4). The greater the difference between the electronegativities of the bonded atoms, the more polar is the bond. What numerical difference in electronegativity between elements is required to form an ionic bond as opposed to a polar covalent bond? A general rule of thumb is that a difference in electronegativity greater than 1.9 will result in an ionic bond. This rule must be used with some caution. If the difference is close to 1.9, be it 1.7 or 2.1, the type of bond cannot be predicted.

Figure 10.4

A representation of covalent, polar covalent, and ionic bonds
The shapes represent the volumes surrounding the nucleus within which
electrons are located. In hydrogen the volume is symmetrical around two
hydrogen nuclei. Note that in HCl the electron space around the hydrogen is
less because the electron pair in the bond is more closely associated with the
chlorine. In sodium chloride the electron transfer is complete and each ion is
spherical.

Only for a large difference or a small difference in electronegativities can the type of bond be predicted with confidence. It should be noted that regardless of electronegativity differences, all compounds of hydrogen and nonmetals have polar covalent bonds.

Example 10.5
_____

What type of bond exists in the molecule HF? Draw the Lewis structure for the molecule. How many unshared pairs of electrons are present in the molecule?

Solution   The hydrogen atom has a $1s^1$ electron configuration and requires one electron to achieve the helium electron configuration $1s^2$. The fluorine atom has a $2s^2 2p^5$ electron configuration and requires one electron to achieve a neon valence shell electron configuration $2s^2 2p^6$. If the two atoms contribute one electron each, a covalent bond results. However, the electrons must be shared unequally because the atoms are different, the shared pair associating more closely with the fluorine. The Lewis structure is

$$H\!-\!\overset{\displaystyle ..}{\underset{\displaystyle ..}{F}}:$$

The fluorine atom retains six of its valence shell electrons in the form of three unshared pairs of electrons.

_____

[*Additional examples may be found in 10.31–10.34 at the end of the chapter.*]

10.7 Coordinate Covalent Bonds

A **coordinate covalent bond** is formed when both electrons of the bonding electron pair shared by the two atoms are provided by one atom. However, once formed, a coordinate covalent bond is indistinguishable from any other covalent bond.

One example of the formation of a coordinate covalent bond is the reaction of ammonia with a hydrogen cation (proton) to form the ammonium ion.

$$
\begin{array}{ccc}
& H & H \\
& | & | \\
H\!-\!N: + H^+ \longrightarrow & H\!-\!\overset{+}{N}\!-\!H \\
& | & | \\
& H & H
\end{array}
$$

covalent bonds ⟶   ⟵ coordinate covalent bond

In ammonia, the nitrogen atom shares three of its five valence electrons with three hydrogen atoms. The remaining two valence shell electrons are not involved in bonding. This pair of electrons, an unshared pair of electrons, is used in a coordinate covalent bond to the proton in forming the ammonium ion. The proton has no electrons in its valence shell and needs the two electrons to form a covalent bond with nitrogen. Once the coordinate covalent bond is formed, it is indistinguishable from the other three covalent bonds.

Consider the structures of hypochlorous acid (HOCl) and chlorous acid ($HClO_2$). The chlorine atom in hypochlorous acid has three unshared pairs of electrons. It can share one pair with an oxygen atom, which only has six valence shell electrons and requires two electrons from chlorine to obtain a Lewis octet. The bond between chlorine and oxygen is a coordinate covalent bond.

$$\text{H}-\overset{\cdot\cdot}{\underset{\cdot\cdot}{\text{O}}}-\overset{\cdot\cdot}{\underset{\cdot\cdot}{\text{Cl}}}: \qquad \text{H}-\overset{\cdot\cdot}{\underset{\cdot\cdot}{\text{O}}}-\overset{\cdot\cdot}{\text{Cl}}-\overset{\cdot\cdot}{\underset{\cdot\cdot}{\text{O}}}: \quad \text{coordinate covalent bond}$$

hypochlorous acid   chlorous acid

Example 10.6

Consider the following structures of PCl_3 and $POCl_3$ and indicate how the combination of PCl_3 with an oxygen atom results in a bond.

$$:\overset{\cdot\cdot}{\underset{\cdot\cdot}{\text{Cl}}}-\underset{\underset{\displaystyle :\overset{\cdot\cdot}{\underset{\cdot\cdot}{\text{Cl}}}:}{|}}{\text{P}}-\overset{\cdot\cdot}{\underset{\cdot\cdot}{\text{Cl}}}: \qquad :\overset{\cdot\cdot}{\underset{\cdot\cdot}{\text{Cl}}}-\overset{\overset{\displaystyle :\overset{\cdot\cdot}{\text{O}}:}{|}}{\underset{\underset{\displaystyle :\overset{\cdot\cdot}{\underset{\cdot\cdot}{\text{Cl}}}:}{|}}{\text{P}}}-\overset{\cdot\cdot}{\underset{\cdot\cdot}{\text{Cl}}}:$$

Solution The oxygen atom, which has only six valence shell electrons, requires two electrons from the phosphorus atom to form a bond. The oxygen atom shown at the top of the representation of $POCl_3$ is bonded to the phosphorus atom by a coordinate covalent bond.

[*Additional examples may be found in 10.35 and 10.36 at the end of the chapter.*]

10.8 Writing Lewis Structures

In order to study chemistry, it is important that you be able to write Lewis structures of molecules and polyatomic ions easily. The conventions that are used are

1. Pairs of electrons are placed about atoms to indicate unshared pairs of electrons.
2. Dashes between atoms are used to indicate bonding pairs of electrons.

With the exception of hydrogen, most common elements require a Lewis octet of electrons in the valence shell. The octet may consist of any combination of bonding electrons and unshared pairs of electrons. The hydrogen atom only requires a total of two electrons to fill its valence shell.

For many simple molecules, Lewis structures can be written by considering how the individual atoms may form an octet of electrons by sharing relatively few electrons. However, some general guidelines are useful in writing more complex molecules. These guidelines are given in Table 10.1.

Table 10.1   **Guidelines for Writing Lewis Structures**

1. For a neutral molecule determine the sum of the number of valence electrons to be shown in the structure of the substance. Simply add the number of valence electrons of each atom in the molecule

2. If the structure is an ion, sum the number of valence electrons of the atoms and add or subtract the number of electrons necessary to form the ion. If the ion has a negative charge, add electrons to the total in sufficient number to produce the charge. If the ion has a positive charge, subtract electrons from the total in sufficient number to produce the charge.

3. Place the atoms about each other in the required skeletal arrangement so that the proper atoms may be bonded to each other. (Usually you need additional information to write this arrangement.)

4. Connect the bonded atoms by a single covalent bond. Check that the number of bonds does not exceed the accepted maximum value for each atom. Hydrogen only forms one bond.

5. Determine the number of electrons represented in the single bonds depicted by step 4. (Multiply the number of bonds by two because there are two electrons per single covalent bond.)

6. Subtract the number of electrons obtained in step 5 from the total number of electrons that must be present in the molecule (see step 1) or the ion (see step 2). Divide that number by two to obtain the number of unshared pairs of electrons that may be used.

7. Determine the number of pairs of electrons required to allow every atom to attain a Lewis octet, with the exception of hydrogen which requires one pair of electrons. If that number of pairs equals the number determined in step 6, distribute the electrons to form the octets about each atom.

8. If there are too few pairs calculated in step 6 to provide the necessary number of unshared pairs of electrons, it will be necessary to change single bonds to double bonds or a triple bond. If the deficiency is a single electron pair then change one single bond to a double bond. If the deficiency is two electron pairs then it may be necessary to form two double bonds or one triple bond.

Example 10.7

Draw the Lewis structure for H_2O, given the fact that each hydrogen atom is bonded to the oxygen atom.

Solution   The total number of valence electrons is $2(1) + 6 = 8$ for the two hydrogen atoms and the one oxygen atom. The arrangement of atoms with the necessary bonds between the hydrogen atom and the oxygen atom is

H—O
|
H

Four electrons are used in bonding. Thus, the remaining four electrons may be used as two electron pairs. Examining the structure, we can see that each hydrogen atom has two electrons that it is sharing with the oxygen atom in a bonded pair. The oxygen atom has four electrons present in two bonded pairs. Thus, the oxygen atom requires four electrons in the form of two unshared pairs of electrons to form a Lewis octet. This number is exactly equal to the number of electrons available. The structure is

$$H{-}\overset{\cdot\cdot}{\underset{|}{O}}:$$
$$\underset{H}{|}$$

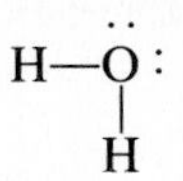

Example 10.8

Draw the Lewis structure for CCl_4, given the fact that each chlorine atom is bonded to the central carbon atom.

Solution The total number of valence electrons is $4(7) + 4 = 32$ for the four chlorine atoms and the one carbon atom. The arrangement of atoms with the necessary bonds between the carbon atom and the chlorine atoms is

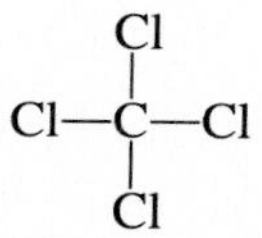

$$\overset{\displaystyle Cl}{\underset{\displaystyle Cl}{Cl{-}\overset{|}{C}{-}Cl}}$$

Eight electrons are used in forming the four bonds. The remaining $32 - 8 = 24$ electrons may be used as 12 electron pairs. Examining the structure, we can see that the carbon atom has eight electrons that it is sharing in four bonds with the four chlorine atoms. Each chlorine atom has only two electrons present in one bonded pair. Thus, each chlorine atom requires six additional electrons in the form of three unshared pairs of electrons to form a Lewis octet. This number is $4(3) = 12$ electron pairs, which is exactly equal to the number of electrons available. The structure is

$$\overset{\displaystyle :\overset{\cdot\cdot}{Cl}:}{\underset{\displaystyle :\overset{\cdot\cdot}{\underset{\cdot\cdot}{Cl}}:}{:\overset{\cdot\cdot}{\underset{\cdot\cdot}{Cl}}{-}\overset{|}{C}{-}\overset{\cdot\cdot}{\underset{\cdot\cdot}{Cl}}:}}$$

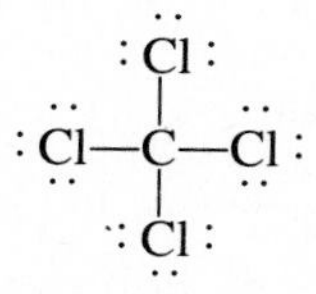

[*Additional examples may be found in 10.39 and 10.40 at the end of the chapter.*]

Example 10.9

Draw the Lewis structure for the hydroxide ion, OH^-.

Solution The total number of valence electrons is $6 + 1 + 1 = 8$ electrons for the one oxygen atom, one hydrogen atom, and one extra electron for the negative charge of the ion. The arrangement of atoms is

$$H{-}O$$

Two electrons are used in forming the one bond. The remaining $8 - 2 = 6$ electrons may be used as three electron pairs. The hydrogen atom has the necessary two electrons in the form of one covalent bond. The oxygen atom has only two of the necessary eight electrons to form the Lewis octet. The six electrons needed can be provided as three electron pairs, which are available according to the calculation above. The structure is

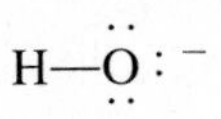

$$H-\ddot{\underset{\cdot\cdot}{O}}:^{-}$$

Example 10.10

Draw the Lewis structure for the sulfite ion, SO_3^{2-}. The three oxygen atoms are bonded to the central sulfur atom.

Solution The total number of valence electrons is $3(6) + 6 + 2 = 26$ electrons for the three oxygen atoms, the one sulfur atom, and the two electrons necessary to give the -2 charge. The arrangement of atoms with the necessary bonds between oxygen and sulfur is

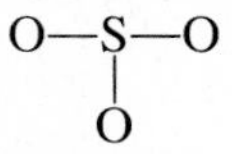

$$\begin{array}{c} O-S-O \\ | \\ O \end{array}$$

Six electrons are used in forming the three bonds. The remaining $26 - 6 = 20$ electrons may be used as ten electron pairs. The sulfur atom has six electrons that it is sharing in three bonds with the three oxygen atoms. The sulfur atom requires two electrons in the form of one unshared pair of electrons to form a Lewis octet. Each oxygen atom has only two electrons present in one bonded pair. Thus each oxygen atom requires six additional electrons as three unshared pairs of electrons to form a Lewis octet. The number of electron pairs required by the three oxygen atoms is $3(3) = 9$. The total number of electron pairs required is ten, which corresponds to the number of electron pairs available.

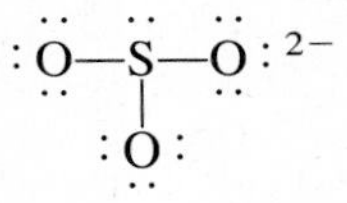

$$\begin{array}{c} :\ddot{O}-\ddot{S}-\ddot{O}:^{2-} \\ | \\ :\ddot{O}: \end{array}$$

[Additional examples may be found in 10.41 and 10.42 at the end of the chapter.]

Example 10.11

Write the Lewis structure for formaldehyde (H_2CO) given the fact that the hydrogen atoms and oxygen atom are bonded to the carbon atom.

Solution The total number of valence electrons is $2(1) + 4 + 6 = 12$ electrons for the two hydrogen atoms, one carbon atom, and one oxygen atom. The arrangement of atoms with the necessary bonds is

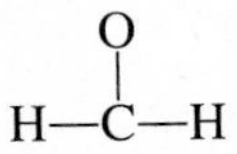

$$\begin{array}{c} O \\ | \\ H-C-H \end{array}$$

Six electrons are used in forming the three bonds. The remaining $12 - 6 = 6$ electrons may be used as three electron pairs. Each of the two hydrogen atoms has the required two electrons present in the form of a covalent bond. The carbon atom has six electrons that it is sharing in three bonds, and it requires two additional electrons or one unshared pair of electrons. The oxygen atom has only two electrons present in one bonded pair. Thus the oxygen atom requires six electrons as three unshared pairs of electrons to form a Lewis octet. The total number of electron pairs required is $3 + 1 = 4$, which exceeds by one the number available. Thus a double bond must be used. The structure is

$$\ddot{:}\!\overset{\displaystyle :\ddot{O}:}{\underset{\displaystyle H-C-H}{\|}}$$

Note that with two single bonds to hydrogen and a double bond to oxygen the carbon atom requires no unshared pairs of electrons.

Example 10.12

Write the Lewis structure for carbon disulfide (CS_2) given the fact that each sulfur is bonded to the central carbon atom.

Solution The total number of valence electrons is $2(6) + 4 = 16$ electrons for the two sulfur atoms and the one carbon atom. The arrangement of atoms with the necessary bonds between the carbon atom and the sulfur atoms is

S—C—S

Four electrons are used in forming the two bonds. The remaining $16 - 4 = 12$ electrons may be used as six electron pairs. The carbon atom has four electrons that it is sharing in two bonds with the two sulfur atoms. Thus the carbon atom requires four additional electrons in two unshared pairs of electrons to form a Lewis octet. Each sulfur atom has only two electrons present in one bonded pair. Thus each sulfur atom requires six electrons in the form of three unshared pairs of electrons to form a Lewis octet. This number is $2(3) = 6$. A total of $6 + 2 = 8$ electron pairs are required if each atom is to have a Lewis octet in a structure containing only single bonds. The deficiency of $8 - 6 = 2$ electron pairs must be made up by two double bonds. Using double bonds between carbon and sulfur the structure is

$$:\!\ddot{S}\!=\!C\!=\!\ddot{S}\!:$$

Note that using two double bonds corresponds to $2(4) = 8$ electrons or four pairs of electrons, and carbon does not require any additional electrons. Each sulfur atom requires four additional electrons or two unshared pairs of electrons.

[*Additional examples may be found in 10.45 and 10.46 at the end of the chapter.*]

10.9 Resonance Structures

There are cases in which a single Lewis structure for a molecule based on the Lewis octet rule is not entirely correct. By that we mean that some of the properties of the molecule are different from those expected from the Lewis structure.

Let's consider the molecule sulfur dioxide (SO_2). A structure based on the Lewis octet rule and using the rules used in the examples in the previous section requires a double bond between sulfur and one of the oxygen atoms. The other oxygen atom is bonded to sulfur by a single bond.

$$\ddot{O}=S-\ddot{O}\!:$$

The sulfur atom and each oxygen atom has a total of eight electrons and thus each has a Lewis octet. The actual sulfur dioxide molecule has properties that indicate the two sulfur–oxygen bonds are identical in every way. Therefore, the Lewis structure showing single and double bonds does not describe the molecule accurately. The real structure of sulfur dioxide can be better represented as a hybrid of two Lewis structures, neither of which is completely correct.

$$:\ddot{O}=S-\ddot{O}\!:\qquad :\ddot{O}-S=\ddot{O}\!:$$

Since the actual sulfur–oxygen bonds are equal, they must be intermediate between single and double bonds. The problem is that we cannot simply represent the actual molecule with a single Lewis structure. A double-headed arrow between two Lewis structures then is used to indicate that the actual structure is similar in part to the two simple structures, but lies somewhere between them. The individual Lewis structures are called **contributing** or **resonance structures.**

$$:\ddot{O}=S-\ddot{O}\!: \longleftrightarrow :\ddot{O}-S=\ddot{O}\!:$$

10.10 Molecular Geometries and Polarities

Up to this point, molecules have been represented in only two dimensions, but molecules are actually three-dimensional. The three-dimensional structures of molecules are important and affect both physical and chemical properties of even simple molecules such as water and carbon dioxide.

The shape of a molecule is given by the positions of the nuclei of atoms in space. Only the positions of the nuclei need to be indicated to describe the molecular shape (Figure 10.5). A molecule of two atoms (diatomic) must necessarily have its nuclei along a line as in the cases of Cl_2 and HCl. Such molecules are called **linear.** Molecules composed of three atoms, such as H_2O or CO_2, must have all atoms in a plane because a plane can be made to contain three points in space. The angle formed by the two bonds to the central atom is used to describe the molecule. A bond angle of 180° results in a **linear** molecule. Any other angle results in a **bent** or **angular** molecule. Water is a bent or angular molecule with a H—O—H bond angle of 104.5° whereas carbon dioxide is a linear molecule, that is the O=C=O bond angle is 180°.

Molecules composed of more than three atoms could not only be linear or planar but many other shapes are also possible (Figure 10.5). Formaldehyde (H_2CO) has all four atoms in a plane with 120° bond angles for both the H—C=O and H—C—H bonds. The arrangement of three atoms about a central atom at 120° to each other in a plane is called **trigonal planar.** In contrast, ammonia has a central nitrogen atom bonded to three hydrogen atoms; the hydrogen atoms are in a common plane but not with the nitrogen atom. The shape of the molecule is described as **trigonal pyramidal.** The H—N—H bond angles in ammonia are all 107°. Molecules containing four atoms bonded to a central atom exist in a variety of shapes. One such shape is described as **tetrahedral.** Methane (CH_4) is a tetrahedral molecule in which the carbon atom is at the center

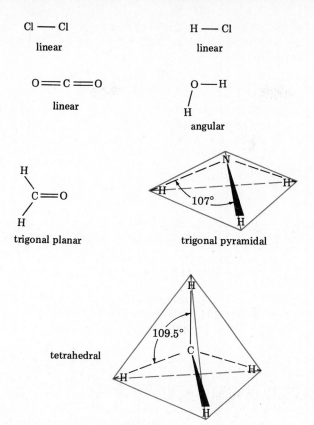

Figure 10.5

Shapes of simple molecules

of the tetrahedron with the four hydrogen atoms at the corners. The H—C—H bond angles are all 109.5°.

Now let's see how the shape of molecules is related to the polarity of molecules. A polar covalent bond exists when there is a difference in electronegativity between the bonded atoms, as in the case of H—Cl, which is a polar molecule. This polarity can be represented by an arrow. The cross on the tail of the arrow indicates the positive end of the bond and the arrow head the negative end.

$$\overset{\longmapsto}{\text{H—Cl}}$$

When two or more polar bonds are present in a molecule, the molecule may be polar or nonpolar depending on its geometry. In order to understand this idea, consider the arrows used to indicate bond polarity as equivalent to forces. If two or more forces are applied on a common object, the object may or may not move. For example, if equal forces in opposite directions exist, the forces cancel each other out and the object will not move.

$$\longleftarrow + \cdot +\longrightarrow \qquad \text{no net force or movement}$$
$$+\longrightarrow \cdot \longleftarrow + \qquad \text{no net force or movement}$$

If the forces are not in opposite directions or some suitable balanced arrangement, the object will move. If the two forces are at an angle other than 180°, there is a net force moving the object in a specific direction.

$$\swarrow \searrow \quad \begin{array}{c} \text{net downward} \\ \text{force} \end{array} \quad \downarrow \qquad \nearrow \nwarrow \quad \begin{array}{c} \text{net upward} \\ \text{force} \end{array} \quad \uparrow$$

In carbon dioxide (CO_2) the C=O bonds are polar with the polarity directed from the carbon atom toward the more electronegative oxygen atoms. The two bonds are located along a common line because the molecule is linear. As a result, the polarity of the bonds cancel each other and the molecule has no net polarity. Thus, it is possible to have polar bonds in a molecule that is itself nonpolar.

$$\overset{\longleftarrow\;+\qquad +\;\longrightarrow}{O\!=\!C\!=\!O}$$

The water molecule is bent, and the polarity of the bonds do not cancel each other out. Water is a polar molecule in which the negative end of the dipole is directed toward the oxygen atom.

$$\begin{array}{c} O \\ H \quad H \end{array} \qquad \nearrow \nwarrow \quad \begin{array}{c} \text{net polarity} \\ \text{toward oxygen} \end{array} \quad \uparrow$$

In ammonia (NH_3) the N—H bonds are polar with the polarity of the bonds directed from each hydrogen atom toward the nitrogen atom. The net result is a polar molecule with the negative end directed toward the nitrogen atom.

$$\begin{array}{c} N \\ H \quad \backslash H \\ H \end{array} \qquad \nearrow \!\!\nwarrow\!\!\nwarrow \quad \begin{array}{c} \text{net polarity} \\ \text{toward nitrogen} \end{array} \quad \uparrow$$

For carbon tetrachloride, (CCl_4) which is a tetrahedral molecule, the four C—Cl bonds are polar with the negative end directed toward chlorine; however, the four equal bonds in a tetrahedral arrangement result in a net nonpolar molecule.

$$\begin{array}{c} Cl \\ C \\ Cl \quad \backslash Cl \\ Cl \end{array} \qquad \begin{array}{c} \uparrow \\ \swarrow \downarrow \searrow \end{array}$$

Example 10.13

Is acetylene (C_2H_2) a polar or nonpolar molecule, given the fact that all four atoms are arranged along a common line? Explain your answer based on the anticipated polarity of each bond in the molecule.

Solution The structure of acetylene was given in Section 10.5.

$$H\!-\!C\!\equiv\!C\!-\!H$$

The electronegativities of hydrogen and carbon are 2.1 and 2.5, respectively. Thus the H—C bond should be weakly polar with the negative end directed toward the carbon atom. The carbon–carbon triple bond must be nonpolar because it involves identical atoms. For a linear arrangement, the polarities of the two C—H bonds must cancel each other out, and the molecule is predicted to be nonpolar.

$$\overset{\longmapsto \quad \longleftarrow}{H-C\equiv C-H}$$

[*Additional examples may be found in 10.51 and 10.52 at the end of the chapter.*]

10.11 Valence-Shell Electron-Pair Repulsion Theory

Why is the triatomic water molecule angular whereas the triatomic carbon dioxide molecule is linear? Why does any molecule have its characteristic shape? A simple concept that can be used to account for molecular structure is based on the idea that electron pairs located about a central atom should repel each other. The concept, called the **valence-shell electron-pair repulsion** (abbreviated VSEPR) theory, predicts that electron pairs should be arranged as far apart as possible. Thus, two electron pairs should be arranged at 180° to each other (Figure 10.6); three pairs should be at 120° in a common plane; for four electron pairs, the arrangement is tetrahedral, with angles of 109.5°, and not in a square planar arrangement in which the angles would be only 90°. A good analogy for the concept of maximum separation of electron pairs involves balloons. Think of the axis going through a balloon from the tied end to the opposite curved surface (Figure 10.7). Now imagine four balloons tied together closely at one point. The balloons would arrange themselves to be as uncrowded as possible. That arrangement is not planar because the angles between the balloons would be only 90°. In a tetrahedron, each angle between balloons is 109.5°. Thus, conceptually we can regard four bonded pairs of electrons like four equivalent balloons.

In carbon dioxide, formaldehyde, and methane, all of the valence electrons about the central

Number of electron pairs	Arrangement of electron pairs	Term	Bond angle
2	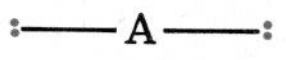 :—A—:	linear	180°
3	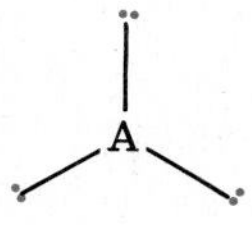 A (trigonal)	trigonal planar	120°
4	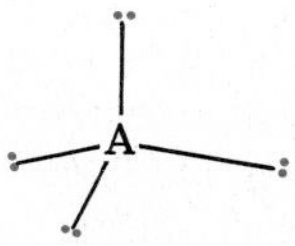 A (tetrahedral)	tetrahedral	109.5°

Figure 10.6

Arrangement of electron pairs according to VSEPR theory
The electron pairs around a central atom, designated as A, are arranged to be at the greatest distance from each other.

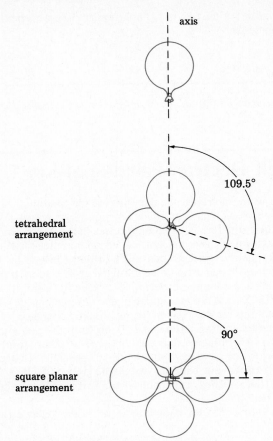

axis

109.5°

tetrahedral
arrangement

90°

square planar
arrangement

Figure 10.7

A balloon analogy for the VSEPR theory
*The tetrahedral arrangement causes each
balloon to be at the maximum distance from
every other balloon. The angle between the
axes of any two balloons is 109.5°.*

atom are involved in bonds. Carbon dioxide has two double bonds; each double bond is separated by the maximum distance, and the resulting angle between the bonds is 180°. Formaldehyde has a double bond and two single bonds to the central carbon atom; these bonds correspond to three regions containing electrons, and they are arranged in a trigonal planar arrangement with bond angles of 120°. In methane, there are four bonded electron pairs, and they are predicted to be located in a tetrahedral arrangement. Each H—C—H bond angle is predicted to be 109.5°, in agreement with the experimental value.

Now let's consider molecules that have both bonded and nonbonded pairs of electrons in the valence shell of the central atom. Water and ammonia have shapes described as angular and trigonal pyramidal, respectively, but there are some important similarities. Both water and ammonia have four electron pairs about the central atom. Some of the electron pairs of the central atom form bonds to other atoms, and others are unshared electron pairs (Figure 10.8).

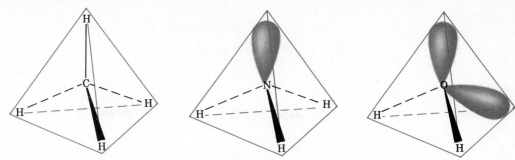

Figure 10.8

Location of electron pairs in some simple molecules
All electron pairs in methane, ammonia, and water are directed to the corners of a tetrahedron.
Note, however, that ammonia is still described as a trigonal pyramidal molecule and water as
an angular molecule.

Methane also has four pairs of electrons about the central atom, but they are all bonded pairs. Water and ammonia do not have tetrahedral shapes like methane. Shape is defined by a geometric figure that describes the positions of the nuclei. The VSEPR theory describes the distribution of electron pairs, including the nonbonded pairs. Therefore despite the fact that four pairs of electrons in both ammonia and water are distributed in a tetrahedral arrangement, ammonia and water are pyramidal and bent molecules, respectively.

Ammonia has three bonding pairs and one nonbonded pair of electrons. The bond angles are the result of the bonding pairs of electrons. However, the nonbonded electron pair affects the positions of the bonded pairs of electrons. The H—N—H bond angles of 107° are somewhat less than the 109.5° tetrahedral bond angles of methane. This smaller angle can be explained by assuming that the nonbonded electron pair is more spread out (a larger balloon) and pushes the bonding electron pairs closer together.

Water has two bonded pairs and two nonbonded pairs of electrons. The two nonbonded pairs exert an even greater effect on the bonded pairs and reduce the bond angle still farther to 104.5°.

Example 10.14

Describe the expected shape of hydrogen cyanide (HCN) given the fact that there is a single bond between the carbon atom and the hydrogen atom and a triple bond between the carbon atom and the nitrogen atom.

Solution The electronic arrangement of the molecule based on the rules given in Section 10.8 is

H—C≡N:

The central carbon atom has a single bond and a triple bond. The electrons in these two bonds will be separated at a maximum distance when they are at a 180° angle to each other. The molecule is predicted to be linear.

[Additional examples may be found in 10.53–10.56 at the end of the chapter.]

Summary

The Lewis theory of bonding involves the valence electrons of atoms and the tendency of atoms to lose, gain, or share electrons to form a noble gas electron configuration. The basic premise of the Lewis theory is that atoms tend to acquire eight electrons in their valence shell.

When the Lewis theory is used with combinations of metals and nonmetals having sufficiently large differences in electronegativity, it predicts the formation of the correct ions and derived formula units. The metal tends to lose enough electrons to achieve an inert gas configuration of lower atomic weight; the nonmetal tends to gain electrons to achieve an inert gas configuration of higher atomic weight.

Covalently bonded molecules usually can be represented with Lewis structures by sharing valence electrons in bonds. The electrons are distributed about the molecular framework according to a set of rules. The total of the nonbonded electrons and bonded electrons associated with each atom is eight.

For some molecules two or more plausible Lewis structures may be drawn. In those cases the true structure is said to be a resonance hybrid represented as a combination of the Lewis structures.

The valence-shell electron-pair repulsion (VSEPR) theory can be used to predict the shapes of molecules. The shape of a molecule depends on the geometrical distribution of bonding and nonbonded pairs of electrons, which tend to achieve maximum separation from each other. The shape of the molecule is defined in terms of the geometric shape described by the positions of the atoms.

New Terms

An **angular (bent)** molecule is a triatomic molecule with atoms arranged at an angle other than 180°.

Bonding electrons are electrons shared between atoms.

A **coordinate covalent bond** between two atoms is formed by the contribution of a pair of electrons from just one of the atoms.

A **covalent bond** is formed by the sharing of a pair of electrons between two atoms.

A **dipole** is a pair of opposite charges of equal magnitude at a distance from each other.

A **double bond** is formed by the sharing of two pairs of electrons between two atoms.

An **ionic (electrovalent) bond** results from the transfer of electrons from a metal to a nonmetal.

Ionic compounds are neutral collections of oppositely charged ions.

The **ionic radius** is the radius of a spherical ion present in ionic compounds.

The **Lewis octet rule** refers to the eight electrons in the valence shell of an atom or ion.

In a **linear** molecule all the atoms of the molecule are arranged along a common line.

Lone pair electrons are valence shell electrons associated with an atom but not involved in bonding.

Nonbonding electrons are valence shell electrons associated with an atom but not involved in bonding.

A **polar covalent bond** is formed by sharing electrons between two atoms of unequal electronegativity.

Resonance or **contributing structures** are two or more plausible Lewis structures used when no single structure can accurately represent the molecule.

A **tetrahedral** molecule has an atom located in the center of a tetrahedron with four atoms bonded to the central atom located at the corners of the tetrahedron.

In a **trigonal planar** molecule three atoms are arranged around a central atom with all atoms in a common plane and all bond angles at 120°.

In a **trigonal pyramidal** molecule the central atom is bonded to three other atoms so that a three sided pyramid is formed. The three atoms bonded to the central atom are in a common plane.

A **triple bond** is formed by the sharing of three pairs of electrons between two atoms.

An **unshared electron pair** is a pair of valence shell electrons associated with an atom but not involved in bonding.

The **valence-shell electron-pair repulsion (VSEPR) theory** relates the shape of molecules to the distribution of electron pairs about a central atom.

Exercises

Terminology

10.1 Describe the use of the VSEPR theory to predict the geometric shapes of molecules.

10.2 Explain what physical characteristics of compounds lead to the classification of bonds into two types.

Lewis Octet Theory of Ions

10.3 Consider the electron configurations of Group IIA elements. What type of ion should be formed by these elements based on the Lewis octet theory?

10.4 Consider the electron configurations of Group VIA elements. What type of ion should be formed by these elements based on the Lewis octet theory?

10.5 Scandium forms a 3+ ion in ionic compounds such as $ScCl_3$. Explain why this ion is formed.

10.6 Zinc forms a 2+ ion by losing its $4s^2$ electrons rather than any of its $3d^{10}$ electrons. Suggest a reason for this fact.

Electron Configuration of Ions

10.7 Write out the entire electron configuration of each of the following ions.
(a) H^+ (b) Ca^{2+} (c) Li^+
(d) Na^+ (e) K^+ (f) Mg^{2+}

10.8 Write out the entire electron configuration of each of the following ions.
(a) Rb^+ (b) Sr^{2+} (c) Cs^+
(d) Ba^{2+} (e) Sc^{3+} (f) Ti^{4+}

10.9 Write out the entire electron configuration of each of the following ions.
(a) H^- (b) F^- (c) Cl^-
(d) N^{3-} (e) S^{2-} (f) O^{2-}

10.10 Write out the entire electron configuration of each of the following ions.
(a) As^{3-} (b) Br^- (c) P^{3-}
(d) Se^{2-} (e) I^- (f) Te^{2-}

10.11 Write the electron configuration of only the valence shell electrons of each ion in Exercise 10.7.

10.12 Write the electron configuration of only the valence shell electrons of each ion in Exercise 10.8.

10.13 Write the electron configuration of only the valence shell electrons of each ion in Exercise 10.9.

10.14 Write the electron configuration of only the valence shell electrons of each ion in Exercise 10.10.

10.15 Write the electron-dot symbols of the ions in Exercise 10.7.

10.16 Write the electron-dot symbols of the ions in Exercise 10.8.

10.17 Write the electron-dot symbols of the ions in Exercise 10.9.

10.18 Write the electron-dot symbols of the ions in Exercise 10.10.

Ionic Bonds

10.19 Write the formula for the ionic compound that results from the combination of each pair of ions.
(a) Na^+ and Cl^- (b) Ca^{2+} and Cl^-
(c) Mg^{2+} and O^{2-} (d) Mg^{2+} and S^{2-}
(e) Al^{3+} and F^- (f) Na^+ and Se^{2-}

10.20 Write the formula for the ionic compound that results from the combination of each pair of ions.
(a) Ca^{2+} and I^- (b) Na^+ and N^{3-}
(c) K^+ and Br^- (d) Cs^+ and Te^{2-}
(e) Ca^{2+} and N^{3-} (f) Rb^+ and P^{3-}

10.21 Write the formula for the ionic compound that results from the reaction of each pair of atoms.
(a) Li and F   (b) Mg and Br   (c) Li and O
(d) Mg and S   (e) Al and F   (f) Na and Se

10.22 Write the formula for the ionic compound that results from the reaction of each pair of atoms.
(a) Ca and I   (b) Na and N   (c) K and S
(d) Ba and F   (e) Mg and Cl   (f) Cs and P

10.23 Scandium combines with chlorine to form the ionic compound $ScCl_3$. Using the Lewis octet rule, explain this observation.

10.24 Zinc forms a +2 ion. Using the Lewis octet theory, explain this observation.

Ionic Radii

10.25 What generalization can be made about the size of an anion compared to the atom of an element?

10.26 What generalization can be made about the size of a cation compared to the atom of an element?

10.27 Which is larger in each of the following? Explain why.
(a) Mg or Mg^{2+} (b) K or K^+ (c) Al or Al^{3+}
(d) Br or Br^- (e) S or S^{2-} (f) N or N^{3-}

10.28 Which is larger in each of the following? Explain why.
(a) O^{2-} or F^- (b) Na^+ or Mg^{2+} (c) S^{2-} or Cl^-
(d) Al^{3+} or Mg^{2+} (e) Mg^{2+} or Ca^{2+} (f) Br^- or I^-

Covalent Bonds

10.29 Why can only covalent bonds result from bonding of identical atoms?

10.30 Write Lewis structures for the following covalent molecules.
(a) H_2 (b) I_2 (c) F_2
(d) Br_2 (e) Cl_2 (f) N_2

Polar Covalent Bonds

10.31 How can one use electronegativity values to predict the direction of the polarity of a bond?

10.32 How might the order of the polarity of a series of bonds be predicted?

10.33 Indicate which element will be the positive end of the dipole when the following pairs of elements are bonded to each other.
  (a) H and Br   (b) Br and Cl   (c) H and O
  (d) O and F   (e) O and Cl   (f) Si and F

10.34 Indicate which element will be the positive end of the dipole when the following pairs of elements are bonded to each other.
  (a) N and H   (b) N and F   (c) P and Cl
  (d) P and F   (e) I and Br   (f) C and Cl

Coordinate Covalent Bonds

10.35 Locate the coordinate covalent bonds present in each of the following compounds.
  (a) HOCl   (b) $HClO_2$ (c) $HBrO_3$
  (d) HIO_4 (e) H_2SeO_4 (f) H_3PO_4

10.36 Locate the coordinate covalent bonds present in each of the following ions.
  (a) PH_4^+ (b) ClO_3^- (c) BrO_2^-
  (d) ClO_4^- (e) SeO_4^{2-} (f) PO_4^{3-}

Writing Lewis Structures

10.37 Using the Lewis octet theory, describe the Br_2 molecule.

10.38 Using the Lewis octet theory, describe the ICl molecule.

10.39 Write Lewis structures for the following covalent molecules.
  (a) H_2S (b) PH_3 (c) CF_4
  (d) H_2Se (e) CBr_4 (f) $SiCl_4$

10.40 Write Lewis structures for the following covalent molecules.
  (a) SiH_4 (b) PCl_3 (c) SbH_3
  (d) H_2Te (e) NF_3 (f) $GeBr_4$

10.41 Write Lewis structures for the following ions.
  (a) SH^- (b) PH_4^+ (c) H_3O^+
  (d) CN^- (e) SO_4^{2-} (f) NH_4^+

10.42 Write Lewis structures for the following ions.
  (a) OCl^- (b) SO_3^{2-} (c) BrO_3^-
  (d) IO_3^- (e) ClO_4^- (f) NH_2^-

10.43 Based on locations in the periodic table, predict the compound that will be formed between antimony and hydrogen. Draw the Lewis structure.

10.44 Based on locations in the periodic table, predict the compound that will be formed between silicon and chlorine.

10.45 The geometric arrangement of atoms in several molecules is given below. Using dashes to represent bonds or shared electron pairs and dots for lone pair electrons, draw Lewis structures for the molecules.

```
        F              H            H  H
(a)  F  C  F   (b)  H  C  O  (c) H  C  N  H
        F              H  H           H

    Cl                         H            O  H
(d)    C  O    (e) H  C  N  (f)    C  N
    Cl                      H
```

10.46 The geometric arrangement of atoms in several molecules is given below. Using dashes to represent bonds or shared electron pairs and dots for lone pair electrons, draw Lewis structures for the molecules.

```
        F                H              H
(a)  Cl  C  Cl   (b)  H  C  S   (c) H  C  P  H
       Br                              H  H

       O                  H              O
(d) H  N  C  N  H  (e) H  C  N  H  (f) H  C
       H  H                            O  H
```

Resonance

10.47 The nitrate ion has identical bonds between each of the three oxygen atoms and the central nitrogen atom. Draw possible Lewis structures for this ion.

10.48 The nitrite ion has identical bonds between each of the two oxygen atoms and the central nitrogen atom. Draw possible Lewis structures for this ion.

10.49 The carbonate ion has identical bonds between each of the three oxygen atoms and the central carbon atom. Draw possible Lewis structures for this ion.

10.50 Sulfur trioxide is a planar molecule with each of the three oxygen atoms bonded to the central sulfur atom by equivalent bonds. Draw possible Lewis structures for this molecule.

Polarity of Molecules

10.51 Indicate whether you expect the following molecules to be polar or nonpolar.
  (a) H_2S (b) PH_3 (c) CF_4
  (d) H_2Se (e) CBr_4 (f) $SiCl_4$

10.52 Indicate whether you expect the following molecules to be polar or nonpolar.
  (a) SiH_4 (b) PCl_3 (c) SbH_3
  (d) H_2Te (e) NF_3 (f) $GeBr_4$

Molecular Shapes

10.53 What types of bond angles are present in each of the compounds listed below? What is the geometry of each molecule?

(a) SiH_4 (b) CCl_4 (c) $SiCl_4$

(d) NF_3 (e) HOCl   (f) OF_2

10.55 Predict the shapes of the following ions according to the VSEPR theory.

(a) SO_4^{2-} (b) CO_3^{2-} (c) NH_4^+

(d) H_3O^+ (e) ClO_4^- (f) NH_2^-

10.54 What types of bond angles are present in each of the compounds listed below? What is the geometry of each molecule?

(a) NCl_3 (b) CF_4 (c) GeF_4

(d) PF_3 (e) HOBr   (f) OCl_2

10.56 Predict the shape of each of the following by the VSEPR theory.

(a) SO_3^{2-} (b) ClO_2^- (c) ClO_3^-

(d) $GeCl_3^-$ (e) PO_4^{3-} (f) BrO_3^-

Additional Exercises

10.57 What group in the periodic table gave Lewis the idea that an octet of electrons is achieved when ions are formed?

10.59 Determine the number of unshared pairs of valence electrons present in each of the following elements.

(a) H_2 (b) I_2 (c) F_2

(d) Br_2 (e) Cl_2 (f) N_2

10.61 Tin forms compounds similar to those of carbon because both elements are in Group IVA. What compound should result from tin and hydrogen? Draw the Lewis structure.

10.63 Indicate whether you expect the following molecules to be polar or nonpolar.

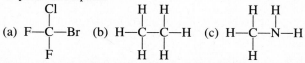

10.65 The molecule BF_3 does not have a nonbonded electron pair on boron. Draw a Lewis structure of this molecule. Based on this structure, explain why all four atoms are in a single plane. What is the expected F—B—F bond angle?

10.67 Predict the shape of ethylene in terms of the VSEPR theory (see Section 10.3).

10.58 Explain why carbon doesn't form C^{4+} or C^{4-} ions.

10.60 How does a covalent coordinate bond differ from a covalent bond?

10.62 Ozone, O_3, is an angular molecule with identical bonds from each terminal oxygen atom. Draw possible Lewis structures for this substance. What are the limitations of these structures?

10.64 What types of bond angles are present in each of the compounds listed below? What is the geometry of each molecule?

(a) CF_4 (b) NF_3 (c) OF_2

(d) HOCl   (e) GeH_4 (f) SbH_3

10.66 Draw a Lewis structure for $BeCl_2$. The molecule does not have any nonbonded electron pairs on beryllium. Would this molecule be expected to be linear or angular?

10.68 Predict the shape of each of the following about the central carbon atom. Note that nonbonded electron pairs are not indicated for any atom.

(a) F—C—F with Cl above and Cl below   (b) H—C—O with H H below   (c) H—C—N—H with H, H, H

(d) Cl, Cl, C=O   (e) H—C≡N   (f) H—C=O with O—H

11 | Nomenclature of Inorganic Compounds

Learning Objectives

After studying Chapter 11 you should be able to

1. Determine the oxidation number of an element in a compound or ion.

2. Give the names of binary compounds of metals and nonmetals and indicate the oxidation state of the metal where necessary.

3. Give the names of binary compounds of nonmetals, indicating the number of each type of atom in the compound.

4. Give the names of ternary compounds containing polyatomic ions.

5. Give the names of acids and bases.

6. Give the formula of a salt formed from a given acid and base.

7. Recognize mixed salts, acidic salts, and basic salts.

11.1 Common and Systematic Names

Nomenclature means naming, and the subject of this chapter is how to name compounds. The chapter is restricted to **inorganic compounds;** that is, compounds that (with a few exceptions) do not contain carbon. The nomenclature of organic compounds is covered in Chapters 19 and 20.

The arbitrary names of compounds that are unrelated to chemical composition are called their **common names.** Water is an example of a common name; it does not provide any information about either the type or number of atoms contained in the molecule. If every known substance were assigned a common name, we would have to learn an immense number of one-to-one relationships between a name and a substance. Since the number of known substances is currently about 7 million, it would be impossible to cope with such a system of naming compounds.

Common chemical names have distinct limitations, but they are prevalent in chemistry. Many compounds were named early in chemical history, before their composition was understood. Even after the composition of these compounds was established, their common names lingered on as part of the language of chemistry. Both table sugar and sucrose are used to describe the common sweetener. However, the systematic name that provides details of the composition of sucrose is α-D-glucopyranosyl-β-D-fructofuranoside.

Common names are often used by industry and by nonchemical specialists because many systematic chemical names are too long or technical to use conveniently. In addition, alternative colloquial terms are used such as ''hypo,'' a compound used in photographic development, which is sodium thiosulfate. Abbreviations using only a few letters are used often because the systematic names may be too long to say easily. For example, the compound 1,1,1-trichloro-2,2-bis(*p*-chlorophenyl)ethane is known by the abbreviation DDT. A list of some substances in common use along with their common names is given in Table 11.1.

Table 11.1   **Common Substances and Common Names**

Name	Formula
alumina	Al_2O_3
lime	CaO
limestone	$CaCO_3$
slaked lime	$Ca(OH)_2$
gypsum	$CaSO_4 \cdot 2H_2O$
plaster of paris	$CaSO_4 \cdot \frac{1}{2}H_2O$
muriatic acid	HCl
water	H_2O
oil of vitriol	H_2SO_4
potash	K_2CO_3
milk of magnesia	$Mg(OH)_2$
epsom salts	$MgSO_4 \cdot 7H_2O$
ammonia	NH_3
sal ammoniac	NH_4Cl
salt	$NaCl$
lye	$NaOH$
baking soda	$NaHCO_3$
hypo	$Na_2S_2O_3$

Despite some advantages of common names, chemists find them unsatisfactory. Chemists prefer **systematic names,** which contain information about the composition of each substance. With a systematic name, the reader learns about the composition of the substance by merely glancing at the words even if the reader is encountering the name for the first time.

The Commission on the Nomenclature of Inorganic Chemistry of the International Union of Pure and Applied Chemistry (IUPAC) first met in 1921 and continues to meet constantly to update its work. This commission of IUPAC devises names for inorganic chemicals to be used internationally.

The systematic names of inorganic compounds are established by a set of rules so that every substance can be uniquely named from its formula and each name provides information about the substance's composition. In the systematic name, the positive portion is named first. This may be a metal cation (such as Na^+), a positive polyatomic ion (such as ammonium, NH_4^+), or the least electronegative of the several nonmetals that may be present in the substance. The more negative portion is then named second. The negative portion may be a simple anion (such as Cl^-), a negatively charged polyatomic ion (such as nitrate, NO_3^-), or the more electronegative element.

Since compounds are formed by the loss, gain, or sharing of electrons between atoms, information about electronic redistribution in compounds is often given in systematic names. This information, given as oxidation numbers, forms part of the rules of nomenclature. Therefore we will discuss oxidation numbers in the next section.

In order to present the rules of nomenclature, the compounds will be divided into classes consisting of binary (two elements) compounds, ternary (three elements) compounds, acids, bases, and salts. Each class of compounds will be discussed in separate sections.

11.2 Oxidation Numbers

The loss or gain of electrons in ionic compounds or the partial gain or loss of electrons by atoms in polar covalent bonds is indicated in terms of a bookkeeping system that uses oxidation numbers. The **oxidation number** of an atom is a positive or negative integer assigned to describe an element as a free atom, an ion, or as part of a polyatomic ion or molecule. The rules for the assignment of oxidation numbers are listed in Table 11.2.

As a reference point, the oxidation number of an element, regardless of whether it is monatomic, diatomic, and so on, is set at zero. The covalent bond that exists in elements such as H_2, F_2, P_4, and S_8 is the result of equal sharing of electrons. No atom that has a covalent bond with an identical atom gains or loses electrons.

For monoatomic ions in an ionic compound, the oxidation number is equal to the charge of the cation or the anion that results from the transfer of electrons away from or to a neutral atom. When an electron is removed from an atom as with Na to yield Na^+, the oxidation number of the cation is $+1$. When an electron is gained as in the case of Cl to yield Cl^-, the oxidation number is -1. These oxidation numbers provide information about the compounds that may be formed. One Na^+ with an oxidation number of $+1$ combines with one Cl^- with an oxidation number of -1. In any ionic compound, the sum of the oxidation numbers must be numerically equal to zero, since the compound is electrically neutral overall.

The sharing of electrons in polar covalent bonds is unequal. As a result, one atom gains a partial negative charge at the expense of another atom, which then has a partial positive charge. By convention, the oxidation number of atoms bonded by polar covalent bonds is based on the arbitrary assignment of the electrons in the bond to the more electronegative atom. Thus, some

Table 11.2   Rules for Assignment of Oxidation Number

1. The oxidation number of an element in its uncombined state is zero.

2. The algebraic sum of the oxidation numbers in an ionic or covalent compound is zero.

3. The oxidation number of a monoatomic ion is the same as the charge of the ion.

4. The algebraic sum of the oxidation numbers in a polyatomic ion equals the charge of the ion.

5. Metals generally have positive oxidation numbers in the combined state.

6. Negative oxidation numbers in covalent compounds of two unlike atoms are assigned to the more electronegative atom.

7. Most hydrogen compounds contain hydrogen with a +1 oxidation number.

8. In most oxygen compounds, the oxidation number is −2.

9. The oxidation number of fluorine, chlorine, bromine, and iodine is −1 except when combined with a more electronegative element.

10. Sulfides have an oxidation number of −2.

---

atoms are viewed as having lost electrons and others as having gained electrons, although no electron transfer actually has occurred.

The most electronegative element in a polar covalent bond is arbitrarily assigned the negative oxidation number because it has the greatest attraction for the electrons in the bond. In carbon tetrachloride (CCl_4) there are four polar covalent bonds between the central carbon atom and the individual chlorine atoms. Each chlorine atom is assigned a −1 oxidation number because chlorine is more electronegative than carbon. Each chlorine atom has a tendency to partially gain an electron from the carbon atom in each polar covalent bond. Carbon is assigned a +4 oxidation number because it has a tendency to partially lose four electrons in bonding to the four chlorine atoms. Since the number of electrons partially gained or lost in polar covalent bonds must be equal, the sum of the oxidation numbers of atoms in a molecule must equal zero.

Example 11.1

What is the oxidation number of sulfur in SO_3?

Solution   The sum of the oxidation numbers of a neutral compound must be zero according to rule 2 of Table 11.2. Abbreviating oxidation number as ox # we write

ox # of sulfur + 3(ox # of oxygen) = 0

Since the oxidation number of oxygen is −2 according to rule 8 of Table 11.2, the oxidation number of sulfur may be calculated algebraically.

ox # of sulfur +3(−2) = 0

ox # of sulfur −6 = 0

ox # of sulfur = +6

Example 11.2

What is the oxidation number of nitrogen in $NaNO_2$?

Solution The sum of the oxidation numbers of sodium, nitrogen, and oxygen must be equal to zero (rule 2 of Table 11.2).

$$ox\ \#\ of\ sodium + ox\ \#\ of\ nitrogen + 2(ox\ \#\ of\ oxygen) = 0$$

The oxidation number of oxygen is -2 (rule 8). The oxidation number of the sodium ion is positive since it is a metal (rule 5). Sodium is known to exist only as the $+1$ ion, and its oxidation number is equal to that charge (rule 3).

$$(+1) + (ox\ \#\ of\ nitrogen) + 2(-2) = 0$$
$$ox\ \#\ of\ nitrogen +1 - 4 = 0$$
$$ox\ \#\ of\ nitrogen = +3$$

[*Additional examples may be found in 11.21–11.26 at the end of the chapter.*]

Example 11.3

What is the oxidation number of manganese in the permanganate ion, MnO_4^-?

Solution The sum of the oxidation numbers of a polyatomic ion is equal to the charge of the ion (rule 4 of Table 11.2).

$$ox\ \#\ of\ manganese + 4(ox\ \#\ of\ oxygen) = -1$$

Since the oxidation number of oxygen is -2 (rule 8), the oxidation number of manganese can be calculated.

$$ox\ \#\ of\ manganese + 4(-2) = -1$$
$$ox\ \#\ of\ manganese = -1 + 8 = +7$$

Example 11.4

What is the oxidation number of carbon in the oxalate ion, $C_2O_4^{2-}$?

Solution The sum of the oxidation numbers of a polyatomic ion is equal to the charge of the ion (rule 4 of Table 11.2).

$$2(ox\ \#\ of\ carbon) + 4(ox\ \#\ of\ oxygen) = -2$$

Since the oxidation number of oxygen is -2 (rule 8), the oxidation number of carbon can be calculated.

$$2(ox\ \#\ of\ carbon) + 4(-2) = -2$$
$$2(ox\ \#\ of\ carbon) = -2 + 8$$
$$(ox\ \#\ of\ carbon) = \frac{-2 + 8}{2} = +3$$

Note that in this calculation the oxidation number of an individual carbon atom in the ion is determined.

[Additional examples may be found in 11.27 and 11.28 at the end of the chapter.]

11.3 Binary Compounds Containing Metals of Fixed Oxidation Number

Binary compounds consist of two elements. A number of metals that form binary compounds with nonmetals have only one oxidation number. For example, sodium has an oxidation number of +1 in NaCl, and calcium has an oxidation number of +2 in CaO. Since this information becomes general knowledge as one studies chemistry, there is no need to indicate the oxidation number in the name. In naming a binary compound of a metal and a nonmetal, the metal is named first and the name of the nonmetal with the ending *-ide* follows. Examples of the names of binary compounds containing a metal are listed in Table 11.3. The names of the negative ions derived from the nonmetals are listed in Table 11.4. Note that compounds may contain more than one atom of the same element, but as long as they contain only two elements, they are considered binary. The number of atoms of a given element is not indicated in the name.

The majority of the metals having fixed oxidation numbers are members of Group IA and Group IIA, and the compounds that they form with nonmetals are ionic. Both groups of elements are electropositive; those of Group IA have a +1 oxidation number, whereas those of Group IIA have a +2 oxidation number. Binary compounds of hydrogen with metals of Group IA and Group IIA contain hydrogen as the hydride ion (H^-) and are also ionic compounds.

Table 11.3   Binary Compounds of Metals with Single Oxidation Numbers

Formula	Name	Oxidation number of the metal
NaCl	sodium chloride	+1
Na_2O	sodium oxide	+1
KBr	potassium bromide	+1
K_2S	potassium sulfide	+1
AgI	silver iodide	+1
Ag_2O	silver oxide	+1
MgSe	magnesium selenide	+2
Mg_3P_2	magnesium phosphide	+2
CaF_2	calcium fluoride	+2
CaH_2	calcium hydride	+2
CaTe	calcium telluride	+2
$AlCl_3$	aluminum chloride	+3
Al_2S_3	aluminum sulfide	+3

Table 11.4 **Names of Common Anions**

Element	Anion	Stem name	Anion name
Br	Br^-	brom	bromide
Cl	Cl^-	chlor	chloride
F	F^-	fluor	fluoride
I	I^-	iod	iodide
H	H^-	hydr	hydride
O	O^{2-}	ox	oxide
S	S^{2-}	sulf	sulfide
Se	Se^{2-}	selen	selenide
Te	Te^{2-}	tellur	telluride
N	N^{3-}	nitr	nitride
P	P^{3-}	phosph	phosphide
As	As^{3-}	arsen	arsenide

Writing the name of a binary compound of a metal cation with a single oxidation number and a nonmetal anion is fairly easy. Writing the formula from a name is somewhat more demanding, as you must know the oxidation numbers of the metal cation and the nonmetal anion. Only then can you determine the number of each ion required to produce a neutral compound. Thus calcium, which exists as Ca^{2+}, requires two bromide ions, which exist as Br^-. Calcium bromide must then be $CaBr_2$. If you do not know the oxidation numbers of the ions, you cannot write the correct formula. Examples of writing formulas from chemical names are given in Table 11.5. In each case, you should verify that the formula is correct. Note that the sum of the number of cations multiplied by their oxidation number and the number of anions multiplied by their oxidation number must equal zero. For $CaBr_2$, the relationship is

$$\text{(number of } Ca^{2+})(\text{ox \# of } Ca^{2+}) + \text{(number of } Br^-)(\text{ox \# of } Br^-) = 0$$

$$(1)(+2) + (2)(-1) = 0$$

Table 11.5 **Names and Formulas for Selected Binary Compounds**

Name	Cation	Anion	Formula
lithium fluoride	Li^+	F^-	LiF
potassium chloride	K^+	Cl^-	KCl
calcium bromide	Ca^{2+}	Br^-	$CaBr_2$
barium iodide	Ba^{2+}	I^-	BaI_2
aluminum bromide	Al^{3+}	Br^-	$AlBr_3$
sodium selenide	Na^+	Se^{2-}	Na_2Se
barium sulfide	Ba^{2+}	S^{2-}	BaS
aluminum oxide	Al^{3+}	O^{2-}	Al_2O_3
strontium telluride	Sr^{2+}	Te^{2-}	SrTe
sodium nitride	Na^+	N^{3-}	Na_3N
strontium phosphide	Sr^{2+}	P^{3-}	Sr_3P_2
magnesium nitride	Mg^{2+}	N^{3-}	Mg_3N_2
aluminum nitride	Al^{3+}	N^{3-}	AlN
sodium hydride	Na^+	H^-	NaH
calcium hydride	Ca^{2+}	H^-	CaH_2

Example 11.5

Name the compound Ca_3P_2 and determine the oxidation number of the phosphorus based on the oxidation number of the calcium ion ($+2$).

Solution The formula contains only two elements and represents a binary compound. Calcium is the least electronegative element and is written first in the name as calcium. The more electronegative element is phosphorus and is written as phosphide. Thus the name of Ca_3P_2 is calcium phosphide.

The calcium ion has a $+2$ charge and its oxidation number is $+2$. For the three calcium ions represented in the formula, the total positive charge is $3(+2) = +6$. The total negative charge carried by the phosphide ions must be -6. Since there are two phosphide ions, the charge of a single phosphide ion is $(-6)/2 = -3$. Therefore, the oxidation number of phosphorus is -3.

[*Additional examples may be found in 11.29 and 11.30 at the end of the chapter.*]

Table 11.6 Common Oxidation Numbers and the Periodic Table

Group number	Common oxidation number
IA	+1
IIA	+2
IIIA	+3
IVA	+4
	or
	−4
VA	−3
VIA	−2
VIIA	−1

The oxidation numbers of the metals of the representative elements are predictable from the group number of the periodic table (Table 11.6). Remember that the group number of the representative elements indicates the number of electrons in the valence shell. For the metals, the loss of these electrons results in an ion whose positive charge corresponds to the group number. That positive charge is equal to the oxidation number. The nonmetals are electronegative and tend to gain electrons to form anions. The charge of the anion is derived by subtracting 8 from the group number. The oxidation number of the anion is equal to its negative charge.

11.4 Binary Compounds Containing Metals with Several Oxidation Numbers

There are many more metals that form multiple cations than there are that form single cations. Iron may exist as Fe^{2+} and Fe^{3+} as in $FeCl_2$ and $FeCl_3$, respectively; titanium may exist with oxidation numbers of $+3$ or $+4$ as in $TiCl_3$ and $TiCl_4$, respectively. In naming compounds containing such metals, the oxidation number of the metal must be indicated because it dictates the number of anions with which the metal combines.

Two systems are in concurrent use to name compounds in this category. In the older system, metals that have two common oxidation numbers are designated by modifying a stem of the metal name with different suffixes. The suffix **-ous** designates the lower oxidation state and the suffix **-ic** the higher oxidation state. Thus, the chloride of titanium with the formula $TiCl_3$ is titanous chloride, whereas $TiCl_4$ is titanic chloride. The stem name of some metals are derived from the name used to form the chemical symbol. Thus the stems cupr-, ferr-, and stann- are derived from cuprum (Cu), ferrum (Fe), and stannum (Sn), respectively. The compound $FeCl_2$ is ferrous chloride, whereas $FeCl_3$ is ferric chloride. Examples of the usage of the stem and the *-ous* and *-ic* suffixes are given in Table 11.7.

The IUPAC recommends the use of the newer **Stock system** of nomenclature. In this system the oxidation state of the metal is indicated by a Roman numeral in parentheses immediately following the name of the metal. The two chlorides of iron are iron(II) chloride and iron(III)

Table 11.7   Names of Ions of Metals with Two Oxidation States

Ion	Old name	Stock name
Cu^+ or Cu_2^{2+}	cuprous	copper(I)
Cu^{2+}	cupric	copper(II)
Fe^{2+}	ferrous	iron(II)
Fe^{3+}	ferric	iron(III)
Mn^{2+}	manganous	manganese(II)
Mn^{3+}	manganic	manganese(III)
Hg_2^{2+}	mercurous	mercury(I)
Hg^{2+}	mercuric	mercury(II)
Pb^{2+}	plumbous	lead(II)
Pb^{4+}	plumbic	lead(IV)
Sn^{2+}	stannous	tin(II)
Sn^{4+}	stannic	tin(IV)
Ti^{3+}	titanous	titanium(III)
Ti^{4+}	titanic	titanium(IV)

chloride. The Roman numeral does not indicate the number of units of the nonmetal combined with the metal, but only the oxidation state of the metal. The oxides of iron, FeO and Fe_2O_3, are iron(II) oxide and iron(III) oxide, respectively. Examples of both the older name and the Stock name are given in Table 11.8 for a number of compounds. Note that the endings *-ous* and *-ic* do not indicate to the reader the oxidation state of the metal. That the term *-ous* means +1 for copper and mercury but +2 for iron or tin must be memorized. Similarly, you must know that the term *-ic* means +2 for copper, +3 for iron, and +4 for tin. The Stock system gives you this information directly.

Table 11.8   Common and Systematic Names of Compounds of Metals with Two Oxidation States

Formula	Older name	Stock name
FeO	ferrous oxide	iron(II) oxide
Fe_2O_3	ferric oxide	iron(III) oxide
$SnCl_2$	stannous chloride	tin(II) chloride
$SnCl_4$	stannic chloride	tin(IV) chloride
Hg_2Cl_2	mercurous chloride	mercury(I) chloride
$HgCl_2$	mercuric chloride	mercury(II) chloride
Hg_2O	mercurous oxide	mercury(I) oxide
HgO	mercuric oxide	mercury(II) oxide
Cu_2S	cuprous sulfide	copper(I) sulfide
$CuBr_2$	cupric bromide	copper(II) bromide
Cu_3P_2	cupric phosphide	copper(II) phosphide
PbF_2	plumbous fluoride	lead(II) fluoride
PbO_2	plumbic oxide	lead(IV) oxide

Example 11.6

What is the formula of tin(IV) oxide? What is the alternate older name?

Solution   The Roman numeral IV following the name of the metal tin indicates the oxidation number of tin is +4 in the compound. Since the oxide ion has a −2 charge, there must be two oxide ions to give a total −4 charge to balance the charge of the tin. Thus, the formula of tin(IV) oxide is SnO_2. The tin is in the higher of its two oxidation states, and the alternate name of SnO_2 is stannic oxide.

Example 11.7

What is the formula of bismuth(III) sulfide?

Solution   The Roman numeral III following bismuth indicates that the oxidation number of bismuth is +3. The oxidation number of the sulfide ion is −2. For each bismuth ion there are required 1.5 sulfide ions in order for the sum of the oxidation numbers to be zero. However, the formula $BiS_{1.5}$ is not acceptable and must be multiplied by two to give the correct formula Bi_2S_3. The total of the oxidation numbers of the two bismuth ions is +6, which balances the total of −6 for the sum of the oxidation numbers of the three sulfide ions.

[*Additional examples may be found in 11.31–11.34 and 11.39–11.42 at the end of the chapter.*]

11.5   Binary Compounds of Nonmetals

With only a few exceptions, the names of binary compounds containing two nonmetals consist of the names of the two elements with the element listed second having an **-ide** ending. In binary compounds of two nonmetals, there are no positive or negative ions because the compounds are covalent substances. The preferred sequence established by IUPAC for selecting which element is named first is B > Si > C > P > N > H > S > I > Br > Cl > O > F for the common elements you will encounter. The element to the left is named first when combined with an element to the right in the series. For example, in a compound of sulfur and oxygen, sulfur would preceed oxygen in the name.

The number of atoms of each element in a molecule of the compound is indicated by a Greek prefix preceding that element. The prefixes *mono-, di-, tri-, tetra-, penta-, hexa-, hepta-, octa-, nona-,* and *deca-* are used for the numbers one through ten. The prefix *mono-* is only used if an ambiguous name would otherwise result. Examples of these names are given in Table 11.9.

Example 11.8

Assign the name for the compound represented by BrF_5.

Solution   Bromine precedes fluorine in the preferred elemental list and is named first. The five fluorine atoms must be indicated by penta-. Thus the systematic name is bromine pentafluoride.

[*Additional examples may be found in 11.35 and 11.36 at the end of the chapter.*]

Table 11.9   **Names of Binary Compounds of Nonmetals**

Name	Formula	Name	Formula
carbon monoxide	CO	carbon tetrachloride	CCl_4
carbon dioxide	CO_2	nitrogen oxide	NO
sulfur dioxide	SO_2	dinitrogen oxide	N_2O
sulfur trioxide	SO_3	dinitrogen trioxide	N_2O_3
disulfur dichloride	S_2Cl_2	dinitrogen pentoxide	N_2O_5
sulfur tetrafluoride	SF_4	iodine monochloride	ICl
sulfur hexafluoride	SF_6	iodine trichloride	ICl_3
phosphorus tribromide	PBr_3	iodine pentachloride	ICl_5
phosphorus pentachloride	PCl_5	diiodine heptoxide	I_2O_7

11.6   Ternary Compounds

Ternary compounds contain three elements. These compounds most frequently consist of a metal ion or hydrogen combined with a negatively charged polyatomic containing two elements. Many of these polyatomic ions contain oxygen (Table 11.10). In order to name ternary compounds it is necessary to know the names of the polyatomic ions.

The polyatomic ions containing oxygen commonly have the suffixes *-ate* and *-ite*. The suffix does not indicate the number of oxygen atoms in the polyatomic ion, but the ion with the -ate suffix contains more oxygen than that with the -ite suffix. Thus nitrate and nitrite are NO_3^- and NO_2^-, respectively; sulfate and sulfite are SO_4^{2-} and SO_3^{2-}, respectively.

Some elements form more than two polyatomic ions with oxygen. Therefore, more endings than *-ate* and *-ite* are required. For example, chlorine and oxygen form four different polyatomic ions. Table 11.10 lists perchlorate (ClO_4^-), chlorate (ClO_3^-), chlorite (ClO_2^-), and hypochlorite (ClO^-). The prefix per- is used to mean over or larger. Thus, perchlorate has one more oxygen atom than the number of oxygen atoms in chlorate. The prefix hypo- means under or smaller. Thus hypochlorite has one less oxygen atom than the chlorite ion. These prefixes are also used with other similar ions such as hypoiodite for IO^-, and perbromate for BrO_4^-.

Only two, commonly occurring, negatively charged polyatomic ions do not use the *-ate* or *-ite* ending. These are hydroxide for OH^- and cyanide for CN^-, and they use the *-ide* ending, which is the ending used for the negative ion of a binary compound.

The ammonium ion (NH_4^+) is the most frequently encountered positive polyatomic ion. A closely related ion, PH_4^+, is called the phosphonium ion, but is less frequently found in compounds. The hydronium ion (H_3O^+) is present in aqueous solutions of acids (Chapter 16).

The procedure for naming ternary compounds is the same as for binary compounds. The only difference is that the name of the negative polyatomic ion replaces the name of the monatomic negative ion. Either the older system or the Stock convention may be used for the positive part of the compound. Several examples of naming ternary compounds are listed in Table 11.11. You should verify that the formulas are correct. Thus, copper(II) phosphate contains three double charged copper ions, which are electrically balanced by two triply charged phosphate ions.

Table 11.10
Polyatomic Ions

Formula	Name
NO_3^-	nitrate
NO_2^-	nitrite
SO_4^{2-}	sulfate
SO_3^{2-}	sulfite
$S_2O_3^{2-}$	thiosulfate
PO_4^{3-}	phosphate
CO_3^{2-}	carbonate
OH^-	hydroxide
CN^-	cyanide
CrO_4^{2-}	chromate
$Cr_2O_7^{2-}$	dichromate
MnO_4^-	permanganate
ClO_4^-	perchlorate
ClO_3^-	chlorate
ClO_2^-	chlorite
ClO^-	hypochlorite
NH_4^+	ammonium
PH_4^+	phosphonium
H_3O^+	hydronium

Table 11.11   **Names of Compounds Containing Polyatomic Ions**

Formula	Name	Formula	Name
KNO_2	potassium nitrite	$CaCr_2O_7$	calcium dichromate
$Ca(NO_3)_2$	calcium nitrate	$Al_2(SO_4)_3$	aluminum sulfate
$LiClO$	lithium hypochlorite	$BaSO_3$	barium sulfite
$NaClO_2$	sodium chlorite	$Na_2S_2O_3$	sodium thiosulfate
$Ba(ClO_3)_2$	barium chlorate	$CsOH$	cesium hydroxide
$AgClO_4$	silver perchlorate	KCN	potassium cyanide
$Na_3(PO_4)_2$	sodium phosphate	$Mg(OH)_2$	magnesium hydroxide
$Cu_3(PO_4)_2$	copper(II) phosphate	$Al(OH)_3$	aluminum hydroxide
$KMnO_4$	potassium permanganate	NH_4Cl	ammonium chloride
$PbCO_3$	lead(II) carbonate	$(NH_4)_2CO_3$	ammonium carbonate
$CuCN$	copper(I) cyanide	$(NH_4)_3PO_4$	ammonium phosphate

Example 11.9

What is the name of the compound represented by $Ba(NO_2)_2$?

Solution   The elemental symbol Ba represents the element barium, which always has an oxidation number of +2 in compounds. It is not necessary or permissible to indicate the oxidation number of elements with single values of oxidation numbers when writing a name. The complex ion NO_2^- represented within parentheses is called nitrite. The subscript 2 following the parenthesis represents the two nitrite ions that are necessary to balance the +2 charge of the barium ion. The name of the compound is barium nitrite.

Example 11.10

Write the formula for gallium(III) sulfate.

Solution   Although you may not have encountered the element gallium in learning the elements and their symbols, you can look up the symbol and find that it is Ga. The Roman number III following Ga indicates that the element has a +3 oxidation number in the compound. The polyatomic sulfate ion, SO_4^{2-}, has a −2 charge. In order to balance the charges of the cations and anions it is necessary to have two gallium ions and three sulfate ions. The formula is $Ga_2(SO_4)_3$.

[*Additional examples may be found in 11.41 and 11.42 at the end of the chapter.*]

Example 11.11

What is the oxidation number of polonium in $Po(CO_3)_2$?

Solution   The carbonate ion, CO_3^{2-}, contained in the compound has a −2 charge. In order for the compound to be electrically neutral the polonium ion must have a charge that balances the 2(−2) of the two carbonate ions. The oxidation number must be +4.

[*Additional examples may be found in 11.45–11.48 at the end of the chapter.*]

11.7 Acids

Hydrogen is less electropositive than many metals but is more electropositive than the simple anions and negative polyatomic ions. Compounds in which hydrogen is the more electropositive element are called acids. These compounds were briefly discussed in Section 6.9. The properties of acids and bonding characteristics of acids are completely different than those of compounds containing metals. For example, HCl is a gaseous compound containing a covalent bond, whereas NaCl is an ionic solid. We discuss acids in Chapter 16.

In the gaseous or liquid state, the acids are covalent compounds and are named as hydrogen compounds. For the binary acids, HCl is hydrogen chloride and H_2S is hydrogen sulfide (Table 11.12). In water, the acids undergo a chemical reaction called ionization. The compounds separate into positive and negative ions as the hydrogen atom gives up its electron to the more negative element. The H^+ ion combines with water to form the hydronium ion, H_3O^+.

$$HCl + H_2O \longrightarrow Cl^- + H_3O^+$$

Because the chemical identity of acids in aqueous solution is different than that of the pure acid, different names are used for aqueous acids. In water solution, the binary acids are named by the prefix **hydro-** followed by the anion name in which -*ide* has been replaced by -*ic* with the word *acid* after it. Thus hydrogen chloride in water is hydrochloric acid. Other examples are given in Table 11.12.

Most ternary acids are commercially available only in aqueous solutions. Thus the compounds are most commonly known by the name of the aqueous solution. Only the name of the polyatomic negative ion is used, and the suffixes -*ate* and -*ite* are replaced by -*ic acid* and -*ous acid,* respectively. The name does not include the *hydro-* prefix and all reference to hydrogen is dropped. Thus, HNO_3 and HNO_2 in aqueous solution are nitric acid and nitrous acid, respec-

Table 11.12 **Names of Acids**

Formula	Name	Name in aqueous solution
	BINARY ACIDS	
HF	hydrogen fluoride	hydrofluoric acid
HCl	hydrogen chloride	hydrochloric acid
HBr	hydrogen bromide	hydrobromic acid
HI	hydrogen iodide	hydroiodic acid
H_2S	hydrogen sulfide	hydrosulfuric acid
	TERNARY ACIDS	
HNO_3	hydrogen nitrate	nitric acid
HNO_2	hydrogen nitrite	nitrous acid
H_2SO_4	hydrogen sulfate	sulfuric acid
H_2SO_3	hydrogen sulfite	sulfurous acid
H_3PO_3	hydrogen phosphite	phosphorous acid
H_3PO_4	hydrogen phosphate	phosphoric acid
$HClO_4$	hydrogen perchlorate	perchloric acid
$HClO_3$	hydrogen chlorate	chloric acid
$HClO_2$	hydrogen chlorite	chlorous acid
$HClO$	hydrogen hypochlorite	hypochlorous acid

tively. For acids containing elements where more than two polyatomic ions are known, the prefixes *per-* and *hypo-* are used. Examples of the names of ternary acids are listed in Table 11.12.

Example 11.12

What is the name of HBrO?

Solution   The compound is ternary and contains hydrogen. The related anion, obtained by removing H^+, is OBr^-, and is called hypobromite by analogy with the name hypochlorite used for OCl^-. By replacing the *-ite* with *-ous,* the acid is named hypobromous acid.

[*Additional examples may be found in 11.51 and 11.52 at the end of the chapter.*]

11.8   Bases

Inorganic **bases** contain the hydroxide ion in combination with a metal ion. The properties of these compounds, first presented in Section 6.9, are discussed in detail in Chapter 16. Although these bases are not binary compounds, they are named using the *-ide* ending based on the name for the hydroxide ion (OH^-). Several examples of bases are listed in Table 11.13. In cases where the metal can exist in a variety of oxidation states, the name must indicate the oxidation state of the metal in that compound.

Example 11.13

Classify and name the compound $Fe(OH)_2$.

Solution   The compound contains the hydroxide ion and is recognized as a base. Iron can exist in two common oxidation states. Since the hydroxide ion has a -1 charge, the iron must have a $+2$ charge to balance the two hydroxide ions. Iron with a $+2$ charge is called ferrous or iron(II). The name is either ferrous hydroxide or iron(II) hydroxide.

[*Additional examples may be found in 11.49 and 11.50 at the end of the chapter.*]

Table 11.13   **Names of Bases**

Formula	Name
LiOH	lithium hydroxide
NaOH	sodium hydroxide
KOH	potassium hydroxide
$Ca(OH)_2$	calcium hydroxide
$Mg(OH)_2$	magnesium hydroxide
$Fe(OH)_3$	iron(III) hydroxide
$Al(OH)_3$	aluminum hydroxide

11.9 Salts

A *salt* is formed when one or more of the hydrogen ions of an acid reacts with one or more hydroxide ions of a base. In the case of hydrogen chloride and sodium hydroxide, the salt formed is sodium chloride.

$$NaOH + HCl \longrightarrow NaCl + H_2O$$

The reaction of an acid and a base to form a salt and water is called *neutralization* (Chapter 16).

All of the binary and ternary compounds containing a metal and a nonmetal or a metal and a polyatomic ion, other than hydroxide, that have been discussed in previous sections are salts. In each case they could be produced from the reaction of an acid and a base. The metal is derived from the base and the nonmetal or polyatomic ion from the acid. Thus, potassium bromide can be formed from the reaction of potassium hydroxide and hydrogen bromide. Similarly, calcium nitrate is the product when calcium hydroxide and nitric acid react, and sodium sulfate is formed from sodium hydroxide and sulfuric acid.

$$KOH + HBr \longrightarrow KBr + H_2O$$
$$Ca(OH)_2 + 2 HNO_3 \longrightarrow Ca(NO_3)_2 + 2 H_2O$$
$$2 NaOH + H_2SO_4 \longrightarrow Na_2SO_4 + 2 H_2O$$

Example 11.14

What acid and base are required to produce $Ca_3(PO_4)_2$ in a neutralization reaction? Write a balanced equation showing the reaction of the acid and base.

Solution The calcium ion can be obtained from calcium hydroxide, which is $Ca(OH)_2$. The polyatomic ion is PO_4^{3-} and can be obtained from the acid H_3PO_4, which is phosphoric acid. The balanced equation is

$$3 Ca(OH)_2 + 2 H_3PO_4 \longrightarrow Ca_3(PO_4)_2 + 6 H_2O$$

[*Additional examples may be found in 11.55 and 11.56 at the end of the chapter.*]

Salts that contain one or more hydrogen atoms that were originally a part of the acid are called **acidic salts.** For example, sulfuric acid may react completely with sodium hydroxide to produce sodium sulfate when both hydrogen atoms react; however, it is also possible that only one hydrogen atom may react, producing sodium hydrogen sulfate (or sodium bisulfate). The acidic salt contains the HSO_4^- anion.

$$NaOH + H_2SO_4 \longrightarrow NaHSO_4 + H_2O$$

Other hydrogen-containing anions are listed in Table 11.14. Acids such as phosphoric acid (H_3PO_4), which contain more than one hydrogen atom can form several acidic salts. To name these anions and the salts that contain them, a Greek prefix such as mono-, di-, or tri- is used to denote the number of hydrogen atoms present.

Salts that contain one or more hydroxide ions that were originally part of a base are called **basic salts** or hydroxy salts. Calcium hydroxide may react with nitric acid to yield calcium

Table 11.14 **Names of Polyatomic Ions Containing Hydrogen**

Ion	Name
HSO_4^-	hydrogen sulfate, bisulfate
HSO_3^-	hydrogen sulfite, bisulfite
HCO_3^-	hydrogen carbonate, bicarbonate
$H_2PO_4^-$	dihydrogen phosphate
HPO_4^{2-}	hydrogen phosphate
$H_2PO_3^-$	dihydrogen phosphite
HPO_3^{2-}	hydrogen phosphite

nitrate if both hydroxide ions react. However, reaction of only one hydroxide ion produces calcium hydroxynitrate.

$$Ca(OH)_2 + HNO_3 \longrightarrow Ca(OH)NO_3 + H_2O$$

The name hydroxynitrate is misleading. There is no polyatomic ion known as hydroxynitrate in this compound; rather, two distinctly different ions, the hydroxide ion and the nitrate ion, are present. Note that one of the anions is put inside parentheses to indicate two ions rather than one are present.

Salts that contain two or more different cations are called **mixed salts.** For these salts each of the cations is named and then the name of the nonmetal anion or negative polyatomic ion is appended. Greek prefixes are used if more than one of the same cation is present. Thus $NaKSO_4$ is sodium potassium sulfate and $K_2NH_4PO_4$ is dipotassium ammonium phosphate.

Summary

Common names of compounds are unrelated to chemical composition, whereas systematic names indicate the elemental composition. Systematic names are based on the rules of the IUPAC.

In naming binary compounds, the metal or electropositive portion of the substance is named first and then the name of the nonmetal or more electronegative element follows. If the metal is one with several possible oxidation numbers, the oxidation number must be indicated by use of an -ous or -ic suffix or by using the Stock system.

Ternary compounds frequently contain a polyatomic anion, which is named as a unit. The prefixes per- and hypo- and the suffixes -ate and -ite are used to name polyatomic ions of an element and oxygen.

Compounds in which hydrogen is the more electropositive element are called acids. Acids are named according to whether they are pure compounds or in aqueous solution. Bases contain a hydroxide ion in combination with a metal ion.

Salts are formed when one or more hydrogen ions of an acid react with one or more hydroxide ions of a base. If a hydrogen ion remains, the salt is an acidic salt. If a hydroxide ion remains, the salt is a basic salt. Compounds containing two different metal ions or ammonium ions are called mixed salts.

New Terms

Acidic salts contain one or more hydrogen atoms originally present in an acid.

-ate is the ending used to designate one of a pair of polyatomic ions containing the larger number of oxygen atoms.

Basic salts contain one or more hydroxide ions originally present in a base.

Binary compounds contain only two elements.

Common names are arbitrary names that are unrelated to composition.

hydro- is used in a name to indicate a binary acid in water solution.

hypo- is the prefix used for the polyatomic ion containing a smaller number of oxygen atoms than the polyatomic ion named with an -ite ending.

-ic is the ending formerly used to designate the higher oxidation state of metal ions in compounds.

-ic acid is used to name acids containing polyatomic anions with an -ate ending.

-ide is the ending used to designate the more negative element in binary compounds.

Inorganic compounds are compounds that do not contain carbon with the exception of carbonates or cyanides.

-ite is the ending used to designate one of a pair of polyatomic ions containing the smaller number of oxygen atoms.

IUPAC or the Internation Union of Pure and Applied Chemistry has devised a system for naming inorganic compounds.

Mixed salts contain two or more different cations.

-ous is the ending formerly used to designate the lower oxidation state of metal ions in compounds.

-ous acid is used to name acids containing polyatomic anions with an -ite ending.

Nomenclature means the naming of compounds.

The **oxidation number** is a positive or negative integer assigned to describe an element as a free atom, an ion, or as part of a polyatomic ion or molecule.

per- is the prefix used for the polyatomic ion containing a larger number of oxygen atoms than the polyatomic ion named with an -ate ending.

Systematic names contain information about the composition of substances.

The **Stock** system is now used to designate the oxidation state of metals in compounds.

Ternary compounds contain three elements.

Exercises

Definitions and terms

11.1 Indicate the usage of each of the following prefixes or suffixes and give one name using the term.
(a) -ate (b) -ite
(c) -ous (d) -ic
(e) -ide (f) -ic acid
(g) -ous acid (h) hydro-
(i) hypo- (j) per-

11.2 Define each of the following terms.
(a) salt (b) acid
(c) acidic salt (d) base
(e) basic salt (f) mixed salt
(g) binary compound (h) ternary compound
(i) -ate and -ite names (j) -ous and -ic names
(k) Stock system (l) IUPAC

Names of ions

11.3 Name each of the following ions.
(a) Cl^- (b) F^- (c) Br^-
(d) I^- (e) O^{2-} (f) S^{2-}

11.4 Name each of the following ions.
(a) H^- (b) Cl^- (c) Sr^{2+}
(d) Na^+ (e) Li^+ (f) Ag^+

11.5 Name each of the following ions.
(a) Mg^{2+} (b) Ca^{2+} (c) K^+
(d) Al^{3+} (e) Hg^{2+} (f) Se^{2-}

11.6 Name each of the following ions.
(a) P^{3-} (b) N^{3-} (c) Cd^{2+}
(d) Rb^+ (e) I^- (f) As^{3-}

11.7 Name each of the following anions.
(a) ClO^- (b) MnO_4^- (c) BrO_4^-
(d) ClO_2^- (e) SO_4^{2-} (f) SO_3^{2-}

11.8 Name each of the following anions.
(a) PO_4^{3-} (b) ClO_3^- (c) CO_3^{2-}
(d) ClO_4^- (e) IO^- (f) BrO_4^-

11.9 Name each of the following anions.
(a) BrO^- (b) NO_2^- (c) IO_2^-
(d) CrO_4^{2-} (e) NO_3^- (f) OH^-

11.10 Name each of the following anions.
(a) BrO_2^- (b) CO_3^{2-} (c) IO_3^-
(d) $Cr_2O_7^{2-}$ (e) $S_2O_3^{2-}$ (f) CN^-

11.11 Name each of the following cations.
(a) Cu^+ (b) Fe^{3+} (c) Hg^{2+}
(d) Hg_2^{2+} (e) Fe^{2+} (f) Cu^{2+}

11.12 Name each of the following cations.
(a) Sn^{2+} (b) Pb^{4+} (c) Pb^{2+}
(d) Sn^{4+} (e) Ti^{3+} (f) Ti^{4+}

Formulas of ions

11.13 Write the symbols of each of the following cations.
(a) sodium ion (b) cuprous ion
(c) ferric ion (d) potassium ion
(e) stannic ion (f) mercurous ion

11.14 Write the symbols of each of the following cations.
(a) cupric ion (b) mercuric ion
(c) calcium ion (d) aluminum ion
(e) barium ion (f) stannous ion

11.15 Write the symbols of each of the following cations.
(a) lithium   (b) silver
(c) magnesium   (d) ferrous
(e) titanic   (f) cesium

11.17 Write the symbols of each of the following ions.
(a) bromide   (b) sulfite
(c) chlorate   (d) bicarbonate
(e) nitrate   (f) perchlorate

11.19 Write the symbols of each of the following ions.
(a) chlorite   (b) chromate
(c) fluoride   (d) carbonate
(e) hypobromite   (f) dichromate

11.16 Write the symbols of each of the following cations.
(a) strontium   (b) rubidium
(c) titanous   (d) zinc
(e) plumbic   (f) manganous

11.18 Write the symbols for each of the following ions.
(a) hydroxide   (b) bisulfite
(c) chloride   (d) phosphate
(e) nitrite   (f) hydrogen sulfite

11.20 Write the symbols of each of the following ions.
(a) permanganate   (b) dihydrogen phosphite
(c) hypobromite   (d) thiosulfate
(e) periodate   (f) hydrogen sulfite

Oxidation Numbers

11.21 Calculate the oxidation number of the indicated element in each of the following.
(a) S in H_2S (b) Sc in ScF_3
(c) Ti in TiO_2 (d) Si in SiO_2
(e) N in Na_3N (f) Mn in $MnCl_2$

11.23 Calculate the oxidation number of the indicated element in each of the following.
(a) Se in H_2Se (b) Ti in $TiCl_3$
(c) Ce in $CeCl_4$ (d) Os in OsO_4
(e) Cu in CuO (f) Xe in XeF_2

11.25 Calculate the oxidation number of the indicated element in each of the following.
(a) P in H_3PO_4 (b) P in H_3PO_3
(c) S in $Na_2S_2O_4$ (d) Cr in $Na_2Cr_2O_7$
(e) Mn in $KMnO_4$ (f) S in H_2SO_4

11.27 Calculate the oxidation number of the indicated element in each of the following.
(a) S in SO_3^{2-} (b) S in SO_4^{2-}
(c) N in NO_2^- (d) N in NO_3^-
(e) Cl in ClO_4^- (f) N in NH_4^+

11.22 Calculate the oxidation number of the indicated element in each of the following.
(a) Te in H_2Te (b) Cr in CrO_3
(c) Ge in GeO_2 (d) C in CO_2
(e) N in NF_3 (f) Tl in $TlCl$

11.24 Calculate the oxidation number of the indicated element in each of the following.
(a) Cu in Cu_2O (b) Fe in Fe_2O_3
(c) Cr in Cr_2O_3 (d) P in P_4O_{10}
(e) N in N_2O_3 (f) As in As_2O_5

11.26 Calculate the oxidation number of the indicated element in each of the following.
(a) Xe in $XeOF_4$ (b) P in $H_5P_3O_{10}$
(c) N in $H_2N_2O_2$ (d) Cl in $NaClO_3$
(e) Al in $Al_2(SO_4)_3$ (f) Cu in $NaCuCl_3$

11.28 Calculate the oxidation number of the indicated element in each of the following.
(a) Mn in MnO_4^- (b) Cr in $Cr_2O_7^{2-}$
(c) P in $P_2O_7^{4-}$ (d) Se in SeO_3^{2-}
(e) Mo in MoO_4^{2-} (f) V in VO_4^{3-}

Naming Compounds

11.29 Write the systematic name for each of the following compounds.
(a) Na_2S (b) BaO
(c) Al_2O_3 (d) $CoBr_2$
(e) CdF_2 (f) Mg_3N_2
(g) K_3P (h) $ZnSe$

11.31 Write the systematic name for each of the following compounds.
(a) PbI_2 (b) HgF_2
(c) Fe_2O_3 (d) $TiCl_3$
(e) $FeBr_2$ (f) CeO_2
(g) SnO (h) Au_2O_3

11.30 Write the systematic name for each of the following compounds.
(a) SrI_2 (b) Cs_2Te
(c) Al_2S_3 (d) Ca_3N_2
(e) Rb_2O (f) $LiBr$
(g) AlN (h) $MgCl_2$

11.32 Write the systematic name for each of the following compounds.
(a) $SnCl_4$ (b) Hg_2O
(c) $CrCl_3$ (d) $AuCl_3$
(e) $TiCl_4$ (f) PbO_2
(g) $MnCl_2$ (h) OsO_4

11.33 Write the systematic name for each of the following compounds.

(a) Li_2CO_3 (b) $AgIO_3$

(c) $Al_2(SO_4)_3$ (d) $Zn(CN)_2$

(e) $Ba(NO_3)_2$ (f) $Na_2Cr_2O_7$

(g) SrF_2 (h) Rb_2SO_3

11.35 Write the systematic name for each of the following compounds.

(a) CCl_4 (b) P_4O_{10} (c) OCl_2

(d) N_2O_3 (e) PCl_3 (f) CO

11.37 Determine the oxidation number of manganese in each of the following compounds and name each compound.

(a) Mn_2O_7 (b) MnO_2 (c) MnO (d) $MnCl_2$

Formulas of Compounds

11.39 Write the formula for each of the following compounds.

(a) calcium iodide   (b) lithium sulfide

(c) magnesium nitride   (d) aluminum sulfide

(e) sodium oxide   (f) barium selenide

(g) potassium phosphide   (h) strontium bromide

11.41 Write the formula for each of the following compounds.

(a) gold(III) oxide   (b) copper(II) sulfide

(c) tin(II) bromide   (d) iron(III) selenide

(e) ferrous oxide   (f) stannic chloride

(g) cupric nitrate   (h) mercuric bromide

11.43 Indicate why each of the following formulas is incorrect.

(a) $NaCl_2$ (b) Fe_3O_2

(c) $Cu(NO_4)_2$ (d) H_2ClO_4

(e) H_2SO_2 (f) $Ba_2(PO_4)_3$

(g) $NaHPO_4$ (h) $Al(CN)_2$

11.45 The formula of an iodate of lanthanum is $La(IO_3)_2$. What is the formula of the sulfate of lanthanum with the same oxidation number?

11.47 The formula of an hydroxide of yttrium is $Y(OH)_3$. What is the formula of the sulfide of yttrium with the same oxidation number?

Acids and Bases

11.49 Write the formula for the hydroxide of each of the following metal ions.

(a) Li^+ (b) Sr^{2+}

(c) Al^{3+} (d) Zn^{2+}

(e) Ni^{2+} (f) Cs^+

11.51 Name each of the following acids.

(a) HNO_2 (b) H_2SO_4

(c) H_3PO_4 (d) H_2CO_3

(e) HF (f) $HBrO_2$

Salts

11.53 The name of $Na_2C_2O_4$ is sodium oxalate. What is the formula of the acid required to form this compound by reaction with sodium hydroxide? Suggest a name for the acid.

11.34 Write the systematic name for each of the following compounds.

(a) $Hg(CN)_2$ (b) Hg_2CrO_4

(c) $TlNO_2$ (d) $CuNO_3$

(e) $Fe_2(SO_4)_3$ (f) NH_4IO_4

(g) $Cr_2(SO_4)_3$ (h) $Au_2(SO_3)_3$

11.36 Write the systematic name for each of the following compounds.

(a) CO_2 (b) SO_2 (c) CBr_4

(d) PCl_5 (e) NF_3 (f) C_3O_2

11.38 Determine the oxidation number of molybdenum in each of the following compounds and name each compound.

(a) $MoCl_4$ (b) Mo_2O_5 (c) Mo_2S_3 (d) $MoCl_3$

11.40 Write the formula for each of the following compounds.

(a) potassium bisulfate   (b) calcium iodate

(c) zinc dichromate   (d) lithium bicarbonate

(e) magnesium phosphate   (f) aluminum sulfate

(g) sodium hypobromite   (h) barium cyanide

11.42 Write the formula for each of the following compounds.

(a) iron(II) sulfite   (b) tin(II) nitrate

(c) mercury(II) chromate   (d) silver(I) permanganate

(e) ferric perchlorate   (f) mercurous sulfate

(g) stannous bromite   (h) mercuric nitrate

11.44 Indicate why each of the following formulas is incorrect.

(a) $Ca(HSO_4)_3$ (b) CO_3

(c) MgO_2 (d) $KCrO_4$

(e) Al_3O_2 (f) AgO_2

(g) H_2PO_4 (h) LiS_2O_3

11.46 The formula of a hydroxide of zirconium is $Zr(OH)_4$. What is the formula of the sulfate of zirconium with the same oxidation number?

11.48 The formula of an oxide of gold is Au_2O_3. What is the formula of the cyanide of gold with the same oxidation number?

11.50 What is the oxidation number of the metal in each of the following bases?

(a) $Zr(OH)_4$ (b) $Ga(OH)_3$

(c) $RbOH$ (d) $Sr(OH)_2$

(e) $Mn(OH)_2$ (f) $Y(OH)_3$

11.52 Write the formula of each of the following acids.

(a) perbromic acid   (b) nitrous acid

(c) phosphoric acid   (d) chloric acid

(e) chlorous acid   (f) sulfuric acid

11.54 The name of $Ca_3(BO_3)_2$ is calcium borate. What is the formula of the acid required to form this compound by reaction with calcium hydroxide? Suggest a name for the acid.

11.55 Write a balanced equation for the reaction of each of the following acids and bases. What is the name of the salt formed? Assume that complete neutralization occurs.
(a) $H_2SO_4 + KOH$
(b) $H_3PO_4 + Ca(OH)_2$
(c) $HCl + Al(OH)_3$
(d) $HNO_3 + Fe(OH)_3$

11.57 Name each of the following compounds.
(a) $MgNH_4PO_4$ (b) $AlNH_4(SO_4)_2$
(c) $KAl(SO_4)_2$ (d) $CaNH_4PO_4$

11.59 What is the oxidation number of vanadium in $NH_4V(SO_4)_2$?

11.61 What is the value of x in the formula $NH_4Cr(SO_4)_x$ if chromium has a +3 oxidation number.

11.56 Write a balanced equation for the reaction of each of the following acids and bases. What is the name of the salt formed? Assume that complete neutralization occurs.
(a) $HBr + Ba(OH)_2$
(b) $H_2SO_4 + Ca(OH)_2$
(c) $H_3PO_4 + LiOH$
(d) $HClO_4 + Mg(OH)_2$

11.58 Write the formula of each of the following.
(a) sodium aluminum sulfate
(b) manganese(II) ammonium phosphate
(c) gallium(III) ammonium sulfate
(d) sodium potassium sulfate

11.60 What is the oxidation number of iron in $NH_4Fe(SO_4)_2$?

11.62 What is the value of x in the formula $KNa(SO_4)_x$?

Additional Exercises

11.63 Name each of the following ions.
(a) Ca^{2+} (b) Ga^{3+} (c) Br^-
(d) Ba^{2+} (e) Zn^{2+} (f) Cs^+

11.65 Name each of the following anions.
(a) BrO_3^- (b) NO_3^- (c) IO^-
(d) MnO_4^- (e) NH_4^+ (f) IO_4^-

11.67 Write the symbols of each of the following ions.
(a) bromite (b) nitrite
(c) sulfate (d) hypoiodite
(e) hydrogen sulfate (f) cyanide

11.69 Write the systematic name for each of the following compounds.
(a) SO_3 (b) CS_2 (c) NO_2
(d) NCl_3 (e) P_4O_6 (f) SiF_4

11.71 Determine the oxidation number of ruthenium in each of the following compounds and name each compound.
(a) RuF_5 (b) RuO_4 (c) $RuCl_3$ (d) RuO_2

11.73 Write the formula for each of the following compounds.
(a) manganese(III) sulfate (b) thallium(III) sulfide
(c) cadmium telluride (d) gold(III) bromide

11.75 Indicate why each of the following names is improper.
(a) chlorite acid (b) monocarbon monoxide
(c) lead chloride (d) sulfur oxide

11.77 The formula of a fluoride of niobium is NbF_5. What is the formula of the oxide of niobium with the same oxidation number?

11.79 The formula of a sulfate of dysprosium is $Dy_2(SO_4)_3$. What is the formula of the phosphate of dysprosium with the same oxidation number?

11.64 Name each of the following cations.
(a) Ce^{3+} (b) Ce^{4+} (c) Sc^{3+}
(d) Ni^{2+} (e) Mn^{2+} (f) Cr^{2+}

11.66 Name each of the following anions.
(a) HSO_4^- (b) HCO_3^- (c) $H_2PO_4^-$
(d) HSO_3^- (e) $H_2PO_3^-$ (f) HPO_4^{2-}

11.68 Write the symbols of each of the following ions.
(a) chlorate (b) nitrate
(c) sulfite (d) hypobromite
(e) hydrogen sulfite (f) hydroxide

11.70 Write the systematic name for each of the following compounds.
(a) SO_2 (b) SeO_2 (c) N_2O_4
(d) PCl_5 (e) IF_5 (f) $GeCl_4$

11.72 Determine the oxidation number of cobalt in each of the following compounds and name each compound.
(a) $CoCl_2$ (b) Co_2O_3 (c) CoS (d) CoF_3

11.74 Write the formula for each of the following compounds.
(a) copper(I) sulfate (b) thallium(I) chloride
(c) copper(II) thiosulfate (d) mercury(II) chlorate

11.76 Indicate why each of the following names is improper.
(a) iron oxide (b) nitrate acid
(c) sodium bicarbonide (d) copper(IV) oxide

11.78 The formula of an oxide of gallium is Ga_2O_3. What is the formula of the perchlorate of gallium with the same oxidation number?

11.80 The formula of a carbonate of yttrium is $Y_2(CO_3)_3$. What is the formula of the nitrate of yttrium with the same oxidation number?

11.81 Name each of the following acids.
    (a) H_2SO_3 (b) HNO_3
    (c) HCl (d) $HClO_3$
    (e) H_3PO_3 (f) HIO_4

11.82 Write the formula of each of the following acids.
    (a) bromous acid   (b) phosphoric acid
    (c) sulfurous acid   (d) nitric acid
    (e) iodic acid   (f) hypochlorous acid

11.83 Calculate the oxidation number of the indicated atom in each of the following.
    (a) Cr in Cr_2O_3 (b) N in N_2H_4
    (c) P in P_4O_6 (d) Si in Si_2Cl_6
    (e) N in N_2O_5 (f) Al in Al_2O_3

11.84 Calculate the oxidation number of the indicated atom in each of the following.
    (a) Ge in GeO_2 (b) P in P_2H_4
    (c) N in N_2O_5 (d) Si in Si_2H_6
    (e) P in PCl_5 (f) Os in OsO_4

11.85 Calculate the oxidation number of the indicated element in each of the following.
    (a) Mn in MnF_6^{2-} (b) V in $V_{10}O_{28}^{6-}$
    (c) Mo in $Mo_8O_{26}^{4-}$ (d) W in $W_{12}O_{42}^{12-}$
    (e) Sb in $SbCl_6^{2-}$ (f) Ni in $NiCl_4^{2-}$

11.86 Calculate the oxidation number of the indicated element in each of the following.
    (a) S in HSO_4^- (b) P in $H_2PO_3^-$
    (c) S in HSO_3^- (d) C in HCO_2^-
    (e) C in CH_3O^- (f) P in $H_2P_2O_2^{2-}$

11.87 Calculate the oxidation number of carbon in each of the following.

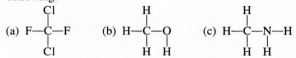

11.88 Calculate the oxidation number of carbon in each of the following.

(a) $H_2C{=}O$ (b) $H{-}C{\equiv}N$ (c) $H{-}C{\overset{O-H}{=}O}$

12 | The Gaseous State

Learning Objectives

After studying Chapter 12 you should be able to

1. Convert units of pressure.

2. Calculate changes in gas volume resulting from pressure changes and vice versa.

3. Calculate changes in gas volume resulting from temperature changes and vice versa.

4. Calculate changes in gas pressure resulting from temperature changes and vice versa.

5. Calculate changes in gas pressure, temperature, or volume resulting from changes in the other two variables.

6. Relate gas volume to the number of moles of a gas using the concept of molar volume.

7. Use the ideal gas law to relate the number of moles of a gas and the pressure, volume, and temperature of a gas.

8. Relate the total pressure of a gas mixture to the partial pressures of the individual gases.

9. Use the law of effusion to calculate the molecular weight of a gas based on its rate of effusion.

12.1 Nature of a Gas

We can see both the liquid and solid states of matter but, with rare exceptions, gases are colorless and are not visible. In addition, we can feel liquids and solids, but unless the wind is blowing we do not "feel" gases. However, we can make observations that tell us that the gaseous state does contain matter. The odors of spring flowers or the smell of decaying matter are both the result of gases reaching our nose. We can also see the result of gases moving about as the wind blows against the branches of a tree. The force exerted by a moving gas against an object can be every bit as strong as that exerted by a moving solid object. Strong winds can knock down trees or buildings.

Gases fill any space available to them and can be compressed into a smaller volume by applying pressure as, for example, oxygen into a hospital oxygen tank or air into a scuba tank. A gas within a tank contains matter that exerts a pressure on the walls of the container. If too much gas is put into a container such as a car tire, the resulting pressure can rupture the tire walls, which will result in an explosion and the escape of the gas.

The properties of gases are affected by temperature. Their volumes change with temperature, expanding when heated and contracting when cooled. The pressures of gases are affected by temperature as occurs in changing weather conditions.

The mathematical relationships among the properties of gases were established by experiments and stated in a series of gas laws in the eighteenth century. In this chapter the gas laws are outlined and methods of calculating the volume of a given mass of a gas at any pressure and temperature are given.

The behavior of gases as given by the gas laws can be rationalized by a model known as the kinetic-molecular theory of matter. This model, which is discussed in this chapter, is then used to explain the properties of liquids and solids in the next chapter.

12.2 Pressure

Pressure is a force per unit area on an object. Let's consider some examples of pressure. The pressure that we exert on our shoes depends on our mass and the area of the soles of our shoes. For two individuals of the same mass who wear different sizes of shoes, the individual with the larger shoe size exerts less force per unit area on the soles of the shoes, although the total force involved is the same in both cases. The effectiveness of snowshoes on snow or water skis on water is due to the distribution of force over a wide area to give a low pressure.

The submicroscopic particles of a gas in a container exert a pressure against a wall by colliding with it (Figure 12.1). This "pushing" against the wall can be measured by a pressure gauge that also experiences the pressure caused by the collisions.

Most people do not realize that they live under gas pressure. We live at the bottom of an ocean of air and are under pressure much like the pressure on a diver in the ocean. However, since we have lived under air pressure all of our lives, we don't feel it. The pressure of our atmosphere is the result of the weight of air from the outer edge of the atmosphere down to the surface of the earth. The weight of a column of air above an area of 1 square inch is 14.7 pounds at sea level. We say the pressure is 14.7 lb/in^2 or 14.7 psi. At higher altitudes the pressure is less because the column of air from that point on the earth's surface to the edge of the atmosphere weighs less (Figure 12.2).

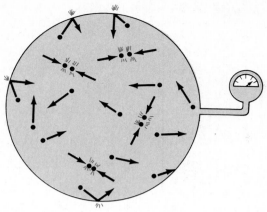

Figure 12.1

Pressure and the molecular motion of a gas
The molecules of a gas are moving and colliding with each other as well as with the walls of the container. The pressure on the walls of the container is the result of the impacts of molecules on the walls.

column of air

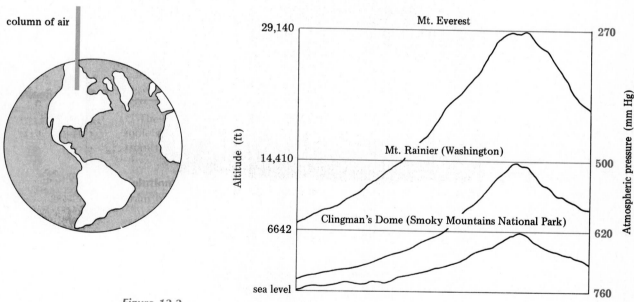

Figure 12.2

Atmospheric pressure
A column of air from the outer atmosphere to a given point on earth contains gases that exert a force equal of 14.7 lb on each square inch. This pressure is 1 atm or 760 mm Hg. At higher altitudes the pressure is less since the mass of the column of air from that point to the outer atmosphere is less. Examples of pressures on some mountains are given at the right.

Some organisms live under extremely high pressures. A fish at a depth of 2000 ft in the ocean lives under a pressure of 1000 psi. The body fluids of this fish must exert an outward pressure equal to the ocean pressure in order to survive. When a scuba diver descends to 120 ft, the pressure of the water is about 59 psi. The diver experiences a total pressure of about 74 psi, because the total pressure is equal to that of the water plus the pressure of the atmosphere.

12.3 Measurement of Pressure

The **barometer**, a standard device for measuring the pressure of the atmosphere, was invented by Evangelista Torricelli in the seventeenth century. It consists of a long tube that is closed at one end. The tube is filled with mercury and inverted in a vessel of mercury. If the tube is more than 76 cm long, part of the mercury will run out of the tube when it is inverted, leaving a vacuum in the upper section of the tube. A column of mercury approximately 76 cm high will remain in the tube (Figure 12.3).

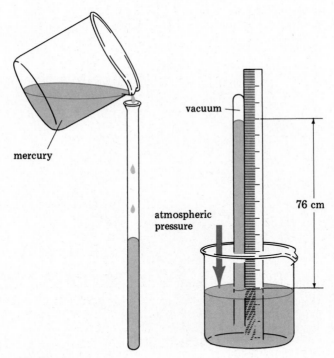

vacuum

mercury

atmospheric
pressure

76 cm

Figure 12.3

The barometer
A barometer is constructed by filling a glass tube with mercury, inverting it, and immersing it in a beaker of mercury. The column of mercury will drop to about 76 cm Hg.

A column of mercury remains in the tube of the barometer because the atmosphere exerts a pressure on the surface of the mercury. This pressure, transmitted through the mercury to the base of the column, supports the mercury in the tube. There is a pressure at the base of the mercury column due to the mass of the mercury in the column. The mercury falls until the pressure exerted by its weight equals the atmospheric pressure.

Pressure should be expressed in units of force per unit area. In the barometer, the downward force exerted by the mercury column is proportional to the mass of the liquid supported in the column, which, in turn, is proportional to the height of the column. Therefore, the pressure can be expressed in centimeters or millimeters of mercury. It is understood that pressure is not the same thing as length, but by convention the term centimeter of mercury (cm Hg) or millimeter of mercury (mm Hg) indicates the pressure or force per unit area exerted by the mass of mercury contained in a stated length of a mercury column.

Atmospheric pressure depends on altitude and the local weather, and it fluctuates from day to day. At 0°C at sea level, the average atmospheric pressure supports a column of mercury 760 mm high. This pressure is called a **standard atmosphere** or 1 atm. The unit of pressure, 1 mm Hg, is also called 1 torr; therefore, the standard pressure (1 atm) is 760 torr.

Example 12.1

The atmospheric pressure in Mexico City one day is 0.750 atm. What is the pressure in torr?

Solution First determine two possible conversion factors relating atmosphere and torr.

$$\frac{1\ atm}{760\ torr} \quad \text{and} \quad \frac{760\ torr}{1\ atm}$$

The second factor cancels the given unit (atm) and gives the required value in torr.

$$0.750\ \cancel{atm} \times \frac{760\ torr}{1\ \cancel{atm}} = 570\ torr$$

Example 12.2

The pressure on the surface of Venus is about 1.0×10^2 atm. What is the pressure in cm Hg?

Solution First determine two possible conversion factors relating atmosphere and cm Hg.

$$\frac{76\ cm\ Hg}{1\ atm} \quad \text{and} \quad \frac{1\ atm}{76\ cm\ Hg}$$

The first factor cancels the given unit of atm and gives the required value in units of cm Hg.

$$1.0 \times 10^2\ \cancel{atm} \times \frac{76\ cm\ Hg}{1\ \cancel{atm}} = 7.6 \times 10^3\ cm\ Hg$$

[Additional examples may be found in 12.11 and 12.12 at the end of the chapter.]

12.4 Kinetic Molecular Theory of Matter

The concept that atoms or molecules in gases are moving is known as the **kinetic molecular theory.** By extension, it can also apply to the liquid and solid states. The phenomenon of pressure, as well as other properties of gases, can be explained by the kinetic molecular theory.

The assumptions made in the kinetic theory of matter are summarized as follows.

1. Gases are composed of atoms or molecules that are widely separated from one another. The space occupied by the atoms or molecules is extremely small compared with the space between them.
2. The atoms or molecules are moving rapidly and randomly in straight lines. Their direction is maintained until they collide with a second atom or molecule or with the walls of the container.
3. There are no attractive forces between molecules or atoms of a gas.
4. Collisions of molecules or atoms do not result in any change in the total energy of the gas, although a transfer of energy between the molecules or atoms may occur in the collision.
5. In a specific gas, individual atoms or molecules move at different speeds. However, the average velocity and average kinetic energy remain constant. As the temperature increases, the average velocity increases and therefore the average kinetic energy also increases. The average kinetic energy is directly proportional to the temperature on the Kelvin scale.

A hypothetical gas that conforms to all of the assumptions of the kinetic theory is an **ideal gas.** Thus, ideal gas particles have a very small volume compared to the total volume of the gas sample. Furthermore, the gas particles are not attracted to each other, and could never be condensed to form a liquid. However, all real gases, including components of our atmosphere such as oxygen and nitrogen, can be converted to liquids. **Real gases** are any gaseous substances that actually exist.

Under conditions of high temperature and low pressure, most real gases behave as ideal gases. At low pressures such as those that exist at high altitudes, the gas particles are far apart, as given by assumption 1 of the kinetic theory. At high temperatures the gas particles have high kinetic energies and are moving so rapidly that the attractive forces existing in real gases can exert very little influence. In this chapter we will assume an ideal gas behavior for all problems. However, we will keep in mind that at high pressures and low temperatures, all real gases can be condensed to liquids. Thus, they do not behave ideally under these conditions, and the laws we study in this chapter must be modified to be useful at high pressures and low temperatures.

12.5 Boyle's Law

In experiments with vacuums and vacuum pumps in 1662, Robert Boyle studied the relationship between volume and pressure in gases. He measured what he termed "the spring of air," the pressure with which a gas sample pushes back when it is compressed. Boyle found a relationship between gas volume and pressure that has been verified many times with a variety of gases. **Boyle's law** states that at constant temperature, the volume of a given quantity of gas varies inversely with the pressure exerted on it. Thus, if the pressure of a given volume of gas is doubled, the volume will be decreased by one half (Figure 12.4). Conversely, if the pressure is decreased by one half, the volume doubles. A mathematical expression of Boyle's law is

$$V = k \times \frac{1}{P}$$

where V stands for volume, P for pressure, and k is a proportionality constant. The value of k depends on the temperature and the quantity of gas being studied. An alternative expression, obtained by multiplying both sides of the equation by P, is

$$PV = k$$

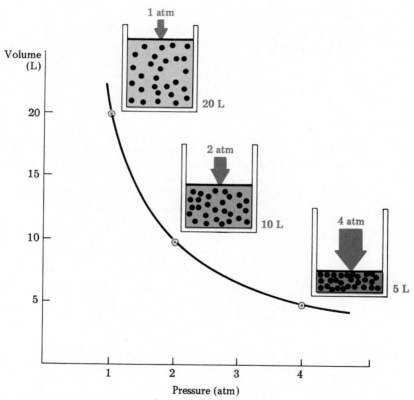

Figure 12.4

An illustration of Boyle's law
As the applied pressure given on the horizontal axis increases, the volume given on the vertical axis decreases until the internal pressure is equal to the applied pressure. As the applied pressure is increased, the number of molecules per unit volume increases. The internal pressure is the result of the increased number of collisions of the molecules with the walls of the container.

Therefore, the product of the volume and pressure of a gas is constant if the quantity of the gas sample and the temperature remain unchanged. Thus for any two sets of pressures and volumes we can restate Boyle's law as

$$P_2 \times V_2 = k = P_1 \times V_1$$

where the subscripts 1 and 2 refer to the two sets of experimental conditions. An example of some data illustrating Boyle's law is given in Figure 12.4.

For a change in pressure, the new volume may be calculated from the rearranged equation.

$$V_2 = V_1 \times \frac{P_1}{P_2} \qquad \text{or} \qquad V_2 = V_1 \times P_{factor}$$

Similarly, if a new pressure must be calculated to produce a desired volume change, then

$$P_2 = P_1 \times \frac{V_1}{V_2} \qquad \text{or} \qquad P_2 = P_1 \times V_{factor}$$

In either a pressure or a volume calculation, the necessary volume factor or pressure factor may be determined by recalling the inverse proportionality of Boyle's law. Thus, if a pressure is increased, the volume must decrease, and only a pressure factor less than unity will give the correct answer.

Example 12.3 _____

An oxygen cylinder for medical use contains 35.5 L at 25°C and 149 atm. What volume is this in liters at 1.00 atm at the same temperature? Assume ideal gas behavior.

Solution Since the temperature for the sample of the gas is unchanged in the described process, we may use Boyle's law. First arrange the information in a clear form so that you can analyze how to solve the problem.

$P_1 = 149$ atm $V_1 = 35.5$ L
$P_2 = 1.00$ atm $V_2 = ?$ L
a pressure decrease   volume must be increased

Direct substitution into the Boyle's law equation will provide the required volume. However, you can also solve the problem by noting that the pressure has decreased from 149 to 1.00 atm and that the volume must increase by a pressure factor containing these two numbers.

$$V_2 = V_1 \times \frac{P_1}{P_2} \qquad \text{or} \qquad V_1 \times P_{factor} \qquad \text{where } P_{factor} > 1$$

$$V_2 = 35.5 \text{ L} \times \frac{149 \text{ atm}}{1.00 \text{ atm}}$$

$$V_2 = 35.5 \text{ L} \times 149$$

$$V_2 = 5.2895 \times 10^3 \text{ L}$$

The answer expressed to the required three significant figures is 5.29×10^3 L.

_____

Example 12.4

The volume of one cylinder in an automobile is 0.45 L. The cylinder containing oxygen and gasoline vapor at 1.0 atm is compressed to 0.075 L. Assuming that the temperature is constant, what pressure is required for this change?

Solution Since the temperature and the sample of the gas are unchanged, we may use Boyle's law. Arrange the information so that you can analyze the problem.

$$V_1 = 0.45 \text{ L} \qquad\qquad P_1 = 1.0 \text{ atm}$$
$$V_2 = 0.075 \text{ L} \qquad\qquad P_2 = ? \text{ atm}$$

volume is decreased   pressure must increase

The required pressure can be obtained by direct substitution into the Boyle's law equation. In addition, you should note that the volume has decreased from 0.45 L to 0.075 L, and the pressure must increase by a volume factor composed of these two numbers.

$$P_2 = P_1 \times \frac{V_1}{V_2} \qquad \text{or} \qquad P_2 = P_1 \times V_{\text{factor}} \quad \text{where } V_{\text{factor}} > 1$$

$$P_2 = 1.0 \text{ atm} \times \frac{0.45 \text{ L}}{0.075 \text{ L}}$$

$$P_2 = 6.0 \text{ atm}$$

[*Additional examples may be found in 12.17–12.22 at the end of the chapter.*]

12.6 Charles's Law

Gases expand when heated under constant pressure and contract when cooled. In 1787, the French physicist J. A. C. Charles observed this relationship between volume and temperature at constant pressure. From the data of Charles and other scientists, **Charles's law** states that at constant pressure, the volume of a fixed mass of a gas is directly proportional to the absolute (Kelvin) temperature. An example of such data is given in Figure 12.5.

The mathematical expression with a proportionality constant, k, is

$$V = kT$$

The proportionality constant is dependent only on the pressure and the sample of gas considered. Dividing both sides of the equation by T gives

$$\frac{V}{T} = k$$

Since the constant remains unchanged, the following relationship must be obeyed under any two sets of experimental conditions.

$$\frac{V_2}{T_2} = k = \frac{V_1}{T_1}$$

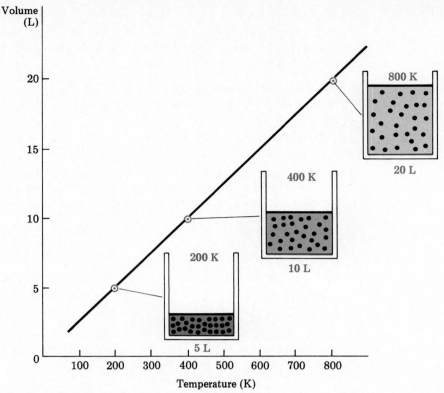

Figure 12.5

An illustration of Charles's law
As the temperature given on the horizontal axis increases, the volume given on the vertical axis increases. The increased kinetic energy of the molecules causes them to move faster. As a result the number of impacts increases, as well as the force of each impact. Therefore the contained volume increases.

For a change in temperature, the new volume may be calculated by the use of a temperature factor.

$$V_2 = V_1 \times \frac{T_2}{T_1} \quad \text{or} \quad V_2 = V_1 \times T_{\text{factor}}$$

Alternatively, if it is necessary to calculate the temperature required to change the volume, we use a volume factor. Remember, in the use of either formula, the temperature must be in kelvins.

$$T_2 = T_1 \times \frac{V_2}{V_1} \quad \text{or} \quad T_2 = T_1 \times V_{\text{factor}}$$

Flight in hot air balloons is possible because of the increase in volume caused by heating air. In fact, Charles rose to a height of about 5 miles in his hot air balloon in 1783. The increase in volume of a gas caused by an increase in temperature results in a decrease in density of the air in the balloon, which then "floats" in the surrounding denser, cooler air.

Example 12.5

A child's balloon is filled with helium to give a volume of 5.0 L in an air conditioned store at 21°C. The child takes the balloon outdoors, where the air temperature is 38°C. What happens to the balloon?

Solution   Since the atmospheric pressure and quantity of the gas remain constant, Charles's law applies to the solution of the problem. Arrange the data in tabular form and calculate the temperature in kelvins.

$$T_1 = 21 + 273 = 294 \text{ K} \qquad\qquad V_1 = 5.0 \text{ L}$$
$$T_2 = 38 + 273 = 311 \text{ K} \qquad\qquad V_2 = ? \text{ L}$$

a temperature increase   volume must be increased

Direct substitution into the Charles's law equation will give the unknown volume. However, note that the volume must increase as a consequence of the increase in the temperature. Therefore, the volume is multiplied by a temperature factor to increase the volume.

$$V_2 = V_1 \times \frac{T_2}{T_1} \quad \text{or} \quad V_2 = V_1 \times T_{\text{factor}} \quad \text{where } T_{\text{factor}} > 1$$

$$V_2 = 5.0 \text{ L} \times \frac{311 \text{ K}}{294 \text{ K}}$$

$$V_2 = 5.289115 \text{ L}$$

The correct answer expressed to the required two significant figures is 5.3 L.

[*Additional examples may be found in 12.23–12.28 at the end of the chapter.*]

12.7   Gay-Lussac's Law

Gay-Lussac's law states that at constant volume the pressure of a fixed mass of a gas is directly proportional to the absolute (Kelvin) temperature. Mathematically stated, the direct proportionality between pressure and temperature is given by

$$P = kT$$

The value of k depends on the volume of the sample and the quantity of gas. An alternative expression is obtained by dividing both sides of the equation by T.

$$\frac{P}{T} = k$$

Therefore, the quotient of the pressure divided by the temperature will be a constant for a given volume of a gas.

$$\frac{P_2}{T_2} = k = \frac{P_1}{T_1}$$

For a change in temperature, the effect on the pressure may be calculated from the following equation.

$$P_2 = P_1 \times \frac{T_2}{T_1} = P_1 \times T_{\text{factor}}$$

Similarly, for a change in pressure, the related necessary temperature change may be calculated from an alternative rearranged equation.

$$T_2 = T_1 \times \frac{P_2}{P_1} = T_1 \times P_{\text{factor}}$$

In either case, you only need to remember that there is a direct proportion between the Kelvin temperature and the pressure.

Heating a gas in a closed container may result in an explosion if the container cannot withstand the pressure. For this reason, spray cans of deodorants, paints, and so on, have a warning label indicating that the can should not be heated and should be stored at temperatures below a stated limit.

Example 12.6

What will happen to a spray can containing only the propellant at a pressure of 1.1 atm at 23°C if it is thrown into a fire at 475°C?

Solution Gay-Lussac's law conditions are stated and the data may be arranged as follows, after converting Celsius temperature to kelvins.

$$T_1 = 296 \text{ K} \qquad\qquad P_1 = 1.1 \text{ atm}$$
$$T_2 = 748 \text{ K} \qquad\qquad P_2 = ? \text{ atm}$$
a temperature increase   pressure must increase

The pressure increases by a temperature factor. Therefore, we write

$$P_2 = P_1 \times \frac{T_2}{T_1} \qquad \text{or} \qquad P_2 = P_1 \times T_{\text{factor}} \quad \text{where } T_{\text{factor}} > 1$$

$$P_2 = 1.1 \text{ atm} \times \frac{748 \text{ K}}{296 \text{ K}}$$

$$P_2 = 2.779729 \text{ atm}$$

The correct answer expressed to the required two significant figures is 2.8 atm. The pressure in the can increases if the can does not rupture.

[Additional examples may be found in 12.29–12.34 at the end of the chapter.]

12.8 Combined Gas Law

Both Boyle's and Charles's laws may be combined into one mathematical expression, which then also gives the Gay-Lussac law.

$$\frac{PV}{T} = k$$

Therefore, for a fixed sample of a gas, it follows that

$$\frac{P_2V_2}{T_2} = k = \frac{P_1V_1}{T_1}$$

This equation need not be memorized because any variable can be changed by the factors of the other two variables.

$$P_2 = P_1 \times V_{factor} \times T_{factor}$$
$$V_2 = V_1 \times P_{factor} \times T_{factor}$$
$$T_2 = T_1 \times P_{factor} \times V_{factor}$$

In order to solve problems in which two variables change, it is only necessary to consider separately what effect each variable will have. Thus, for a volume to change as a consequence of pressure and temperature changes, you determine first what effect the pressure change will have on the volume. If the pressure increases, the volume will decrease, and vice versa. After the proper pressure factor has been determined, the temperature factor can be considered. If the temperature decreases, the volume will decrease, and vice versa. Application of the pressure factor and the temperature factor gives the final correct answer.

Example 12.7

A high altitude weather balloon is filled with 2.0×10^4 L of helium at 20°C and 730 mm Hg. What volume will the balloon occupy at an altitude where the pressure is 69 mm Hg and the temperature is −45°C?

Solution   Convert Celsius temperature to Kelvin. Then arrange the data as

$$V_1 = 2.0 \times 10^4 \text{ L} \qquad V_2 = ? \text{ L}$$
$$T_1 = 293 \text{ K} \qquad T_2 = 228 \text{ K}$$
$$P_1 = 730 \text{ mm Hg} \qquad P_2 = 69 \text{ mm Hg}$$

The volume can be calculated by substitution into the combined gas law equation. However, you should also note the effect of the change of both pressure and temperature on the volume using the appropriate factors. Because of the decrease in temperature, the volume must decrease by a factor of 228/293. Because of the pressure decrease, the volume must increase by a factor of 730/69.

$$V_2 = V_1 \times \frac{T_2}{T_1} \times \frac{P_1}{P_2} \qquad \text{or} \qquad V_2 = V_1 \times T_{factor} \times P_{factor}$$

$$V_2 = 2.0 \times 10^4 \text{ L} \times \frac{228 \text{ K}}{293 \text{ K}} \times \frac{730 \text{ mm Hg}}{69 \text{ mm Hg}}$$

$$V_2 = 1.646535 \times 10^5 \text{ L}$$

The correct answer expressed to the required two significant figures is 1.6×10^5 L.

[Additional examples may be found in 12.35–12.40 at the end of the chapter.]

12.9 Avogadro's Hypothesis

The behavior of gases as shown in the preceding sections does not depend on the nature of the gas. The gas laws are due to a common feature of all gases. **Avogadro's hypothesis** (1811) states that equal volumes of gases under the same conditions of temperature and pressure contain the same number of submicroscopic particles. For example, equal volumes of hydrogen and oxygen at 1 atm pressure and 0°C (273 K), called standard temperature and pressure or **STP,** contain the same number of molecules (Figure 12.6).

Avogadro's hypothesis can be used to determine the relative weight of atoms and molecules. Because equal volumes of gases under the same conditions of temperature and pressure contain the same number of submicroscopic particles, the weight of the gas particles must be in the same ratio as the gas densities. The densities of the monatomic gases helium and argon at standard temperature and pressure are 0.179 and 1.79 g/L, respectively, and indicate that one argon atom is ten times as massive as one helium atom.

The concept of the mole and Avogadro's hypothesis allow the determination of the molecular weight of an unknown gaseous compound. The volume occupied by 1 mole (32 g) of oxygen molecules at standard temperature and pressure has been determined to be 22.4 L. Since equal volumes of gases contain the same number of particles, 22.4 L must be the volume occupied by 1 mole of any gaseous substance at standard temperature and pressure. Thus 22.4 L at STP is the **molar volume** of a gas.

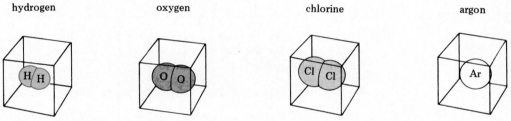

Figure 12.6

Avogadro's hypothesis and the volume of a gas

The number of molecules in a defined volume does not depend on the nature of the gas. Thus the volume occupied by equal numbers of hydrogen, oxygen, or chlorine molecules and argon atoms is the same under the same conditions of pressure and temperature.

Example 12.8

Calculate the density of CO_2 at STP.

Solution   The molecular weight (mol wt) of CO_2 is

$$\text{mol wt } CO_2 = \text{at wt C} + 2(\text{at wt of O})$$
$$= 12.0 \text{ amu} + 2(16.0 \text{ amu})$$
$$= 44.0 \text{ amu}$$

Therefore, 22.4 L of CO_2 at STP must have a mass of 44.0 g. The density of CO_2 can be calculated as follows.

$$CO_2 \text{ density} = \frac{44.0 \text{ g}}{22.4 \text{ L}} = 1.964285 \text{ g/L}$$

The correct answer expressed to the required three significant figures is 1.97 g/L.

Example 12.9

A 1.12 L sample of butane gas at 0.200 atm and 0°C weighs 0.58 g. What is the molecular weight of the gas?

Solution   It is first necessary to calculate what the volume would be at STP in order to make use of the molar volume quantity of 22.4 L/mole. The gas is already at standard temperature. Only the effect of pressure must be calculated, using Boyle's law.

$$V_2 = \frac{P_1}{P_2} \times V_1 \qquad \text{or} \qquad V_2 = V_1 \times P_{factor}$$

$$V_2 = 1.12 \text{ L} \times \frac{0.200 \text{ atm}}{1.00 \text{ atm}} = 0.224 \text{ L}$$

The volume is larger at the lower pressure. Now the number of moles present in the sample can be calculated.

$$\text{number of moles} = 0.224 \text{ L} \times \frac{1 \text{ mole}}{22.4 \text{ L}} = 0.0100 \text{ mole}$$

Once the number of moles present in the sample is known, the mass of the sample can be used to calculate the mass of a mole.

$$\text{mass of 1 mole} = \frac{0.58 \text{ g}}{0.0100 \text{ mole}} = 58 \text{ g/mole}$$

The molecular weight is 58 amu.

[*Additional examples may be found in 12.41–12.50 at the end of the chapter.*]

12.10   Ideal Gas Law

Now that Avogadro's hypothesis has been discussed, the significance of a constant value of k for the same initial volumes of various gases under the same conditions can be understood. The same initial volumes, according to Avogadro's hypothesis, contain the same number of atoms or molecules.

The gas laws can be combined as a more general expression, known as the **ideal gas law equation.** The term n represents the number of moles of a gas, and R is a proportionality constant.

$$\frac{PV}{T} = k = nR \qquad \text{or} \qquad PV = nRT$$

The proportionality constant R is called the **universal gas constant.** It can be evaluated from the knowledge that 1 mole of a gas at standard temperature (273 K) and pressure (1 atm) occupies 22.4 L.

$$R = \frac{PV}{nT} = \frac{1 \text{ atm} \times 22.4 \text{ L}}{1 \text{ mole} \times 273 \text{ K}} = 0.0821 \frac{\text{L atm}}{\text{mole K}}$$

To three significant figures the value for R is 0.0821 and the units are (L atm mole^{-1} K^{-1}), read as "liter atmosphere per mole per Kelvin."

Example 12.10

A mass of 1.34 g of a gas occupies 2.00 L at 91°C and 0.500 atm. What is the molecular weight of the gas?

Solution The number of moles in this sample is

$$n = \frac{PV}{RT} = \frac{0.500 \text{ atm} \times 2.00 \text{ L}}{0.0821 \text{ L atm mole}^{-1} \text{ K}^{-1} \times 364 \text{ K}} = 0.0334623 \text{ mole}$$

Rounding to the required three significant figures the 0.0335 mole of the gas has a mass of 1.34 g. Thus the mass of 1 mole is

$$\frac{1.34 \text{ g}}{0.0335 \text{ mole}} = 40.0 \text{ g/mole}$$

[*Additional examples may be found in 12.51–12.58 at the end of the chapter.*]

12.11 Stoichiometry and Gas Laws

In Chapter 7 you learned how to do stoichiometry problems by using the mole ratio obtained from a balanced chemical equation. If the mass of a substance was given, you converted the mass to moles by dividing by the molecular weight and then used the mole ratio to obtain the number of moles of the desired substance. If the mass of the desired substance was required, you converted the moles to grams by multiplying by the molecular weight. Now that you know the molar volume of a gas as well as the ideal gas equation, stoichiometric calculations can be done using the volume of gases.

Volume–Volume Stoichiometry Problems

Since chemical reactions are often carried out under conditions of constant pressure and temperature, the volume of the gases used is proportional to the number of moles (Avogadro's hypothesis). Therefore the coefficients of the equation that give the relative number of moles of reactants and products also give the relative volumes of gaseous reactants and products. For example, the reaction to form 2 moles of water requires 2 moles of hydrogen and 1 mole of oxygen (Figure 12.7).

$$2 \text{ H}_2(g) + \text{O}_2(g) \longrightarrow 2 \text{ H}_2\text{O}(g)$$

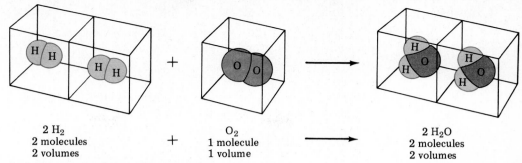

2 H$_2$
2 molecules
2 volumes

O$_2$
1 molecule
1 volume

2 H$_2$O
2 molecules
2 volumes

Figure 12.7

Stoichiometry and the volume of gases
According to Avogadro's hypothesis, equal volumes of gaseous hydrogen, oxygen, and water contain the same number of molecules under the same conditions of pressure and temperature. Thus the volumes of the gases involved in a chemical reaction will stand in the same ratio as the coefficients of the balanced chemical equation.

Example 12.11

Methyl alcohol is produced industrially from carbon monoxide and hydrogen at 375°C and 3000 psi using a catalyst of chromium(III) oxide and zinc oxide. How many liters of gaseous methyl alcohol can be produced from 2.0×10^3 L of hydrogen gas, assuming that sufficient carbon monoxide is available.

$$CO(g) + 2 H_2(g) \longrightarrow CH_3OH(g)$$

Solution Since all of the reactants and products are gaseous under the reaction conditions, the coefficients of the equation not only give the relative number of moles of reactants and products, but the relative volumes as well. Thus, the coefficients can be used as a volume ratio as well as a mole ratio.

$$2.0 \times 10^3 \text{ L H}_2 \times \frac{1 \text{ volume CH}_3\text{OH}}{2 \text{ volumes H}_2} = 1.0 \times 10^3 \text{ L CH}_3\text{OH}$$

Note that there was not any need to convert the stated conditions to STP or to use the ideal gas law equation. The concept of Avogadro applies to any samples of gases under the same set of conditions.

[*Additional examples may be found in 12.59 and 12.60 at the end of the chapter.*]

Mole–Volume Stoichiometry Problems

In this type of stoichiometry problem at least one of the substances is a gas. If the known quantity is in mole units, the unknown quantity may be calculated in volume units; if the known is in volume units, the unknown may be calculated in mole units. In either case you may have to use the fact that 1 mole of any gas at STP occupies 22.4 L. If the conditions are not at STP, then you will have to use the ideal gas law equation to calculate the number of moles of the substance for which the volume is given or required.

Example 12.12

How many liters of oxygen gas at STP can be produced from the reaction of 1.00 mole of potassium nitrate according to the following equation?

$$2 \text{ KNO}_3(s) \longrightarrow 2 \text{ KNO}_2(s) + \text{O}_2(g)$$

Solution Since the known quantity is given in moles, we can calculate the moles of the desired substance, oxygen, by using the mole ratio.

$$1.00 \text{ mole KNO}_3 \times \frac{1 \text{ mole O}_2}{2 \text{ moles KNO}_3} = 0.500 \text{ mole O}_2$$

Now the 0.500 mole of oxygen can be converted into volume units. Since the conditions are STP, we can use the fact that 1 mole of any gas at STP occupies 22.4 liters.

$$0.500 \text{ mole O}_2 \times \frac{22.4 \text{ L}}{1 \text{ mole}} = 11.2 \text{ L O}_2$$

Example 12.13

How many moles of potassium chlorate are required to produce 34.0 mL of oxygen gas at 740 mm Hg and 27°C according to the following equation?

$$2 \text{ KClO}_3(s) \longrightarrow 2 \text{ KCl}(s) + 3 \text{ O}_2(g)$$

Solution It is first necessary to determine how many moles of oxygen gas are required based on the experimental conditions given. The ideal gas law provides a way of calculating the number of moles, given the volume, pressure, and temperature. Remember that the pressure must be expressed in atm and the temperature in Kelvin.

$$n = \frac{PV}{RT} = \frac{(740/760 \text{ atm}) \times (0.0340 \text{ L})}{(0.0821 \text{ L atm mole}^{-1} \text{ K}^{-1}) \times (300 \text{ K})} = 1.34\underline{4} \times 10^{-3} \text{ mole}$$

Now the number of moles of potassium chlorate required to produce the calculated number of moles of oxygen may be determined by using the mole ratio obtained from the coefficient of the equation.

$$1.34\underline{4} \times 10^{-3} \text{ mole O}_2 \times \frac{2 \text{ moles KClO}_3}{3 \text{ moles O}_2} = 8.960 \times 10^{-4} \text{ mole KClO}_3$$

The correct answer expressed to the required three significant figures is 8.96×10^{-4} mole.

[*Additional examples may be found in 12.61 and 12.62 at the end of the chapter.*]

Volume–Mass Stoichiometry Problems

At this point you should be quite familiar with the use of volumes of gases in solving stoichiometry problems. We now have one last type of stoichiometry problem to examine. You will be given the known quantity in mass units and asked to calculate the unknown in volume units or be given the known in volume units and asked to calculate the unknown in mass units. You have to be able to convert mass into moles or vice versa. In addition, you must be able to convert the volume of a gas into moles or vice versa.

Example 12.14

How many liters of hydrogen gas will be produced at STP from the reaction of 3.91 g of potassium with sufficient water?

$$2 \text{ K(s)} + 2 \text{ H}_2\text{O(l)} \longrightarrow 2 \text{ KOH(aq)} + \text{H}_2\text{(g)}$$

Solution First the mass of potassium must be converted into moles.

$$3.91 \text{ g K} \times \frac{1 \text{ mole K}}{39.1 \text{ g K}} = 0.100 \text{ mole K}$$

Now the coefficient of the equation can be used to calculate the moles of hydrogen gas produced by the given quantity of potassium.

$$0.100 \text{ mole K} \times \frac{1 \text{ mole H}_2}{2 \text{ moles K}} = 0.0500 \text{ mole H}_2$$

Since the experimental conditions are at STP, the number of moles of hydrogen gas can be converted into volume by using the molar volume.

$$0.0500 \text{ mole H}_2 \times \frac{22.4 \text{ L H}_2}{1 \text{ mole H}_2} = 1.12 \text{ L H}_2$$

Example 12.15

What mass of copper(II) oxide will react with 6.20 L of hydrogen gas measured at 730 mm Hg and 100°C?

$$\text{CuO(s)} + \text{H}_2\text{(g)} \longrightarrow \text{Cu(s)} + \text{H}_2\text{O(g)}$$

Solution The volume of hydrogen given must first be converted into its equivalent number of moles using the ideal gas law equation, since the conditions are not at STP.

$$n = \frac{PV}{RT} = \frac{(730/760 \text{ atm}) \times (6.20 \text{ L})}{(0.0821 \text{ L atm mole}^{-1} \text{ K}^{-1}) \times (373 \text{ K})} = 1.94\underline{5} \times 10^{-1} \text{ mole}$$

Now the number of moles of copper(II) oxide that will react can be calculated using the mole ratio obtained from the coefficients of the equation.

$$1.94\underline{5} \times 10^{-1} \text{ mole H}_2 \times \frac{1 \text{ mole CuO}}{1 \text{ mole H}_2} = 1.94\underline{5} \times 10^{-1} \text{ mole CuO}$$

Finally, the number of grams of CuO can be calculated by multiplying by the formula weight of CuO.

$$1.94\underline{5} \times 10^{-1} \text{ mole CuO} \times \frac{79.54 \text{ g CuO}}{1 \text{ mole CuO}} = 15.470530 \text{ g CuO}$$

The correct answer expressed to the three significant figures required in this problem is 15.5 g.

[*Additional examples may be found in 12.63 and 12.64 at the end of the chapter.*]

12.12 Dalton's Law of Partial Pressures

John Dalton, of atomic theory fame, also studied the properties of gases. His **law of partial pressures** states that in a gas mixture the sum of the partial pressures of all of the gases is equal to the total pressure. Dalton's law expressed mathematically is

$$P_{total} = P_1 + P_2 + P_3 + \ldots$$

where the subscripted P values are the partial pressures of the gases in the mixture. The **partial pressure** of a gas is equal to the pressure it would exert if it were the only gas present under the experimental conditions.

Gases are commonly collected by water displacement in college chemistry laboratories (Figure 12.8). One such gas is oxygen. Potassium chlorate, when heated, decomposes to produce potassium chloride and oxygen.

$$2 \text{ KClO}_3 \longrightarrow 2 \text{ KCl} + 3 \text{ O}_2$$

The oxygen, which is collected by the downward displacement of water from an inverted jar, is not pure but contains some water vapor. The pressure of the collected gas is equal to the external atmospheric pressure when the level of the water inside the jar is equal to the level in the large water bath. That pressure is a sum of the partial pressures of the two gases in the jar, which are water vapor and oxygen.

$$P_{total} = P_{atm} = P_{O_2} + P_{H_2O}$$

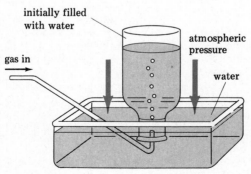

Figure 12.8
Dalton's law of partial pressures
The total pressure of a gas collected over water is equal to the sum of the partial pressure of the gas and the partial pressure of the water vapor.

In order to determine the pressure of the oxygen gas, it is necessary to subtract the partial pressure of the water vapor at the temperature of the experiment.

Example 12.16

A 200 mL sample of oxygen is collected at 26°C over water. The vapor pressure of water is 25 mm Hg at 26°C, and the atmospheric pressure is 750 mm Hg at the time of the experiment. Calculate the partial pressure of oxygen, and determine the volume that the dry oxygen would occupy at 26°C and 750 mm Hg.

Solution The partial pressure of oxygen is determined by use of the law of partial pressures.

$$P_{total} = P_{atm} = P_{O_2} + P_{H_2O}$$
$$750 \text{ mm Hg} = P_{O_2} + 25 \text{ mm Hg}$$
$$P_{O_2} = 750 \text{ mm Hg} - 25 \text{ mm Hg} = 725 \text{ mm Hg}$$

Now the problem may be treated using Boyle's law. If 200 mL of dry oxygen at 725 mm Hg was subjected to a new pressure of 750 mm Hg, the volume would decrease.

$$V_2 = 200 \text{ mL} \times \frac{725}{750} = 193 \text{ mL}$$

This means that only 193 mL of the 200 mL is actually oxygen.

[*Additional examples may be found in 12.69 and 12.70 at the end of the chapter.*]

Another illustration of the law of partial pressures is the air of our atmosphere. Air consists mostly of nitrogen and oxygen and, in lesser amounts, water vapor and carbon dioxide. The total pressure exerted by air is a sum of the partial pressures of the individual gases. At 1 atm, the partial pressures of nitrogen, oxygen, and water vapor are approximately 595, 159, and 6 mm Hg, respectively; that of carbon dioxide is lower. The 159 mm Hg pressure of oxygen is 20.9% of the total atmospheric pressure. At higher altitudes, the total pressure of air decreases but the percent composition remains the same. Thus the partial pressures of all the components of air decrease. For example, at an altitude of 1 mile, the total pressure is 630 mm Hg, and the partial pressures of nitrogen, oxygen, and water vapor are 494, 132, and 4 mm Hg, respectively.

Our bodies function efficiently if the partial pressure of oxygen is about 160 mm Hg. Thus, an individual accustomed to living at sea level cannot operate as efficiently and tires more easily at higher altitudes. At the higher altitude the percent of oxygen is still about 21%, but the partial pressure is less than 160 mm Hg. As a result of the lower partial pressure of oxygen at higher altitudes, the red blood cells absorb a smaller amount of oxygen. However, if one stays at the high altitude for a time, the body adapts by producing more red blood cells. A high partial pressure of oxygen can be detrimental to the body. For example, newborn infants who are placed in oxygen atmospheres for a long period of time may develop damage to the retinal tissue and become either partially or totally blind.

Deep sea divers or scuba divers breathe air at a pressure greater than atmospheric pressure. If compressed air is used, the high partial pressure of nitrogen can cause nitrogen narcosis. This condition causes the diver to become ''intoxicated,'' and as a result, the diver may develop irrational behavior that could cause death. Because helium does not have a narcotic effect on the body, helium is often used to dilute the oxygen for diving. The percent oxygen used in the mixture is less than 21%, because the pressure of the gas mixture is greater than 1 atm. However, the partial pressure of the oxygen is maintained at approximately 160 mm Hg.

12.13 Graham's Law of Effusion

Effusion is a process in which gas molecules pass through a small opening in a container from a region of high pressure to one of low pressure (Figure 12.9). Graham examined the rate of this process in 1829 and observed that under the same conditions the relative rates of effusion of two gases at the same temperature are inversely proportional to the square root of their densities. Since the densities of gases are directly proportional to their atomic or molecular weights, **Graham's law** may be stated: The rate of effusion of a gas is inversely proportional to the square root of the atomic or molecular weight. For two gases the mathematical relationship is

$$\frac{r_1}{r_2} = \frac{\sqrt{m_2}}{\sqrt{m_1}} = \sqrt{\frac{m_2}{m_1}}$$

where r_1 and r_2 represent the rates of effusion of the two gases and m_1 and m_2 are the corresponding atomic or molecular weights.

Example 12.17

A compound of oxygen and sulfur effuses at a rate that is one quarter that of helium. What is the molecular weight of the compound?

Solution Since the compound effuses at a slower rate than helium, it must have a molecular weight larger than that of helium. Substitute the given values into the formula, and solve the equation by squaring both sides.

$$\frac{r_{He}}{r_2} = \frac{1}{0.25} = \sqrt{\frac{m_2}{m_1}} = \sqrt{\frac{m_2}{4 \text{ amu}}}$$

$$4^2 = 16 = \frac{m_2}{4 \text{ amu}}$$

$$m_2 = 16 \times 4 \text{ amu} = 64 \text{ amu}$$

[*Additional examples may be found in 12.71–12.78 at the end of the chapter.*]

The fifth postulate of the kinetic molecular theory can be used to derive Graham's law. At the same temperature the average kinetic energies (KE) of all gases are equal. Since the average kinetic energy is defined as $\frac{1}{2} mv^2$, where v is the average velocity, we can write

$$KE_1 = KE_2$$
$$\tfrac{1}{2} m_1 v_1{}^2 = \tfrac{1}{2} m_2 v_2{}^2$$

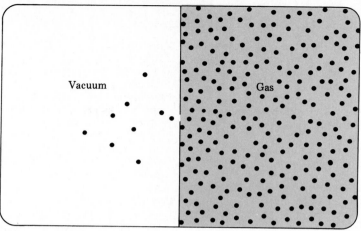

Figure 12.9

Effusion of a gas
Gas molecules move from a high-pressure region to a low-pressure region through a small hole.

Rearranging the equation and solving for v_1/v_2 gives

$$\frac{v_1{}^2}{v_2{}^2} = \frac{m_2}{m_1}$$

$$\frac{v_1}{v_2} = \sqrt{\frac{m_2}{m_1}}$$

Recognizing that the rate at which gas effuses is directly proportional to the average velocity, this equality is Graham's law.

$$\frac{v_1}{v_2} = \frac{r_1}{r_2} = \sqrt{\frac{m_2}{m_1}}$$

Summary

Gases may be described by the pressure, volume, temperature, and amount of the gas. The gas pressure is measured by comparing it to the pressure exerted by a column of mercury. Some of the units of pressure are atmospheres, cm Hg, and mm Hg (or torr).

The theoretical basis for understanding the behavior of gases is the kinetic molecular theory. Different gases at the same temperature move with the same average kinetic energy. Real gases behave according to the theory only at high temperatures and low pressures. Under these conditions, gas molecules are widely separated and experience only low attractive forces.

Relationships between the pairs of the four gas variables (P, V, T, and n), the other two remaining constant, are defined by Boyle's law, which relates pressure and volume; by Charles's law, which relates volume and temperature, by Gay-Lussac's law, which relates pressure and temperature; and by Avogadro's hypothesis, which relates volume and the amount of the gas.

The individual gas laws can be combined into one comprehensive gas law. By using a constant called the universal gas law constant, a general expression known as the ideal gas law is obtained. Stoichiometric calculations can be done with gas volumes using either the molar volume of the gas at STP, or the ideal gas law equation under any condition of temperature and pressure.

Dalton's law of partial pressures relates the properties of gas mixtures. The partial pressure of a gas can be treated using any of the gas laws. Graham's law of effusion relates the rate of effusion of a gas and its molecular weight.

New Terms

One **atmosphere,** the pressure exerted by the atmosphere under normal conditions, supports a column of mercury 760 mm high.

Avogadro's hypothesis states that equal volumes of gases under the same conditions contain the same number of atoms or molecules.

A **barometer** is a device to measure atmospheric pressure.

Boyle's law states that the volume of a fixed amount of a gas at constant temperature varies inversely with pressure.

Charles's law states that the volume of a fixed amount of a gas at constant pressure is directly proportional to the absolute temperature.

Dalton's law of partial pressures states that the sum of the partial pressures of the gases in a mixture equals the total pressure of the mixture.

Effusion is a process whereby gaseous material is transferred through a small opening from a region of high pressure to a region of low pressure.

Gay-Lussac's law states that at constant volume, the pressure of a fixed amount of a gas is directly proportional to the absolute temperature.

Graham's law states that the rate of effusion of a gas is inversely proportional to the square root of the atomic or molecular weight of the gas.

An **ideal gas** is a gas whose behavior is predicted by the ideal gas law and that conforms to the kinetic molecular theory.

The **ideal gas law equation** relates the pressure, volume, temperature, and number of moles of a gas.

The **kinetic molecular theory** is a model that describes gas behavior based on the concept of an average kinetic energy that is directly proportional to the absolute temperature.

The **molar volume** is 22.4 moles/liter at STP.

A **partial pressure** is the pressure exerted by a single gas in a gaseous mixture.

Pressure is the force per unit area. In gases the pressure is measured in terms of the height of a column of mercury that can be maintained by the gas.

A **real gas** is any gaseous substance that actually exists.

STP or **standard temperature and pressure** is 0°C (273 K) and 1 atm.

A **torr** is 1 mm Hg.

The **universal gas constant,** R, is a proportionality constant of the ideal gas law equation.

Exercises

Terminology

12.1 How is a real gas different from the ideal gas?

12.2 Describe the kinetic molecular theory.

Pressure

12.3 Describe why a "pressure group" is effective in politics.

12.5 How many mm Hg equal 1 torr?

12.7 How many torr are equal to 1 atm?

12.9 Explain why the pressure on Mt. McKinley might be only 350 torr.

12.11 Convert each of the following pressures into the indicated unit.

    (a) 380 mm Hg into atm   (b) 0.500 atm into cm Hg
    (c) 0.100 atm into torr   (d) 190 torr into cm Hg
    (e) 13.5 psi into atm   (f) 12.8 psi into cm Hg

12.13 Using a density of water equal to 1.00 g/cm^3, calculate the pressure exerted by water in psi at a depth of 100 ft in a lake.

12.4 What is atmospheric pressure? Is it a constant value?

12.6 How many cm Hg are equal to 1 atm?

12.8 How many mm Hg are equal to 1 atm?

12.10 Why did the astronauts wear protective suits when they were on the moon?

12.12 Convert each of the following pressures into the indicated unit.

    (a) 0.750 atm into mm Hg   (b) 38 cm Hg into atm
    (c) 520 torr into atm   (d) 19.0 cm Hg into torr
    (e) 0.90 atm into psi   (f) 57 cm Hg into psi

12.14 Using a density of air equal to 1.2 g/L, calculate the pressure at the bottom of a mine shaft at 500 m on a day when the pressure at the top of the shaft is 760 mm Hg.

Kinetic Molecular Theory

12.15 A sample of a gas is cooled from 80°C to 40°C. How is the average kinetic energy affected?

12.16 A sample of a gas is heated from −73°C to 127°C. How is the average kinetic energy affected?

Boyle's Law

12.17 A sample of a gas at 600 torr and 25°C occupies 300 mL. What volume will the gas occupy at 800 torr and 25°C?

12.18 Calculate the volume of a gas in liters at 1.0 atm if it occupies 2.0 L at 158 cm Hg.

12.19 A sample of a gas at 0.5 atm and 100°C occupies 2 L. What pressure is necessary to cause the gas to occupy 0.5 L?

12.20 Calculate the pressure required to convert 500 mL of a gas at 0.200 atm into a volume of 10 mL.

12.21 Gases in cylinders used in chemistry labs may be at 2000 psi. What volume would 10 L of such a compressed gas occupy at 1 atm?

12.22 A scuba tank has a volume of 13.2 L. What volume of air in liters at 720 torr is required to fill the tank to a pressure of 150 atm assuming no change in temperature?

Charles's Law

12.23 A sample of a gas at −91°C and 1 atm occupies 2.0 L. What volume will the gas occupy at 0°C at the same pressure?

12.24 A 2.0 L sample of air at −23°C is warmed to 100°C. What is the new volume if the pressure remains constant?

12.25 A sample of a gas at 300 K and 700 torr occupies 60 mL. What temperature is necessary to increase the volume to 75 mL at the same pressure?

12.26 In order to change the volume of a 1200 mL sample of a gas at 250 K to 300 mL, what temperature must be reached, assuming the pressure is kept constant?

12.27 A 300 mL sample of helium at 127°C is cooled to 150 K. What is the volume if the pressure is kept constant?

12.28 A 150 cm^3 sample of argon gas exerts a pressure of 820 mm Hg at 25°C. At what temperature will the volume decrease to 125 cm^3 if the pressure is kept constant?

Gay-Lussac's Law

12.29 A sample of a gas in a rigid container is heated from 273 K to 273°C. If the gas was initially at 1 atm, what is the final pressure?

12.30 The pressure in an aerosol container is 2.7 atm at 27°C. What will be the pressure if the temperature is lowered to −73°C?

12.31 A sample of gas in a rigid container at −23°C is at a pressure of 500 torr. What temperature will be necessary to change the pressure to 800 torr?

12.32 The pressure in an automobile tire is 30 psi at 27°C at the start of a trip. At the end of a trip the pressure is 33 psi. What is the temperature of the air in the tire?

12.33 A sample of a gas in a container of fixed volume at 28°C is at a pressure of 0.50 atm. What temperature (°C) is required to give a pressure of 0.75 atm?

12.34 A 0.50 L container contains nitrogen at a pressure of 800 torr and a temperature of 0°C. The container can only withstand a pressure of 2.5 atm. What is the highest temperature (°C) to which the container may be heated?

Combined Gas Law

12.35 A gas occupies 250 mL at 700 torr and 300 K. What volume will the gas occupy at 350 torr and 450 K?

12.36 At STP a gas has a volume of 5.0 L. What volume will the gas occupy at 80°C and 800 torr?

12.37 A gas occupies 800 mL at 1 atm and 250 K. At what pressure will the gas occupy 400 mL at 500 K?

12.38 A 2.0 L sample of air at −50°C has a pressure of 700 torr. What will be the new pressure if the temperature is raised to 50°C and the volume is increased to 4.0 L?

12.39 A gas occupies 2 L at 127°C and 2 atm. At what temperature will the gas occupy 3 L at 1 atm?

12.40 A gas occupies 1 L at 27°C and 0.5 atm. At what temperature will the gas occupy 0.5 L at 1 atm?

Avogadro's Hypothesis and Molar Volume

12.41 What statement can be made about the number of molecules contained in a 2 L sample of H_2 molecules at STP and the number of atoms in a 2 L sample of helium at STP?

12.42 Why are the densities of gases under the same pressure and temperature different?

12.43 Using the molar volume calculate the volume occupied by 1.0 g of hydrogen gas at STP. Compare that volume to the volume occupied by 16.0 g of oxygen gas at STP.

12.44 Using the molar volume, determine which sample, 10 g of neon or 10 g of argon, occupies the larger volume at STP.

12.45 Convert each of the following samples to STP and determine how many moles of a gas are present in each sample.
(a) 1.12 L at 2.00 atm and 0°C
(b) 2.99 L at 1.00 atm and 91°C
(c) 560 mL at 5.00 atm and 182°C
(d) 22.4 L at 273°C and 2 atm

12.46 Convert each of the following samples to STP and determine how many molecules are present in each sample.
(a) 2.24 L of H_2 at 2.00 atm and 0°C
(b) 5.98 L of O_2 at 1.00 atm and 91°C
(c) 280 mL of CO_2 at 5.00 atm and 182°C
(d) 22.4 L of CH_4 at 2 atm and 273°C

12.47 How many molecules of sulfur dioxide (SO_2) are contained in a 1 mole sample of the gas at STP? How many molecules are in 1 L at STP?

12.48 How many molecules of carbon dioxide (CO_2) are contained in a 1 mole sample of the gas at STP? How many molecules are in 1 L at STP?

12.49 A 1.12 L sample of a gaseous compound has a mass of 2.2 g at STP. Using the molar volume, calculate the mass of a molecule of this compound in atomic mass units.

12.50 A 4.48 L sample of a gaseous compound has a mass of 0.80 g at STP. Using the molar volume, calculate the molecular weight of this compound in atomic mass units.

Ideal Gas Law

12.51 A hydrocarbon weighing 0.185 g occupies 110 cm^3 at 25°C and 74 cm Hg. What is the molecular weight of the gas?

12.53 A 5.6 L sample of a gas at 182°C and 38 cm Hg has a mass of 4.8 g. What is the mass of a mole of this gas?

12.55 A small cylinder of helium for use in chemistry lectures has volume of 334 mL. How many moles of helium are contained in the cylinder at a pressure of 150 atm and 25°C?

12.57 Calculate the density of C_2H_6 at 0°C and 2 atm.

12.52 A 51 mL sample of a diatomic gas at 10°C and 725 torr weighs 58 mg. What is the atomic weight of the gas?

12.54 How many moles of Cl_2 are contained in a 10.3 L tank at 25°C if the pressure is 725 torr?

12.56 A 25 L cylinder contains 140 g of nitrogen gas at 20°C. How many grams of nitrogen must be released to reduce the pressure to 1.5 atm?

12.58 Calculate the density of CH_4 at 273°C and 2 atm.

Gas Stoichiometry

12.59 What volume of hydrogen (in L) at STP is required to react with 10 L of C_2H_2 according to the following equation?

$$2 H_2(g) + C_2H_2(g) \longrightarrow C_2H_6(g)$$

12.61 What volume of hydrogen (in L) at STP would be required to react with 0.100 mole of nitrogen to form ammonia?

$$N_2(g) + 3 H_2(g) \longrightarrow 2 NH_3(g)$$

12.63 What mass of magnesium is required to produce 5.15 L of hydrogen at STP by the following reaction?

$$Mg(s) + 2 HCl(aq) \longrightarrow MgCl_2(aq) + H_2(g)$$

12.60 What volume of oxygen at STP is required for the complete combustion of 5.0 L of butane (C_4H_{10})?

$$2 C_4H_{10}(g) + 13 O_2(g) \longrightarrow 8 CO_2(g) + 10 H_2O(g)$$

12.62 What volume of hydrogen (in L) at STP is required to produce 0.80 mole of HCl by the following reaction?

$$H_2(g) + Cl_2(g) \longrightarrow 2 HCl(g)$$

12.64 What volume of oxygen at STP is required for the complete combustion of 1.14 g of octane (C_8H_{18}) according to the following equation?

$$2 C_8H_{18}(l) + 25 O_2(g) \longrightarrow 16 CO_2(g) + 18 H_2O(g)$$

Dalton's Law of Partial Pressures

12.65 A mixture of three gases has the following partial pressures: oxygen, 100 torr; nitrogen, 300 torr; hydrogen, 150 torr. What is the total pressure of the mixture? If carbon dioxide were added until the pressure reached 750 torr, what would be the partial pressure of the carbon dioxide?

12.67 A mixture of 15 g Ne and 30 g Ar occupies 4.0 L at 10.4 atm and 75°C. What is the partial pressure of neon?

12.69 If 300 mL of nitrogen gas at 760 torr was bubbled through water at 26°C and collected, what would be the volume of the wet gas sample?

12.66 The partial pressure of nitrogen on Mt. McKinley is 288 torr on a day when the atmospheric pressure is 0.480 atm. What is the partial pressure of oxygen?

12.68 The U.S. Navy developed an undersea habitat whose atmosphere contains 79% helium, 17% nitrogen, and 4% oxygen by volume. What is the partial pressure of oxygen if the pressure within the habitat is 6.0 atm?

12.70 If 250 mL of argon gas at 740 torr was bubbled through water at 26°C and collected, what would be the volume of the wet gas sample?

Graham's Law of Effusion

12.71 Calculate the relative rate of effusion of O_2 compared to O_3.

12.73 Calculate the relative rate of effusion of SO_2 compared to CO_2.

12.75 Heavy water, D_2O (mol wt = 20.03) can be separated from ordinary water by the difference in the relative rates of effusion of the molecules in the gas phase. Calculate the relative rate of effusion of H_2O to D_2O.

12.77 In an effusion apparatus, H_2 is found to effuse at the rate of 5.9 milliliters per second (mL/s). Another gas in the same apparatus effuses at the rate of 0.55 mL/s. What is the molecular weight of the gas?

12.72 Calculate the relative rate of effusion of HCl compared to NH_3.

12.74 Calculate the relative rate of effusion of CO_2 compared to CO.

12.76 Under standard conditions, the density of He is 0.179 g/L and of Xe is 5.86 g/L. In an apparatus, helium effuses at the rate of 18 mL/min. At what rate will xenon effuse in the same apparatus?

12.78 It takes 14.5 s for 0.1 mL of nitrogen to effuse from a porous container. It takes 46 s for 0.2 mL of an unknown gas to effuse under the same conditions from the porous container. What is the molecular weight of the gas?

Additional Exercises

12.79 Explain why the mercury does not run out of a barometer.

12.80 How does the location beneath the ocean affect the pressure on an organism?

12.81 The pressure of gaseous mercury above liquid mercury is 0.0012 torr at 20°C. What is this pressure in atmospheres?

12.82 The pressure of gaseous gallium above liquid gallium at 1350°C is 1.3×10^{-3} atm. What is this pressure in mm Hg?

12.83 Calculate the mass of mercury in a barometer tube having a cross section of 1.0 cm^2 at 1 atm. The density of mercury is 13.6 g/cm^3.

12.84 What height of a column of water (cm) can be supported in a barometer tube on a day when the atmospheric pressure is 730 torr. Neglect the vapor pressure of water. The density of water and mercury are 1.00 g/cm^3 and 13.6 g/cm^3, respectively.

12.85 A gas bubble has a volume of 7.8 mL when released by a scuba diver at the bottom of a lake at 120 ft where the water pressure is 4.0 atm. What is the volume of the bubble when it reaches the surface of the lake where the pressure is 1.00 atm? Assume that the temperature is constant.

12.86 A weather balloon is inflated with helium. The balloon has a volume of 100 m^3, and it must be inflated to a pressure of 0.10 atm. If 40 L gas cylinders of helium at a pressure of 150 atm are used, how many cylinders are needed? Assume that the temperature is constant.

12.87 A sample of neon that occupies 125 cm^3 at 730 mm Hg and 30°C is heated to 60°C at constant pressure. What is the new volume?

12.88 An aerosol container has a pressure of 3.0 atm at 27°C. The can is left in the sun at 100°F. What is the pressure in the can?

12.89 A weather balloon is filled with helium to a volume of 30.0 L at 20°C and 1.0 atm. In the stratosphere the temperature and pressure are −23°C and 3.0×10^{-3} atm, respectively. What will be the volume in the stratosphere?

12.90 Gas evolved in the fermentation of sugar in making wine at home one day occupies a volume of 0.75 L at 20°C and 720 mm Hg. What volume would the gas occupy at 25°C and 1.00 atm?

12.91 The ozone molecules in the stratosphere absorb ultraviolet radiation from the sun. The pressure of ozone (O_3) is 1.5×10^{-7} atm and the temperature is −20°C. How many molecules are in 1.0 L under these conditions?

12.92 A pressure of 1.0×10^{-3} torr is obtained using a vacuum pump. Calculate the number of molecules in 1.0 mL of gas at this pressure at 20°C.

12.93 The concentration of carbon monoxide in a city one day is 100 parts per million (ppm). Calculate the partial pressure of the carbon monoxide if the atmospheric pressure is 730 mm Hg.

12.94 A 1.50 L container of H_2 at 765 mm Hg and 25°C is connected to a 2.50 L container of He at 740 mm Hg and 25.0°C. What is the total pressure after the gases have mixed if the temperature remains at 25°C?

12.95 A sheet of aluminum is placed in pure oxygen at 1.0 atm in a sealed 1.00 L container at 25°C. One hour later, the pressure is 0.9 atm. How many grams of oxygen have reacted with the aluminum?

12.96 A sample of phosphorus is placed in a mixture of nitrogen and oxygen at 1.0 atm in a sealed 1.00 L container at 25°C. After a period of time, the pressure is 0.8 atm. How many grams of oxygen have reacted with the phosphorus?

12.97 Two identical balloons are filled, one with helium and one with nitrogen at the same temperature. If the nitrogen balloon leaks at the rate of 64 mL per hour, what will be the rate of leakage from the helium balloon?

12.98 Two identical porous containers are filled, one with hydrogen and one with carbon dioxide at the same temperature. If 1.6 mL of carbon dioxide leaks from the container in one day, how much hydrogen will leak in one day?

13 | Liquids and Solids

Learning Objectives

After studying Chapter 13 you should be able to

1. Describe the liquid and solid states using the kinetic molecular theory.
2. List the equilibria between the states of matter.
3. Use Le Châtelier's principle to explain the effects of pressure on changes of state.
4. Relate vapor pressure and temperature to the boiling point of a liquid.
5. Describe the three types of intermolecular forces and give examples of their effects on the physical properties of matter.
6. Give a summary of the unique properties of water.
7. Name hydrates of ionic compounds.
8. Calculate the percent composition of hydrates.

13.1 Equilibrium and the States of Matter

In this chapter and Chapter 15, which deals with chemical reactions, the concept of a dynamic equilibrium is presented. A **dynamic equilibrium** is a condition in which two or more opposing processes occur at the submicroscopic level at the same time and are in balance. As a result, on the macroscopic level a system at equilibrium appears to be static because there is no net observable change.

Consider your finances as an example of equilibrium. If the deposits and withdrawals to and from your checking account are controlled to give a constant balance, you are in financial equilibrium. Your checking account balance remains unchanged and appears to be static, but the same dollars are not always there. Your finances are dynamic.

$$\text{student's wallet} \underset{\text{withdrawal}}{\overset{\text{deposit}}{\rightleftharpoons}} \text{bank}$$

The concept of equilibrium is important in describing and understanding the liquid and solid states as compared to the gaseous state. In our model of the gaseous state the attractive forces between molecules are unimportant because the molecules are widely separated. In the liquid and solid states the molecules are closer together, and there is a balance between the forces that attract molecules to each other and their average kinetic energy, which keeps them moving apart. This balance is affected by both temperature, which changes the kinetic energy, and pressure, which controls how close the molecules are to each other. We will study the equilibrium between the liquid state and the gaseous state first in this chapter and then consider the equilibrium between the liquid and solid states. We will use the equilibrium concepts developed for physical processes in discussing chemical processes in subsequent chapters.

Condensation, the conversion of a gas into a liquid, occurs when a gas is cooled to a certain temperature at the prevailing atmospheric pressure. The reverse of condensation, **vaporization,** is the conversion of a liquid into a gas. According to the kinetic molecular theory, the average kinetic energy of a gas is proportional to the temperature and the molecules move more slowly as a gas is cooled. At a temperature when the average kinetic energy approaches the average forces of attraction between molecules, the molecules tend to stick together to form a liquid. These attractive forces are always present, but they cannot cause the formation of a liquid if the kinetic energy is too high. The nature of these attractive forces and how they affect the vapor pressure, boiling point, and heat of vaporization of liquids is discussed in this chapter.

Since the attractive forces between molecules hold them closer together in a liquid than they are in the gas phase, the density of a liquid is higher than that of a gas. The balance between attractive forces and the kinetic energy of the molecules is sufficient for a degree of cohesion but not sufficient to form a rigid structure. Consequently the cluster of moving liquid molecules are sufficiently attracted to one another to have a definite volume but not held firmly enough to have a definite shape. This concept explains why a liquid can be poured and will assume the shape of its container.

The further cooling of a liquid continues to decrease the kinetic energy of the atoms or molecules and allows the attractive forces between neighboring particles to exert a greater influence. Eventually cooling results in the formation of crystals in which the motion of the particles is restricted to vibrating about fixed positions. The molecules are held in a rigid structure with a definite volume and a definite shape. The conversion of a liquid into a solid is called **freezing.** The reverse of the process, conversion of a solid into a liquid, is **melting.** The melting point and heat of fusion of solids is discussed in this chapter.

13.2 Vapor Pressure of Liquids

You are probably familiar with the phenomenon of evaporation, in which matter is transferred from the liquid phase to the gas phase. In a liquid, individual particles are traveling at different speeds. Particles with high velocity may have enough kinetic energy to exceed the attractive forces of their neighbors and break away to enter the gas phase. The departure of particles of high kinetic energy results in a decrease in the average kinetic energy of the remaining particles. As a result, the temperature of the liquid decreases.

Any individual who has exercised and perspired or who has stood in a breeze immediately after emerging from a swimming pool is well acquainted with the cooling effect of evaporation. The water molecules that leave the surface of your skin most readily are the most energetic particles, and the remaining liquid is cooler.

If a heat source is available and the rate of evaporation is slow, the temperature of a liquid will be maintained during evaporation. For example, a glass of water in a room will evaporate slowly without any noticeable cooling (Figure 13.1). As the most energetic particles leave the liquid phase, heat is transferred from the surroundings to the liquid, and the temperature is maintained.

Evaporation of matter from the liquid phase is less likely as the temperature is decreased because the average kinetic energy of the particles is less. This is easily verified by experience. For example, it is more difficult to dry clothes outdoors on a cool day than on a hot day.

At the same temperature, different liquids evaporate at different rates. For example, gasoline evaporates faster than lubricating oil. Because the average kinetic energies of the particles in two different liquids at the same temperature are identical, this implies that the escaping tendency of the molecules depends on the attractive forces between neighboring molecules. If the

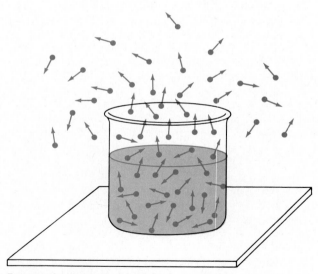

Figure 13.1

Evaporation of a liquid
The molecules that enter the gas phase tend to wander away from the beaker and not return to the liquid phase. As a consequence, eventually all of the liquid will evaporate.

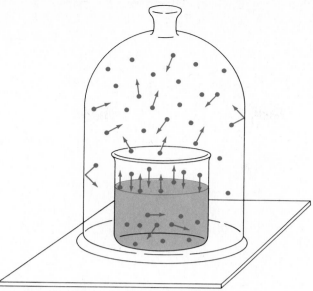

Figure 13.2

Equilibrium and vapor pressure of a liquid
Molecules of liquid leave the liquid phase and enter the
gas phase, while molecules from the gas phase return to
the liquid phase. The pressure exerted by the gaseous
molecules in equilibrium with the liquid is the vapor
pressure.

attractive forces between neighboring molecules are large, the escaping ability of the particles is smaller.

When a liquid is in a closed container, particles still leave the liquid phase and enter the gas phase. As the vapor particles become more numerous in the gas phase, they are more likely to collide with the liquid surface and return to the liquid phase. Eventually a balance is achieved. The rates at which the particles leave and return to the liquid phase become equal (Figure 13.2). When such a balance occurs, the system is in equilibrium and no change is observed on a macroscopic level, although evaporation and condensation are still occurring at the submicroscopic level.

$$\text{liquid} \underset{\text{condensation}}{\overset{\text{evaporation}}{\rightleftharpoons}} \text{vapor}$$

The particles in the gas phase exert a pressure like those of any gas. The pressure of a gas in equilibrium with its liquid phase is called the **vapor pressure** of the liquid. The vapor pressure is a physical property that indicates the escaping tendency of the liquid. Like other physical properties, the vapor pressure is a function of the temperature.

At the same temperature, although the average kinetic energies of various liquids are equal, all liquids do not have the same vapor pressure. Recall that the kinetic energy is equal of $\frac{1}{2}mv^2$, where m and v are the mass and average velocity, respectively, of the molecules. Therefore, heavy molecules have low velocities and light molecules have high velocities. It then might be expected that light molecules would have high vapor pressures, and heavy molecules would

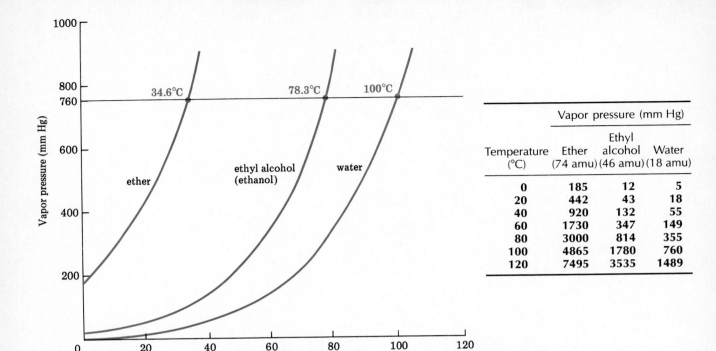

Figure 13.3
Vapor pressure of liquids and temperature

Temperature (°C)	Vapor pressure (mm Hg)		
	Ether (74 amu)	Ethyl alcohol (46 amu)	Water (18 amu)
0	185	12	5
20	442	43	18
40	920	132	55
60	1730	347	149
80	3000	814	355
100	4865	1780	760
120	7495	3535	1489

have low vapor pressures. Such a trend is observed for hexane (C_6H_{14}), octane (C_8H_{18}), and decane ($C_{10}H_{22}$) with vapor pressures 400, 50, and 9 mm Hg, respectively. However, there are many substances for which this type of correlation is not observed.

Figure 13.3 shows a graph of vapor pressures of diethyl ether ($C_4H_{10}O$), ethyl alcohol (C_2H_5OH), and water. At any temperature, the order of vapor pressures is diethyl ether > ethyl alcohol > water. The molecular weights of diethyl ether, ethyl alcohol, and water are 74, 46, and 18 amu, respectively. Clearly, molecular weight cannot be used to explain the observed vapor pressures. Based on molecular weight, the order of increasing vapor pressure would be diethyl ether < ethyl alcohol < water. The observed order of vapor pressures can be explained by considering the attractive forces between particles that affects the escape of molecules from the liquid phase. Water molecules attract one another very strongly, and their escaping tendency is low in spite of the low molecular weight. We will return to discuss this very common but unusual liquid in later sections. Ethyl alcohol also has large attractive forces between molecules, and it has a much lower vapor pressure than diethyl ether, which has a higher molecular weight.

Example 13.1

The vapor pressures of methyl alcohol (CH_3OH) and methyl iodide (CH_3I) at 25°C are 120 and 400 mm Hg, respectively. Interpret this data in terms of the intermolecular forces that are present in each compound.

Solution The molecular weights of methyl alcohol and methyl iodide are 32 and 142 amu, respectively. In spite of the larger molecular weight for CH_3I, this compound has the higher vapor pressure. Alternatively one could say that in spite of the lower molecular weight of methyl alcohol, this compound has the lower vapor pressure. Methyl alcohol has the larger intermolecular forces.

[*Additional examples may be found in 13.13–13.16 at the end of the chapter.*]

13.3 Boiling Point of Liquids

When heating a liquid there is a specific temperature at which it eventually undergoes a very pronounced transformation. Bubbles are formed throughout the liquid, rise rapidly to the surface, burst, and release vapor in large quantities. This process is called **boiling,** and the temperature at which it occurs is called the boiling point of the liquid.

At the **boiling point** the vapor pressure of the liquid equals the atmospheric pressure. For water, the temperature at which the vapor pressure equals 760 mm Hg is 100°C. Water, or any other liquid, can boil at many different temperatures if the external pressure is changed appropriately. The vapor pressure of water at 80°C is 355 mm Hg. If we lived on a planet with a normal atmospheric pressure of 355 mm Hg, water would boil at 80°C. At this temperature, the vapor pressure of the liquid would equal the atmospheric pressure. To avoid ambiguity, the standard or **normal boiling point** is defined as the boiling point at 1 standard atmosphere.

The fact that liquids boil at lower temperatures at lower pressures can be confirmed by anyone who cooks food in boiling water at high altitudes where the atmospheric pressure is less than

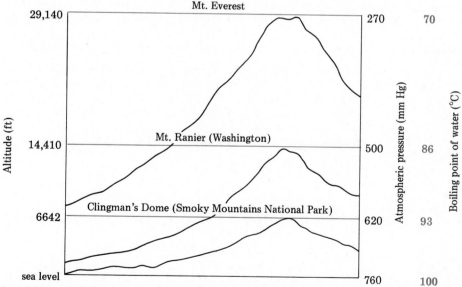

Figure 13.4

The effect of pressure on the boiling point of water
At higher altitudes the atmospheric pressure decreases. When the vapor pressure of water is equal to the atmospheric pressure, the water boils.

Table 13-1   **Boiling Point and Heat of Vaporization of Some Liquids**

Substance	Boiling point (K)	Molar mass	Heat of vaporization (cal/g)	Heat of vaporization (kcal/mole)
argon	87	39.9	39.1	1.56
krypton	121	83.8	25.8	2.16
xenon	166	131.3	23.0	3.02
methane	109	16.0	122	1.95
ammonia	240	17.0	329	5.60
water	373	18.0	540	9.73

1 atm. In Figure 13.4, the effect of altitude on the boiling point of water is illustrated. Water, or any liquid at its boiling point in an open container, stays at the temperature of boiling, and no amount of heat can raise the temperature above the boiling point. Food requires longer to cook at a lower boiling point, and cooking time is twice as long at an altitude of 7000 ft as at sea level. Conversely, food can be cooked faster at the high pressures in a pressure cooker. When the pressure gauge is set for 5 psi above atmospheric pressure, the boiling point of the water in the pressure cooker is about 108°C. Under these conditions, food cooks twice as fast as at 100°C.

Energy is required to maintain boiling and to transfer matter from the liquid to the gaseous phase. The heat energy added does not increase the temperature of the liquid at the boiling point; it maintains the average kinetic energy of the liquid at a high enough level that the most energetic particles can continue to escape. The energy required to transform 1 g of a substance at its boiling point from a liquid into a gas is called its **heat of vaporization.** The heat of vaporization of water is 540 cal/g, a value that is one of the largest for common substances (Table 13.1) and reflects the strong attractive forces between water molecules.

Example 13.2

The heats of vaporization of ethyl alcohol and diethyl ether are 204 and 84 cal/g, respectively. The molecular weights of ethyl alcohol and diethyl ether are 46 and 74, respectively. Calculate and compare the heat of vaporization of 1 mole of each liquid. What do these values indicate about the relative strengths of the intermolecular forces in these two compounds?

Solution   The heat of vaporization per mole can be obtained by multiplying the heat of vaporization per gram by the molecular weight.

$$\frac{204 \text{ cal}}{1 \text{ g}} \times \frac{46 \text{ g}}{1 \text{ mole}} = 9.4 \times 10^3 \text{ cal/mole} \quad \text{(ethyl alcohol)}$$

$$\frac{84 \text{ cal}}{1 \text{ g}} \times \frac{74 \text{ g}}{1 \text{ mole}} = 6.2 \times 10^3 \text{ cal/mole} \quad \text{(diethyl ether)}$$

On a mole basis, the heat of vaporization of ethyl alcohol is larger than that of diethyl ether. Thus the intermolecular forces in ethyl alcohol must be larger than for diethyl ether.

[*Additional examples may be found in 13.25–13.28 at the end of the chapter.*]

13.4 Solids

When heat energy is added to a solid, the temperature increases until the solid starts to melt. That temperature at which the added heat energy is used only to melt the solid without raising the temperature of the solid or liquid is called the **melting point.** At the melting point, the solid and liquid states exist in equilibrium. Particles from the solid state, which are in ordered arrangements, escape and enter the more random and mobile liquid state, while particles from the liquid may be deposited on the surface of the solid.

$$\text{solid} \underset{\text{freezing}}{\overset{\text{melting}}{\rightleftharpoons}} \text{liquid}$$

The effect of pressure on the melting point is not as dramatic as the effect of pressure on the boiling point. For most solids, the melting point of a substance increases slightly with pressure. Water is not typical; its melting point decreases slightly with increasing pressure. The decrease in melting point is approximately 0.01°C/atm. When skaters skate on the surface of the ice, the pressure exerted by the narrow edge of a hollow-ground skate blade melts the ice to provide water as a lubricant for the skate blade.

The amount of heat energy required to transform 1 g of a solid into a liquid at the melting point is called the **heat of fusion.** The heat of fusion of water is 80 cal/g. Like the heat of vaporization for water, this value is higher than that for many solids and is yet another indication that water has strong attractive forces between neighboring molecules. The melting point of a solid can be considered an indication of its intermolecular attractive forces. For substances of similar molecular weight, those with the higher melting points have the stronger intermolecular forces.

Example 13.3

The heat of fusion of carbon tetrachloride (CCl_4) is 4.2 cal/g. Calculate the heat of fusion per mole and compare this value to the heat of fusion per mole for water. What can be concluded from these values?

Solution The heat of fusion per mole can be calculated by multiplying the heat of fusion per gram by the molecular weight.

$$\frac{80 \text{ cal}}{1 \text{ g}} \times \frac{18 \text{ g}}{1 \text{ mole}} = 1.4 \times 10^3 \text{ cal/mole} \quad \text{(water)}$$

$$\frac{4.2 \text{ cal}}{1 \text{ g}} \times \frac{154 \text{ g}}{1 \text{ mole}} = 6.5 \times 10^2 \text{ cal/mole} \quad \text{(carbon tetrachloride)}$$

That the heat of fusion per mole of water is greater than that for carbon tetrachloride is indicative that the intermolecular attractive forces of water are larger than those of carbon tetrachloride.

[*Additional examples may be found in 13.35–13.38 at the end of the chapter.*]

13.5 Le Châtelier's Principle and Changes of State

The French chemist Henri Le Châtelier suggested, in 1888, a simple generalization about how systems at equilibrium are affected by changes in conditions. If an external force is applied to a system at equilibrium, the system will readjust to reachieve equilibrium by reducing the stress or change applied to it. **Le Châtelier's principle** is one that you should find acceptable based on your observations. Consider how you react to changes in temperature. If the air temperature drops, you experience a change that your body attempts to counteract. Your body responds by shivering in the cold in order to produce heat. It would not respond by perspiring when it is cold. If the air temperature increases, your body cools itself by perspiring. It isn't reasonable to expect the body to produce heat when it is hot.

Le Châtelier's principle explains the effect of pressure on the boiling point of a liquid and the melting point of a solid. Placing pressure on matter should make it tend to occupy the smallest volume possible. In the case of boiling a liquid to form a gas, there is a very large increase in volume. Applying pressure causes the equilibrium system of a liquid and its vapor at its normal boiling point to shift toward the liquid state, which has a lower volume. Therefore, in order to boil a liquid at higher pressures, higher temperatures are necessary.

$$\text{liquid} \underset{\text{increased pressure}}{\overset{\text{decreased pressure}}{\rightleftharpoons}} \text{gas}$$
$$\text{small volume} \qquad\qquad \text{large volume}$$

Most substances are about 10% more dense as solids than as liquids. In melting most solids, applying pressure causes an equilibrium system of liquid and solid to shift toward the solid state because the solid has the smaller volume. Therefore, in order to melt most solids under pressure, higher temperatures are necessary. Increased pressure increases the melting point. Only small increases in melting points are usually observed due to the small difference in volume between solids and liquids.

$$\text{liquid} \underset{\text{decreased pressure}}{\overset{\text{increased pressure}}{\rightleftharpoons}} \text{solid}$$
$$\text{less dense} \qquad\qquad \text{more dense}$$

The abnormal behavior of water results from the lower density of ice compared to water. Solid water has a larger volume than liquid water (Section 13.7). Thus there is an increase in volume in going from water to ice, and an increase in pressure shifts the ice–water equilibrium toward water. Increased pressure decreases the melting point of water.

$$\text{water} \underset{\text{increased pressure}}{\overset{\text{decreased pressure}}{\rightleftharpoons}} \text{ice}$$
$$\text{smaller volume} \qquad\qquad \text{larger volume}$$

13.6 Intermolecular Forces

The bonds that hold atoms or ions together in compounds are **intramolecular forces.** The prefix intra- means within, as in intramural sports within a college. In order to break a chemical bond, large amounts of energy must be added to overcome the intramolecular forces between atoms. To break hydrogen molecules into hydrogen atoms requires 104 kcal/mole.

There are weaker forces that exist between individual molecules. The forces between molecules are **intermolecular forces,** and they are responsible for holding molecules close together in the liquid and solid states. These forces are usually less than 10 kcal/mole. Intermolecular forces are of three types: London forces, dipole–dipole forces, and hydrogen-bonding forces.

London Forces

The gaseous elements and nonpolar compounds can be liquefied by decreasing the temperature, indicating that there are intermolecular forces even between nonpolar molecules and atoms in elements such as neon and helium. What is responsible for these forces?

On the average, the electrons in a nonpolar molecule or atom are distributed uniformly around the nucleus. However, at some instant the electrons may be distributed closer to one nucleus in a molecule or toward one side of an atom. At that instant a **temporary dipole** is present (Figure 13.5). A temporary dipole exerts an influence on nearby molecules or atoms. The ease with which an electron cloud can be distorted by nearby charges is called **polarizability.** The effect of the temporary dipole is to polarize neighboring molecules, resulting in an **induced dipole.** The attractive forces between a temporary dipole and an induced dipole are called **London forces.**

The strength of London forces depends on the number of electrons in a molecule or atom. The size and shape of a molecule are also important. The more electrons there are and the farther away they are from the nucleus, the more easily distorted or polarizable are the electrons in the molecule.

Consider the boiling points of the nonpolar molecules chlorine, bromine, and iodine given in Figure 13.6. The boiling points increase with increasing molecular weight, and a plot of boiling point versus molecular weight for these elements gives a straight line. The relationship shown is

temporary   induced
dipole   dipole

Figure 13.5

London attractive forces

The electron distribution in an atom becomes distorted, and a temporary dipole results. An adjacent atom then has a dipole induced by the movement of electrons to result in a net attraction between atoms.

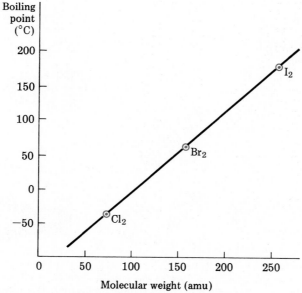

Figure 13.6

Correlation between boiling point and molecular weight

the result of increased London forces due to more polarizable electrons of the larger atoms in the molecules. These forces tend to hold the molecules together and therefore the temperature required to boil the substance is increased.

Example 13.4

The boiling points of liquid krypton and liquid argon are −152.9 and −185.8°C, respectively. Suggest a reason for these facts and predict the boiling point of xenon.

Solution The atomic numbers of krypton and argon are 36 and 18, respectively. The electrons in krypton are in higher energy levels and farther from the nucleus than the electrons in argon. As a result, the electrons in krypton are more polarizable and larger London forces result. One would predict the boiling point of xenon to be still higher. It is −108.1°C.

[Additional examples may be found in 13.45–13.50 at the end of the chapter.]

Dipole–Dipole Forces

Molecules with a negative end and a positive end are polar and tend to associate closely, with the positive end of one molecule, attracting the negative end of another molecule. The physical properties of polar molecules reflect this association (Figure 13.7).

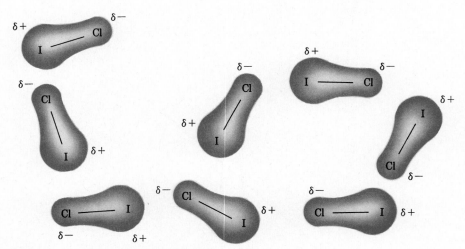

Figure 13.7

Dipole–dipole intermolecular forces
The positive iodine atom is attracted to the negative chlorine atom of a neighboring ICl molecule. The dipole is the result of a difference in the electronegativities of the two atoms.

The molecular weight of ICl (162 amu) is close to the molecular weight of Br_2 (160 amu). Yet the boiling point of ICl is 97°C, 38°C higher than that of Br_2. If the boiling point of ICl were plotted on the graph given in Figure 13.6, the point would be above the line. Why does ICl boil at a higher temperature than Br_2? In Br_2 the bonding pair of electrons is shared equally in a nonpolar covalent bond between the two atoms. In ICl, the bonding pair of electrons is attracted toward chlorine, the more electronegative element, and the bond in ICl is polar covalent. The molecules are more strongly attracted to one another by intermolecular forces between dipoles. In order to boil ICl, it is necessary to heat the molecules to a higher temperature than that needed for a nonpolar molecule of comparable molecular weight.

Example 13.5

The boiling points of nitrogen (N_2) and carbon monoxide (CO) are 77 and 83 K, respectively. Suggest a reason for this order of boiling points.

Solution   The molecular weights of the two substances are the same. Nitrogen is a nonpolar molecule, and the only intermolecular forces are of the London type. Carbon monoxide is a slightly polar compound as a result of the difference in electronegativity between carbon and oxygen. The dipole–dipole forces cause an additional attraction between the positive carbon atom and the negative oxygen atom of a neighboring molecule.

$$^{\delta+} : C \equiv O : {}^{\delta-} \qquad : N \equiv N :$$

[*Additional examples may be found in 13.51 and 13.52 at the end of the chapter.*]

Hydrogen-Bonding Forces

Compounds containing hydrogen bonded to fluorine, oxygen, and nitrogen have much higher intermolecular forces than expected. This is best seen in Figure 13.8, which gives the boiling points of the hydrogen compounds of the elements in Groups IVA, VA, VIA, and VIIA.

The boiling point of SnH_4 is higher than other compounds of Group IVA because of the large London forces. As the molecules become smaller, the electrons of the atoms are less polarizable and the attractive forces are decreased. Methane, CH_4, has the lowest boiling point.

In the compounds of the Groups VA, VIA and VIIA elements there is an obvious anomaly. The compounds NH_3, H_2O, and HF all have much higher boiling points than expected. The three molecules NH_3, H_2O, and HF have two structural features in common. Each has at least one hydrogen atom joined by a polar covalent bond to an electronegative atom. Each has at least one unshared pair of electrons (Figure 13.9). The very small hydrogen atom has a partial positive charge because the bonding electrons are attracted toward an electronegative element. Consequently, an intermolecular attraction exists between hydrogen and an electron pair on a neighboring molecule. The "bridging" of a hydrogen atom between two electronegative atoms is called a **hydrogen bond.** Hydrogen bonds are quite strong, on the order of 3–10 kcal/mole.

The order of decreasing hydrogen bond strength with the electronegative atoms might be expected to be F > O > N. However, in Figure 13.8 it appears that the anticipated order is

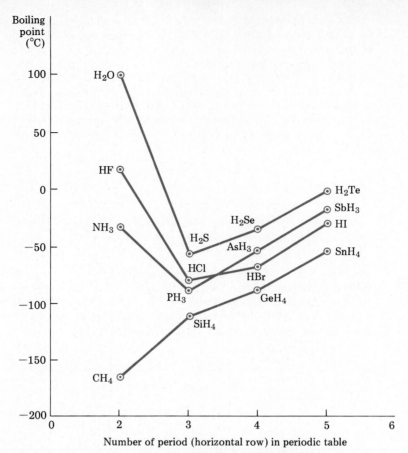

Illustration of the effect of hydrogen bonding on boiling points
*The boiling points of compounds within a family increase with
increasing molecular weight for elements in periods 3, 4, and 5. Only
CH₄ of Group IV is a part of this trend. The elements fluorine, oxygen,
and nitrogen form compounds with hydrogen that boil considerably
higher than anticipated on the basis of London forces. The special force
of attraction in these compounds is called a hydrogen bond.*

incorrect. HF boils at a lower temperature than H₂O. Actually the hydrogen bonds in HF are
stronger than those in H₂O. The reason for the apparent anomaly is that although the hydrogen
bond to fluorine is strong, the HF molecule can only form one hydrogen bond per molecule
(Figure 13.9). The fluorine atom has three unshared electron pairs, but only one hydrogen atom.
In the case of water, there are two hydrogen atoms and two unshared electron pairs per molecule, and the extent of aggregation of H₂O molecules can be greater than that of HF molecules.
Ammonia (NH₃) has three hydrogen atoms, but only one electron pair to share. Water has the
proper balance of hydrogen atoms and electron pairs to form the maximum number of hydrogen
bonds per mole.

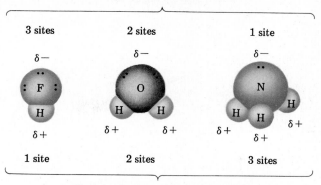

3 sites   2 sites   1 site

1 site   2 sites   3 sites

Figure 13.9   Sites to associate with electron pairs

Molecules that form hydrogen bonds
Hydrogen fluoride is limited in forming hydrogen bonds by its single hydrogen atom in spite of its three electron pairs. Ammonia is limited in forming hydrogen bonds by its single electron pair in spite of its three hydrogen atoms. Only water has an equal number of hydrogen atoms and electron pairs.

Example 13.6

The boiling points of ethyl alcohol and dimethyl ether are 78.5 and $-24°C$, respectively. The compounds have the same molecular formula, C_2H_6O, but have different structures. Explain the large difference in the boiling points of the two compounds.

ethyl alcohol   dimethyl ether

Solution Ethyl alcohol has a —OH group as part of its structure. This group can form hydrogen bonds via its hydrogen atom or the unshared pairs of electrons on the oxygen atom. Dimethyl ether cannot form hydrogen bonds to itself because it does not have a positively charged hydrogen atom attached to oxygen. The hydrogen bonding in ethyl alcohol is responsible for the higher boiling point.

[*Additional examples may be found in 13.55–13.60 at the end of the chapter.*]

13.7 Water—An Unusual Compound

At 100°C the density of water is 0.958 g/mL. The density increases as the temperature decreases until at 4°C the density of water is 1.0000 g/mL, which is its maximum density. Then below 4°C the density decreases until the density of 0.99987 g/mL is reached at 0°C (Figure 13.10).

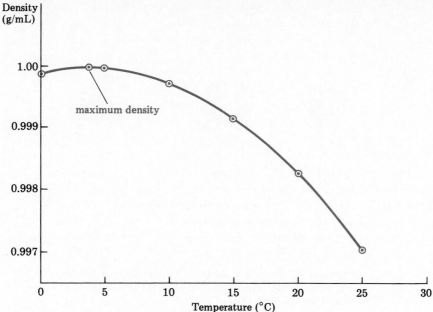

Figure 13.10

Density of water and temperature
The density of water is at a maximum at 4°C. The density decreases from 4°C to 0°C.

Ice at 0°C has a density of 0.917 g/mL and is less dense than liquid water. For that reason ice displaces its own weight and floats in water. The volume of ice that remains floating above the surface of the water is approximately 8%. Since 92% of a mass of ice is below the surface, the visible part of an iceberg is indeed only "the tip of the iceberg."

The facts that ice is less dense than water and that the maximum density of liquid water occurs at 4°C rather than 0°C have profound ecological significance. As the temperature of the water near the surface of a lake is lowered in cold weather, the cooler and more dense water tends to sink toward the bottom of the lake. The warmer water in the lake rises to the top since it is less dense. This circulation of water continues until the overall water temperature is 4°C. Further cooling results in a lower density and the cooler water remains at the surface and no longer sinks to the bottom of the lake. Thus the warmest water is 4°C at the lake bottom.

On top of the lake, the surface water eventually freezes. Since ice is less dense than water, the ice remains on the surface. If ice were more dense than water, the ice would sink to the bottom. A continued cold spell then would cause the lake to freeze from the bottom to the top and living organisms in the lake would not survive. The lower density of ice results in a sheet of ice that protects the water beneath its surface from further decreases in temperature. The ice thickness will slowly increase, but the water temperature beneath the ice is maintained comfortably enough for aquatic life to survive.

The change in the density of water at temperatures below 4°C and the fact that ice is less dense than liquid water are due to the nature of the hydrogen bonding in water. Most liquid materials become more dense as the temperature decreases and become still more dense when the solid state is formed. For most liquids the decrease in temperature lowers the kinetic energy

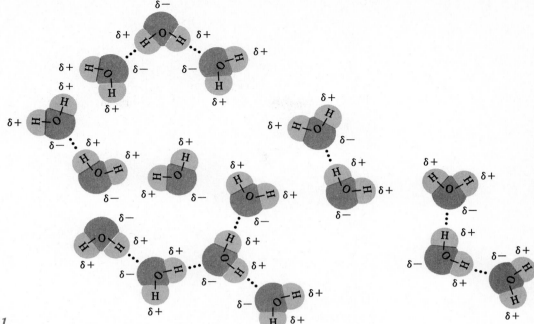

Figure 13.11

Hydrogen bonding in liquid water
The δ+ and δ− charges represent the polarity of the hydrogen–oxygen bonds. In liquid water the molecules can move about in a somewhat restricted manner. The hydrogen bonds of clumps of associated water molecules are formed and broken easily. As the temperature is lowered, the kinetic energy decreases and the liquid volume contracts. The density then increases.

and allows more of the matter to become aggregated in "clumps" of molecules. In the solid state the molecules get as close as possible and leave a minimum of free space.

In water the aggregation as a result of hydrogen bonding has some unique structural requirements. As long as the "clumps" are small (Figure 13.11), as is the case at higher temperatures, water behaves like other liquids and becomes more dense as the temperature decreases. However, at 4°C a pattern of molecules emerges as the number of hydrogen bonds is increased. Then, as freezing occurs, the pattern of hydrogen-bonded molecules crystallizes. The pattern is an open cage-like network. "Holes" in the cages then mean that the density of ice is less than that of water (Figure 13.12).

13.8 Hydrates

Hydrates are crystalline compounds that contain water in definite proportions by weight. Examples of some common hydrates are listed in Table 13.2. Each hydrate is known by a common name, but the systematic name is one that indicates the number of molecules of water contained per formula unit. The designations for one through ten water molecules involve the prefixes

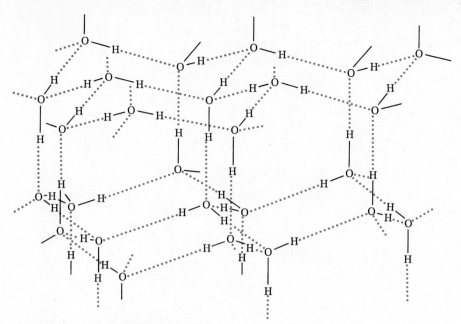

Figure 13.12

Structure of ice and hydrogen bonds
The three-dimensional structure shows each covalent oxygen–hydrogen bond as a solid line. The hydrogen bonds are shown by colored dotted lines. Each oxygen atom in the structure has two covalent bonds and two hydrogen bonds. Each hydrogen atom is covalently bonded to one oxygen atom and hydrogen-bonded to another oxygen atom. The void or space created by the structure accounts for the lower density of ice than of water.

Table 13.2   **Some Common Hydrates**

Formula	Common name	Systematic name
$CuSO_4 \cdot 5H_2O$	blue vitriol	copper(II) sulfate pentahydrate
$Na_2SO_4 \cdot 10H_2O$	Glauber's salt	sodium sulfate decahydrate
$KAl(SO_4)_2 \cdot 12H_2O$	alum	potassium aluminum sulfate dodecahydrate
$MgSO_4 \cdot 7H_2O$	epsom salts	magnesium sulfate heptahydrate
$CaSO_4 \cdot 2H_2O$	gypsum	calcium sulfate dihydrate

mono- through *deca-* given in Section 11.5. Other prefixes include *undeca-, dodeca-, trideca-, tetradeca-, pentadeca-* for 11 through 15 units.

Example 13.7

Name the hydrate given by the formula $FeBr_3 \cdot 6H_2O$.

Solution   The iron is in the $+3$ oxidation state because it is combined with three bromide ions.

Without the molecules of water, the compound is iron(III) bromide. There are six water molecules in the hydrate that are indicated in the complete name iron(III) bromide hexahydrate.

[*Additional examples may be found in 13.69–13.72 at the end of the chapter.*]

Example 13.8

Write the formula for barium perchlorate trihydrate.

Solution   The barium ion is Ba^{2+} and the perchlorate ion is ClO_4^-. The compound barium perchlorate is $Ba(ClO_4)_2$. The three water molecules indicated by the trihydrate are added to the formula for the salt to give $Ba(ClO_4)_2 \cdot 3H_2O$.

[*Additional examples may be found in 13.67 and 13.68 at the end of the chapter.*]

Water in hydrates may be incorporated in three ways to yield the definite proportion by weight, which is a characteristic of hydrates. These are

1. Coordinate covalent bonding to the metal cation.
2. Hydrogen bonding with the anion.
3. Crystal entrapment.

Examples of coordinate covalent bonding to the metal cation include all four water molecules of $BeSO_4 \cdot 4H_2O$ and four of the five water molecules of $CuSO_4 \cdot 5H_2O$. Coordination of six water molecules to the metal cation is very common in transition metal ions. Examples include $NiSO_4 \cdot 6H_2O$ and six of the seven water molecules of $CoSO_4 \cdot 7H_2O$. Thus in these two sulfate salts, the metal ions exist as $Ni(H_2O)_6^{2+}$ and $Co(H_2O)_6^{2+}$.

Hydrogen bonding to the anion generally involves a smaller number of water molecules. In both $CuSO_4 \cdot 5H_2O$ and $CoSO_4 \cdot 7H_2O$, one of the water molecules is hydrogen bonded to the sulfate ion.

Crystal entrapment involves the location of water molecules at definite positions within the crystal. In the mixed salt $KAl(SO_4)_2 \cdot 12H_2O$, six of the water molecules are coordinated to aluminum and six water molecules are spaced in the lattice in positions surrounding the potassium ion. The water molecules around potassium are not coordinate covalent bonded.

Example 13.9

Calculate the percent water contained in $LiI \cdot 3H_2O$ to four significant figures.

Solution   The percent composition of water is determined by the same procedures used in Chapter 5 when calculating the percent composition of the elements in a compound. First determine the formula weight of the compound.

formula weight = (at wt Li) + (at wt I) + 3(mol wt H_2O)

formula weight = 6.94 + 126.90 + 3(18.016) = 187.89

The mass of 1 mole of the hydrate is 187.89 g. The mass of 3 moles of water in the hydrate to four significant figures is 54.05 g. The percent of water is obtained by dividing the mass of the 3 moles of water by the mass of one mole of the compound.

$$\% \text{ water} = \frac{54.05 \text{ g water}}{187.89 \text{ g hydrate}} \times 100 = 28.766832\%$$

The correct answer expressed to the required four significant figures is 28.77%.

[*Additional examples may be found in 13.73–13.78 at the end of the chapter.*]

Example 13.10

A 16.20 g sample of a hydrate of strontium iodide is heated to give 12.30 g of the anhydrous salt. Using the formula $SrI_2 \cdot xH_2O$, what is the value of x?

Solution The number of moles of SrI_2 in the hydrate is given by dividing the mass of the dry salt by the formula weight of the dry salt.

$$\text{moles } SrI_2 = \frac{12.30 \text{ g } SrI_2}{341.43 \text{ g/mole}} = 0.036025 \text{ mole } SrI_2$$

The mass of water in the hydrate is obtained by the difference of the original mass and that of the dry salt. The number of moles of water is obtained by dividing the mass of water by the molecular weight.

$$\text{moles } H_2O = \frac{3.90 \text{ g water}}{18.016 \text{ g/mole}} = 0.2165 \text{ mole } H_2O$$

The ratio of the number of moles of water to moles of SrI_2 is determined as follows:

$$\frac{0.2165 \text{ mole } H_2O}{0.036025 \text{ mole } SrI_2} = 6.01 \text{ moles } H_2O/\text{mole } SrI_2$$

The formula of the hydrate is $SrI_2 \cdot 6H_2O$.

[*Additional examples may be found in 13.79 and 13.80 at the end of the chapter.*]

Efflorescence

The loss of water of hydration to air is called **efflorescence.** Most hydrates will release some or all of their water upon heating. When the water is given off, the crystalline character of the salt changes. For example, blue vitriol, a blue salt of formula $CuSO_4 \cdot 5H_2O$, produces a white salt, $CuSO_4$, upon heating.

$$CuSO_4 \cdot 5H_2O \xrightarrow{\Delta} CuSO_4 + 5 H_2O(g)$$
$$\text{blue} \qquad\qquad \text{white}$$

The hydrates have vapor pressures due to the water present. This vapor pressure increases as the temperature increases. If a hydrate has a vapor pressure that is higher than the partial pressure of water in the air, the hydrate will lose all or part of its water to the air. Glauber's salt loses all of its water rapidly when exposed to air if the partial pressure of water vapor is less than

14 mm Hg. Blue vitriol is unstable if the partial pressure of water vapor is less than 8 mm Hg; it undergoes transformations successively to yield $CuSO_4 \cdot 3H_2O$ and eventually $CuSO_4 \cdot H_2O$. Washing soda loses nine water molecules to yield $Na_2CO_3 \cdot H_2O$.

Deliquescence

Some substances when exposed to air with a high vapor pressure of water will absorb water to form hydrates. Thus any substance that effloresces to yield a lower hydrate can absorb water to reform the higher hydrate. Any substance that absorbs water from the air is **hygroscopic.** Certain substances will remove so much water from the air that the solid disappears and becomes dissolved completely in a water solution. A substance that absorbs water from the air to form a solution is **deliquescent.** An example of a deliquescent substance is anhydrous calcium chloride, which first forms $CaCl_2 \cdot 6H_2O$; this, in turn, absorbs more water to form the solution. Sodium hydroxide is also deliquescent.

13.9 Hard Water

Limestone ($CaCO_3$) is an abundant mineral in the earth's crust. Most groundwater comes in contact with it and dissolves small amounts of the calcium ion. Groundwater containing dissolved calcium ion (Ca^{2+}) is called **hard water.** The presence of this ion in the water supply leads to many household and industrial problems. The calcium ion forms precipitates when heated or when soap is added.

When soap, which is the sodium salt of a complex acid called stearic acid ($C_{17}H_{35}CO_2H$), is added to a solution containing calcium ions, an insoluble salt called calcium stearate is formed. Its more familiar name is scum or bathtub ring.

$$Ca^{2+} + 2\,C_{17}H_{35}CO_2^- \longrightarrow Ca(C_{17}H_{35}CO_2)_2$$

If enough soap is added, eventually all the calcium ion will be precipitated, and the residual soluble sodium salt will be an effective cleansing agent. However, the precipitate is unsightly, and hard water is uneconomical to use if the concentration of calcium ions is too high.

When heated, hard water forms a precipitate of calcium carbonate ($CaCO_3$), which creates industrial problems.

$$Ca^{2+} + 2\,HCO_3^- \rightleftharpoons CaCO_3 + CO_2 + H_2O$$

If the bicarbonate ion HCO_3^- is present, the preceding reaction can be used to remove the calcium ion before the water is used. Hard water containing the bicarbonate ion therefore is called **temporary hard water.** However, this process is impractical for large-scale industry. Both temporary and permanently hard water can be softened by the addition of sodium carbonate (Na_2CO_3), commonly called washing soda. The carbonate ion (CO_3^{2-}) precipitates out the calcium ions.

$$Ca^{2+} + CO_3^{2-} \rightleftharpoons CaCO_3$$

Temporary hardness can be eliminated by the addition of a base that reacts with bicarbonate in an acid–base reaction to produce carbonate ions, which then precipitate the calcium ions.

$$HCO_3^- + OH^- \rightleftharpoons H_2O + CO_3^{2-}$$

The use of silicate minerals called zeolites to soften water by an ion-exchange process has provided a remarkably convenient way of removing calcium ions and replacing them by ions such as sodium ions. The ion-exchanger material consists of a large, covalently bonded, solid substance containing many negatively charged sites. The electrical neutrality of the substance is maintained by the presence of sodium ions. When water containing calcium ions is passed through the ion exchanger, the sodium ions are replaced by calcium ions. The hard water is softened and now contains sodium ions.

Summary

When the temperature of a gas is lowered, the average kinetic energy of the molecules decreases, and a point is reached at which intermolecular forces cause condensation to form a liquid. As the temperature is further lowered, the liquid solidifies into a rigid structure of arranged molecules.

A liquid has a characteristic vapor pressure that depends on the temperature. When the vapor pressure equals the atmospheric pressure, the liquid boils. The normal boiling point of a liquid is the temperature at which the vapor pressure equals 1 atm. The energy required to transform 1 g of a liquid to a gas at its boiling point is the heat of vaporization of the liquid.

The temperature at which a solid is converted to a liquid in equilibrium is called the melting point. Pressure exerts little effect on the melting point. The energy required to transform 1 g of a solid into a liquid is the heat of fusion of the substance.

Intermolecular forces are of three types: London, dipole–dipole, and hydrogen bonding. London forces, the attraction of a temporary dipole and an induced dipole, are the consequence of the polarizability of electrons in atoms and molecules. Dipole–dipole forces, the attraction of the permanent dipole of molecules,

are the consequence of the unequal distribution of electrons in the bonds between atoms of different electronegativity. Hydrogen bonds are the attraction between a bonded hydrogen atom and an electron pair on a neighboring molecule. Compounds of nitrogen, oxygen, and fluorine bonded to hydrogen form hydrogen bonds. The unusual properties of water are the result of hydrogen bonds.

Le Châtelier's principle provides an explanation for the physical changes of the states of matter. Systems at equilibrium change to a new equilibrium condition to reduce the stress caused by an external change. Under pressure, the gaseous state is converted to the liquid state, which has the smaller volume. The equilibrium between the liquid and the solid states is less sensitive to pressure because there is only a small difference in the volume of the two states. Water is an anomalous liquid; it has a larger volume in the solid state than in the liquid state.

Water is an unusual compound as a result of its ability to form networks of hydrogen bonds. Water forms crystalline compounds called hydrates that contain water and another substance in definite proportions by weight.

New Terms

Boiling is a process that occurs when the vapor pressure of a liquid is equal to atmospheric pressure.

At the **boiling point** the vapor pressure of a liquid equals atmospheric pressure.

Condensation is a process in which a gas is converted to a liquid.

A **deliquescent** substance absorbs water from the air to form a solution.

A **dynamic equilibrium** is a condition in which two or more opposing processes occur at the same time and are in balance.

Efflorescence is the loss of water of hydration to the atmosphere.

An **equilibrium** is a state in which opposing processes are in balance.

Freezing is a process in which a liquid is converted to a solid. This process occurs at the same temperature at which a solid melts.

Hard water contains dissolved calcium ions.

Heat of fusion is the amount of energy released as 1 g of a solid is converted into a liquid at its melting point.

Heat of vaporization is the amount of energy required to convert 1 g of a liquid to a gas at its boiling point.

A **hydrate** is a compound that contains water in definite proportions by weight.

A **hydrogen bond** is an intermolecular attraction between an electropositive hydrogen atom and a nonbonded electron pair of an electronegative atom of a neighboring molecule.

A **hygroscopic** substance absorbs water from the air.

An **induced dipole** is a separation of charge within a molecule caused by a temporary dipole in the vicinity.

Intermolecular forces are forces of attraction between separate molecules.

Intramolecular forces are forces that hold atoms together in compounds.

Le Châtelier's principle states that a system will readjust to reduce the stress placed on it.

London forces are a type of intermolecular force involving temporary and induced dipoles.

Melting is the conversion of a solid into a liquid.

The **melting point** is the temperature at which a solid is converted into a liquid.

Normal boiling point is the temperature at which the vapor pressure of a liquid equals 1 atm.

Polarizability indicates the ease with which an electron cloud can be distorted by nearby charges.

A **temporary dipole** is a separation of charge produced momentarily in an otherwise nonpolar substance.

Temporary hard water contains bicarbonate ions.

Vaporization is the process of converting molecules from the liquid to the gaseous state.

Vapor pressure is the pressure of a vapor in equilibrium with its liquid form.

Exercises

Terminology

13.1 Using the kinetic molecular theory, distinguish between liquids and solids.

13.3 Distinguish between boiling point and normal boiling point.

13.2 To which states of matter do the heat of fusion and heat of vaporization refer?

13.4 How are vapor pressure and boiling point related?

Evaporation of Liquids

13.5 What factors control the drying of clothes at temperatures above the freezing point of water?

13.7 The rate of evaporation of water from a beaker inside a larger sealed container at constant temperature decreases with time. Explain this phenomenon and contrast it with the rate of evaporation from the same beaker in a room at constant temperature.

13.9 Two liquids A and B are placed in similar separate beakers on a laboratory table. Liquid B decreases in volume much faster than liquid A. Explain why the difference is observed.

13.6 Explain the difference felt by a swimmer emerging from the water on a windy day versus a calm day, assuming the temperature is the same.

13.8 A beaker of water in a closed room at constant temperature does not decrease in volume on a given day. Why does the water not evaporate?

13.10 Which substance, ethyl alcohol or diethyl ether, will evaporate at the faster rate at 25°C?

Vapor Pressure of Liquids

13.11 The space above the column of mercury in a barometer is considered to be a vacuum. Is this idea strictly correct?

13.13 The vapor pressure of $SbCl_3$ and $SbBr_3$ are 10 and 100 mm Hg respectively at 145°C. Which substance should evaporate at the faster rate at 145°C?

13.15 The vapor pressures of CS_2 and CSSe at 28°C are 400 and 100 mm Hg, respectively. Which substance has the larger intermolecular forces?

13.12 Why is mercury used in a barometer rather than another liquid such as water?

13.14 The vapor pressure of chloroform at 43°C is 400 mm Hg, and the vapor pressure of carbon tetrachloride is 400 mm Hg at 58°C. Which substance will evaporate at the faster rate at 43°C?

13.16 The vapor pressures of $GeCl_4$ and $GeBr_4$ are 760 and 40 mm Hg, respectively at 84°C. Which substance has the larger intermolecular forces?

Boiling Points of Liquids

13.17 What is the highest temperature at which water vapor will condense to yield water under 1 atm pressure? At what temperature will water boil under a pressure of 148.9 cm Hg?

13.19 The vapor pressure of $SnCl_4$ at 92°C is 400 mm Hg. The vapor pressure of SnI_4 is 400 mm Hg at 315°C. which substance should have the higher boiling point?

13.21 The boiling points of $SiCl_4$ and SiF_4 at 1 atm are 56.8 and −94.8°C, respectively. What do these values indicate about intermolecular forces in these two compounds?

13.18 Explain why the time required to fry meat at a high altitude is no different than at sea level, whereas the time required to boil potatoes varies significantly as a function of altitude.

13.20 The vapor pressures of methanol at 50 and 65°C are 400 mm Hg and 760 mm Hg, respectively. Methanol is boiling one day at 60°C. What can one conclude about the atmospheric pressure that day?

13.22 The boiling point of CS_2 is 46.5°C. What can be said about the intermolecular forces in CS_2 compared to CO_2, which is a gas at 46.5°C?

Heat of Vaporization

13.23 Calculate the amount of heat required to convert 10 g of water at its normal boiling point into steam.

13.24 How many calories are released by 10 g of steam when it condenses to water at 100°C?

13.25 The heats of vaporization of CCl_4 and CBr_4 are 53.7 and 32.8 cal/g, respectively. Calculate the molar heats of vaporization. Which compound has the larger intermolecular forces?

13.26 The heats of vaporization of CH_3OH and CH_3SH are 294 and 132 cal/g, respectively. Calculate the molar heats of vaporization. Which compound has the larger intermolecular forces?

13.27 The molar heat of vaporization of CH_3CH_2OH is 9673 cal/mole. Calculate the heat energy required to vaporize 4.6 g of this compound.

13.28 The molar heat of vaporization of CH_3CH_2SH is 6728 cal/mole. Calculate the heat energy required to vaporize 3.6 g of this compound.

Melting Points of Solids

13.29 A highly trained figure skater skates on only a small portion of the edge of a blade and may exert 300 atm pressure. What is the melting point of the ice under this pressure?

13.30 A speed skating blade is long but very thin. Explain why this design is chosen.

13.31 Mercury thermometers cannot be used below −39°C. Explain why.

13.32 If the density of a solid is less than that of the corresponding liquid, what will be the effect of increased pressure on the melting point?

Heat of Fusion

13.33 How many calories are required to melt 1000 g of ice?

13.34 How many calories must be removed from water at 0°C to freeze 100 g of water?

13.35 The heat of fusion of CH_4 is 13.99 cal/g. What is the molar heat of fusion?

13.36 The heat of fusion of CH_3OH is 23.7 cal/g. What is the molar heat of fusion?

13.37 The heat of fusion of CH_3Br is 15.05 cal/g. What is the molar heat of fusion?

13.38 The heats of fusion of HF and HI are 54.7 and 5.4 cal/g, respectively. Calculate the molar heats of fusion. Based on this data, which of the compounds has the larger intermolecular forces?

Le Châtelier's Principle

13.39 Explain the prediction of the effect of pressure on boiling point using Le Châtelier's principle.

13.40 Explain the effect of pressure on the melting point of solids using Le Châtelier's principle.

13.41 The melting point of CH_4 is −182.5°C at 1 atm. Assuming that solid methane is more dense than liquid methane, how will be the melting point of methane be affected by 100 atm pressure?

13.42 The melting point of bismuth is 271.3°C. The density of liquid bismuth is greater than solid bismuth at the melting point. How will the melting point be affected by increased pressure.

Intermolecular Forces

13.43 List and describe the three types of intermolecular forces.

13.44 If water were a "normal" compound, what would be its expected boiling point based on London and dipole–dipole forces?

13.45 The boiling points of propane and butane are −42 and −0.5°C, respectively. Explain this order.

propane   butane

13.46 The compounds CF_4 and CCl_4 are tetrahedral molecules. Neither is polar. Explain why the boiling points of CF_4 and CCl_4 are −129 and 76.8°C, respectively.

13.47 The boiling points of $SiCl_4$ and SiF_4 are 56.8 and −94.8°C, respectively, at 1 atm. What type of intermolecular forces are responsible for the observed boiling points?

13.48 The boiling point of CS_2 is 46.5°C, whereas CO_2 is a gas at room temperature. What type of intermolecular forces are responsible for the observed boiling points.

13.49 The compounds BBr_3 and BCl_3 are trigonal planar in shape. The vapor pressures of BBr_3 and BCl_3 at 13°C are 40 and 760 mm Hg, respectively. Which type of intermolecular forces are responsible for the observed vapor pressures.

13.50 The compounds CS_2 and CO_2 are linear in shape. The molar heats of vaporization of CS_2 and CO_2 are 6787 and 5539 cal/mole, respectively. Which type of intermolecular forces are responsible for the observed heats of vaporization?

13.51 The boiling points of $CH_3CH_2CH_3$ and CH_3CH_2F are −42.1 and −32°C, respectively. Which type of intermolecular forces are responsible for the observed boiling points?

13.52 Draw Lewis structures of CH_3OH and CH_3SH and determine their shapes. The molar heats of vaporization of CH_3OH and CH_3SH are 9380 and 6330 cal/mole, respectively. Which type of intermolecular forces are responsible for the observed heats of vaporization?

13.53 Draw Lewis structures of CH_3OCH_3 and CH_3SCH_3 and determine their shapes. The molar heats of vaporization of CH_3OCH_3 and CH_3SCH_3 are 5409 and 6742 cal/mole, respectively. Which type of intermolecular forces are responsible for the observed heats of vaporization?

13.54 The boiling points of ethyl alcohol and ethylene glycol (used in antifreeze) are 78.5 and 190°C, respectively. Explain the high boiling point of ethylene glycol.

ethyl alcohol   ethylene glycol

13.55 Predict which of the following compounds would have the higher boiling point. On what type of intermolecular forces did you base your prediction?

ethyl alcohol   ethyl mercaptan

13.56 Should methane or ethane have the higher boiling point?

methane   ethane

13.57 Which of the following has the lower normal boiling point?

13.58 Which of the following has the lower normal boiling point?

13.59 Give the number of sites per molecule where hydrogen atoms may be donated in hydrogen-bond formation and the number of sites where electron pairs may be supplied for hydrogen-bond formation.

13.60 Give the number of sites per molecule where hydrogen atoms may be donated in hydrogen-bond formation and the number of sites where electron pairs may be supplied for hydrogen-bond formation.

Water

13.61 Describe the special intermolecular forces that exist in water.

13.62 Why does water have a relatively low vapor pressure and a high boiling point?

13.63 Why does ice float on water?

13.64 Explain why the maximum density of water allows water at 4°C to exist at the bottom of a lake.

13.65 Why is the lower density of ice over water important to aquatic life?

13.66 Use the structure of ice to explain its being lower in density than water.

Hydrates

13.67 Write the formula of each of the following hydrates.
(a) sodium dichromate dihydrate
(b) strontium bromide hexahydrate

13.68 Write the formula of each of the following hydrates.
(a) aluminum iodide hexahydrate
(b) barium perchlorate trihydrate

(c) aluminum bromate nonahydrate
(d) barium hydroxide octahydrate
(e) calcium bromide hexahydrate

13.69 Name each of the following hydrates.
(a) $Al(BrO_3)_3 \cdot 9H_2O$ (b) $BaBr_2 \cdot 2H_2O$
(c) $AlI_3 \cdot 6H_2O$ (d) $(NH_4)_3PO_4 \cdot 3H_2O$
(e) $(NH_4)_2SO_3 \cdot H_2O$ (f) $K_2S \cdot 5H_2O$

13.71 Name each of the following hydrates.
(a) $Sr(BrO_3)_2 \cdot H_2O$ (b) $NaBr \cdot 2H_2O$
(c) $K_2CO_3 \cdot 2H_2O$ (d) $NiSO_3 \cdot 6H_2O$
(e) $SnCl_4 \cdot 4H_2O$ (f) $CoF_2 \cdot 4H_2O$

13.73 Calculate the percent water in each of the following hydrates.
(a) $AlBr_3 \cdot 15H_2O$ (b) $BaCl_2 \cdot 2H_2O$
(c) $NH_4CoPO_4 \cdot H_2O$ (d) $Ba(ClO_4)_2 \cdot 3H_2O$
(e) $Al(NO_3)_3 \cdot 9H_2O$ (f) $NH_4MgAsO_4 \cdot 6H_2O$

13.75 Thorium(IV) sulfate forms tetra-, hexa-, octa- and nonahydrates. Calculate the percent water in each hydrate.

13.77 Iron(II) sulfate forms tetra-, penta-, and heptahydrates. Calculate the percent water in each hydrate.

13.79 Given the percent water in each of the following hydrates, determine the formula of the hydrate.
(a) $YBr_3 \cdot xH_2O$ (33.0%)
(b) $Zn_3(PO_4)_2 \cdot xH_2O$ (27.2%)
(c) $Zr(SO_4)_2 \cdot xH_2O$ (20.3%)

13.81 Describe the different ways in which water may be found in hydrates.

13.83 A sample of a hydrate slowly becomes wet and eventually becomes a liquid. Explain why.

(c) barium bromide dihydrate
(d) cadmium permanganate hexahydrate
(e) calcium fluoride tetrahydrate

13.70 Name each of the following hydrates.
(a) $CdBr_2 \cdot 4H_2O$ (b) $Ca(ClO)_2 \cdot 3H_2O$
(c) $ZnF_2 \cdot 4H_2O$ (d) $SnCl_2 \cdot 2H_2O$
(e) $TlCl_3 \cdot H_2O$ (f) $NiBr_2 \cdot 3H_2O$

13.72 Name each of the following hydrates.
(a) $NaOCl \cdot 5H_2O$ (b) $KF \cdot 2H_2O$
(c) $Ni(IO_3)_2 \cdot 4H_2O$ (d) $MnBr_2 \cdot 4H_2O$
(e) $Sr(NO_2)_2 \cdot H_2O$ (f) $Cs_2S \cdot 4H_2O$

13.74 Calculate the percent water in each of the following hydrates.
(a) $CoF_2 \cdot 4H_2O$ (b) $Cd(ClO_3)_2 \cdot 2H_2O$
(c) $Ce(IO_3)_2 \cdot 2H_2O$ (d) $Ca(NO_3)_2 \cdot 4H_2O$
(e) $Y_2(SO_4)_3 \cdot 8H_2O$ (f) $ZnCr_2O_7 \cdot 3H_2O$

13.76 Cerium(III) sulfate forms penta-, octa-, and nonahydrates. Calculate the percent water in each hydrate.

13.78 Rhodium(III) sulfate forms tetra-, dodeca-, and pentadecahydrates. Calculate the percent water in each hydrate.

13.80 Given the percent water in each of the following hydrates, determine the formula of the hydrate.
(a) $NaMnO_4 \cdot xH_2O$ (27.2%)
(b) $MgCO_3 \cdot xH_2O$ (51.7%)
(c) $Mg(BrO_3)_2 \cdot xH_2O$ (27.8%)

13.82 A sample of a hydrate which is known to be efflorescent is exposed to air in a room over a period of time and remains unchanged. Explain why.

13.84 A sample of a known deliquescent substance remains unchanged while exposed to the air in a room. Explain why.

Additional Exercises

13.85 Why does rubbing alcohol feel cold when rubbed on your skin?

13.87 The vapor pressure of liquid A at 25°C is equal to that of liquid B at 50°C. If the molecular masses of A and B are identical, which of the two liquids has the stronger intermolecular attractive forces?

13.89 The vapor pressures of PBr_3 and PCl_3 are 35 mm Hg and 760 mm Hg, respectively, at 74°C. Which substance has the larger intermolecular forces?

13.91 The vapor pressure of PBr_3 at 150°C is 400 mm Hg, whereas PCl_3 at 57°C has a 400 mm Hg vapor pressure. Which compound has the higher normal boiling point?

13.86 Steam at 100°C causes more severe burns than water at 100°C. Why?

13.88 Is the vapor pressure of water at 25°C at an altitude of 10,000 ft in the mountains less than, equal to, or more than the vapor pressure of water at sea level at 25°C.

13.90 The vapor pressures of chloroform at 43 and 61°C are 400 mm and 760 mm Hg, respectively. The vapor pressures of carbon tetrachloride at 58 and 77°C are 400 mm and 760 mm Hg, respectively. Which will have the higher boiling point at 700 mm Hg?

13.92 The vapor pressures of PBr_3 at 150°C is 400 mm Hg, whereas PCl_3 at 57°C has a vapor pressure of 400 mm Hg. At what temperature will the compounds boil at 400 mm Hg?

290

13.93 Calculate the amount of heat required to convert 1 g of water from 0°C to water at 100°C. (See Section 3.5) Compare this value with the amount of heat required to convert 1 g of water at 100°C to steam at 100°C.

13.94 The molar heat of vaporization of CCl_4 is 8270 cal/mole. Calculate the heat energy required to vaporize 1.54 g of this compound.

13.95 The heats of vaporization of CH_3F and CH_3Cl are 117 and 46.6 cal/g respectively. Calculate the molar heats of vaporization. Which compound has the larger intermolecular forces?

13.96 The compounds PBr_3 and PCl_3 are trigonal pyramidal in shape. The vapor pressure of PBr_3 at 150°C is 400 mm Hg, whereas PCl_3 has a 400 mm Hg vapor pressure at 57°C. Which type of intermolecular forces are responsible for the observed vapor pressures.

13.97 The boiling points of CH_3CH_3 and CH_3F at 1 atm are −88.6°C and −78.2°C, respectively. Which type of intermolecular forces are responsible for the observed vapor pressures?

13.98 Why doesn't methane form hydrogen bonds?

13.99 Which of the following can form intermolecular hydrogen bonds in the pure liquid?

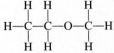

13.100 Which of the following can form intermolecular hydrogen bonds in the pure liquid?

13.101 Give the number of sites per molecule where hydrogen atoms may be donated in hydrogen-bond formation and the number of sites where electron pairs may be supplied for hydrogen-bond formation.

13.102 Give the number of sites per molecule where hydrogen atoms may be donated in hydrogen-bond formation and the number of sites where electron pairs may be supplied for hydrogen-bond formation.

13.103 Write the formula of each of the following hydrates.
(a) calcium chloride monohydrate
(b) zinc iodate dihydrate
(c) calcium nitrate trihydrate
(d) strontium chloride dihydrate
(e) aluminum iodide hexahydrate

13.104 Write the formula of each of the following hydrates.
(a) ytterbium(III) chloride hexahydrate
(b) thorium(IV) chloride dihydrate
(c) terbium(III) chloride hexahydrate
(d) samarium(III) bromide trihydrate
(e) scandium(III) sulfate hexahydrate

13.105 Calculate the percent water in each of the following hydrates.
(a) $SnCl_4 \cdot 4H_2O$ (b) $Na_2CrO_4 \cdot 10H_2O$
(c) $Th(NO_3)_4 \cdot 12H_2O$ (d) $Sc(NO_3)_3 \cdot 4H_2O$
(e) $SrI_2 \cdot 6H_2O$ (f) $RbAl(SO_4)_2 \cdot 12H_2O$

13.106 Calculate the percent water in each of the following hydrates.
(a) $TlAl(SO_4)_2 \cdot 12H_2O$ (b) $SrBr_2 \cdot 6H_2O$
(c) $Na_2Cr_2O_7 \cdot 2H_2O$ (d) $RbCr(SO_4)_2 \cdot 12H_2O$
(e) $Sm_2(SO_4)_3 \cdot 8H_2O$ (f) $NaH_2PO_4 \cdot H_2O$

13.107 Given the mass of a hydrate and the amount of water contained in the hydrate, calculate the formula of the hydrate.
(a) 3.71 g of $LaCl_3 \cdot xH_2O$ contains 1.26 g of H_2O
(b) 9.34 g of $FePO_4 \cdot xH_2O$ contains 1.80 g of H_2O
(c) 7.23 g of $Sr(BrO_3)_2 \cdot xH_2O$ contains 0.36 g of H_2O

13.108 Given the mass of a hydrate and the amount of water contained in the hydrate, calculate the formula of the hydrate.
(a) 21.5 g of $AlBr_3 \cdot xH_2O$ contains 10.8 g of H_2O
(b) 3.48 g of $Th(NO_3)_4 \cdot xH_2O$ contains 1.08 g of H_2O
(c) 11.40 g of $Na_2CrO_4 \cdot xH_2O$ contains 6.00 g of H_2O

14 | Solutions

Learning Objectives

After studying Chapter 14 you should be able to

1. Identify solutions involving all three states of matter.
2. Calculate solution concentrations using percent, parts per million, molarity, and molality.
3. Perform calculations based on the preparation of dilute solutions.
4. Distinguish between electrolytes and nonelectrolytes.
5. Explain the effects of pressure, temperature, and the solvent on solubility.
6. Calculate the effect of solute on the boiling point and freezing point of a solution.
7. Calculate the effect of solute on the osmotic pressure of a solution.

14.1 Solvents and Solutes

As indicated in Chapter 3, very few materials that you are familiar with are pure substances; most are mixtures—either homogeneous or heterogeneous. Homogeneous mixtures, more commonly known as solutions, have components that are uniformly distributed. Only homogeneous mixtures will be considered in this chapter. Examples of solutions include the air that we breathe, the body fluids that support our life, and the many metal alloys that we use everyday.

The substance present in the largest quantity in a solution is the **solvent;** the substance dissolved in the solvent is called the **solute.** The solvent may be a gas, liquid, or solid (Figure 14.1). We will briefly discuss each type of solvent before we consider in detail the more common liquid solutions in which water is the solvent. A solution of a solute in water is called an **aqueous solution.** The oceans are an aqueous solution of many substances as are the body fluids which sustain our life.

All gases that do not react with each other will mix with each other in all proportions. In gaseous solutions, the individual molecules are relatively far apart (Figure 14.1) and move independently. Air is the most common example of a gaseous solution. Dry air contains approximately 78% nitrogen molecules, 21% oxygen molecules, and 1% argon atoms.

Gases, liquids, and solids may dissolve in liquid solvents. In liquid solutions the molecules of solute and solvent are close together (Figure 14.1). Oxygen dissolved in water, a solution of a gas in a liquid, maintains aquatic life. Another example of a solution of a gas in a liquid is carbon dioxide in carbonated beverages. Two common examples of liquids that dissolve in water are ethyl alcohol and acetic acid. Ethyl alcohol is a component of all alcoholic beverages; acetic acid is present in vinegar. Both salt and sugar are well-known examples of solids that will dissolve in water.

Although solid solutions might appear to be less common, there are, in fact, numerous examples of this type of solution. Many metals dissolve in one another to form solid solutions called **alloys** (Figure 14.1). Examples of such alloys are brass (zinc and copper) and sterling silver (silver and copper). A less common example of a solid solution is vinertia alloy (67% cobalt, 27% chromium, 6% molybdenum) used to replace body joints such as hips and knees.

In this chapter we will discuss concentrations, which are a quantitative method of expressing the amount of solute in a solution. We will also consider the process of forming a solution and the submicroscopic nature of the dissolved solute. Then the properties of the resulting solutions—freezing point, boiling point, and osmotic pressure—will be discussed.

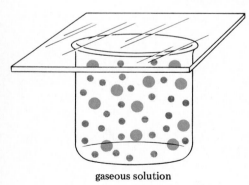

gaseous solution

liquid solution

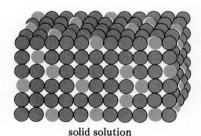

solid solution

Figure 14.1
Solutions and the states of matter
Solutes may be dissolved in solvents in any of the three states of matter.

14.2 Concentrations

The composition of a solution is described in terms of concentration. The **concentration** is a measure of the amount of solute in a given amount of solution or solvent. A **concentrated solution** contains a large amount of solute per given amount of solvent or solution. A **dilute solution** contains a small amount of solute per given amount of solvent or solution. There is no sharp line between these two qualitative descriptions of solutions; however, one would judge a solution of 1 g of sugar in 1 L of water to be dilute, whereas a solution of 100 g of sugar in 1 L of water would be concentrated.

When a few milligrams of sodium chloride is added to a test tube containing 10 mL of water, only a small amount of stirring is required to dissolve the solid. If more and more sodium chloride is added, a point is reached at which no additional salt will dissolve. At this point, when added solute remains undissolved, the solution is **saturated.** As long as more solute will dissolve, the solution is **unsaturated.** Thus, the term saturated means ''full,'' whereas unsaturated means that the ''solution could hold more.''

Sodium chloride is quite soluble, and 36 g is required to saturate 100 mL of water at room temperature. On the other hand, only 9.0×10^{-5} g of silver chloride is required to saturate 100 mL of water. The term saturated should not be confused with concentrated, nor should unsaturated be confused with dilute. A solution may be ''full'' but yet contain very little material. Solutions of 9.0×10^{-5} g of sodium chloride and 9.0×10^{-5} g of silver chloride in 100 mL of water are both dilute solutions. Only the silver chloride solution is saturated!

The qualitative terms dilute versus concentrated or unsaturated versus saturated are descriptively useful. However, quantitative measures of concentrations are needed for many purposes. For example, the effective administration of a medicine usually requires a prescribed amount of the therapeutic agent. Quantitative expressions of the amount of solute in a solvent or solution are required in chemical laboratories, industries, hospitals, and pharmacies.

A variety of methods are used to express concentration. Each method has been chosen for convenience under some particular set of circumstances. These include percent concentration, molarity, and molality. Normality, another unit of concentration, will be discussed in Chapter 16.

14.3 Percent Concentration

The concentration of a solution is easily expressed using the percent of the solute in the solution. Since percent means parts per hundred, a 3% solution means 3 parts of solute out of a total of 100 parts of solution. There are, of course, 97 parts of solvent present to make the 100 parts. It is not necessary that the percent be an integer. In a 0.9% solution, there are 0.9 parts of the solute and 99.1 parts of the solvent. Percent concentrations may be expressed with units of mass or volume. Three concentrations commonly used are weight to weight, weight to volume, and volume to volume.

The weight to weight percent (w/w %) is given by the following equation. The weights are usually given in grams.

$$\text{w/w \%} = \frac{\text{weight of solute}}{(\text{weight of solute} + \text{weight of solvent})} \times 100$$

Example 14.1

What is the w/w % of a solution of 5.0 g of sugar in 45.0 g of water?

Solution Both quantities are in the same units and may be substituted directly into the percentage equation for w/w %.

$$\% \text{ sugar solution} = \frac{5.0 \text{ g sugar}}{5.0 \text{ g sugar} + 45.0 \text{ g water}} \times 100$$

$$\% \text{ sugar solution} = 10\%$$

[*Additional examples may be found in 14.12 at the end of the chapter.*]

A common way of expressing percent concentrations in clinical situations is weight to volume percent (w/v %), which compares the weight of the solute to the total volume of the solution.

$$\text{w/v } \% = \frac{\text{grams of solute}}{100 \text{ mL of solution}} \times 100$$

For example, the normal saline solution used to dissolve drugs for intravenous therapy is 0.9% aqueous sodium chloride. The solution is prepared by dissolving 0.9 g of NaCl in water to give 100 mL of solution. To prepare a liter (1000 mL) of normal saline, 9 g of sodium chloride is weighted out (Figure 14.2) and enough water then is added to make 1000 mL of solution.

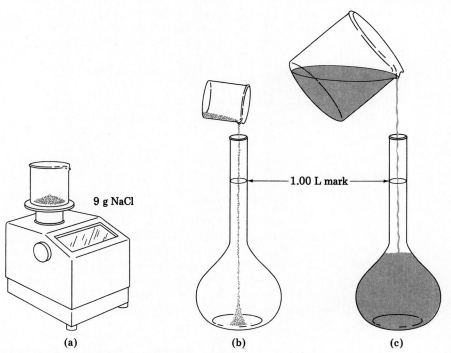

9 g NaCl

1.00 L mark

(a)   (b)   (c)

Figure 14.2

Preparation of 0.9% NaCl solution
(a) Weigh 9 g of sodium chloride. (b) Place in 1.00 L flask. (c) Add water to the mark while shaking slightly until the NaCl dissolves completely.

Example 14.2

A 2.00 mL sample of blood plasma contains 6.80 mg of sodium ions. Calculate the concentration in milligrams per 100 milliliters (mg/100 mL).

Solution First we calculate the weight to volume ratio in the units given.

$$\frac{6.80 \text{ mg sodium}}{2.00 \text{ mL plasma}} = 3.40 \text{ mg/mL}$$

In a 100 mL volume there would be 100 times the mass of sodium ions.

$$3.40 \text{ mg/mL} \times 100 \text{ mL} = 340 \text{ mg}$$

Therefore, there is 340 mg of sodium ions in 100 mL of blood plasma, and the concentration is 340 mg/100 mL. The unit mg/100 mL can be used to report sodium ion concentration in the blood, although the equivalent milligrams per deciliter (mg/dL) is now the preferred unit.

Example 14.3

How many grams of sodium chloride are present in 250 mL of a salt solution that is 0.90% sodium chloride?

Solution A 0.90% solution contains 0.90 g of sodium chloride for every 100 mL of solution. Therefore, the amount of sodium chloride in any quantity of sodium chloride solution may be calculated by multiplying the volume by the concentration.

$$250 \text{ mL solution} \times \frac{0.90 \text{ g NaCl}}{100 \text{ mL solution}} = 2.2 \text{ g NaCl}$$

[*Additional examples may be found in 14.9 and 14.10 at the end of the chapter.*]

Volume to volume percent (v/v %) is useful when dealing with liquid solutes. The equation is as follows.

$$\text{v/v \%} = \frac{\text{volume of solute}}{\text{total volume of solution}} \times 100$$

The volumes are usually given in milliliters. Note that the v/v % concentration expresses directly the milliliters of solute per 100 mL of solution. This method is used to express the concentration of alcohol in beverages. A wine that is 12% alcohol contains 12 mL of alcohol in 100 mL of solution. **Proof,** a term that is also used to express the alcohol content in alcoholic beverages, is equal to twice the v/v % of the alcohol. An alcoholic beverage that is 40% alcohol is 80 proof. Since pure alcohol is 100%, it is 200 proof.

Example 14.4

A solution is prepared by dissolving 200 mL of ethyl alcohol in sufficient water to produce 1 L of solution. What are the v/v % concentration and proof of the solution?

Solution   Both volumes must be expressed in the same units.

$$\frac{200 \text{ mL of ethyl alcohol}}{1000 \text{ mL of solution}} \times 100 = 20.0\% \text{ alcohol}$$

The proof is twice that of the concentration or 40 proof.

[Additional examples may be found in 14.11 at the end of the chapter.]

Parts per Million

Some solutes, such as pollutants in air or water, are in such small amounts that decimal percentage concentrations would have to be used. A unit that avoids the use of decimals is parts per million (ppm). Like percent concentration, the ppm unit may refer to w/w, w/v, or v/v. One **ppm** is one part of solute per million parts of solution. For example, a pollutant in a water source might be present at 0.0005 g per 100 mL of solution. The percent would then be 0.0005%. However, given in parts per million, the concentration would be

$$\frac{0.0005 \text{ g}}{100 \text{ mL}} \times \frac{10^4}{10^4} = \frac{5 \text{ g}}{1,000,000 \text{ mL}} = 5 \text{ ppm}$$

Example 14.5

The Federal Food and Drug Administration (FDA) has set the human tolerance for mercury in fish at 0.5 ppm. A 5.0 g sample of fish contains 20 μg of mercury. Calculate the concentration of mercury in parts per million, and determine if the FDA tolerance is exceeded.

Solution   First convert the 20 μg into grams so that the mass of mercury and that of the fish are in the same units. The answer (in ppm) will be on a weight to weight basis.

$$20 \, \mu g \times \frac{1 \text{ g}}{10^6 \, \mu g} = 20 \times 10^{-6} \text{ g} = 2.0 \times 10^{-5} \text{ g}$$

Now determine the ratio of the number of grams of mercury to the grams of fish. Multiplication by 10^6 gives the number of grams of mercury per million grams of fish or parts per million.

$$\frac{2.0 \times 10^{-5} \text{ g Hg}}{5.0 \text{ g fish}} \times 10^6 = 4.0 \text{ ppm}$$

The concentration exceeds the tolerance limit of the FDA.

[Additional examples may be found in 14.13–14.16 at the end of the chapter.]

14.4 Molarity

Although percent composition solutions are easy to prepare, they have limitations in chemistry. Solutions that have the same percent composition contain the same weight of solute in a specified volume of a solution. However, they usually do not contain the same number of moles because the molecular weights of the solutes are not usually the same.

In chemistry the number of moles in a sample is important. The number of moles is related to the number of molecules and it is this quantity that is used in discussing chemical reactions. Therefore, a concentration known as molarity is convenient for chemistry.

Molarity is defined as the number of moles of solute per liter of solution. The abbreviation for molarity is *M*.

$$\text{molarity} = \frac{\text{moles of solute}}{\text{L of solution}} = M$$

To prepare a 1 *M* solution, it is necessary to add sufficient solvent to 1 mole of the solute to make a total volume of solution of exactly 1 L. A 1.00 *M* NaCl solution is prepared by dissolving 58.5 g of NaCl in sufficient water to give 1.00 L of solution (Figure 14.3).

The molarity of any solution can be calculated by dividing the number of moles of solute by the number of liters of solution. A solution of 5.85 g of NaCl in 500 mL of solution is 0.200 *M*.

$$5.85 \text{ g NaCl} \times \frac{1 \text{ mole NaCl}}{58.5 \text{ g NaCl}} = 0.100 \text{ mole NaCl}$$

$$500 \text{ mL} \times \frac{1 \text{ L}}{1000 \text{ mL}} = 0.500 \text{ L}$$

$$\frac{0.100 \text{ mole NaCl}}{0.500 \text{ L}} = 0.200 \text{ M NaCl}$$

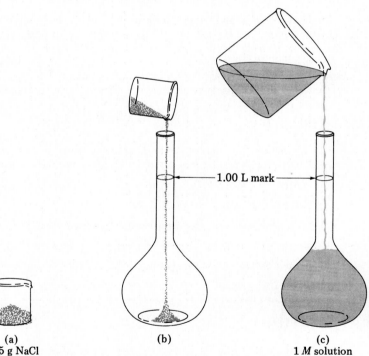

— 1.00 L mark —

(a)   (b)   (c)

Figure 14.3   58.5 g NaCl   1 *M* solution

Preparation of a 1 M solution of NaCl
(a) Weigh 58.5 g (1 mole) of sodium chloride. (b) Place in 1.00 L flask.
(c) Add water and mix thoroughly until 1 L of solution results.

Molarity is a very convenient concentration unit because it gives information about the number of moles contained in a given volume of solution or the volume of solution required to provide a desired number of moles. Volumes are easily measured with volumetric laboratory glassware such as a graduated cylinder or pipet. Thus the required mass of a solute can be provided by delivering a calculated volume of solution.

The number of moles of solute contained in a volume of solution is given by

volume of solution × molarity = moles of solute

$$\text{liters} \times \frac{\text{moles}}{\text{liter}} = \text{moles}$$

The volume of solution required to provide a stated number of moles of solute is

$$\text{moles of solute} \times \frac{1}{\text{molarity}} = \text{volume}$$

$$\text{moles} \times \frac{\text{liter}}{\text{mole}} = \text{liters}$$

The connection between volume of a solution and the number of moles of solute is summarized as follows

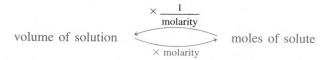

Example 14.6

How many grams of sodium hydroxide are required to produce 250 mL of a 0.20 M solution?

Solution   First the number of moles of NaOH required for this volume and molarity must be determined, and then the number of grams may be calculated. The number of moles is obtained by multiplying the molarity by the volume. The volume must be converted into liters.

$$250 \text{ mL} \times \frac{1 \text{ L}}{1000 \text{ mL}} \times \frac{0.20 \text{ mole}}{1 \text{ L}} = 0.050 \text{ mole}$$

The number of grams is obtained by multiplying the number of moles by the molecular weight.

$$0.050 \text{ mole} \times \frac{40.0 \text{ g}}{\text{mole}} = 2.0 \text{ g}$$

[*Additional examples may be found in 14.21 and 14.22 at the end of the chapter.*]

Example 14.7

Calculate the number of liters of a 0.10 M glucose solution required to provide 90 g of glucose, $C_6H_{12}O_6$.

Solution   First the number of moles contained in 90 g must be calculated.

$$90 \text{ g} \times \frac{1 \text{ mole}}{180 \text{ g}} = 0.50 \text{ mole}$$

Now the volume that contains 0.50 mole of glucose may be calculated by using the inverse factor of the molarity.

$$0.50 \text{ mole} \times \frac{1 \text{ L}}{0.10 \text{ mole}} = 5.0 \text{ L}$$

[*Additional examples may be found in 14.23 and 14.24 at the end of the chapter.*]

Example 14.8

The maximum permissible amount of Cd^{2+} in drinking water is 0.010 mg/L. What is this concentration in molarity units?

Solution   First calculate the number of grams of Cd^{2+} in the 1 L.

$$0.010 \text{ mg } Cd^{2+} \times \frac{1 \text{ g}}{1000 \text{ mg}} = 1.0 \times 10^{-5} \text{ g } Cd^{2+}$$

Now the number of moles of Cd^{2+} can be calculated by multiplying by the reciprocal of the molar mass of the element.

$$1.0 \times 10^{-5} \text{ g } Cd^{2+} \times \frac{1 \text{ mole}}{112.4 \text{ g}} = 8.8\underline{9} \times 10^{-8} \text{ mole}$$

The molarity is calculated from the number of moles of solute and the volume of solution.

$$\frac{8.8\underline{9} \times 10^{-8} \text{ mole}}{1.0 \text{ L}} = 8.8\underline{9} \times 10^{-8} M \text{ Cd}^{2+}$$

The correct answer to the required two significant figures is $8.9 \times 10^{-8} M$.

[*Additional examples may be found in 14.19 and 14.20 at the end of the chapter.*]

14.5   Dilution of Solutions

Sometimes it is necessary to prepare a solution of a desired concentration by diluting a more concentrated solution. In this process, the addition of solvent increases the volume and, as a result, decreases the concentration of the solute. However, the total number of moles of solute is unchanged. Since the product of the volume and molarity gives the number of moles we may write

$$V_{\text{initial}} \times M_{\text{initial}} = \text{moles} = V_{\text{final}} \times M_{\text{final}}$$

In doing problems using this formula you should be careful to distinguish between the total final volume and the volume of solvent that must be added to dilute the solution. For aqueous solutions

$$V_{final} = V_{initial} + V_{solvent}$$

Example 14.9

If 1.00 L of $AgNO_3$ is made by adding 900 mL of water to 100 mL of 2.0 M $AgNO_3$, what is the molarity of the dilute solution?

Solution The number of moles of $AgNO_3$ is obtained by multiplying the volume by the molarity of the initial solution.

$$0.100 \, \cancel{L} \times \frac{2.0 \text{ moles}}{1 \cancel{L}} = 0.20 \text{ mole}$$

The volume of the diluted solution is 1.00 L because the total volume is the sum of 100 mL and 900 mL. The number of moles of $AgNO_3$ in the solution is the same as in the concentrated solution. Thus the concentration is calculated by dividing the number of moles by the final volume.

$$\frac{0.20 \text{ mole}}{1.00 \text{ L}} = 0.20 \, M$$

[*Additional examples may be found in 14.29 and 14.30 at the end of the chapter.*]

Example 14.10

How many milliliters of water must be added to 50 mL of a 0.40 M NaCl solution to obtain a 0.050 M solution?

Solution The final volume of the dilute solution can be obtained by using the formula

$$V_{initial} \times M_{initial} = \text{moles} = V_{final} \times M_{final}$$
$$0.050 \, L \times 0.40 \, M = V_{final} \times 0.050 \, M$$
$$0.40 \, L = V_{final}$$

The volume of the solvent to be added can be obtained by subtracting the final volume from the initial volume.

$$V_{solvent} = V_{final} - V_{initial}$$
$$V_{solvent} = 0.40 \, L - 0.050 \, L$$
$$V_{solvent} = 0.35 \, L$$

The 0.35 L required is equal to 350 mL.

[*Additional examples may be found in 14.31 and 14.32 at the end of the chapter.*]

14.6 Molality

Molality is defined as the number of moles of solute per kilogram of solvent. The abbreviation for molality is m, not to be confused with M which is used for molarity.

$$\text{molality} = \frac{\text{moles of solute}}{\text{kg of solvent}}$$

It is also important to remember that this unit of concentration refers to the mass of the solvent, whereas molarity involves the volume of solution.

Although molarity and molality are different concentration units, their values are similar for aqueous solutions. Consider the preparation of a $1.00\ M$ solution of sucrose ($C_{12}H_{22}O_{11}$) which is made by dissolving 342 g of sucrose in sufficient water in a volumetric flask to make 1.00 L of solution. Is the volume of water added equal to 1.00 L? No, because the sucrose molecules constitute some fraction of the final volume of the solution. About 900 mL are added but we do not measure this quantity. The definition of molarity involves the final volume of the solution.

To prepare a $1.00\ m$ solution of sucrose, the 342 g sample is placed in a flask or beaker and exactly 1.00 kg of water is added. This quantity of water is equal to 1.00 L because the density of water is 1.0 g/mL. Now let's determine the molarity of this $1.00\ m$ solution. The number of moles is 1.00 and we need to know the volume of the solution. The volume of the solution is 1.11 L! Why is this quantity larger than the 1.00 L of water that we originally added? The final volume of the solution is equal to both the volume occupied by the sucrose molecules as well as that occupied by the water molecules. The solution is $0.900\ M$.

$$\frac{1.00\ \text{mole}}{1.11\ \text{L}} = 0.900\ \text{M}$$

For very dilute aqueous solutions, where the volume occupied by the solute molecules is small, the values of molality and molarity become very similar. In dilute aqueous solutions, essentially 1.00 L or 1.00 kg of solvent is required to make 1.00 L of solution. However, the difference between molality and molarity will be greater for solvents whose densities are very different from 1 g/mL regardless of the concentration of solute.

Example 14.11

What are the molarity and molality of a solution prepared by placing 0.143 g of vitamin A ($C_{20}H_{30}O$) in a volumetric flask and adding enough CCl_4 to make 1.00 L of solution. The density of CCl_4 is 1.59 g/mL.

Solution   The molecular weight of vitamin A is 286 g/mole.

$$0.143\ g \times \frac{1\ \text{mole}}{286\ g} = 5.00 \times 10^{-4}\ \text{mole}$$

There is 5.00×10^{-4} mole of vitamin A in the sample. The molarity is 5.00×10^{-4} because the volume of the solution is 1.00 L.

$$\frac{5.00 \times 10^{-4}\ \text{mole}}{1.00\ \text{L}} = 5.00 \times 10^{-4}\ M$$

In order to calculate the molality we need to use the mass of solvent added. Since the volume of the vitamin A is small, the 1.00 L of solution results from adding 1.00 L of CCl_4 solvent.

$$1.59 \text{ g/mL} \times 1000 \text{ mL} = 1.59 \times 10^3 \text{ g}$$

The mass of the solvent is 1.59×10^3 g or 1.59 kg.

The molality is calculated by dividing the number of moles by the 1.59 kg.

$$\frac{5.00 \times 10^{-4} \text{ mole}}{1.59 \text{ kg}} = 3.144654 \times 10^{-4} \, m$$

The correct answer to the three significant figures required in the problem is $3.14 \times 10^{-4} \, m$.

[*Additional examples may be found in 14.33–14.38 at the end of the chapter.*]

14.7 Electrolytes and Nonelectrolytes

Some compounds, when dissolved in water, form a solution that conducts electricity, whereas solutions of other compounds do not conduct electricity. Thus there is something to be learned about the nature of the solute present in various aqueous solutions. Very pure water is essentially a nonconductor. This can be illustrated by the use of the apparatus shown in Figure 14.4. Two electrodes, not in direct contact with each other, are immersed in water. If water could conduct electricity between the two electrodes, the light bulb would glow. Because the light does not glow, we can conclude that pure water is not a conductor.

Any substance that forms a solution in water that conducts electricity is called an **electrolyte.** When a solution of sodium chloride is used in the apparatus in Figure 14.4, the light glows brightly. The explanation for this phenomenon was provided by the Swedish scientist Svante Arrhenius in 1884. He suggested that substances whose aqueous solutions conduct electricity form ions in solution. The cations in the solution migrate to the negative electrode, called the **cathode,** while the anions migrate to the positive electrode, called the **anode.**

When ionic substances are dissolved in water, their ions are separated from the crystal structure by the water molecules. The separation of ions already present in another phase is called **dissociation.** There are covalent substances that produce ions when dissolved in water. The formation of ions upon dissolution in water is called **ionization.** For example, the gaseous covalent molecule hydrogen chloride yields ions when dissolved in water. The ions formed are given by the following equation.

$$HCl + H_2O \longrightarrow H_3O^+ + Cl^-$$

Any substance that does not conduct a current when dissolved in water is called a **nonelectrolyte.** Ordinary cane sugar, known as sucrose, $C_{12}H_{22}O_{11}$, is a nonelectrolyte. Sucrose does not contain ions in the solid state, and therefore dissociation does not occur. Furthermore, water cannot produce ions from sucrose. Another common substance that is a nonelectrolyte is ethyl alcohol, C_2H_5OH.

Electrolytes are divided into two classes called strong electrolytes and weak electrolytes. **Strong electrolytes** dissociate or ionize completely and thus enable the ready passage of electricity in the device shown in Figure 14.4, causing the light bulb to glow brightly. The degree of brightness is related to the number of ions present in solution. All ionic substances that are

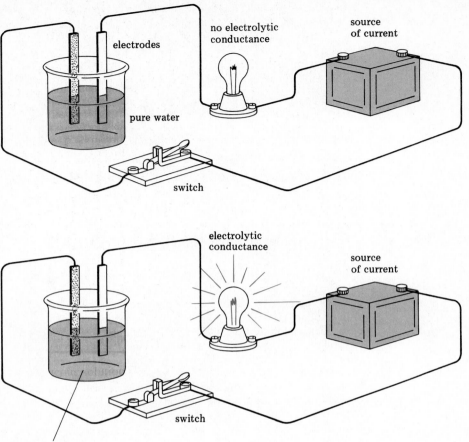

Figure 14.4 sodium chloride
$(Na^+ \ Cl^-)$ + water

Conductivity of a solution
The current flows only if ions exist in solution. Substances that provide ions in solution are called electrolytes.

soluble in water are strong electrolytes. Some covalent compounds ionize completely in water. Because these compounds produce ions that cause the bulb to glow brightly, they are also called strong electrolytes. Other covalent substances cause only a dull glow because they do not ionize completely in water. These substances are called **weak electrolytes.**

14.8 Solubility

Solubility is a measure of the amount of solute that can dissolve in a solvent. The process that occurs when a solute dissolves in a solvent can be described by the kinetic molecular theory. In an ionic solid such as sodium chloride, the sodium and chloride ions are attracted to each other.

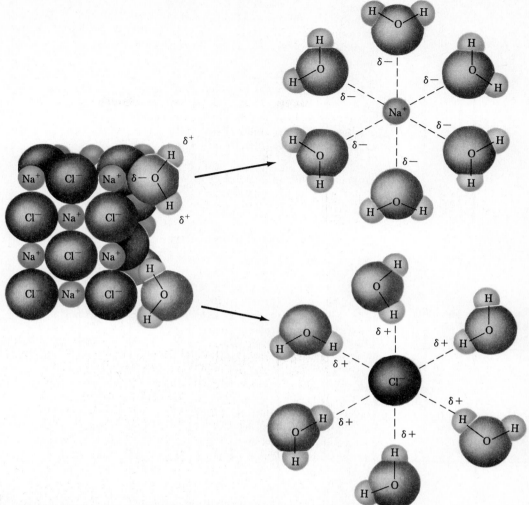

Figure 14.5

Dissolving an ionic compound in water
The attractive forces between ions and the polar water molecules are sufficient in soluble compounds to overcome the attractive forces within the crystal. The cations are approached by the negative part of the water molecule and enter solution to be surrounded by water molecules. The anions are approached by the positive part of the water molecule and enter solution to be surrounded by water molecules.

In the process of dissolving in water, these ions are separated from one another. This separation occurs when the polar water molecules approach and eventually surround the ions (Figure 14.5). The ions are removed from the site of the sodium chloride crystal by the water molecules. The electronegative oxygen of water is oriented toward the positive sodium ion, whereas the electropositive hydrogen atoms of water are oriented toward the negative chloride ion.

The amount of sodium chloride that dissolves in water depends on the temperature. If 30 g of sodium chloride is placed in 100 g of water at 25°C, the solid dissolves completely. If another 6 g of sodium chloride is added, it too dissolves, although at a somewhat slower rate. Now, if more sodium chloride is added, the solid remains at the bottom of the container. At this point, the solution is saturated. Thus, 36 g of sodium chloride in 100 mL of water at 25°C is a saturated solution.

Although no further change is apparent in a saturated solution in contact with undissolved solid, two processes are constantly taking place. Some of the solid is still dissolving in the solvent, while some of the dissolved solute is precipitating. There is a state of equilibrium. Thus, a saturated solution is one in which the rate of dissolution of the solute equals the rate of crystallization of the solute. The conditions that determine this equilibrium are temperature, pressure, and the identities of the solute and solvent.

$$\text{solvent} + \text{solute} \underset{\text{precipitation}}{\overset{\text{dissolution}}{\rightleftharpoons}} \text{solution}$$

Effect of Pressure on Solubility

As was pointed out in Chapter 12, there are large changes in volume that occur with pressure changes in the gaseous state. Similarly, the solubility of gaseous solutes in liquids show a large dependence on pressure. The solubility of a gas is directly proportional to the partial pressure of that gas above the surface of the solution. This relationship is known as Henry's law.

The solubility of carbon dioxide in water demonstrates Henry's law. All carbonated beverages are bottled under a pressure of carbon dioxide. When the bottle is opened, there is a low partial pressure of carbon dioxide above the liquid, since there is little carbon dioxide in the atmosphere. As a result, the solubility of carbon dioxide decreases and the solution effervesces as the carbon dioxide bubbles off. Table 14.1 lists the solubilities for some common gases in water. The values are for a pressure of 1 atm of the specific gas. Thus, the solubility of oxygen is 4.0×10^{-2} g/L at 25°C under a pressure of 1 atm of oxygen. Since the partial pressure of oxygen in air is 0.2 atm, the solubility of oxygen is only 8×10^{-3} g/L. The amount of oxygen this solubility could provide in blood would be insufficient to support human life; however, hemoglobin molecules in our blood cells chemically bind oxygen molecules, and this chemical process greatly increases the amount of oxygen available to the tissues.

Table 14.1 Solubility of Gases in 1 L of Water at 0°C and 1 atm

Gas	Mass (g)	Moles
He	0.0017	0.00042
N_2	0.029	0.0010
O_2	0.070	0.0021
CO_2	3.3	0.076
Ar	0.10	0.0025

Effect of Temperature on Solubility

The solubility of many gases in water decreases with increasing temperature. As water is warmed, the dissolved air forms bubbles and escapes from the liquid. Many industrial plants and power plants use water for cooling, and the hot water is passed into lakes and streams. As these waterways warm up, oxygen is less soluble, and the water is said to be thermally polluted. Fish cannot adapt to rapid temperature changes. An increase of 10°C in the temperature approximately doubles the metabolic rate. The increased metabolism requires more oxygen at a time when the solubility of oxygen has been decreased by the higher temperature.

Quite often the solubility of solids in water increases with increasing temperature. For example, the solubility of potassium chloride (KCl) increases from 28 g/100 g of water at 0°C to 57 g/100 g of water at 100°C. Similarly, the solubility of sodium chloride (NaCl) increases from 36 g/100 g to 39 g/100 g of water over the same temperature range. The solubilities of some salts, such as cerium sulfate, $Ce_2(SO_4)_3$, decrease with increasing temperature (Figure 14.6).

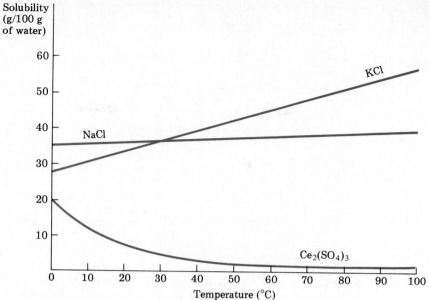

Figure 14.6

Solubility of solids in water as a function of temperature

Effect of Solvent Polarity on Solubility

A maxim of the chemistry laboratory is that "likes dissolve likes." This generalization is reasonable, since molecules of solute that are similar to molecules of solvent are expected to be better able to coexist. Water, which is a polar solvent, is a good solvent for polar solutes, ionic compounds, and substances that can produce ions in water. Carbon tetrachloride (CCl_4), a nonpolar substance, will not dissolve ionic compounds such as sodium chloride. However, fats and waxes readily dissolve in this nonpolar solvent because they are relatively nonpolar substances.

Liquids that dissolve in each other in all proportions are **miscible.** Liquids that do not dissolve in each other are immiscible. Ethyl alcohol is miscible with water, but carbon tetrachloride is immiscible with water. The miscibility of ethyl alcohol with water can be rationalized based on its structure.

$$\begin{array}{ccc} \text{H} & \text{H} & \\ | & | & \\ \text{H——C——C——O} \\ | & | & | \\ \text{H} & \text{H} & \text{H} \end{array}$$

The structural feature —O—H resembles water. Ethyl alcohol is polar, as is water. The nonbonded electron pairs on the oxygen atom in ethyl alcohol will form hydrogen bonds with water, thus making it soluble.

14.9 Colligative Properties

The properties of a solvent containing a solute are different from those of the pure solvent. The properties that depend on the number of particles dissolved and not on their chemical identity are known as **colligative properties.** Some of these colligative properties are vapor pressure, boiling point, freezing point, and osmotic pressure.

A solution containing a nonvolatile solute has a lower vapor pressure than the pure solvent. This lowered vapor pressure is a consequence of the solute molecules or ions occupying positions on the surface of the solution. For example, dissolved sodium chloride in water decreases the escaping tendency of water to the gaseous phase. The decrease in vapor pressure is related to the concentration of the solute particles present in the solution.

Example 14.12

The vapor pressure of a 0.5 M solution of NaCl in water is the same as the vapor pressure of a 1.0 M solution of sucrose. Explain why.

Solution   A 0.5 M solution of NaCl in water contains 0.5 mole of Na^+ per liter and 0.5 mole of Cl^- per liter. Thus, the solution contains the same total number of particles as are in a 1.0 M solution of sucrose. The vapor pressure of the solvent depends only on the number of solute particles present in solution.

[*Additional examples may be found in 14.53–14.56 at the end of the chapter.*]

Boiling Points of Solutions

The decreased vapor pressure of a solution of a nonvolatile solute means that it will require a higher temperature to raise the vapor pressure to atmospheric pressure. Thus the boiling point of the solvent must be increased (Figure 14.7). There is a direct relationship between the escaping tendency of the solvent molecules (vapor pressure) and the number of solute particles. Therefore the increase in the boiling point of the solvent also is directly proportional to the concentration of solute.

The elevation of the boiling point of a solution caused by the addition of sufficient nonelectrolyte to produce a 1 m solution is called the **boiling point elevation constant.**

change in boiling point = $\Delta T_b = K_b m$

ΔT_b is the change in the boiling point, K_b is the boiling point elevation constant, and m is the molality of the solution. In the case of water a 1.0 m solution of a nonelectrolyte such as sugar boils at 100.52°C at atmospheric pressure. The boiling point elevation constant is 0.52°C/m. A solution containing 1.0 mole of sugar in 400 g of water will boil at 101.3°C because the solution is 2.5 m.

change in boiling point = $\Delta T_b = K_b m$

$\Delta T = (0.52°C/m)(2.5\ m) = 1.3°C$

$T_{\text{boiling}} = 100.0°C + \Delta T = 101.3°C$

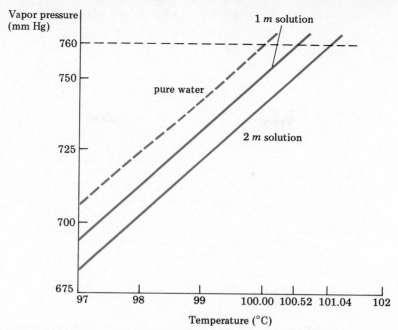

Figure 14.7

Effect of dissolved solute on the vapor pressure and boiling point of water

Example 14.13

What is the molality of a sucrose solution whose boiling point at 1 atm is 100.39°C?

Solution   The boiling point elevation of 0.39°C is less than 0.52°C which is obtained for a 1 *m* solution. Therefore the molality must be less than 1 *m*. The concentration can be obtained by direct substitution into the boiling point elevation equation.

$\Delta T_b = K_b m$

0.39°C = (0.52°C/*m*) (molality)

$0.39°C \times \dfrac{1\ m}{0.52°C}$ = molality

0.75 *m* = molality

[*Additional examples may be found in 14.57–14.60 at the end of the chapter.*]

A 1.0 *m* solution of sodium chloride in water does not boil at 100.52°C but at approximately 101°C. The boiling point elevation is twice that expected on the basis of the sugar example. The reason for the higher boiling point elevation in the case of sodium chloride is that a 1 *m* solution contains 2 moles of particles: 1 mole of sodium ions and 1 mole of chloride ions. Remember that the boiling point elevation constant actually is based on the number of moles of particles present in solution. Thus the boiling point elevation of a solution of $CaCl_2$ of a given molality is three times that of a solution of sucrose of the same molality since 3 moles of ions are present in solution per mole of compound.

$$CaCl_2(s) \longrightarrow Ca^{2+}(aq) + 2\ Cl^-(aq)$$

Freezing Point of Solutions

In both liquid–vapor and liquid–solid transformations, the escaping tendency of a solvent decreases with the addition of a solute. The escaping tendency of solvent molecules from the solid state to the liquid state is not affected by the addition of solute because only solvent molecules are present in the solid state. Thus if solute is added to an equilibrium mixture of solid and liquid, the conversion of solid into liquid is unaffected, but the conversion of liquid into solid is impeded. As a consequence more solid will melt than liquid will form solid. In order to equalize the relative escaping tendencies of solvent between the two phases, the temperature must be lowered. At some lower temperature, the equilibrium between liquid and solid phases will be reestablished, and this temperature is the freezing point of the solution. The depression of the freezing point of a 1 m solution of a nonelectrolyte is called the **freezing point depression constant.**

$$\text{freezing point depression} = \Delta T_f = K_f m$$

Where ΔT_f is the change in the freezing point, K_f is the freezing point depression constant, and m is the molality of the solution. A solution of 1 mole of a nonelectrolyte in 1 kg of water depresses the freezing point of water to $-1.86°C$. The freezing point depression constant for water is $-1.86°C/m$.

Just as the boiling point elevation of a solution depends on the number of moles of solute per 1000 g of solvent, the freezing point depression of a solution also depends on the number of particles present in solution. For an ionic substance such as sodium chloride the freezing point of a 1 m solution reflects the presence of 2 moles of ions that are present in each mole of compound.

Example 14.14

A 0.80 sample of a nonelectrolyte is dissolved in 20.0 g of water, and the freezing point of the resultant solution is $-0.93°C$. What is the molecular weight of the substance?

Solution   The freezing point of the solution indicates that the molality of the solution is less than 1 m. The calculated molality is 0.50.

$$\Delta T_f = (-1.86°C/m)\ (\text{molality})$$
$$-0.93°C = (-1.86°C/m)\ (\text{molality})$$
$$0.50\ m = \text{molality}$$

The amount of substance required to form the solution corresponds to 0.50 mole/1 kg of water. For the 20.0 g of water the number of moles is calculated by multiplying the molality by the mass of the solvent expressed in kilograms.

$$0.50\ \text{mole/kg} \times 0.0200\ \text{kg} = 0.010\ \text{mole}$$

Therefore the 0.80 g of compound is 0.010 mole.

$$\frac{0.80 \text{ g}}{0.010 \text{ mole}} = 80 \text{ g/mole}$$

The mass of 1 mole is 80 g.

[*Additional examples may be found in 14.61–14.64 at the end of the chapter.*]

There are many practical applications of the freezing point depression of liquids. In the northern states salt is commonly spread on snow and ice in order to melt them. In the presence of salt, snow and ice cannot exist at 0°C. Of course this method of melting ice is ineffective if the temperature of the ice is below that at which the freezing point of water can be depressed by the addition of salt. Another common example of freezing point depressions of liquids is the use of antifreeze in car radiators. The antifreeze consists of ethylene glycol ($C_2H_6O_2$), which is very soluble in water. The addition of antifreeze prevents the water from freezing at temperatures above that determined by the concentration of antifreeze.

Osmosis

The process of transferring water molecules through a semipermeable membrane is called **osmosis.** A **semipermeable membrane** is a membrane through which water molecules can pass but other molecules cannot. The phenomenon of osmosis and its related osmotic pressure is another example of a colligative property. Osmosis is extremely important in cells.

If a solution such as sugar in water is separated from pure water by a semipermeable membrane, the volume of the solution increases while the volume of the pure water decreases (Figure 14.8). The water remains pure because sugar molecules cannot pass through the membrane. The water molecules can cross the membrane and do so in both directions.

The tendency of water molecules to pass out of the solution side to the pure water side of the membrane is diminished by the presence of solute molecules. Since there is no such restriction on passage of water molecules toward and through the membrane from the pure water side of the membrane, a net transfer of water molecules occurs. The result of the transfer is to dilute the sugar solution.

Eventually no further net transfer of water molecules occurs, and the level of the solution remains constant at some height above that of the water. The difference in the levels of the solution and the pure solvent is related to the net tendency of water to go through the membrane to dilute the solution. There is a "back pressure" due to the height of the column of liquid on the solution side of the membrane. In effect the column of liquid provides a counterpush to the tendency of the water molecules to push across the membrane.

If pressure is mechanically applied on the side of the tube containing the solution, the net flow of water molecules can be reversed, and the height of the two columns of liquid can be made equal. The pressure required to maintain equal levels of the water and the solution is called the **osmotic pressure.**

The greater the concentration of the solution, the higher is the osmotic pressure. A 0.2 *M* sugar solution has twice the osmotic pressure of a 0.1 *M* sugar solution. However, the osmotic pressure depends not only on the molarity of the solution but also on the number of moles of solute particles present. For example, a 0.1 *M* NaCl solution has the same osmotic pressure as a 0.2 *M* sugar solution. The reason for the phenomenon is that a 0.1 *M* NaCl solution is 0.1 *M* in sodium ions and 0.1 *M* in chloride ions. Thus, in 1 L of solution there is 0.2 mole of ions. In a sugar solution the sugar is present as covalent molecules. A 0.2 *M* sugar solution has 0.2 mole of molecules per liter.

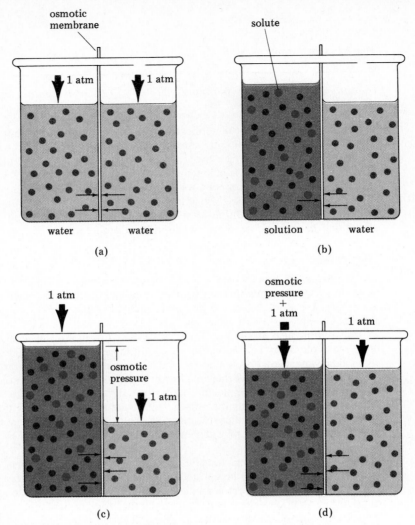

Figure 14.8

Osmotic pressure of a solution

(a) Equal rates of transfer of water result in an equilibrium between the volumes of pure water on both sides of the membrane. (b) The rate of transfer of water from the solution side is decreased because of the presence of solute. The rate of transfer from the pure water is unchanged. A net transfer of water results, and the level of water increases on the solution side of the membrane. (c) When the level of the liquid on the solution side is sufficient to reestablish the balance of rates of transfer from one side to the other, a new equilibrium results. The difference in the levels of the two liquids is due to the osmotic pressure. (d) By applying pressure on the solution side equal to the osmotic pressure the liquid levels are made equal again.

The relationship between osmotic pressure, π, and the molarity of the solute particles is similar to the ideal gas law.

$$\pi V = nRT$$

In this equation n is the number of moles of all solute particles; V is the volume of the solution. The equation can be rearranged to obtain the quotient n/V.

$$\pi = \frac{n}{V} \times RT$$

The quotient n/V corresponds to the molarity of all solute particles. For a molecular substance, the quotient is equal to the molarity of the solution. For an ionic compound it is equal to the sum of the molarities of all ions.

Physiologists use the osmole and a related term, osmolarity, to calculate osmotic pressure. An **osmole** is the molecular or formula weight of the substance in grams divided by the number of particles provided by the material in solution. Thus 58.5 g of NaCl is equal to 1 mole but 2 osmoles.

$$\frac{58.5 \text{ g/mole}}{2 \text{ moles ions/mole NaCl}} = 29.2 \text{ g/mole ion} = 29.2 \text{ g/osmole}$$

A $1.00\ M$ solution of NaCl contains 58.5 g NaCl/L. A 1.00 osmolar solution of NaCl contains 29.2 g NaCl/L and is $0.50\ M$ in Na^+ and $0.50\ M$ in Cl^-. The concentration as osmolarity can be used directly in place of n/V in the osmotic pressure equation where

$$R = 0.0821 \frac{\text{L atm}}{\text{osmole K}}$$

Example 14.15

What is the osmotic pressure of a physiological saline solution that is $0.15\ M$ NaCl at 37°C?

Solution The osmolarity of the solution is twice the molarity, since 1 mole of NaCl yields 2 moles of ions in solution. Remember that temperatures must be in kelvins.

$$\frac{n}{V} = 2(0.15 \text{ mole/L}) = 0.30 \text{ osmole/L}$$

$$\pi = 0.30 \frac{\text{osmole}}{\text{L}} \times 0.0821 \frac{\text{L atm}}{\text{osmole K}} \times 310 \text{ K} = 7.6353 \text{ atm}$$

The correct answer to the required two significant figures is 7.6 atm.

Example 14.16

The osmotic pressure of a solution containing 5.0 g of insulin per liter of solution is 16.3 mm Hg at 25°C. Calculate the approximate molecular weight of insulin. Insulin is a nonelectrolyte.

Solution The osmotic pressure in atmospheres is first calculated.

$$16.3 \text{ mm Hg} \times \frac{1 \text{ atm}}{760 \text{ mm Hg}} = 0.0214\underline{4} \text{ atm}$$

The osmolarity of the solution is calculated as follows.

$$n/V = \frac{\pi}{RT} = \frac{0.02144 \text{ atm}}{0.0821\dfrac{\text{L atm}}{\text{osmole K}} \times 298 \text{ K}} = 8.76\underline{3}\times10^{-4} \text{ osmolar}$$

The 5.0 g sample in the 1 L is equal to $8.76\underline{3}\times10^{-4}$ mole, and thus the molecular weight can be calculated.

$$\frac{5.0 \text{ g}}{8.76\underline{3}\times10^{-4} \text{ mole}} = 5.705808 \times 10^3 \text{ g/mole}$$

The correct answer to the required two significant figures is 5.7×10^3 g/mole.

[*Additional examples may be found in 14.67 and 14.68 at the end of the chapter.*]

Solutions of different osmolarity separated by a semipermeable membrane will undergo a net flow of water. When the osmotic pressures of solutions are unequal, the tendencies of water to cross the membrane are not equal. Water will flow from less concentrated to more concentrated solutions.

Osmotic pressure differences account for the wilting of plants. For example, if the dressing on a salad has an osmotic pressure higher than the fluid in the lettuce cells, the cells will lose the water that gives the lettuce its crispness. Similarly, a flower placed in a solution will wilt as it loses water from its cells. However, if it is then placed in pure water, it will regain its shape.

Summary

In describing a solution (homogeneous mixture), we indicate the relative amounts of solute and solvent. Concentrations can be qualitatively described as dilute or concentrated. A variety of quantitative expressions for concentration are used. Percent concentration units include weight/weight, weight/volume, and volume/volume. Parts per million (ppm) may be used for very dilute solutions. Molarity and molality are concentration units that provide direct information about the number of moles of solute.

Solutes in aqueous solution may be electrolytes or nonelectrolytes. Electrolytes are substances that are ionic or are converted into ions when dissolved in water. Nonelectrolytes are substances that exist as molecules in solution.

Solubility is the limit of the amount of solute that can dissolve in a solvent. The extent to which a solute dissolves in a solvent depends on the structure of the solvent and the solute, temperature, and pressure.

Properties of a solution that depend on the number of solute particles in solution and not on their identity are called colligative properties. These properties include lowering the vapor pressure, raising the boiling point, decreasing the freezing point, and increasing the osmotic pressure of a solvent.

New Terms

An **aqueous solution** is a mixture of a solute in water.

An **alloy** is a solid solution of metals.

The **boiling point elevation constant** relates the effect of a 1 molal solution of a nonelectrolyte on the boiling point of the solvent.

A **colligative property** is a property of a solution that depends on the number of dissolved particles.

A **concentrated** solution is a solution that has a high concentration of solute.

The **concentration** of a solution is a measure of the amount of solute in a given amount of solvent or solution.

A **dilute** solution is a solution that has a low concentration of solute.

Dissociation is the separation of ions in water.

Electrolytes are ionic compounds or compounds that form ions when dissolved in water.

The **freezing point depression constant** relates the effect of a 1 molal solution of a nonelectrolyte on the freezing point of the solvent.

Ionization is the formation of ions upon dissolution in water.

Miscible liquids dissolve in each other in all proportions.

Molality is equal to the ratio of moles of solute per kilogram of solvent.

Molarity is equal to the ratio of moles of solute per liter of solution.

A **nonelectrolyte** is a neutral covalent compound that does not form ions in solution and does not conduct electricity.

An **osmole** is the molecular weight in grams divided by the number of particles formed in solution.

Osmosis is the net flow of solvent molecules through a semipermeable membrane.

Osmotic pressure is the pressure required to stop the net transfer of a solvent across a semipermeable membrane.

A **ppm** is a part per million, a unit of concentration.

Proof, a concentration unit used for alcohol solutions, is twice the v/v % concentration.

A **saturated** solution is a solution in which additional solute will not dissolve.

A **semipermeable membrane** is a material that allows only water to pass through it.

Solubility is a measure of the amount of solute that can dissolve in a solvent.

The **solute** is the minor component of a solution.

The **solvent** is the major component of a solution.

A **solution** is a homogeneous mixture of two or more substances.

Strong electrolytes dissociate or ionize completely in water.

An **unsaturated** solution is a solution that has a lower concentration of a solute than its solubility.

Weak electrolytes do not dissociate or ionize completely in water.

Exercises

Terminology

14.1   Explain the difference between the terms solute and solvent.

14.3   Explain how a saturated solution may be considered to be a dilute solution.

14.2   Explain the difference between the terms saturated and unsaturated.

14.4   Explain why some covalent substances are electrolytes, whereas others are nonelectrolytes.

Solvents and Solutes

14.5   What is the solute in vinegar?

14.7   The solubility of sodium chloride is 36 g/100 mL at 0°C. A solution contains 6 g of NaCl in 20 mL at 0°C. Is the solution saturated?

14.6   What is the solvent in wine?

14.8   The solubility of sodium nitrate is 81 g/100 mL at 25°C. A solution contains 8.1 g of $NaNO_3$ in 10 mL at 25°C. Is the solution saturated?

Percent Concentrations

14.9   Calculate the w/v % concentration of each of the following.
(a) 10.0 g of NaH_2PO_4 in 1.00 L of solution
(b) 16.0 g of $NaNO_3$ in 400 mL of solution
(c) 1.50 g of NaCl in 50 mL of solution
(d) 4.0 g of $NaHCO_3$ in 5.00 L of solution

14.11   Calculate the v/v % concentration of each of the following.
(a) 100 mL of ethyl alcohol in 500 mL of solution
(b) 3 mL of acetic acid in 60 mL of vinegar solution
(c) 0.003 mL of alcohol in 1 mL of blood

14.13   You breathe 1×10^4 L of air a day. The concentration of sulfur dioxide (SO_2), a pollutant in the air, is 0.1 ppm. What volume of SO_2 molecules will you breathe?

14.15   Some solutions are so dilute that parts per million concentrations are in the 0.001–0.009 ppm range. What type of concentration unit might be used to eliminate the decimal and the zeros?

14.10   Calculate the volume of solution required to provide the indicated mass of solute from the given w/v % solutions.
(a) 18.0 g of NaCl from a 0.900% solution
(b) 30 g of $C_6H_{12}O_6$ from a 5.0% glucose solution
(c) 60.0 g of Na_2CO_3 from a 1.0% solution

14.12   Calculate the number of grams of solute needed to make each of the following w/w % solutions.
(a) 1.0 kg of 5.00% NH_4Cl
(b) 500 g of 3.00% H_2SO_4
(c) 100 g of 1% $C_6H_{12}O_6$ (glucose)

14.14   When the carbon monoxide concentration in air reaches 100 ppm, dizziness and headaches occur. Express this concentration in v/v %.

14.16   A gaseous pollutant is present in 0.2 ppm in air. What volume of pollutant is contained in 10 L of air?

14.17 A person has 0.25% alcohol in his blood. Using v/v % concentration, calculate the volume of alcohol in the blood, assuming a blood volume of 6 L.

14.18 A sherry is 20% alcohol. What concentration unit is being used?

Molarity

14.19 Calculate the molarity of each of the following solutions.
(a) 4.0 g of NaOH in 500 mL of solution
(b) 5.6 g of KOH in 200 mL of solution
(c) 414.0 g of H_2SO_4 in 1 L of solution

14.20 Calculate the molarity of each of the following solutions.
(a) 0.0058 g of $Mg(OH)_2$ in 2000 mL of solution
(b) 2.0 g of NaOH in 2000 mL of solution
(c) 20.7 g of H_2SO_4 in 400 mL of solution

14.21 How many grams of solute are there in each of the following solutions?
(a) 250 mL of 1.00 M glucose, $C_6H_{12}O_6$
(b) 250 mL of 1.00 M sucrose, $C_{12}H_{22}O_{11}$
(c) 1.50 L of 0.250 M NaOH

14.22 How many grams of solute are there in each of the following solutions?
(a) 125 mL of 1.00 M ethyl alcohol (C_2H_6O)
(b) 250 mL of 0.500 M glycerin ($C_3H_8O_3$)
(c) 50 mL of 0.125 M KOH

14.23 What volume will provide the required number of moles of solute with the given concentration of solution?
(a) 0.20 mole of NaOH from 0.50 M NaOH
(b) 1.00 mole of H_2SO_4 from 0.100 M H_2SO_4
(c) 0.010 mole of NaCl from 0.100 M NaCl

14.24 What volume will provide the required number of moles of solute with the given concentration of solution?
(a) 0.10 mole of KOH from 0.25 M KOH
(b) 0.50 mole of H_2SO_4 from 0.200 M H_2SO_4
(c) 0.250 mole of NaCl from 0.05 M NaCl

14.25 A 2.00 mL sample of blood contains 0.490 mg of calcium ions. Calculate the molarity of calcium ions.

14.26 Household ammonia contains 0.85 g NH_3 in 100 mL. What is the molarity of ammonia?

14.27 How many moles of $FeSO_4$ are in 250 mL of drinking water that is 1.5×10^{-6} M $FeSO_4$?

14.28 How many moles of glucose ($C_6H_{12}O_6$) are contained in 200 mL of a 0.30 M glucose solution used in intravenous injection?

Dilution of Solutions

14.29 What will be the molarity of the solution that results from mixing each of the following?
(a) 10 mL of 12 M HCl and 200 mL of H_2O
(b) 25 mL of 16 M HNO_3 and 250 mL of H_2O
(c) 100 mL of 18 M H_2SO_4 and 400 mL of H_2O
(d) 50 mL of 15 M NH_3 and 500 mL of H_2O

14.30 What volume of the indicated concentrated reagent is required to prepare the indicated diluted solution?
(a) 15 M NH_3 to give 100 mL of 1.0 M NH_3
(b) 18 M H_2SO_4 to give 1.00 L of 0.10 M H_2SO_4
(c) 16 M HNO_3 to give 250 mL of 1.0 M HNO_3
(d) 12 M HCl to give 500 mL of 0.5 M HCl

14.31 What volume of water must be added to the indicated sample to give the desired final concentration of solution?
(a) 100 mL of 0.50 M NaOH to give 0.20 M NaOH
(b) 50 mL of 1.0 M NH_3 to give 0.020 M NH_3
(c) 25 mL of 0.20 M NaCl to give 0.010 M NaCl
(d) 200 mL of 4.0 M HCl to give 0.20 M HCl

14.32 What volume of water must be added to the indicated sample to give the desired final concentration of solution?
(a) 25 mL of 2.0 M NH_3 to give 0.010 M NH_3
(b) 50 mL of 1.0 M HCl to give 0.025 M HCl
(c) 100 mL of 1.5 M NaCl to give 0.15 M NaCl
(d) 200 mL of 2.0 M NaOH to give 0.02 M NaOH

Molality

14.33 What is the molality of a solution of 50 g of methyl alcohol (CH_3OH) in 500 g of water?

14.34 What is the molality of a solution of 2.5 g of moth balls ($C_6H_4Cl_2$) in 50 g of benzene (C_6H_6)?

14.35 What is the molality of a solution of 9.6 g of baking soda ($NaHCO_3$) in 100 g of H_2O?

14.36 What is the molality of glucose ($C_6H_{12}O_6$) in spinal fluid which contains 75 mg of glucose per 100 g of water?

14.37 Concentrated hydrobromic acid is 48.0% HBr by mass and has a density of 1.50 g/mL. What is the molality of this solution?

14.38 Concentrated phosphoric acid is 48.0% H_3PO_4 by mass and has a density of 1.70 g/mL. What is the molality of this solution?

Electrolytes

14.39 Hydrogen bromide (HBr) is a covalent molecule, but an aqueous solution of HBr conducts electricity. Write an equation that explains this observation.

14.40 Ethyl alcohol is a nonelectrolyte. What does this tell us about ethyl alcohol?

14.41 What criterion is used to distinguish between strong and weak electrolytes.

14.42 Silver chloride (AgCl) is an ionic compound. A saturated solution of AgCl does not show appreciable conductivity. Explain why.

Solubility

14.43 The solubility of sodium chloride is 36 g/100 mL at 25°C. Is a solution that contains 150 g of NaCl in a liter saturated or unsaturated?

14.45 Why do bubbles escape from a bottle of beer after the cap is removed?

14.47 How is the oxygen content of river water affected by industrial plants that use the water for cooling?

14.49 Ammonia (NH_3) is very soluble in water. Explain why.

14.51 Iodine is more soluble in carbon tetrachloride (CCl_4) than in water. Suggest a reason.

14.44 The solubility of barium sulfate is 0.00024 g/100 mL at 20°C. A solution contains 0.00120 g of $BaSO_4$ in 500 mL. Is the solution saturated or unsaturated?

14.46 Why does an open bottle of soft drink go "flat" at room temperature faster than one stored in a refrigerator?

14.48 Why is the concentration of oxygen in water increased by an increase in the partial pressure of oxygen above the surface of water?

14.50 Some ionic solids are insoluble in water. What does this fact indicate?

14.52 Bromine is less soluble than iodine in CCl_4. Suggest a reason.

Colligative Properties

14.53 Why does fresh water evaporate more rapidly than sea water at the same temperature?

14.55 The freezing points of a 0.10 M KCl solution and a 0.10 M glucose solution are not equal. Explain why.

14.54 How are the boiling point and freezing point of water affected by the concentration of solute?

14.56 Why would $CaCl_2$ be more effective than NaCl in melting ice in the winter?

Boiling Point Elevation

14.57 A 6.0 g sample of a nonelectrolyte is dissolved in 40.0 g of water to produce a solution that boils at 100.52°C. What is the molecular weight of the substance?

14.59 What is the boiling point of a solution of 1.75 g of sucrose ($C_{12}H_{22}O_{11}$) in 10 g of water?

14.58 Urea (NH_2CONH_2) is a nonelectrolyte. What is the boiling point of an aqueous solution of urea containing 3.0 g of urea and 25 g of water?

14.60 A solution of 1.9 g of a nonelectrolyte in 5.0 g of water boils at 101.05°C. What is the molecular weight of the substance?

Freezing Point Depression

14.61 A solution of 6.75 g of a nonelectrolyte in 100 g of water has a freezing point of −0.74°C. What is the molecular weight of the compound?

14.63 What is the freezing point of a solution of 0.010 mole of NaCl in 0.10 kg of water?

14.62 A 1.33 g sample of nicotine in 20 g of water freezes at −0.75°C. What is the molecular weight of nicotine?

14.64 What is the freezing point of a solution of 0.10 mole of $CaCl_2$ in 0.30 kg of water?

Osmosis

14.65 Celery that is kept in a refrigerator for a long time can become limp. The crispness can be restored by putting the celery in water. Explain why.

14.67 Amylose, a soluble starch, has a molecular weight of approximately 30,000 amu. What would be the osmotic pressure of a solution of 2.0 g of amylose dissolved in 1 L of water?

14.66 Salt spread on driveways to melt ice may be injurious to nearby plants and shrubs. Explain why.

14.68 The osmotic pressure of a solution of 10 g of a milk protein in 1 L of water is 5 mm Hg at 25°C. What is the molecular weight of the protein?

Additional Exercises

14.69 Give one example of each of the following solutes in a solvent.
(a) a gas in a gas   (b) a gas in a liquid

14.70 Give one example of each of the following solutes in a solvent.
(a) a liquid in a liquid   (b) a solid in a liquid

14.71 What is the solvent in vinertia alloy?

14.72 What qualitative terms are used to describe the amount of solute dissolved in a solvent?

14.73 Calculate the w/v % concentration of each of the following.
(a) 10.0 g of K_2SO_4 in 1.00 L of solution
(b) 16.0 g of KNO_3 in 500 mL of solution
(c) 1.50 g of KBr in 100 mL of solution
(d) 4.0 g of NaOH in 5.00 L of solution

14.74 Calculate the volume of solution required to provide the indicated mass of solute from the given w/v % solutions.
(a) 4.0 g of NaOH from a 1.2% solution
(b) 20 g of $C_{12}H_{22}O_{11}$ from a 2.0% solution
(c) 30.0 g of KCl from a 2.0% solution

14.75 Calculate the v/v % concentration of each of the following.
(a) 100 mL of glycerin ($C_3H_8O_3$) in 500 mL of solution
(b) 5 mL of wood alcohol (CH_3OH) in 75 mL of solution
(c) 0.04 mL of ethyl alcohol (C_2H_5OH) in 10 mL of blood

14.76 Calculate the number of grams of solute needed to make each of the following w/w % solutions.
(a) 2.0 kg of 2.00% NaCl
(b) 250 g of 4.00% HCl
(c) 200 g of 1% $C_{12}H_{22}O_{11}$ (sucrose)

14.77 A vodka is 100 proof. What is its alcohol concentration?

14.78 A wine is 10% alcohol. What is its proof?

14.79 Calculate the molarity of each of the following solutions.
(a) 4.9 g of H_3PO_4 in 50 mL of solution
(b) 3.65 g of HCl in 100 mL of solution

14.80 Calculate the molarity of each of the following solutions.
(a) 0.98 g of H_3PO_4 in 200 mL of solution
(b) 0.182 g of HCl in 5 mL of solution

14.81 How many grams of solute are there in each of the following solutions?
(a) 0.50 L of 0.14 M $NaHCO_3$
(b) 25.0 mL of 0.100 M NaCl

14.82 How many grams of solute are there in each of the following solutions?
(a) 200 mL of 0.010 M KCl
(b) 250 mL of 0.100 M NaCl

14.83 What volume will provide the required number of moles of solute with the given concentration of solution?
(a) 0.010 mole of NaCl from 0.200 M NaCl
(b) 0.100 mole of glucose from 0.500 M glucose

14.84 What volume will provide the required number of moles of solute with the given concentration of solution?
(a) 0.200 mole of NaCl from 0.10 M NaCl
(b) 0.100 mole of sucrose from 0.200 M sucrose

14.85 Rubbing alcohol contains 60 g of isopropyl alcohol (C_3H_8O) in 100 mL. What is the molarity of isopropyl alcohol?

14.86 A water sample has 0.1 mg of Cd^{2+} in 10 mL of solution. What is the molarity of Cd^{2+}.

14.87 What will be the molarity of the solution that results from mixing each of the following?
(a) 10 mL of 1.0 M NaOH and 90 mL of H_2O
(b) 25 mL of 2.0 M HCl and 75 mL of H_2O
(c) 5 mL of 0.50 M NaCl and 500 mL of H_2O
(d) 50 mL of 0.20 M KCl and 150 mL of H_2O

14.88 What volume of the indicated concentrated reagent is required to prepare the indicated diluted solution?
(a) 2.0 M NaOH to give 100 mL of 0.1 M NaOH
(b) 1.0 M NH_3 to give 500 mL of 0.010 M NH_3
(c) 1.0 M NaCl to give 200 mL of 0.020 M NaCl
(d) 1.0 M KCl to give 2.00 L of 0.0050 M KCl

14.89 What is the molality of ethylene glycol ($C_2H_4O_2$) in antifreeze containing 210 g of ethylene glycol and 200 g of water?

14.90 Rubbing alcohol contains 60 g of isopropyl alcohol (C_3H_8O) in 100 mL. What is the molarity of isopropyl alcohol?

14.91 Is anything observed when a solid is added to a saturated solution of the solid? Is anything actually happening?

14.92 On hot days, fish tend to swim to lower depths in a lake. Recalling the fact that colder water has a higher density, explain why the fish prefer the depths of the lake.

14.93 Explain what would happen to a saturated solution in contact with undissolved solute for each of the following as the temperature is increased from 25 to 100°C.
(a) NaCl   (b) KCl   (c) $Ce_2(SO_4)_3$

14.94 Explain what would happen to a saturated solution at 100°C of each of the following when the temperature is decreased to 75°C.
(a) NaCl   (b) KCl   (c) $Ce_2(SO_4)_3$

14.95 Glycerol has the following structure. Predict its solubility in water and carbon tetrachloride.

14.96 Moth balls have the following structure. Predict its solubility in water and carbon tetrachloride.

14.97 Which of the following solutions will have the lowest vapor pressure?

(a) 0.2 M NaCl

(b) 0.1 M CaCl$_2$

(c) 0.2 M glucose

14.99 Explain what would happen if a 0.4% NaCl solution were separated by a semipermeable membrane from a 0.8% NaCl solution.

14.98 Which of the following solutions will have the highest vapor pressure?

(a) 0.3 M KCl

(b) 0.25 M CaCl$_2$

(c) 0.4 M glucose

14.100 The osmotic pressure of 0.100 g of hemoglobin in 10.0 mL of solution is 2.87 torr at 25°C. Calculate the molecular weight of hemoglobin.

15 Reaction Rates and Equilibrium

Learning Objectives

After Studying Chapter 15 you should be able to

1. Relate the spontaneity of a reaction to enthalpy changes and entropy changes involved in the process.

2. Relate the rate of reactions to the nature of the reactants, concentrations of reactants, temperature, and catalysts.

3. Discuss the concept of activation energy and show how it is related to reaction rate.

4. Express an equilibrium reaction in terms of an equilibrium constant.

5. Perform calculations relating equilibrium concentrations and the equilibrium constant.

6. Demonstrate the use of Le Châtelier's principle in chemical equilibria.

7. Write net ionic equations.

8. Use solubility product constants to calculate concentrations and vice versa.

15.1 Chemical Change

Chemical changes or chemical reactions are what chemistry and indeed even life is all about. From the photosynthesis by plants to the metabolism of food by animals, we are immersed in chemical change. Reading this page involves many complex chemical reactions in our eyes and brain. The transmission of electrical impulses by nerves and muscle contractions also are the results of chemical reactions.

Chemical reactions occur at a variety of rates. The formation of oil over centuries involves slow reactions, whereas the burning of gasoline in a car engine occurs with explosive speed. Chemists are interested in how to change the speed of chemical reactions. The study of the speed or rate of chemical reactions is called **kinetics.** The kinetics of chemical reactions is of concern in industry, where it is important to understand how to rapidly form a desired chemical product in preference to other products.

It is also important to determine under what experimental conditions reactants may be converted quantitatively to products. Chemical reactions are reversible and can result in an equilibrium mixture of reactants and products. An understanding of these equilibrium processes is important in both industrial and biological chemistry. In industry, it is necessary to find those experimental conditions under which as much reactant is converted to product as possible. Not only is the inefficient conversion of chemicals costly, but the unwanted material must be removed to purify the product. In biological chemistry, our concern for equilibrium also has practical consequences. Our equilibrium biochemical processes may be disturbed as a result of sickness, improper diet, or the introduction of foreign chemicals. It is then necessary to understand how to help restore the normal equilibrium state.

Each chemical reaction is a unique event in which certain bonds in specific reactants are broken and certain bonds in specific products are formed. The number of known and potential reactions among the millions of compounds is astronomically large. In order to understand these reactions, it is useful to classify them according to defined similar processes. However, unlike bonding, for which a few classes suffice, chemical reactions involve numerous classes. Even the methods of classification differ within the various specialized fields of chemistry such as organic chemistry and biochemistry. In this chapter we will study one type of reaction— precipitation of insoluble salts. In subsequent chapters acid–base reactions and oxidation– reduction reactions will be discussed.

15.2 Why Do Chemical Reactions Occur?

Although many chemical equations can be written, they do not necessarily represent chemical reactions that actually do occur and convert reactants to products. Reactions that occur without a continued source of energy being supplied are called **spontaneous reactions.** A reaction that does not give the indicated product unless energy is added continuously is **nonspontaneous.** What determines that some reactions are spontaneous, whereas others are not? An analysis of all spontaneous reactions reveals that a combination of two features control whether or not a reaction is spontaneous: enthalpy and entropy.

Enthalpy

All substances contain stored chemical energy in their bonds. When reactants are converted to products, the stored chemical energies are not the same because the number and types of bonds

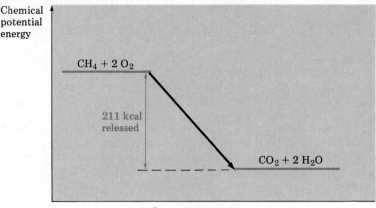

Figure 15.1

Potential energy and the energy of an exothermic reaction
The chemical potential energies of the reactants are higher than those
of the products. When the reaction of the combustion of methane
occurs, the energy difference is released as heat energy.

are altered. If the products of a reaction contain less chemical energy than the reactants, the net difference can be released as heat. A reaction that gives off heat energy is **exothermic** (Figure 15.1). The combustion of methane (natural gas) is exothermic and releases 211 kcal/mole of methane.

$$CH_4 + 2 O_2 \longrightarrow CO_2 + 2 H_2O + 211 \text{ kcal}$$

When the products of a reaction contain more stored chemical energy than the reactants, the net difference must be supplied during the reaction. A reaction that requires or absorbs heat energy is **endothermic.**

Enthalpy is a chemical term used to represent heat energy or some equivalent energy form. The energy difference between the products and the reactants is called a **change in enthalpy,** $\Delta H°$. By convention, the release of energy in an exothermic process is given a negative sign. Thus, the combustion of methane has a $\Delta H°$ of -211 kcal/mole.

$$CH_4 + 2 O_2 \longrightarrow CO_2 + 2 H_2O \qquad \Delta H° = -211 \text{ kcal}$$

The energy released or absorbed in a chemical reaction need not be in the form of heat. Electrical or light energy may be involved. Chemical reactions in a battery produce an electric current. Other reactions, such as the electrolysis of water to produce hydrogen and oxygen, require electrical energy.

Light can be released by some reactions, which is what happens in the firefly. Some reactions occur when light energy is added, as in photosynthesis, in which plants produce carbohydrates from carbon dioxide and water. Light energy equivalent to 686 kcal/mole is required for each mole of glucose formed (Figure 15.2).

$$686 \text{ kcal} + 6 CO_2 + 6 H_2O \longrightarrow C_6H_{12}O_6 + 6 O_2$$

The enthalpy change for endothermic processes is given a positive sign. For the photosynthesis process, $\Delta H° = +686$ kcal/mole of glucose formed.

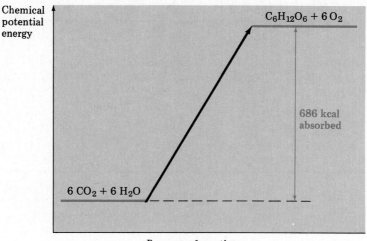

Figure 15.2

Potential energy and the energy of an endothermic reaction
The chemical potential energies of the reactants are lower than those of the products. When the reaction of photosynthesis occurs, energy equal to the net energy difference must be added. This energy is supplied as light energy from the sun.

$$6\,CO_2 + 6\,H_2O \longrightarrow C_6H_{12}O_6 + 6\,O_2 \qquad \Delta H^\circ = +686 \text{ kcal}$$

Reactions that have a negative enthalpy change (exothermic) tend to be spontaneous. Such reactions are considered to be favorable processes; however, it is also necessary to consider the effect of the entropy change on the reaction as well.

Entropy

The second controlling feature of all physical processes and chemical reactions is the tendency to achieve the most random or disordered arrangement possible. The degree of randomness or disorder is called the **entropy** of a system. By definition, the **entropy change, ΔS°**, is positive for increasing disorder.

A positive entropy change acts in opposition to unfavorable enthalpy changes in some reactions. If the increase in the degree of disorder is great, an endothermic process can occur. Conversely, a reaction leading to increased order (a negative entropy change) can occur only in an exothermic reaction.

The temperature at which a reaction occurs also is important in determining the importance of enthalpy and entropy contributions. At absolute zero, most substances are ordered and entropy differences between two or more substances are zero. Therefore, only their relative energies determine their relative stabilities. With an increase in temperature, a variety of molecular motions becomes possible, and the tendency toward disorder varies from substance to substance. The effect of differences in entropy in chemical reactions depends on the temperature. At extremely high temperatures, the entropy differences between substances may play a dominant role in the course of a reaction.

Table 15.1 **Contributions of $\Delta H°$ and $\Delta S°$ to $\Delta G°$**

$\Delta H°$	$\Delta S°$	$\Delta G°$	Result
negative	positive	negative at all temperatures	reaction will proceed
positive	negative	positive at all temperatures	reaction will not proceed
negative	negative	negative if the temperature is sufficiently low	reaction might proceed at sufficiently low temperature
positive	positive	negative if the temperature is sufficiently high	reaction might proceed at sufficiently high temperature

Free Energy

The relationship between enthalpy changes and entropy changes is given by the following expression, in which $\Delta G°$ symbolizes the change in free energy of a system at constant pressure.

$$\Delta H° - T \times \Delta S° = \Delta G°$$

The **free energy change** is a measure of the tendency of a reaction to proceed spontaneously. When $\Delta G°$ is negative, a chemical or physical process occurs spontaneously. The negative enthalpy change contributes toward making $\Delta G°$ negative, whereas a positive entropy change contributes to making $\Delta G°$ negative. Note from the expression that $\Delta H°$ is more important at low temperatures and $\Delta S°$ becomes more important at high temperatures. A summary of the contributions of $\Delta H°$ and $\Delta S°$ to $\Delta G°$ and the resultant prediction of the spontaneity of a chemical reaction is given in Table 15.1.

15.3 Kinetics

As we mentioned briefly before, kinetics is a study of the rates of chemical reactions. The rate of reaction is a measure of how quickly reactants are converted into products. The factors that influence the reaction rate are the structures of the reactants, the temperature, the concentrations of the reactants, and the presence of substances called catalysts.

Reactants and Reaction Rates

The transformation of reactants into products involves the rupture of some bonds and the formation of others. Therefore, the nature of the chemical substances involved is the most important feature controlling the reaction. For example, in the reaction of CO and NO_2 (Figure 15.3), an oxygen atom from NO_2 is transferred to CO. Bonding must occur between carbon and oxygen and a bond must be broken between nitrogen and oxygen.

Even simple molecules can react at dramatically different rates. The colorless compound nitrogen oxide (NO), which is produced in an internal combustion engine, reacts very quickly with oxygen at room temperature to form a reddish brown gas, nitrogen dioxide (NO_2).

$$2\,NO + O_2 \longrightarrow 2\,NO_2$$

Figure 15.3

Reaction of nitric oxide and carbon monoxide
The collision of the two molecules results in the rupture of the bond between one oxygen atom and the nitrogen atom. A bond is formed between the carbon atom and the oxygen atom that was originally bonded to the nitrogen atom.

Oxides of nitrogen, such as nitrogen dioxide, are responsible in part for the smog in some metropolitan areas.

When carbon monoxide (CO) is produced from the incomplete combustion of gasoline in a car engine, it does not react very fast with oxygen at room temperature.

$$2\,CO + O_2 \longrightarrow 2\,CO_2$$

As a consequence, the carbon monoxide level can build up in the atmosphere in urban areas. If the reaction of carbon monoxide with oxygen were as rapid as that of nitrogen oxide, the dangerous effects of carbon monoxide poisoning, such as decreased reaction times, drowsiness, or death would not occur.

Although the chemical equations for the reactions of NO and CO with oxygen appear similar, there is a large difference in the rates of the two reactions. The difference in reactivity is due to differences in the bonding in these two molecules. The carbon atom in CO has an octet of electrons. Nitrogen oxide does not have an octet of electrons around nitrogen, and its reactivity may be attributed to this deficiency of electrons.

$$:C\equiv O: \qquad \cdot \overset{..}{N}=\overset{..}{O}:$$

Concentrations and Reaction Rates

Two common reactions are the burning of wood and the rusting of iron. Both reactions involve heterogeneous mixtures in which oxygen gas reacts with a solid. The reactants must come in contact with each other to react, and the reaction velocity increases with an increase in surface area. If wood is chopped into fine kindling or if iron is ground into powder, more of the solid comes in contact with the oxygen of air and the reaction rate increases.

For reactions in homogeneous mixtures the reactants must similarly also come into contact. As the concentration of reactants in either a gas or liquid is increased, the reaction velocity increases because of the increased probability that the reactant molecules will collide. In a gas, the reaction velocity can be increased either by increasing the amount of reactants in a constant volume or by decreasing the volume (increasing the pressure) containing the reactants. In the liquid phase, reactant concentrations may be increased by adding reactants.

Temperature and Reaction Rates

All reaction velocities increase with a rise in temperature. Some reaction rates are very sensitive to temperature changes, whereas others are only slightly affected. However, a general rule that can be used is that on the average a 10°C rise in temperature doubles the reaction rate.

Chemical reactions in living organisms are affected by changes in temperature. Increased temperature results in an increased metabolic rate. In humans with a fever, the metabolism rate increases by 5% for each degree Fahrenheit above the normal temperature. As a consequence, either food intake must be increased or a weight loss results.

Lowering of the temperature slows down the reaction rates of the body. Animals that hibernate are able to survive because their body temperatures are lowered and as a result their need for food is decreased. A woodchuck's heart rate decreases from about 75 beats a minute in the nonhibernating state to about 5 beats a minute while hibernating. The rates of all chemical reactions dependent on stored body fat and oxygen are decreased. The woodchuck can then survive the hibernation period.

The technique of lowering the body temperature (hypothermia) of surgical patients is used by doctors to enable them to perform certain operations and avoid deterioration of vital tissue. Without oxygen from blood circulation, the brain is irreparably damaged in a few minutes at normal body temperature. However, if the body temperature is lowered from 37 to 20°C, the brain can be deprived of oxygen for about an hour and not be affected. Thus, surgeons can perform operations on a still heart, which can then be restarted after repairs have been made.

Example 15.1

Milk is known to spoil at room temperature but to remain suitable for drinking if stored in a refrigerator. Explain why.

Solution The spoiling of milk involves chemical reactions. These reactions occur faster at room temperature than at the temperature of the refrigerator. However, given sufficient time, even milk stored in a refrigerator eventually will spoil. It is for this reason that expiration dates are placed on milk and milk products such as cottage cheese in grocery stores.

[Additional examples may be found in 15.13–15.16 at the end of the chapter.]

Catalysts and Reaction Rates

A **catalyst** is a substance that increases a reaction rate when it is present in the reaction mixture. A catalyst is said to catalyze the reaction, and its effect is known as catalysis. Catalysts are usually required only in small amounts. The catalyst is present in the same amount before and after the reaction takes place, even though it must interact with the reactant at a given step in the reaction.

The production of oxygen in college chemistry laboratories shows the effect of catalysis. When potassium chlorate ($KClO_3$) is heated, it slowly decomposes into oxygen and potassium chloride (KCl).

$$2 \, KClO_3 \longrightarrow 2 \, KCl + 3 \, O_2$$

If a small amount of manganese dioxide (MnO_2) is added to the potassium chlorate, the rate of decomposition is increased. The manganese dioxide serves as a catalyst in the reaction.

Catalysts in plants and animals allow these organisms to carry out reactions at rates sufficient for the organism to survive. Catalysts in organisms are called **enzymes.** Any one species has many enzymes because each enzyme is highly specific in its catalysis of one reaction.

In the body, glucose is metabolized efficiently at 37°C.

$$C_6H_{12}O_6 + 6 \, O_2 \longrightarrow 6 \, H_2O + 6 \, CO_2$$

The actual conversion occurs in several reactions, each catalyzed by a specific enzyme. Outside the body, the same combustion can occur rapidly only above 600°C. At body temperature, the chemical conversion without enzymes would require months. This rate is too slow to provide the energy necessary to support life.

Enzymes are as common in the plant kingdom as in the animal kingdom. One of the very important processes catalyzed by enzymes is nitrogen fixation. The nitrogen gas in the atmosphere ultimately is the principal natural source of ammonia and other nitrogen compounds essential for protein formation. Nitrogen gas is quite unreactive, although small quantities of nitrogen are converted into nitrogen compounds by lightning discharges. However, under substantially less dramatic conditions, the bacteria in the roots of leguminous plants such as alfalfa, beans, clover, and peas readily convert nitrogen gas into ammonia. These bacteria contain enzymes that catalyze the conversion under mild conditions and provide the largest source of nitrogen compounds to organisms.

15.4 Activation Energy

In a chemical reaction, reactant molecules are converted into product molecules by breaking some bonds and forming other bonds. For bonding changes to occur, reactant molecules must be brought together and collide with considerable energy. Only if the energy is high can the molecules be forced close enough together to overcome the repulsion between the electrons around the nuclei in the reactants. The minimum energy required for a successful collision leading to a chemical reaction is called the **activation energy.** Molecules with less than the activation energy rebound without reaction. Each reaction has its own characteristic activation energy based on the number and type of bonds broken and formed in the reaction.

Because there is a specific activation energy for each reaction, a certain temperature is required for the reaction to occur at an appreciable rate. The temperature controls the kinetic energy of the molecules. As the kinetic energy increases, the chances increase for molecular collisions with energy equal to the activation energy. A chemical reaction then occurs more frequently, and the observed rate of reaction increases.

In the combustion of methane, a flame increases the temperature so that more methane molecules attain the activation energy. This activation energy is shown in the reaction-progress diagram of Figure 15.4. The potential energy of the molecules must be increased by the activation energy to boost them to the top of the energy hump shown. Once the molecules reach this point, they can be converted to products and energy is released. The energy released is equal to the activation energy originally added plus an amount equal to that characteristic for the exothermic reaction. In total, the net release of energy for the reaction is 211 kcal/mole. The energy released in this exothermic reaction is sufficient to continue to increase the kinetic energy of the remaining reactant molecules. The reaction is then self-sustaining and the methane continues to burn.

Endothermic reactions require an input of energy to allow the molecules to undergo a chemical reaction. Consider the photosynthesis of glucose shown in the reaction-progress diagram in Figure 15.5. Light energy is used to activate the molecules. Some energy is released as products are formed, but it is less than the activation energy. The total energy difference is equal to the amount by which the reaction is endothermic. Thus, if the light energy is not continually provided, the photosynthetic reaction ceases.

When the temperature is increased for either an exothermic or an endothermic reaction, the reaction rate increases. This rate increase reflects both the increased frequency of collision of faster-moving molecules and the fact that more molecules have the necessary activation energy for reaction when they do collide.

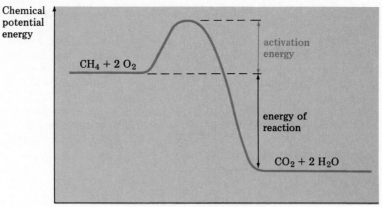

Figure 15.4

Activation energy of an exothermic reaction
After the highest energy point is reached by adding energy, the activation energy and energy of reaction was released. Since the activation energy was needed to cause the reactants to reach the high point of chemical energy, the net energy released is equal to the heat of the reaction.

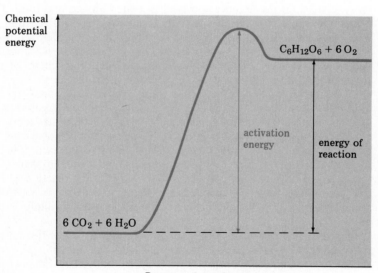

Figure 15.5

Activation energy of an endothermic reaction
After the highest energy point is reached by adding the activation energy, some energy is released. The energy released is less than the activation energy. Thus a net energy input is required for the reaction.

In the preceding section the importance of catalysts in increasing the rate of reaction was noted. The effect of the catalyst is to provide a different path for the progress of the reaction. The path starts at the reactants and concludes at the products. However, the path for the catalyzed reaction has a different activation energy (Figure 15.6). An analogy for a catalyst is that of a pathfinder who locates a different route from one altitude to some other altitude via a lower mountain range.

To illustrate the effect of a catalyst on the path of a reaction, consider the hypothetical reaction of A and B, a reaction with a high activation energy.

$$A + B \longrightarrow X$$

In order for the reaction to occur, the high activation energy must be provided, so the reaction is slow and difficult. However, in the presence of a catalyst such as an enzyme represented by E, the following reactions may occur.

Step 1 $A + E \longrightarrow A\text{---}E$

Step 2 $A\text{---}E + B \longrightarrow X + E$

The enzyme may combine with A in a reaction with a low activation energy. Similarly, the reaction of A---E with B may require little energy. If the activation energy of each step is low, the molecules will be able to react faster via this enzyme-catalyzed pathway than they could without the enzyme.

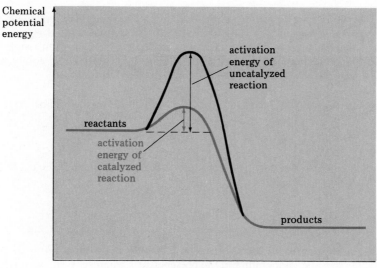

Figure 15.6

Activation energy of a reaction and the effect of catalysis
The catalyst provides an alternate pathway for a reaction. The activation energy for the process is lower than for the uncatalyzed reaction. This lower energy requirement results in a faster reaction.

15.5 Chemical Equilibrium and the Equilibrium Constant

In previous chapters the concept of an equilibrium was presented in terms of physical processes. Thus, water vapor can exist in equilibrium with liquid water. At equilibrium, the rates at which molecules leave and return to the liquid are equal. Similarly, in osmosis, water molecules cross the membrane in both directions until an equilibrium is achieved based on concentration and pressure. In dynamic equilibria, processes occur constantly although there is no net macroscopic change.

Chemical reactions do not proceed in one direction. As a reaction occurs, product molecules are formed and they can collide. Some collisions lead to the formation of the original reactants. Thus, two opposing reactions occur. When the rate at which products form is equal to the rate at which reactants form, an equilibrium is established.

$$\text{reactants} \underset{\text{reverse reaction}}{\overset{\text{forward reaction}}{\rightleftharpoons}} \text{products}$$

For a general equation at equilibrium, where A and B are reactants, X and Y are products, and m, n, p, and q are coefficients,

$$m\text{A} + n\text{B} \rightleftharpoons p\text{X} + q\text{Y}$$

the following expression has been shown experimentally to be a constant at a specific temperature. The brackets indicate molar concentrations.

$$\frac{[\text{X}]^p[\text{Y}]^q}{[\text{A}]^m[\text{B}]^n} = K$$

The numerical value of the expression for the relationship between the concentrations of reactants and products is called the **equilibrium constant** and is given by the symbol K.

The expression of the equilibrium constant for any reaction can be written by inspection of the balanced equation. For example, hydrogen and iodine can combine to form hydrogen iodide.

$$\text{H}_2 + \text{I}_2 \rightleftharpoons 2\,\text{HI}$$

According to the generalized form, the equilibrium constant expression is

$$K = \frac{[\text{HI}]^2}{[\text{H}_2][\text{I}_2]}$$

The concentration of hydrogen iodide is squared because the coefficient of hydrogen iodide in the balanced equation is 2.

Example 15.2

Ammonia is produced commercially by a reaction of nitrogen and hydrogen gases at high pressure and temperature in the presence of a catalyst. Write an equilibrium constant expression for the reaction

$$\text{N}_2 + 3\,\text{H}_2 \rightleftharpoons 2\,\text{NH}_3$$

Solution   First write the concentration of ammonia, the product, in the numerator and square it because the coefficient of NH_3 in the equation is 2.

$$\frac{[NH_3]^2}{?}$$

Next, place the concentration of nitrogen, a reactant, in the denominator. The power of 1 is not indicated.

$$\frac{[NH_3]^2}{[N_2]}$$

Finally, place the concentration of hydrogen, another reactant, in the denominator and cube it because the coefficient of H_2 in the equation is 3.

$$K = \frac{[NH_3]^2}{[N_2][H_2]^3}$$

[*Additional examples may be found in 15.19–15.24 at the end of the chapter.*]

The equilibrium constant gives the relationship between the concentrations of reactants and products. If the concentrations of all reactants and products are known at equilibrium, the equilibrium constant can be calculated by substituting the concentrations into the equilibrium expression. If the equilibrium constant is known, the equilibrium concentration of one substance can be calculated, providing the concentrations of the other substances are known.

Example 15.3

Consider the following reaction at equilibrium at 1000°C where the concentrations are $5.6 \times 10^{-3}\ M$ CO, $5.6 \times 10^{-3}\ M$ H_2O, $4.4 \times 10^{-3}\ M$ CO_2, and $4.4 \times 10^{-3}\ M$ H_2. What is the equilibrium constant?

$$CO(g) + H_2O(g) \rightleftharpoons CO_2(g) + H_2(g)$$

Solution   First write the equilibrium constant expression.

$$\frac{[CO_2][H_2]}{[CO][H_2O]} = K$$

Next, substitute the equilibrium concentrations into the expression and solve the equation.

$$\frac{(4.4 \times 10^{-3}\ M)(4.4 \times 10^{-3}\ M)}{(5.6 \times 10^{-3}\ M)(5.6 \times 10^{-3}\ M)} = K = 0.617346$$

The correct answer to the two required significant figures is $K = 0.62$.

[*Additional examples may be found in 15.25–15.30 at the end of the chapter.*]

Example 15.4

Consider the following equilibrium reaction in which the equilibrium concentrations of H_2 and I_2 are both 0.25 M. What is the equilibrium concentration of HI if the equilibrium constant is 57?

$$H_2(g) + I_2(g) \rightleftharpoons 2\,HI(g)$$

Solution First write the equilibrium constant expression.

$$\frac{[HI]^2}{[H_2][I_2]} = 57$$

Next, substitute the equilibrium concentrations into the expression and solve the equation. Note that it is necessary to take a square root to obtain the concentration of HI.

$$\frac{[HI]^2}{(0.25\ M)(0.25\ M)} = K = 57$$

$$[HI]^2 = 57 \times 0.25\ M \times 0.25\ M = 3.5625\ M^2$$

$$[HI] = 1.887458\ M$$

The correct answer to the required two significant figures is $[HI] = 1.9\ M$.

[*Additional examples may be found in 15.31–15.34 at the end of the chapter.*]

For an equilibrium in which the concentrations of the products raised to the appropriate powers in the numerator are larger than the concentration of the reactants raised to the appropriate powers in the denominator, the equilibrium constant is large. Conversely, for a small equilibrium constant, the numerical quantities in the numerator are less than the numerical quantities in the denominator. A large equilibrium constant such as 10^{10} indicates that there is little reactant left at equilibrium. For an equilibrium constant of 10^{-10}, there is little product present at equilibrium.

Catalysts do not change the position of a chemical equilibrium or the value of the equilibrium constant. The catalyst facilitates both the forward and reverse reactions equally and no position difference results.

Equilibrium constants depend on temperature. For reactions that are exothermic in the forward direction, the equilibrium constant decreases with increasing temperature. For reactions that are endothermic in the forward direction, the equilibrium constant increases with increasing temperature (Table 15.2). Recall that an exothermic reaction has a negative $\Delta H°$; an endothermic reaction has a positive $\Delta H°$. The first reaction listed in the table is exothermic, and the second reaction is endothermic.

Table 15.2 **Effect of Temperature on the Equilibrium Constant**

Exothermic reaction $CO + H_2O \rightleftharpoons CO_2 + H_2$ $\Delta H° = -9900$ cal		Endothermic reaction $N_2 + O_2 \rightleftharpoons 2\ NO$ $\Delta H° = 43{,}200$ cal	
$T(°C)$	K	$T(°C)$	K
600	3.5	1600	2.5×10^{-4}
800	1.23	1800	7.3×10^{-4}
1000	0.60	2000	1.9×10^{-3}
1200	0.35	2200	4.0×10^{-3}

(Exothermic: increase T ↓, decrease K ↓; Endothermic: increase T ↓, increase K ↓)

15.6 Chemical Equilibrium and Le Châtelier's Principle

Le Châtelier's principle can be applied to chemical as well as physical equilibria. A change in the conditions of an equilibrium will cause a readjustment to reestablish the equilibrium. We will consider three types of changes in conditions: concentration, pressure, and temperature. The first two changes do not affect the value of the equilibrium constant; concentration and pressure affect only the individual concentrations of the substances in equilibrium with each other. Temperature changes alter the value of the equilibrium constant as well as the concentrations of the substances in equilibrium. Catalysts have no effect on either the value of the equilibrium constant or the concentrations because they affect the rates of the forward and reverse reactions to an equal extent.

Concentration Changes

If additional reactant is added to a chemical reaction at equilibrium, the concentrations of reactants and products change so as to maintain the equilibrium constant. This is done by decreasing the concentration of the reactants and increasing the concentration of the products. In short, the force imposed on the system by the addition of material is countered by causing that material to react. If a product is removed from a chemical reaction at equilibrium, the reaction will proceed to produce more of that product. Regardless of the condition imposed on the chemical equilibrium, the concentrations ultimately are changed so that the same value of the equilibrium constant is obtained.

Consider the equilibrium that exists between glucose in the blood and glycogen stored in the liver.

$$n \text{ glucose} \rightleftharpoons n \text{ H}_2\text{O} + \text{glycogen}$$

Glycogen is a very large molecule consisting of many covalently bonded glucose molecules. These bonds can be broken with water to give free glucose molecules. At equilibrium we have 65–100 mg of glucose per 100 mL of blood. When we do work, our blood glucose level decreases. One of the ways our system responds is to produce more glucose from stored glycogen in the liver. By removing a substance on the left side of the equation, the reverse or right-to-left reaction is favored. When an excess of glucose is ingested, the forward reaction is favored and the liver converts the excess blood glucose into glycogen.

Example 15.5

Consider the following reactions at equilibrium and determine what changes in the position of the equilibrium are caused by the indicated change in concentration of one of the substances given in parentheses.

(1) $CO(g) + 3 H_2(g) \rightleftharpoons CH_4(g) + H_2O(g)$ (remove CO)
(2) $CO_2(g) + C(s) \rightleftharpoons 2 CO(g)$ (add CO_2)

Solution   For reaction 1 the removal of CO, one of the reactants, will cause the reaction to proceed to the left to produce more CO. For reaction 2, the addition of CO_2, one of the reactants, will cause the reaction to use that CO_2 and proceed to the right.

[*Additional examples may be found in 15.35–15.40 at the end of the chapter.*]

Pressure Changes

Pressure changes in chemical systems also are countered in a way predicted by Le Châtelier's principle. If the pressure of a gas is increased, the system will respond by processes that tend to decrease the pressure. In the hemoglobin–oxyhemoglobin reaction, the amounts of the two compounds are controlled by the pressure of oxygen.

$$Hb + O_2 \rightleftharpoons HbO_2$$

High pressures of oxygen such as those in a hyperbaric chamber cause the hemoglobin (Hb) to be converted into oxyhemoglobin (HbO_2) as the system seeks to decrease the pressure of oxygen. The benefit to an individual is that the increased amount of oxyhemoglobin allows more oxygen to be transported to the cells. Conversely, at lower oxygen concentrations at high altitudes, the amount of oxyhemoglobin is lowered and the body cannot function as efficiently.

Reactions will respond to changes in pressure only if there is a difference in the number of moles of gaseous reactants as compared to gaseous products. Since the volumes of liquids and solids are not affected very much by pressure changes, reactants and products that are not gases are not considered in deciding which way the equilibrium will shift.

Example 15.6

Consider the reactions given below. Will any of the reactions proceed more to the right by increasing the pressure?

(1) $2\,CO(g) + O_2(g) \rightleftharpoons 2\,CO_2(g)$

(2) $2\,NO(g) \rightleftharpoons N_2(g) + O_2(g)$

(3) $CO_2(g) + C(s) \rightleftharpoons 2\,CO(g)$

Solution   In reaction 1 there are 2 moles of gaseous product and 3 moles of gaseous reactants. An increase will cause the reaction to proceed to the material of lower volume, that is, the product side.

In reaction 2 there are 2 moles of gaseous reactant as well as 2 moles of gaseous products. The reaction will be unaffected by pressure changes since there is no difference in the volumes.

In reaction 3 there are two moles of gaseous product but only 1 mole of gaseous reactant. Note that carbon is a solid. The reaction will proceed to the side of lower volume which is the reactant side.

[*Additional examples may be found in 15.41 and 15.42 at the end of the chapter.*]

Temperature Changes

Changes in temperature affect the value of the equilibrium constant. We can understand the change by considering heat as a reactant or product. For a reaction that is exothermic in the forward direction, heat can be treated as a product.

$$A \rightleftharpoons B + heat$$

Recall from our discussion of $\Delta H°$ that an exothermic reaction has a negative enthalpy change. If the system with an exothermic reaction is heated, it will respond by using up the added heat energy; the reverse reaction will be favored, and B will be consumed. Thus, the value of the equilibrium constant is decreased. If the system is cooled, it will react to produce heat, much as our body does; the amount of B will increase as will the equilibrium constant.

For a reaction that is endothermic in the forward direction, heat can be treated as a reactant.

$$A + \text{heat} \rightleftharpoons B$$

If the system with an endothermic reaction is heated, it will respond by using up the added heat energy; the forward reaction will be favored, and B will be produced. Thus, the value of the equilibrium constant is increased. If the system is cooled, it will react to produce heat, the amount of B will decrease as will the equilibrium constant.

Example 15.7

Consider the reactions given below. Which of the two reactions will proceed to the right and increase the equilibrium constant by increasing the temperature?

(1) $\qquad 2\,CO_2(g) \rightleftharpoons 2\,CO(g) + O_2(g) \qquad \Delta H° = 135 \text{ kcal}$

(2) $\quad H_2(g) + F_2(g) \rightleftharpoons 2\,HF(g) \qquad\qquad \Delta H° = -130 \text{ kcal}$

Solution Reaction 1 is endothermic and requires heat energy to proceed to the right. Thus increasing the temperature will increase the extent of the reaction to the right and will increase the equilibrium constant. Reaction 2 is exothermic and releases heat energy. Increasing the temperature will decrease the extent of the forward reaction and increase the extent of the reverse reaction. The equilibrium constant will decrease.

[*Additional examples may be found in 15.43 and 15.44 at the end of the chapter.*]

15.7 Precipitation Reactions

Soluble ionic compounds exist in aqueous solutions as cations and anions. When two solutions containing dissolved ionic compounds are mixed, the ions of both substances are intermingled. If any of these ions can react to form an insoluble substance, that substance will separate from the solution. The combination of ions from a solution to form a solid is called **precipitation.** The insoluble solid that deposits from solution is called a **precipitate.**

When a solution of sodium sulfate is added to a solution of barium chloride, a precipitate of barium sulfate is formed. The other product, sodium chloride, remains in solution.

$$Na_2SO_4(aq) + BaCl_2(aq) \longrightarrow 2\,NaCl(aq) + BaSO_4(s)$$

The equations for precipitation reactions can be written in simpler form by using the ions actually involved in the reaction. Equations that use the formula of the predominant species present in solutions and include only those species that undergo chemical change are called **net ionic equations.**

For the reaction of $BaCl_2$ with Na_2SO_4, we write the formula of each species as it actually exists in solution. Since barium sulfate is not in solution, it is written as $BaSO_4(s)$.

$$2\,Na^+(aq) + SO_4{}^{2-}(aq) + Ba^{2+}(aq) + 2\,Cl^-(aq) \longrightarrow 2\,Na^+(aq) + 2\,Cl^-(aq) + BaSO_4(s)$$

Notice that the sodium ions and chloride ions appear on both sides of the equation and neither reacted. This often is the case in reactions of ionic compounds in solution. Ions that are present in solution but not involved in a reaction are called **spectator ions.**

Table 15.3 **General Rules for the Water Solubilities of Common Ionic Compounds**

1. All common salts of the alkali (IA) metals and of the ammonium (NH_4^+) ion are **soluble.**
2. All nitrates (NO_3^-), chlorates (ClO_3^-), perchlorates (ClO_4^-), and acetates ($C_2H_3O_2^-$) are **soluble.**
3. The chlorides (Cl^-), bromides (Br^-), and iodides (I^-) of most metals are **soluble.** The principal **exceptions** are those of Pb^{2+}, Ag^+, and Hg_2^{2+}.
4. All sulfates (SO_4^{2-}) are **soluble** except for those of Sr^{2+}, Ba^{2+}, Pb^{2+}, and Hg_2^{2+}.
5. All carbonates (CO_3^{2-}), chromates (CrO_4^{2-}), and phosphates (PO_4^{3-}) are **insoluble** except for those of the alkali metals and ammonium salts.
6. The group IA metal hydroxides (OH^-) are **soluble.** The hydroxides of Ca^{2+}, Sr^{2+}, and Ba^{2+} are **moderately soluble.** The rest of the hydroxides are **insoluble.**
7. The sulfides of all metals are **insoluble** except for those of NH_4^+ and the IA and IIA metals.

The net ionic equation for a reaction in solution does not include the spectator ions. Only the ions that react or are produced are used in the equation. Ions that appear on both sides of the equation are canceled. The net ionic equation for the reaction under consideration is

$$Ba^{2+}(aq) + SO_4^{2-}(aq) \longrightarrow BaSO_4(s)$$

In order to predict when precipitation reactions will occur and to write net ionic equations, some general solubility rules are necessary (Table 15.3).

Example 15.8

An aqueous solution of silver nitrate is added to a solution of sodium chloride and a precipitate results. Write the net ionic equation for the reaction.

Solution The possible products are silver chloride and sodium nitrate. From the solubility rules we note that silver chloride is insoluble but sodium nitrate is soluble. The molecular equation is

$$AgNO_3(aq) + NaCl(aq) \longrightarrow AgCl(s) + NaNO_3(aq)$$

According to the rules for expressing ionic equations, the compounds are written as

$$Ag^+(aq) + NO_3^-(aq) + Na^+(aq) + Cl^-(aq) \longrightarrow AgCl(s) + Na^+(aq) + NO_3^-(aq)$$

The Na^+ and NO_3^- ions are not involved in the reaction and are eliminated.

$$Ag^+(aq) + Cl^-(aq) \longrightarrow AgCl(s)$$

[*Additional examples may be found in 15.55–15.58 at the end of the chapter.*]

15.8 Solubility Product Equilibria

The solubility phenomena described in the previous section can be expressed in terms of an equilibrium constant. When a solution is saturated, the rate of dissolution equals the rate of precipitation, and an equilibrium between the dissolved solute and the solid phase exists.

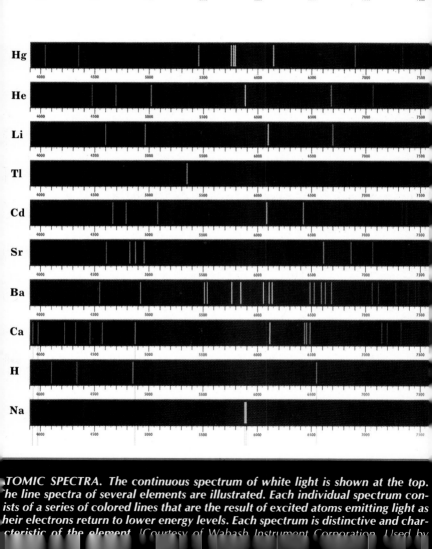

ATOMIC SPECTRA. The continuous spectrum of white light is shown at the top. The line spectra of several elements are illustrated. Each individual spectrum consists of a series of colored lines that are the result of excited atoms emitting light as their electrons return to lower energy levels. Each spectrum is distinctive and characteristic of the element. *[Courtesy of Wabash Instrument Corporation. Used by*

SOLUBILITY OF IODINE. Carbon [disulfide is a]
nonpolar molecule that is more den[se than water. The]
lower layer in the flask is CS₂ and the [upper layer is water.]
Iodine (I₂) is nonpolar and is more so[luble in CS₂ than in]
water as evidenced by the intense pu[rple color of the lower]
layer. [Photograph by Carey B. Van [Loon. From General]
Chemistry, Fourth Edition, by Ralph H[. Petrucci. Copyright]
© 1985 by Ralph H. Petrucci. Reprinte[d with permission of]
Macmillan Publishing Company.]

[SUBLIMATION OF I]**ODINE. Iodine has a vapor pressure at** [room temperature]
[th]**at results in gaseous iodine molecules in** [equilibrium with solid]
[solid] **iodine. At 70°C, as in this experiment, the** [vapor pressure is high]
[high] **enough that the iodine vapor can be seen.** [Deposition, the oppo-]
[oppo]**site process of sublimation, occurs when** [iodine molecules]
[molecules] **contact the cooler walls of the flask.** [Pho-[tograph by Carey B.]
[B. V]**an Loon.** From General Chemistry, Fourth [Edition, by Ralph H.]
[H.] Petrucci. Copyright © 1985 by Ralph H. [Petrucci. Reprinted]
[Reprinted w]**ith permission of Macmillan Publishing** [Company.]

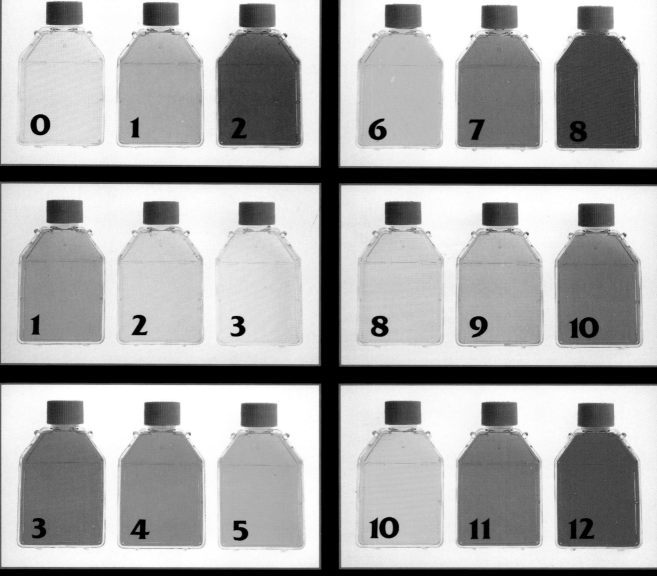

pH INDICATORS. *Indicators are substances whose acid and conjugate base forms have different colors. As a result the color of the indicator depends on the pH of the solution. The indicators shown here and the pH ranges over which they change color are methyl violet (0–2); thymol blue (1–3); methyl orange (3–5); phenol red (6–8); thymol blue (8–10) and alizarin yellow (10–12). In each set of photographs the acidic form of the indicator is on the left; the basic form is on the right.*
[Photographs by Carey B. Van Loon. From General Chemistry, *Fourth Edition,* by Ralph H. Petrucci. Copyright © 1985 by Ralph H. Petrucci. Reprinted with permission of Macmillan Publishing Company.]

ACTIVITY SERIES OF METALS. *Zinc is more active than copper. Zinc will reduce copper ions* (Cu^{2+}) *to produce copper metal and zinc ions* (Zn^{2+}). *In the left photograph the characteristic blue color of an aqueous solution of* Cu^{2+} *is shown; zinc metal is in the small container. Addition of zinc to* $Cu^{2+}(aq)$ *results in an oxidation–reduction reaction to produce a colorless solution of* $Zn^{2+}(aq)$ *shown in the right photograph and a red-brown precipitate of copper metal.* [Photographs by Carey B. Van Loon. From General Chemistry, Fourth Edition, by Ralph H. Petrucci. Copyright © 1985 by Ralph H. Petrucci. Reprinted with permission of Macmillan Publishing Company.]

OXIDATION–REDUCTION REACTIONS. *In most reactions, the oxidizing agent and reducing agent are different compounds. However, ammonium dichromate,* $(NH_4)_2Cr_2O_7$, *pictured on the left, contains both an oxidizing agent* ($Cr_2O_7^{2-}$) *and a reducing agent* (NH_4^+). *Upon heating, a reaction occurs (center) that evolves both heat and light. The products are* N_2, H_2O, *and* Cr_2O_3. *The product on the right is* Cr_2O_3. [Photographs by Carey B. Van Loon. From General Chemistry, Fourth Edition, by Ralph H. Petrucci. Copyright © 1985 by Ralph H. Petrucci. Reprinted with permission of Macmillan Publishing Company.]

$$\text{solid} \underset{\text{precipitation}}{\overset{\text{dissolution}}{\rightleftharpoons}} \text{solution of ions}$$

Consider the slightly soluble ionic compound A_nB_m in a saturated solution. The equilibrium expression is

$$A_nB_m(s) \rightleftharpoons n\,A^{m+} + m\,B^{n-}$$

The equilibrium constant is expressed as

$$K = \frac{[A^{m+}]^n[B^{n-}]^m}{A_nB_m}$$

Since the concentration of the solute A_nB_m in the solid phase is a constant, its concentration is incorporated into the constant that is called the solubility product constant, K_{sp}.

$$K_{sp} = K[A_nB_m] = [A^{m+}]^n[B^{n-}]^m$$

A **solubility product constant** is equal to the product of the concentrations of the ions, each raised to a power that is equal to the coefficient in the equilibrium equation. The value of the solubility product constant depends on the nature of the solute and the temperature of the solvent.

Each solubility product constant has units that depend on the power to which the concentrations of the ions are raised in the equilibrium expression. For the general case cited for A_nB_m, the units are derived as follows.

$$K_{sp} = [\text{mole/liter}]^n[\text{mole/liter}]^m = \frac{\text{mole}^{m+n}}{\text{liter}^{m+n}}$$

Thus the units are mole per liter raised to a power that is the sum of the coefficients of the ions in the equilibrium equation.

A comparison of K_{sp} values for compounds of the same type gives the order of solubilities. The compound with the lowest K_{sp} has the lowest solubility. For example, the K_{sp} values for AgCl and AgBr are 1.6×10^{-10} mole²/liter² and 7.7×10^{-13} mole²/liter². Silver bromide is less soluble than silver chloride. It must be reemphasized that such comparisons are valid only for compounds with the same values of m and n, as only then will the units of K_{sp} be the same. For silver chromate, Ag_2CrO_4, the units of K_{sp} are mole³/liter³ and are not the same as the units for AgCl and AgBr. However, the solubility of Ag_2S could be compared to the solubility of Ag_2CrO_4 by using the K_{sp} values. The solubility product constants of Ag_2CrO_4 and Ag_2S are 9×10^{-12} mole³/liter³ and 1.6×10^{-49} mole³/liter³, respectively. Thus we see that Ag_2S is less soluble than Ag_2CrO_4.

When the product of the concentrations of the ions present in solution, each raised to their respective powers, exactly equals the solubility product constant, a saturated solution is established. If the product of these concentrations, each raised to its respective power, is less than the K_{sp} value, the solution is unsaturated.

Example 15.9

Write the K_{sp} expression for silver bromide.

Solution The formula for silver bromide is AgBr and the equilibrium for the solubility of AgBr is

$$AgBr(s) \rightleftharpoons Ag^+(aq) + Br^-(aq)$$

The coefficients for both the silver ion and the bromide ion are 1. Therefore, the exponents in the K_{sp} expression are also 1.

$$K_{sp} = [\text{Ag}^+][\text{Br}^-]$$

[*Additional examples may be found in 15.59 and 15.60 at the end of the chapter.*]

Example 15.10

Write the K_{sp} expression for iron(III) hydroxide and determine the units of the expression.

Solution The formula for iron(III) hydroxide is Fe(OH)_3 and the equilibrium for its solubility is

$$\text{Fe(OH)}_3(\text{s}) \rightleftharpoons \text{Fe}^{3+}(\text{aq}) + 3 \text{ OH}^-(\text{aq})$$

The coefficient for the hydroxide ion is 3, and therefore the K_{sp} expression must have the same exponent for the concentration of the hydroxide ion.

$$K_{sp} = [\text{Fe}^{3+}][\text{OH}^-]^3$$

The units of the expression are mole/L or M raised to the power of the sum of the exponents for the concentrations of the ions. Thus the units are $\text{mole}^4/\text{liter}^4$ or M^4.

[*Additional examples may be found in 15.61 and 15.62 at the end of the chapter.*]

Example 15.11

The K_{sp} for AgCl is 1.6×10^{-10} $\text{mole}^2/\text{liter}^2$. What is the molarity of a saturated AgCl solution?

Solution Writing the K_{sp} expression and setting it equal to 1.6×10^{-10} $\text{mole}^2/\text{liter}^2$, we have

$$K_{sp} = [\text{Ag}^+][\text{Cl}^-] = 1.6 \times 10^{-10} \text{ mole}^2/\text{liter}^2$$

When silver chloride dissolves in water, the concentration of the silver ion is equal to that of the chloride ion. If we let $x = [\text{Ag}^+]$, then $x = [\text{Cl}^-]$. Substituting into the K_{sp} expression, we get

$$[x][x] = 1.6 \times 10^{-10} \text{ mole}^2/\text{liter}^2$$
$$x^2 = 1.6 \times 10^{-10} \text{ mole}^2/\text{liter}^2$$
$$\sqrt{x^2} = \sqrt{1.6 \times 10^{-10} \text{ mole}^2/\text{liter}^2}$$
$$x = 1.3 \times 10^{-5} \text{ mole/liter}$$

The molarity of dissolved silver chloride is equal to the calculated concentration of silver ions.

[*Additional examples may be found in 15.67 and 15.68 at the end of the chapter.*]

Example 15.12

The solubility of Mn(OH)_2 is 0.0019 g/liter. What is the K_{sp} of Mn(OH)_2?

Solution The formula weight of $Mn(OH)_2$ is 88.9 g/mole. The molarity of a saturated solution of $Mn(OH)_2$ is

$$\frac{0.0019 \text{ g/liter}}{88.9 \text{ g/mole}} = 2.1\underline{3} \times 10^{-5} \text{ mole/liter}$$

The concentration of Mn^{2+} is equal to the molarity of the dissolved compound but the concentration of OH^- is twice this value because 2 moles of OH^- are produced for every mole of $Mn(OH)_2$ that dissolves.

$$Mn(OH)_2 \rightleftharpoons Mn^{2+} + 2\,OH^-$$

The K_{sp} expression is now written and the concentrations substituted into it.

$$K_{sp} = [Mn^{2+}][OH^-]^2$$
$$K_{sp} = [2.1\underline{3} \times 10^{-5}][4.2\underline{6} \times 10^{-5}]^2$$
$$K_{sp} = 3.865439 \times 10^{-14} \text{ mole}^3/\text{liter}^3$$

The correct answer to the two significant figures is $K_{sp} = 3.9 \times 10^{-14}$ mole^3/liter^3.

[Additional examples may be found in 15.63–15.66 at the end of the chapter.]

Summary

A spontaneous reaction has a natural tendency to occur. All spontaneous reactions release free energy by a combination of an enthalpy change and an entropy change.

Reaction rates are determined by the identity of the reactants, concentration, temperature, and catalysts. Increasing the temperature of a reaction increases the reaction rate. Catalysts cause an increase in reaction rates by providing a reaction pathway that can occur at a faster rate.

When the rates of conversion of reactants to products and products to reactants are equal, a chemical equilibrium results. The equilibrium constant gives a measure of the concentrations of reactants and products at equilibrium. Altering the concentration of one or more substances in the reaction causes a shift in the concentrations of all reactants and products to maintain the equilibrium constant. The equilibrium constant is affected by temperature changes, but not by the presence of a catalyst.

Precipitation reactions are the result of the limited solubility of certain combinations of anions and cations. These reactions can be written as net ionic equations. The solubility product constant for a slightly soluble ionic compound is equal to the product of the concentrations of the ions each raised to a power equal to the coefficient in the balanced equilibrium equation.

New Terms

The **activation energy** is the minimum energy required for a reaction to occur.

Catalysis is the increase in the speed of a reaction in the presence of a substance called a catalyst.

A **catalyst** increases the speed of a chemical reaction.

An **endothermic** reaction requires heat energy.

Enthalpy is symbolized by $H°$, and $\Delta H°$ is the energy difference between two states or substances. In chemical reactions the $\Delta H°$ is the heat of the reaction and may be positive or negative.

Entropy is symbolized by $S°$, and $\Delta S°$ is a measure of the change in the degree of disorder in a system. A positive $\Delta S°$ means an increase in disorder.

Enzymes are catalysts in organisms.

An **equilibrium constant** describes numerically the relationship between the concentrations of reactants and products at equilibrium.

An **exothermic** reaction gives off heat energy.

The **free energy change,** symbolized by $\Delta G°$, measures the spontaneity of a chemical reaction. A spontaneous reaction has a negative $\Delta G°$.

Kinetics is the study of the rates of chemical reactions.

A **net ionic equation** gives only the species actually participating in a chemical reaction.

A **nonspontaneous** reaction requires the continuous addition of energy to occur.

A **precipitate** is an insoluble solid deposited from a solution.

A **precipitation reaction** is a process in which a precipitate forms by the combination of ions in solution.

The **solubility product constant** is the equilibrium constant for the dissolution of a slightly soluble salt.

Spectator ions are present in solution but are not involved in a reaction.

Spontaneous reactions occur without an outside source of energy.

Exercises

Terminology

15.1 What does a change in enthalpy for a reaction indicate about the heat evolved from or required for the reaction?

15.3 What is meant by the term spontaneous in reference to chemical reactions?

15.2 What does the change in entropy indicate about a chemical reaction?

15.4 What term is used to indicate the degree of spontaneity of a reaction?

Enthalpy, Entropy, and Free Energy

15.5 What is meant by a spontaneous reaction?

15.7 A reaction has a negative $\Delta S°$. What is meant by this fact?

15.9 What predictions may be made about reactions with the following combinations of $\Delta H°$ and $\Delta S°$?
(a) positive $\Delta H°$ and negative $\Delta S°$
(b) negative $\Delta H°$ and negative $\Delta S°$

15.11 Explain why the energy required in the photosynthesis of glucose is equal to the energy released in the metabolism of glucose.

15.6 Is a reaction with a negative $\Delta H°$ exothermic or endothermic?

15.8 What relationships exist among $\Delta G°$, $\Delta H°$, and $\Delta S°$?

15.10 What predictions may be made about reactions with the following combinations of $\Delta H°$ and $\Delta S°$?
(a) positive $\Delta H°$ and positive $\Delta S°$
(b) negative $\Delta H°$ and positive $\Delta S°$

15.12 Are the products in an exothermic reaction at a higher or lower potential energy than the reactants?

Kinetics

15.13 Why might an industry choose to carry out a reaction at a high temperature in spite of the higher costs of heat?

15.15 Meat will spoil more quickly at room temperature than when kept in a refrigerator. Explain why.

15.17 List three experimental factors that can be used to speed up a reaction.

15.14 Substances burn more rapidly in pure oxygen than in air. Explain why. This phenomenon was responsible for the tragedy during a ground test of an Apollo spacecraft in which three American astronauts died.

15.16 Explain why persons with fevers for a prolonged period must increase their food intake to avoid losing weight.

15.18 Explain why chemical reactions increase in rate as the temperature is increased.

Equilibrium Constants

15.19 Write the equilibrium constant expression for each of the following reactions
(a) $3 O_2 \rightleftharpoons 2 O_3$
(b) $N_2 + 3 H_2 \rightleftharpoons 2 NH_3$
(c) $CH_4 + Cl_2 \rightleftharpoons CH_3Cl + HCl$
(d) $2 CO + O_2 \rightleftharpoons 2 CO_2$

15.21 Write the equilibrium constant expression for each of the following reactions
(a) $CS_2 + 4 H_2 \rightleftharpoons CH_4 + 2 H_2S$
(b) $2 Cl_2 + 2 H_2O \rightleftharpoons 4 HCl + O_2$
(c) $N_2O_4 \rightleftharpoons 2 NO_2$
(d) $N_2 + 2 O_2 \rightleftharpoons 2 NO_2$

15.20 Write the equilibrium constant expression for each of the following reactions
(a) $CO + 3 H_2 \rightleftharpoons CH_4 + H_2O$
(b) $H_2 + Br_2 \rightleftharpoons 2 HBr$
(c) $2 H_2O \rightleftharpoons 2 H_2 + O_2$
(d) $N_2 + O_2 \rightleftharpoons 2 NO$

15.22 Write the equilibrium constant expression for each of the following reactions
(a) $COCl_2 \rightleftharpoons CO + Cl_2$
(b) $NO + Br_2 \rightleftharpoons NOBr_2$
(c) $4 NH_3 + 5 O_2 \rightleftharpoons 4 NO + 6 H_2O$
(d) $2 SO_2 + O_2 \rightleftharpoons 2 SO_3$

15.23 How are the units of an equilibrium constant related to the coefficients of the balanced equation?

15.24 How does the magnitude of the equilibrium constant indicate the amounts of product and reactant present at equilibrium?

Equilibrium Calculations

15.25 Given $[H_2] = 0.0343\ M$, $[I_2] = 0.326\ M$, and $[HI] = 0.826\ M$ for the following reaction at 400°C, calculate the equilibrium constant.

$$I_2(g) + H_2(g) \rightleftharpoons 2\ HI(g)$$

15.26 Given $[CO] = 0.152\ M$, $[H_2] = 0.157\ M$, $[CH_4] = 0.0478\ M$, and $[H_2O] = 0.0478\ M$ for the following reaction at 1200 K calculate the equilibrium constant.

$$CO(g) + 3\ H_2(g) \rightleftharpoons CH_4(g) + H_2O(g)$$

15.27 Given $[N_2O_4] = 0.208\ M$ and $[NO_2] = 3.11 \times 10^{-2}\ M$ for the following reaction at 25°C calculate the equilibrium constant.

$$N_2O_4(g) \rightleftharpoons 2\ NO_2(g)$$

15.28 Given $[COCl_2] = 0.275\ M$, $[Cl_2] = 2.10 \times 10^{-5}\ M$, and $[CO] = 2.10 \times 10^{-5}\ M$ for the following reaction at 125°C calculate the equilibrium constant.

$$COCl_2(g) \rightleftharpoons CO(g) + Cl_2(g)$$

15.29 Given $[CO] = 0.120\ M$, $[H_2] = 0.250\ M$, and $[CH_3OH] = 0.0788\ M$ for the following reaction at 500 K calculate the equilibrium constant.

$$CO(g) + 2\ H_2(g) \rightleftharpoons CH_3OH(g)$$

15.30 Given $[CS_2] = 0.120\ M$, $[H_2] = 0.100\ M$, $[H_2S] = 0.200\ M$, and $[CH_4] = 8.40 \times 10^{-5}\ M$ for the following reaction at 900°C calculate the equilibrium constant.

$$CS_2(g) + 4\ H_2(g) \rightleftharpoons CH_4(g) + 2\ H_2S(g)$$

15.31 The equilibrium constant for the following reaction is 3.93 at 1200 K. A system at equilibrium has $[CO] = 0.0613\ M$, $[H_2] = 0.1839\ M$, and $[CH_4] = 0.0387\ M$. What is the $[H_2O]$?

$$CO(g) + 3\ H_2(g) \rightleftharpoons CH_4(g) + H_2O(g)$$

15.32 The equilibrium constant for the following reaction is 4.66×10^{-3} at 25°C. A system at equilibrium has $[N_2O_4] = 0.208\ M$. What is the $[NO_2]$?

$$N_2O_4(g) \rightleftharpoons 2\ NO_2(g)$$

15.33 The equilibrium constant for the following reaction is 54.8 at 425°C. A system at equilibrium has $[H_2] = 0.0378\ M$ and $[HI] = 0.728\ M$. What is the $[I_2]$?

$$I_2(g) + H_2(g) \rightleftharpoons 2\ HI(g)$$

15.34 The equilibrium constant for the following reaction is 10.5 at 500 K. A system at equilibrium has $[CO] = 0.250\ M$ and $[H_2] = 0.120\ M$. What is the $[CH_3OH]$?

$$CO(g) + 2\ H_2(g) \rightleftharpoons CH_3OH(g)$$

Le Châtelier's Principle

15.35 Consider the following reaction and indicate the effect of each of the changes given.

$$CO(g) + 3\ H_2(g) \rightleftharpoons CH_4(g) + H_2O(g)$$

(a) addition of CO
(b) addition of CH_4
(c) removal of H_2
(d) removal of H_2O

15.36 Consider the following reaction and indicate the effect of each of the changes given.

$$CS_2(g) + 4\ H_2(g) \rightleftharpoons CH_4(g) + 2\ H_2S(g)$$

(a) addition of CS_2
(b) addition of CH_4
(c) removal of H_2
(d) removal of H_2S

15.37 Consider the following reaction and indicate the effect of each of the changes given.

$$CO(g) + 2\ H_2(g) \rightleftharpoons CH_3OH(g)$$

(a) addition of CO
(b) addition of CH_3OH
(c) removal of H_2

15.38 Consider the following reaction and indicate the effect of each of the changes given.

$$N_2(g) + 3\ H_2(g) \rightleftharpoons 2\ NH_3(g)$$

(a) addition of NH_3
(b) addition of H_2
(c) removal of N_2
(d) removal of NH_3

15.39 Consider the following reactions at equilibrium and determine which of the indicated changes will cause the reaction to proceed to the right.
(a) $CO(g) + 3\ H_2(g) \rightleftharpoons CH_4(g) + H_2O(g)$
$\qquad$ (remove H_2O)
(b) $CO_2(g) + C(s) \rightleftharpoons 2\ CO(g)$ $\qquad$ (add CO)
(c) $N_2(g) + 3\ H_2(g) \rightleftharpoons 2\ NH_3(g)$ $\qquad$ (add NH_3)

15.40 Consider the following reactions at equilibrium and determine which of the indicated changes will cause the reaction to proceed to the right.
(a) $2\ CO_2(g) \rightleftharpoons 2\ CO(g) + O_2(g)$ $\qquad$ (add CO)
(b) $CO(g) + 2\ H_2(g) \rightleftharpoons CH_3OH(g)$ $\qquad$ (add CO)
(c) $CO(g) + 3\ H_2(g) \rightleftharpoons CH_4(g) + H_2O(g)$
$\qquad$ (remove CO)

15.41 Consider the reactions given below. In which case(s) will the reaction proceed more to the right by increasing the pressure?
(a) $2 CO(g) + O_2(g) \rightleftharpoons 2 CO_2(g)$
(b) $2 NO(g) \rightleftharpoons N_2(g) + O_2(g)$
(c) $N_2O_4(g) \rightleftharpoons 2 NO_2(g)$

15.43 Consider the reactions below. In which cases is product formation favored by increased temperature?
(a) $CO(g) + 3 H_2(g) \rightleftharpoons CH_4(g) + H_2O(g)$
$\Delta H° = -49.4$ kcal
(b) $CO_2(g) + C(s) \rightleftharpoons 2 CO(g)$ $\Delta H° = +41.2$ kcal
(c) $H_2(g) + I_2(g) \rightleftharpoons 2 HI(g)$ $\Delta H° = -2.2$ kcal

15.45 For which of the following reactions is product formation favored by high pressure and high temperature?
(a) $CO(g) + 3 H_2(g) \rightleftharpoons CH_4(g) + H_2O(g)$
$\Delta H° = -49.4$ kcal
(b) $CO_2(g) + C(s) \rightleftharpoons 2 CO(g)$ $\Delta H° = +41.2$ kcal
(c) $H_2(g) + I_2(g) \rightleftharpoons 2 HI(g)$ $\Delta H° = -2.2$ kcal

15.42 Consider the reactions given below. In which case(s) will the reaction proceed more to the right by increasing the pressure?
(a) $Ni(s) + 4 CO(g) \rightleftharpoons Ni(CO)_4(g)$
(b) $N_2(g) + 3 H_2(g) \rightleftharpoons 2 NH_3(g)$
(c) $I_2(g) + H_2(g) \rightleftharpoons 2 HI(g)$

15.44 Consider the reactions below. In which cases is product formation favored by increased temperature?
(a) $3 O_2(g) \rightleftharpoons 2 O_3(g)$ $\Delta H° = 68.2$ kcal
(b) $2 H_2O(g) \rightleftharpoons 2 H_2(g) + O_2(g)$ $\Delta H° = 116$ kcal
(c) $N_2(g) + 3 H_2(g) \rightleftharpoons 2 NH_3(g)$ $\Delta H° = -21.9$ kcal

15.46 For which of the following reactions is product formation favored by high pressure and low temperature?
(a) $N_2(g) + O_2(g) \rightleftharpoons 2 NO(g)$ $\Delta H° = 43.3$ kcal
(b) $N_2(g) + 2 O_2(g) \rightleftharpoons 2 NO_2(g)$ $\Delta H° = 16.1$ kcal
(c) $N_2(g) + 3 H_2(g) \rightleftharpoons 2 NH_3(g)$ $\Delta H° = -21.9$ kcal

Precipitation Reactions

15.47 Identify which of the following compounds are insoluble in water.
(a) Na_2SO_4 (b) $AgBr$ (c) $(NH_4)_2SO_4$
(d) Na_2CO_3 (e) $CaCl_2$ (f) $NaNO_3$

15.49 Identify which of the following compounds are soluble in water.
(a) $Ca_3(PO_4)_2$ (b) NH_4I (c) $BaSO_4$
(d) $CaCO_3$ (e) $PbCl_2$ (f) Na_2CrO_4

15.51 Predict whether the following pairs of solutions will give a precipitate when mixed.
(a) $NH_4I + AgNO_3$
(b) $K_2SO_4 + Ba(NO_3)_2$
(c) $NH_4Br + NaNO_3$

15.53 Predict whether the following pairs of solutions will give a precipitate when mixed.
(a) $Ba(NO_3)_2 + NaCl$
(b) $RbCl + MgBr_2$
(c) $AgNO_3 + Na_3PO_4$

15.48 Identify which of the following compounds are insoluble in water.
(a) $BaCO_3$ (b) CdS (c) $(NH_4)_2S$
(d) $KClO_3$ (e) $PbSO_4$ (f) $Fe(OH)_2$

15.50 Identify which of the following compounds are soluble in water.
(a) $LiClO_4$ (b) $PbCrO_4$ (c) K_3PO_4
(d) $Zn(NO_3)_2$ (e) $Na_2Cr_2O_7$ (f) PbI_2

15.52 Predict whether the following pairs of solutions will give a precipitate when mixed.
(a) $Pb(NO_3)_2 + NaCl$
(b) $Hg(NO_3)_2 + Na_2S$
(c) $Pb(NO_3)_2 + Na_2CrO_4$

15.54 Predict whether the following pairs of solutions will give a precipitate when mixed.
(a) $NH_4Cl + Na_2S$
(b) $MgCl_2 + KClO_3$
(c) $Zn(NO_3)_2 + K_2CrO_4$

Net Ionic Equations

15.55 Silver bromide is insoluble in water. Based on this fact, write net ionic equations for the expected reaction when solutions of each of the following substances are mixed.
(a) $AgNO_3 + NaBr$
(b) $AgNO_3 + HBr$
(c) $AgNO_3 + NH_4Br$

15.57 Write net ionic equations for the reactions between each of the following sets of reactants.
(a) $NH_4I + AgNO_3$ (b) $K_2SO_4 + Ba(NO_3)_2$
(c) $Pb(NO_3)_2 + NaCl$ (d) $Hg(NO_3)_2 + Na_2S$
(e) $Ba(NO_3)_2 + Na_2CO_3$

15.56 Lead sulfate is insoluble in water. Based on this fact, write net ionic equations for the expected reaction when solutions of each of the following substances are mixed.
(a) $Pb(NO_3)_2 + Na_2SO_4$
(b) $Pb(NO_3)_2 + H_2SO_4$
(c) $Pb(NO_3)_2 + (NH_4)_2SO_4$

15.58 Write net ionic equations for the reactions between each of the following sets of reactants.
(a) $Na_2CrO_4 + BaCl_2$ (b) $(NH_4)_2S + Hg(NO_3)_2$
(c) $BaCl_2 + Na_2SO_4$ (d) $FeCl_2 + KOH$
(e) $ZnCl_2 + Na_2CrO_4$

Solubility Product Constant

15.59 Write the solubility product constant expression for each of the following insoluble salts.
(a) AgBr (b) BaSO$_4$ (c) PbI$_2$
(d) Fe(OH)$_2$ (e) Ag$_2$SO$_4$ (f) Cu$_2$S

15.60 Write the solubility product constant expression for each of the following insoluble salts.
(a) Fe(OH)$_3$ (b) InF$_3$ (c) Ag$_3$PO$_4$
(d) Mg$_3$(PO$_4$)$_2$ (e) Ba$_3$(AsO$_4$)$_2$ (f) Gd$_2$(SO$_4$)$_3$

15.61 What are the units of the solubility product constant expression for each of the following insoluble salts?
(a) AgBr (b) BaSO$_4$ (c) PbI$_2$
(d) Mg$_3$(PO$_4$)$_2$ (e) Ba$_3$(AsO$_4$)$_2$ (f) Gd$_2$(SO$_4$)$_3$

15.62 What are the units of the solubility product constant expression for each of the following insoluble salts?
(a) Fe(OH)$_3$ (b) InF$_3$ (c) Ag$_3$PO$_4$
(d) Fe(OH)$_2$ (e) Ag$_2$SO$_4$ (f) Cu$_2$S

15.63 The solubility of NiCO$_3$ (formula weight 119) is 0.018 g in 1900 mL. What is the K_{sp}?

15.64 The solubility of PbSO$_4$ (formula weight 303) is 0.045 g in 1800 mL. What is the K_{sp}?

15.65 The solubility of BaF$_2$ (formula weight 175) is 0.79 g/L. What is the solubility product?

15.66 The solubility of Ag$_2$CrO$_4$ (formula weight 334) is 0.028 g/L. What is the solubility product?

15.67 How many grams of PbSO$_4$ (formula weight 303) will dissolve in 1.5 L of water? ($K_{sp} = 1.6 \times 10^{-8}\ M^2$)

15.68 How many grams of CaCO$_3$ (formula weight 100) will dissolve in 2.0 L of water? ($K_{sp} = 8.7 \times 10^{-9}\ M^2$)

15.69 What is the molarity of Ba^{2+} ions in a saturated solution of BaF$_2$? ($K_{sp} = 1.7 \times 10^{-6}\ M^3$)

15.70 What is the molarity of F$^-$ ions in a saturated solution of BaF$_2$? ($K_{sp} = 1.7 \times 10^{-6}\ M^3$)

Additional Exercises

15.71 Under what conditions may an endothermic reaction be spontaneous?

15.72 How is the potential energy stored in chemicals? Why does this potential energy differ in two compounds such as H$_2$S and H$_2$O?

15.73 Why is the temperature inside a compost pile higher than that of the surrounding air?

15.74 Individuals who "drown" in very cold water and have stopped breathing sometimes can be revived even after a prolonged period of time under water. Explain why.

15.75 Ripened tomatoes are stored in the refrigerator, whereas unripened tomatoes are left at room temperature. Explain why.

15.76 Certain antibiotic drugs must be stored under refrigeration. Suggest a reason for this requirement.

15.77 Given [N$_2$] = 0.100 M, [H$_2$] = 0.200 M, and [NH$_3$] = 0.236 M for the following reaction at 350°C calculate the equilibrium constant.

$$N_2(g) + 3\,H_2(g) \rightleftharpoons 2\,NH_3(g)$$

15.78 The equilibrium constant for the following reaction is 0.28 at 900°C. A system at equilibrium has [CS$_2$] = 0.120 M, [H$_2$] = 0.100 M, and [CH$_4$] = 8.40 $\times$ 10^{-5} M. What is the [H$_2$S]?

$$CS_2(g) + 4\,H_2(g) \rightleftharpoons CH_4(g) + 2\,H_2S(g)$$

15.79 The equilibrium constant for the following reaction is 70 at 350°C. A system at equilibrium has [N$_2$] = 0.100 M and [H$_2$] = 0.200 M. What is the [NH$_3$]?

$$N_2(g) + 3\,H_2(g) \rightleftharpoons 2\,NH_3(g)$$

15.80 The equilibrium constant for the following reaction is 70 at 350°C. A system at equilibrium has [NH$_3$] = 0.100 M and [H$_2$] = 0.100 M. What is the [N$_2$]?

$$N_2(g) + 3\,H_2(g) \rightleftharpoons 2\,NH_3(g)$$

15.81 Consider the following reaction and indicate the effect of each of the changes given.

$$I_2(g) + H_2(g) \rightleftharpoons 2\,HI(g)$$

(a) addition of HI
(b) addition of H$_2$
(c) removal of I$_2$
(d) removal of HI

15.82 Consider the following reaction. What effect would each of the changes in condition listed have on the reaction?

$$O_3(g) + NO(g) \rightleftharpoons O_2(g) + NO_2(g)$$

(a) addition of O$_3$
(b) addition of O$_2$
(c) removal of NO
(d) removal of NO$_2$

15.83 Consider the following reactions at equilibrium and determine which of the indicated changes will cause the reaction to proceed to the right.

(a) $CO_2(g) + C(s) \rightleftharpoons 2\,CO(g)$ (add CO_2)
(b) $N_2(g) + 3\,H_2(g) \rightleftharpoons 2\,NH_3(g)$ (add NH_3)
(c) $2\,CO_2(g) \rightleftharpoons 2\,CO(g) + O_2(g)$ (add CO)

15.84 Consider the following reactions at equilibrium and determine which of the indicated changes will cause the reaction to proceed to the right.

(a) $CO(g) + 2\,H_2(g) \rightleftharpoons CH_3OH(g)$
                           (remove CH_3OH)
(b) $CS_2(g) + 4\,H_2(g) \rightleftharpoons CH_4(g) + 2H_2S(g)$
                           (remove H_2S)
(c) $CO(g) + 3\,H_2(g) \rightleftharpoons CH_4(g) + H_2O(g)$
                           (add H_2)

15.85 Consider the reactions given below. In which cases will the reaction proceed more to the right by decreasing the pressure?

(a) $4\,HCl(g) + O_2(g) \rightleftharpoons 2\,Cl_2(g) + 2\,H_2O(g)$
(b) $CS_2(g) + 4\,H_2(g) \rightleftharpoons CH_4(g) + 2\,H_2S(g)$
(c) $N_2(g) + O_2(g) \rightleftharpoons 2\,NO(g)$

15.86 Consider the reactions given below. In which case(s) will the reaction proceed more to the right by decreasing the pressure?

(a) $CO(g) + 3\,H_2(g) \rightleftharpoons CH_4(g) + H_2O(g)$
(b) $COCl_2(g) \rightleftharpoons CO(g) + Cl_2(g)$
(c) $CO(g) + 2\,H_2(g) \rightleftharpoons CH_3OH(g)$

15.87 Consider the reactions below. In which cases is product formation favored by increased temperature?

(a) $2\,CO_2(g) \rightleftharpoons 2\,CO(g) + O_2(g)$ $\Delta H° = 136$ kcal
(b) $CO(g) + 2\,H_2(g) \rightleftharpoons CH_3OH$ $\Delta H° = -5.2$ kcal
(c) $N_2(g) + O_2(g) \rightleftharpoons 2\,NO(g)$ $\Delta H° = 43.3$ kcal

15.88 For which of the following reactions is product formation favored by high pressure and high temperature?

(a) $2\,CO_2(g) \rightleftharpoons 2\,CO(g) + O_2(g)$ $\Delta H° = 159$ kcal
(b) $3\,O_2(g) \rightleftharpoons 2\,O_3(g)$ $\Delta H° = 68.2$ kcal
(c) $H_2(g) + F_2(g) \rightleftharpoons 2\,HF(g)$ $\Delta H° = -129$ kcal

15.89 Identify which of the following compounds are insoluble in water.

(a) K_2SO_4 (b) AgI (c) NH_4Cl
(d) Rb_2CO_3 (e) $BaCl_2$ (f) $Sr(NO_3)_2$

15.90 Identify which of the following compounds are insoluble in water.

(a) $SrCO_3$ (b) HgS (c) NH_4SO_4
(d) $NaClO_4$ (e) $ZnSO_4$ (f) $Ba(OH)_2$

15.91 Write a net ionic equation for the reaction between each of the following sets of reactants.

(a) $Pb(NO_3)_2 + NaI$ (b) $Cd(NO_3)_2 + Na_2S$
(c) $NH_4Br + AgNO_3$ (d) $Na_2SO_4 + Ba(NO_3)_2$
(e) $Sr(NO_3)_2 + Na_2CO_3$

15.92 Write a net ionic equation for the reaction between each of the following sets of reactants.

(a) $BaCl_2 + Li_2SO_4$ (b) $FeCl_2 + NaOH$
(c) $Li_2CrO_4 + BaCl_2$ (d) $(NH_4)_2S + Pb(NO_3)_2$
(e) $CdCl_2 + Na_2CrO_4$

15.93 Which of the following when dissolved in water has the highest concentration of magnesium ion?

(a) $MgCO_3$: $K_{sp} = 2.6 \times 10^{-5}\ M^2$
(b) MgF_2: $K_{sp} = 6.5 \times 10^{-9}\ M^3$
(c) $Mg_3(PO_4)_2$: $K_{sp} = 1 \times 10^{-25}\ M^5$

15.94 Which of the following when dissolved in water has the highest concentration of silver ion?

(a) $AgCl$: $K_{sp} = 1.6 \times 10^{-10}\ M^2$
(b) Ag_2CO_3: $K_{sp} = 6.2 \times 10^{-12}\ M^3$
(c) $AgBr$: $K_{sp} = 5.0 \times 10^{-13}\ M^2$

15.95 The solubility of $BaSO_4$ (formula weight 233) is 0.0046 g in 1900 mL. What is the K_{sp}?

15.96 The solubility of $BaCO_3$ (formula weight 197) is 0.020 g in 1400 mL. What is the K_{sp}?

15.97 The solubility of PbF_2 (formula weight 245) is 0.46 g/L. What is the solubility product?

15.98 The solubility of $Cu(IO_3)_2$ (formula weight 413) is 1.3 g/L. What is the solubility product?

15.99 Calculate the molarity of a saturated solution of $NiCO_3$?

$$(K_{sp} = 6.6 \times 10^{-9}\ M^2)$$

15.100 What is the molarity of a saturated solution of $BaCO_3$?

$$(K_{sp} = 8.1 \times 10^{-9}\ M^2)$$

16 Acids and Bases

Learning Objectives

As a result of studying Chapter 16 you should be able to

1. Identify conjugate acid–base pairs according to the Brønsted–Lowry concept.

2. Describe the properties of, and give examples of, strong and weak acids and strong and weak bases.

3. Use acid ionization data to calculate K_a and vice versa.

4. Use K_w to relate hydronium and hydroxide ion concentrations.

5. Calculate the pH of a solution given the hydronium or hydroxide ion concentration.

6. Calculate the pH of a solution of a salt that undergoes hydrolysis.

7. Calculate the pH of a chemical buffer.

8. Calculate the normality of an acid or base solution.

9. Calculate the amount of acid or base present in a sample from titration data.

16.1 Properties of Acids and Bases

In Chapter 11 you learned that acids are compounds in which hydrogen is the more electropositive element. Bases contain a hydroxide ion in combination with a metal ion. In this chapter the concept of acids and bases will be examined in greater detail.

Many common household items are either acids or bases. Citric acid in lemon juice, acetic acid in vinegar, and acetylsalicylic acid in aspirin tablets are acids. Baking soda, household ammonia, antacids, and "lye" are bases. In agriculture, the acidity of the soil is important in determining what crops may be grown. Fertilizers containing acids or bases can be applied to make the soil more suitable for particular crops.

Acid–base reactions are important in the chemistry of your body. Most of the food we eat contains compounds that are acids or can be converted into acids. The metabolism of food produces organic acids that are eventually converted into carbon dioxide and water. As the chemistry of life processes occurs, acids are produced or used, and their concentrations must be precisely controlled. Very slight changes in chemical equilibria involving acids and bases can disrupt physiological reactions. Hemoglobin–oxygen binding in respiration is affected by acid concentration. Increased or decreased respiration affects the acid–base balance of the blood. Gastric acid, a solution of hydrochloric acid, is required for the digestion that occurs in the stomach. However, the excessive production of gastric acid can cause an ulcer. Antacids, used to treat conditions of excess stomach acidity, are bases. The compositions of some common antacids are listed in Table 16.1.

Many acids have a sour taste. Examples include the dilute solutions of citric acid in lemons and acetic acid in vinegar. Most acids that you will encounter in the laboratory are aqueous solutions. Concentrated solutions of acids in the laboratory are very corrosive and should not be tasted. Acids that are accidentally spilled on your skin can cause serious chemical burns because they react with the proteins that make up cell membranes. Any affected area particularly the eye, must be immediately flooded with water to decrease the damage. For this reason, safety goggles must be worn in all chemistry laboratories. An eyewash should also be available.

Table 16.1   Composition of Some Antacids

Compound	Comments
aluminum hydroxide, $Al(OH)_3$	An effective and nonhazardous antacid with no dosage restriction. The compound is combined with other substances in Di-Gel, Gelusil, and Maalox.
dihydroxyaluminum sodium carbonate, $NaAl(OH)_2CO_3$	An effective antacid found in Rolaids. However, usage must be restricted by individuals on a low sodium diet.
calcium carbonate, $CaCO_3$	This antacid, used in Tums and Pepto Bismol, may cause constipation. Excessive usage can lead to a high blood calcium level and cause kidney stones.
magnesium hydroxide, $Mg(OH)_2$	Small dosages are used as an antacid, but larger amounts act as a laxative. As a suspension the compound is known as Milk of Magnesia.
sodium bicarbonate, $NaHCO_3$	Also known as baking soda and found in Alka-Seltzer. It gives fast and effective antacid action. Usage should be restricted by individuals on a low sodium diet.

Most common bases are solid ionic compounds containing hydroxide ions. Examples include sodium hydroxide and potassium hydroxide. Solutions of bases have a slippery feeling and a bitter taste. Bases can react with fats and oils and change them into low-molecular-weight soluble compounds. Thus, solutions of bases are useful as cleaning agents. However, they also react with materials that make up cell membranes, and therefore you should avoid skin contact with them. A fact not generally known is that bases are more destructive to the eyes than acids. Therefore, if a base gets splashed in your face, immediately flood your eyes with water.

16.2 Common Acids and Bases

Although there are many acids and bases, only a few of the more important ones are discussed here. The concentrations of some acids are given in Table 16.2.

Sulfuric acid (H_2SO_4) is the chemical produced in the largest commercial quantity in the world. It is used directly or indirectly in a great number of industrial processes. In fact, one measure of the development of a modern industrial nation is its sulfuric acid production and consumption. Sulfuric acid is the "battery acid" found in automobile batteries. Sulfuric acid is a powerful dehydrating agent; it has the ability to remove water. This dehydration process is very harmful if the acid is spilled on tissue.

A 12 M hydrochloric acid (HCl) solution is made by dissolving gaseous hydrogen chloride in water. The gastric juice in your stomach is 0.1 M hydrochloric acid, and the stomach wall must be protected against even this low concentration by a special mucosal layer. However, excessive secretion of hydrochloric acid, hyperchlorhydria, can cause ulcers over an extended time.

Nitric acid (HNO_3) is a very corrosive acid. If you spill it on your skin, a yellow stain results because the nitric acid reacts with protein molecules to produce colored compounds that then become part of your skin. The color is lost only as newly formed skin replaces the discolored tissue. Nitric acid is used to make products as diverse as fertilizers and explosives.

Phosphoric acid (H_3PO_4) is a thick, viscous liquid that is used in dilute form in soft drinks and in the production of detergents and fertilizers. Phosphoric acid combined with organic molecules plays an important role in the chemistry of cells.

Acetic acid (CH_3CO_2H) is an organic acid in which only one of the four hydrogen atoms is acidic. The hydrogen atoms bonded to carbon are not acidic. Vinegar contains 5% acetic acid.

Table 16.2 **Properties of Concentrated Solutions of Common Acids**

Name	Formula	Concentration		Density (g/mL)
		w/v %	M	
hydrochloric acid	HCl	37	12	1.19
nitric acid	HNO_3	70	16	1.42
sulfuric acid	H_2SO_4	96	18	1.82
phosphoric acid	H_3PO_4	85	15	1.70
acetic acid	CH_3CO_2H	100	17	1.05

The compound also is formed when wine is oxidized by air.

$$H-\overset{\overset{\displaystyle H}{|}}{\underset{\underset{\displaystyle H}{|}}{C}}-\overset{\overset{\displaystyle O}{\|}}{C}-O-H$$

Carbonic acid is formed in an equilibrium reaction with carbon dioxide and water. It plays an important role in respiration and in the acid–base balance of the blood.

$$H_2O + CO_2 \rightleftharpoons H_2CO_3$$

Only about 1% of the carbon dioxide dissolved in water exists as carbonic acid. Carbonic acid is present in carbonated beverages.

Sodium hydroxide ($NaOH$), commonly called lye or caustic soda, is a solid. Both the solid and aqueous solutions of sodium hydroxide must be handled with care. Sodium hydroxide is used in the manufacture of soap, in paper production, textile manufacturing, and many other industrial processes. It is also the active ingredient in some oven and drain cleaners.

Ammonia (NH_3) is produced in industrial quantities second only to sulfuric acid. It is used in the production of fertilizer. Ammonia, which is a gas, dissolves in water to form small amounts of ammonium and hydroxide ions. For this reason ammonia solutions are sometimes called **ammonium hydroxide.** However, the major component of the solution is ammonia.

$$NH_3 + H_2O \rightleftharpoons NH_4^+ + OH^-$$

The 27% solution of ammonia sold commercially will cause skin burns. A 2% solution of ammonia is used as an inhalant to revive people who have fainted. Although less common now than a few years ago, aqueous ammonia may be used to clean items such as windows.

16.3 Brønsted–Lowry Concept of Acids and Bases

Over the years, acids and bases have been classified in a variety of ways. However, for reactions of acids and bases in aqueous solution, the most useful definition is that of **Brønsted and Lowry.** An **acid** is a substance that can donate a proton (H^+). A **base** is a substance that can accept a proton.

In aqueous solutions of acids, the proton is donated to water and the **hydronium ion** (H_3O^+) is formed. For example, when gaseous hydrogen chloride is dissolved in water, virtually all of the HCl molecules transfer their proton to water and a solution of hydronium ions and chloride ions results.

$$HCl + H_2O \longrightarrow H_3O^+ + Cl^-$$

The transfer of a proton from an acid to water is called **ionization.** An illustration of this process is shown in Figure 16.1. In the hydronium ion, the proton can be viewed as attached to oxygen by a nonbonded pair of electrons on oxygen. The terms proton, hydrogen ion, or hydronium ion may be used interchangeably in describing aqueous acid solutions.

All Brønsted–Lowry acids produce hydronium ions in solution but in different amounts. Acids that completely transfer their proton to water are called **strong acids.** Acids that do not

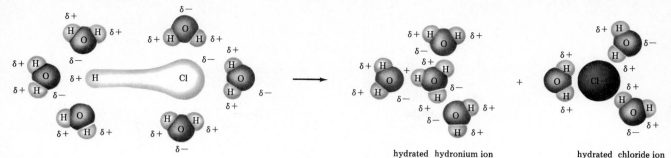

hydrated hydronium ion   hydrated chloride ion

Figure 16.1

Ionization of HCl
The polar covalent bond in HCl becomes stretched and is broken by the action of surrounding water molecules. The proton becomes attached to a water molecule to produce a hydronium ion.

completely transfer their proton to water are **weak acids.** It is the hydronium ion that is responsible for the acidic properties of acids. In fact, the hydronium ion is an acid because it has a proton that it may donate to a base.

The most common base is the hydroxide ion. The hydroxide ion exists as an ion in compounds such as NaOH, KOH, and Mg(OH)$_2$. When dissolved in water, the ions of these compounds are separated from the crystal and distributed in the solution. The hydroxide ion is a base because it can accept a proton from an acid such as the hydronium ion. Ammonia is also a base because it can accept a proton from an acid.

$$NH_3 + H_3O^+ \longrightarrow NH_4^+ + H_2O$$

The reactions of acids and bases are related in an interesting way. When an acid and a base react by the transfer of a proton, another base and acid are produced. This relationship is emphasized by considering a substance and the related substance formed after proton transfer as a conjugate pair. When an acid loses a proton, the species formed is called the **conjugate base** of the acid. The conjugate base of hydrogen chloride is the chloride ion.

$$\underset{\text{acid}}{HCl} + H_2O \longrightarrow H_3O^+ + \underset{\text{conjugate base}}{Cl^-}$$

When a base accepts a proton, the new substance formed is called the **conjugate acid** of the base. The conjugate acid of ammonia is the ammonium ion.

$$\underset{\text{base}}{NH_3} + H_3O^+ \longrightarrow H_2O + \underset{\text{conjugate acid}}{NH_4^+}$$

Substances that can either lose or gain a proton are **amphoteric substances** and can act as either an acid or a base. Water itself is amphoteric. It may function as an acid when, for example, it donates a proton to ammonia. Water can also accept protons when, for example, it reacts with hydrogen chloride.

$$NH_3 + \underset{\text{acid}}{H_2O} \rightleftharpoons NH_4^+ + OH^-$$

$$HCl + \underset{\text{base}}{H_2O} \longrightarrow H_3O^+ + Cl^-$$

Acids that transfer only one proton are **monoprotic.** Hydrochloric acid, nitric acid, and perchloric acid, HClO$_4$, are monoprotic acids. In each case the acids are strong. However, regardless of whether the acid is strong or weak, monoprotic acids yield a 1:1 ratio of hydronium ion to the conjugate base of the acid.

$$H_2O + HCl \longrightarrow H_3O^+ + Cl^-$$
$$H_2O + HNO_3 \longrightarrow H_3O^+ + NO_3^-$$
$$H_2O + HClO_4 \longrightarrow H_3O^+ + ClO_4^-$$

Diprotic acids can transfer two protons to water or to a base. Sulfuric acid is a diprotic acid. It can react to donate two protons to water. Transfer of one proton yields the hydrogen sulfate (or bisulfate) ion (HSO_4^-). The second proton transfer results in the formation of the sulfate ion (SO_4^{2-}). The first transfer is virtually quantitative and for this reason sulfuric acid is a strong acid. The second transfer is incomplete and a substantial amount of HSO_4^- is present at equilibrium.

$$H_2SO_4 + H_2O \longrightarrow H_3O^+ + HSO_4^-$$
$$HSO_4^- + H_2O \rightleftharpoons H_3O^+ + SO_4^{2-}$$

Carbonic acid is a weak diprotic acid that plays an important role in respiration and in the acid–base balance of the blood.

$$H_2CO_3 + H_2O \rightleftharpoons H_3O^+ + HCO_3^-$$
$$HCO_3^- + H_2O \rightleftharpoons H_3O^+ + CO_3^{2-}$$

Triprotic acids can transfer three protons to water or a base. Phosphoric acid is a triprotic acid because it may transfer three protons to water. All steps are equilibrium reactions and phosphoric acid is a weak acid.

$$H_3PO_4 + H_2O \rightleftharpoons H_3O^+ + H_2PO_4^-$$
$$H_2PO_4^- + H_2O \rightleftharpoons H_3O^+ + HPO_4^{2-}$$
$$HPO_4^{2-} + H_2O \rightleftharpoons H_3O^+ + PO_4^{3-}$$

Example 16.1

What are the conjugate acid and the conjugate base of the amphoteric $H_2PO_4^-$ ion?

Solution   When a substance acts as an acid, it donates a proton to a base such as water. The material related to the acid is called the conjugate base. The conjugate base of $H_2PO_4^-$ is HPO_4^{2-}.

$$H_2PO_4^- + H_2O \rightleftharpoons HPO_4^{2-} + H_3O^+$$

When a substance acts as a base, it accepts a proton from an acid such as water. The material related to the base is called the conjugate acid. The conjugate acid of $H_2PO_4^-$ is H_3PO_4.

$$H_2PO_4^- + H_2O \rightleftharpoons H_3PO_4 + OH^-$$

[*Additional examples may be found in 16.21–16.28 at the end of the chapter.*]

16.4   Strengths of Acids and Bases

The words **strong** and **weak** used in describing acids and bases refer to the degree of ionization. These terms do not refer to the concentrations of acids and bases. An acid such as HCl, which is

Table 16.3 Brønsted–Lowry Conjugate Acid–Base Pairs

Acid	Base
$HClO_4$	ClO_4^-
H_2SO_4	HSO_4^-
HCl	Cl^-
HNO_3	NO_3^-
H_3O^+	H_2O
HSO_4^-	SO_4^{2-}
H_3PO_4	$H_2PO_4^-$
HF	F^-
CH_3CO_2H	$CH_3CO_2^-$
H_2CO_3	HCO_3^-
H_2S	HS^-
$H_2PO_4^-$	HPO_4^{2-}
NH_4^+	NH_3
HCO_3^-	CO_3^{2-}
HPO_4^{2-}	PO_4^{3-}
H_2O	OH^-
HS^-	S^{2-}
OH^-	O^{2-}

increasing acid strength → (down the Acid column)

increasing base strength → (down the Base column)

strong, is strong regardless of whether it is in a dilute solution, as in a 0.1 M HCl, or in concentrated solution, as in 12 M HCl. Similarly, a weak acid, such as acetic acid, is weak regardless of its concentration in water.

Strong Acids

When hydrogen chloride is dissolved in water, virtually no covalent HCl remains. The transfer of a proton from HCl to water occurs essentially completely. This fact can be indicated by arrows of unequal length in the equilibrium equation.

$$HCl + H_2O \rightleftharpoons H_3O^+ + Cl^-$$

Since the equilibrium is overwhelmingly to the right, we conclude that HCl has a stronger tendency to donate protons than does H_3O^+ and is therefore a stronger acid than H_3O^+. In addition, the position of the equilibrium reflects the relative abilities of H_2O and Cl^- to accept protons. The water molecule is a stronger base than Cl^-.

From a consideration of the equilibria between acids and bases and their conjugate bases and acids, it can be concluded that a strong acid, with its great tendency to lose protons, is conjugate to a weak base that has a low affinity for protons. The stronger the acid, the weaker is its conjugate base. Strong bases attract protons strongly and are conjugate to weak acids, which do not readily lose protons. The stronger a base, the weaker is its conjugate acid. A list illustrating the relationship between common acid–base pairs is given in Table 16.3.

Weak Acids

Most acids do not transfer their protons completely to water, and few ions are produced. Acetic acid, an organic acid, is a weak acid. At 25°C, a 1 M solution of acetic acid is approximately 0.4% ionized and the concentration of ions is very low. The position of the equilibrium can be indicated by using equilibrium arrows of unequal length.

$$CH_3CO_2H + H_2O \rightleftharpoons H_3O^+ + CH_3CO_2^-$$

Acetic acid is a weaker acid than H_3O^+ and $CH_3CO_2^-$ is a stronger base than H_2O. In this reaction, the equilibrium lies to the side containing the weaker acid and weaker base. This statement is general and quite logical because the proton must reside with the weaker acid or the substance that has the smallest tendency to lose it. Furthermore, the proton remains with the acid because the base on that side of the equation does not have much tendency to remove it.

The strengths of acids are measured by their tendencies to transfer protons to a base, usually water. The order of acid strengths is given by measuring the equilibrium constant for ionization. For an acid with the general formula HA, the equilibrium constant for ionization is obtained from the equation for ionization.

$$HA + H_2O \rightleftharpoons H_3O^+ + A^-$$

$$K = \frac{[H_3O^+][A^-]}{[HA][H_2O]}$$

The concentration of water is about 55 M and is so large compared to that of the other components of the equilibrium that its value changes very little when the acid HA is added. Therefore, it is included in a constant called the **acid ionization constant** K_a.

$$K_a = K[H_2O] = \frac{[H_3O^+][A^-]}{[HA]}$$

The acid ionization constants of some acids and the percent ionization of a 1 M solution are

Table 16.4 **Acidity of Acids (1 M)**

Acid	K_a (mole/L)	Percent ionization
HSO_4^-	1.3×10^{-2}	11
H_3PO_4	7.5×10^{-3}	8.3
HF	6.6×10^{-4}	2.5
CH_3CO_2H	1.8×10^{-5}	0.42
H_2CO_3	4.3×10^{-7}	0.065
H_2S	5.7×10^{-8}	0.024
$H_2PO_4^-$	6.2×10^{-8}	0.025
HCN	4.0×10^{-10}	0.0020
HCO_3^-	5.6×10^{-11}	0.00075
HPO_4^{2-}	2.2×10^{-13}	0.000047
HS^-	1.3×10^{-13}	0.0000036

given in Table 16.4. The larger the value of K_a, the larger is the percent of ionization at the same concentration.

Example 16.2

Lactic acid is a monoprotic organic acid produced in metabolic reactions. A 1.0 M solution of lactic acid is about 1% ionized. Calculate the K_a for lactic acid.

Solution The concentration of hydronium ion will be one hundredth that of the lactic acid. Thus, the concentration of hydronium ion is

$$[H_3O^+] = (0.01)(1.0\,M) = 0.01\,M$$

The process of ionization of a monoprotic acid produces hydronium ions and the conjugate base of the acid in a 1:1 ratio. If the acid is represented as HL, then the concentration of the conjugate base L^- is 0.01 M.

The concentration of lactic acid, which is slightly diminished by ionization, is 0.99 M. The acid ionization constant is then calculated.

$$\frac{[H_3O^+][L^-]}{[HL]} = \frac{[0.01][0.01]}{[0.99]} = 1.010101 \times 10^{-4}$$

The correct answer to two significant figures required in this problem is 1.0×10^{-4}.

[*Additional examples may be found in 16.39 and 16.40 at the end of the chapter.*]

Example 16.3

Formic acid (HCO_2H) is an irritating organic acid produced by red ants and is a component of their sting. The K_a of formic acid, which is monoprotic, is 1.8×10^{-4}. What is the concentration of hydronium ion in a 0.50 M solution? What percent of the acid is ionized?

Solution The concentration of hydronium ion produced by the ionization of formic acid can be set as x. The concentration of the conjugate base of formic acid also must be x because the number of hydronium ions and the conjugate base for a monoprotic acid are equal. The concentration of the covalent formic acid will be $(0.50 - x)\,M$. However, because the ionization constant is small, the value may be approximated as $0.50\,M$. Substituting into the equilibrium constant expression, we have

$$K_a = \frac{[H_3O^+][HCO_2^-]}{[HCO_2H]} = \frac{[x][x]}{0.50\,M} = 1.8 \times 10^{-4}\,M$$

$$x^2 = (1.8 \times 10^{-4}\,M)(0.50\,M)$$

$$x = 9.5 \times 10^{-3}\,M$$

The fraction of the acid ionized is given by the ratio of the concentration of the hydronium ion to the concentration of the acid.

$$\% \text{ ionized acid} = \frac{9.5 \times 10^{-3}}{5.0 \times 10^{-1}} \times 100 = 1.9\%$$

[*Additional examples may be found in 16.41 and 16.42 at the end of the chapter.*]

Strengths of Bases

A **strong base** completely removes the proton of an acid. The most common strong base is the hydroxide ion, which will remove and accept protons from even weak acids such as acetic acid.

$$OH^- + CH_3CO_2H \rightleftharpoons H_2O + CH_3CO_2^-$$

Weak bases do not have a very large attraction for the protons of an acid. Only a small fraction of a weak base in a sample will accept protons at equilibrium. Ammonia is the most common example of a weak base. When ammonia dissolves in water, a low concentration of hydroxide ions is formed as a result of the transfer of a proton from water by ammonia.

$$NH_3 + H_2O \rightleftharpoons NH_4^+ + OH^-$$

16.5   Self-Ionization of Water

In water at 25°C there are both hydronium ions and hydroxide ions as a result of self-ionization (Figure 16.2). In the self-ionization of water, some water molecules are behaving as acids and some are behaving as bases. Note from the equation that for every hydronium ion formed, there must be one hydroxide ion formed. At equilibrium the concentration of H_3O^+ is $1 \times 10^{-7}\,M$, the same as the concentration of OH^-.

The equilibrium constant expression for the self-ionization of water is written as follows.

$$K = \frac{[H_3O^+][OH^-]}{[H_2O]^2}$$

Since the concentration of water is hardly affected by the slight extent of the reaction, it is essentially constant and is included in an **ion product constant** of water, K_w.

$$K[H_2O]^2 = K_w = [H_3O^+][OH^-]$$

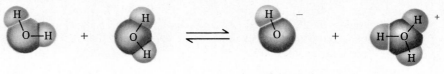

Figure 16.2

Self-ionization of water

A proton is transferred from one water molecule to another to produce equal numbers of hydroxide and hydronium ions. The reaction proceeds to only a limited extent.

The value of K_w is $1.0 \times 10^{-14} M^2$, since the $[H_3O^+]$ and $[OH^-]$ are both $1 \times 10^{-7} M$ at 25°C.

$$K_w = [H_3O^+][OH^-] = [1.0 \times 10^{-7} M][1.0 \times 10^{-7} M] = 1.0 \times 10^{-14} M^2$$

When a strong acid such as HCl is dissolved in water, the hydronium ion concentration becomes very large compared to what it is in pure water. For example, 0.010 M HCl, which contains $1.0 \times 10^{-2} M$ H_3O^+, is 100,000 times as concentrated in H_3O^+ as it is in pure water. An **acidic solution** has a higher concentration of hydronium ions than pure water.

How is the concentration of hydroxide ions affected by the increased concentration of hydronium ions? Le Châtelier's principle gives the answer.

$$2 H_2O \rightleftharpoons H_3O^+ + OH^-$$

As hydronium ions are added, the hydroxide ions will tend to resist this change by reacting with them. The OH^- concentration then must decrease. For $1.0 \times 10^{-2} M$ H_3O^+ solution, the hydroxide ion concentration may be calculated as follows.

$$K_w = [H_3O^+][OH^-] = 1.0 \times 10^{-14} M^2$$
$$(1 \times 10^{-2} M)[OH^-] = 1.0 \times 10^{-14} M^2$$
$$[OH^-] = \frac{1.0 \times 10^{-14} M^2}{1.0 \times 10^{-2} M}$$
$$[OH^-] = 1.0 \times 10^{-12} M$$

Thus, the hydroxide ion concentration is reduced from $1.0 \times 10^{-7} M$ in pure water to $1.0 \times 10^{-12} M$ in a 0.010 M HCl solution.

Now let's consider what occurs in a basic solution. If 0.0010 mole of sodium hydroxide is added to sufficient water to produce 1.0 L of solution, the OH^- concentration from this completely ionized base is $1.0 \times 10^{-3} M$.

How is the concentration of hydronium ions affected by the increased concentration of hydroxide ions? Le Châtelier's principle gives the answer.

$$2 H_2O \rightleftharpoons H_3O^+ + OH^-$$

As hydroxide ions are added, the hydronium ions will tend to resist this change by reacting with them. The H_3O^+ concentration then must decrease. For $1.0 \times 10^{-3} M$ OH^- solution, the hydronium ion concentration may be calculated as follows.

$$[H_3O^+][OH^-] = 1.0 \times 10^{-14} M^2$$
$$[H_3O^+](1.0 \times 10^{-3} M) = 1.0 \times 10^{-14} M^2$$

$$[H_3O^+] = \frac{1.0 \times 10^{-14}\,M^2}{1.0 \times 10^{-3}\,M}$$

$$[H_3O^+] = 1.0 \times 10^{-11}\,M$$

A **basic solution** then has a hydronium ion concentration that is lower than that of pure water.

Example 16.4

The hydronium ion concentration of a blood sample is $4.5 \times 10^{-8}\,M$. What is the value of the hydroxide ion concentration?

Solution   The blood is slightly basic, since the hydronium ion concentration is less than $1.0 \times 10^{-7}\,M$. The hydroxide ion concentration must be greater than $1.0 \times 10^{-7}\,M$. Substitution into the K_w expression gives the expected answer.

$$K_w = [H_3O^+][OH^-] = 1.0 \times 10^{-14}\,M^2$$
$$(4.5 \times 10^{-8}\,M)[OH^-] = 1.0 \times 10^{-14}\,M^2$$
$$[OH^-] = 2.222222 \times 10^{-7}\,M$$

The correct answer to the two required significant figures is $2.2 \times 10^{-7}\,M$.

Example 16.5

Morphine is used medically as a pain reliever. A $0.0050\,M$ solution of morphine has a $[OH^-] = 8.8 \times 10^{-5}\,M$. What is the hydronium ion concentration of the solution?

Solution   The solution is basic, since the hydroxide ion concentration is greater than $1.0 \times 10^{-7}\,M$. The hydronium ion concentration must be less than $1.0 \times 10^{-7}\,M$. Substitution into the K_w expression gives the expected answer.

$$K_w = [H_3O^+][OH^-] = 1.0 \times 10^{-14}\,M^2$$
$$[H_3O^+](8.8 \times 10^{-5}\,M) = 1.0 \times 10^{-14}\,M^2$$
$$[H_3O^+] = 1.136364 \times 10^{-10}\,M$$

The correct answer to the two required significant figures is $1.1 \times 10^{-10}\,M$.

[Additional examples may be found in 16.47 and 16.48 at the end of the chapter.]

16.6 The pH Scale

The pH scale was developed as a more convenient method of expressing the hydronium ion concentration for many common applications involving acids and bases. The **pH** is defined as the negative logarithm of the hydronium ion concentration expressed in molarity.

$$pH = -\log [H_3O^+]$$

Table 16.5   The pH Scale

$[H_3O^+]$	pH	$[OH^-]$	
10^0	0	10^{-14}	
10^{-1}	1	10^{-13}	
10^{-2}	2	10^{-12}	
10^{-3}	3	10^{-11}	acidic solution
10^{-4}	4	10^{-10}	
10^{-5}	5	10^{-9}	
10^{-6}	6	10^{-8}	
10^{-7}	7	10^{-7}	neutral solution
10^{-8}	8	10^{-6}	
10^{-9}	9	10^{-5}	
10^{-10}	10	10^{-4}	
10^{-11}	11	10^{-3}	basic solution
10^{-12}	12	10^{-2}	
10^{-13}	13	10^{-1}	
10^{-14}	14	10^0	

In pure water, the hydronium ion concentration is $1.0 \times 10^{-7}\,M$, and the pH is 7.0.

$$pH = -\log [H_3O^+] = -\log(1.0 \times 10^{-7}) = -(-7.0) = 7.0$$

At higher hydronium ion concentrations, that is, in acidic solution, the pH is smaller. In a $0.010\,M$ HCl solution, the hydronium ion concentration is $1.0 \times 10^{-2}\,M$ and the pH is 2.0.

$$pH = -\log(1.0 \times 10^{-2}) = -(-2.0) = 2.0$$

In basic solutions, the hydronium ion concentration is lower than in pure water. For a $0.0010\,M$ NaOH solution, the hydroxide ion concentration is $1.0 \times 10^{-3}\,M$, and the hydronium ion concentration is $1.0 \times 10^{-11}\,M$. The pH is then 11.0.

$$pH = -\log(1.0 \times 10^{-11}) = -(-11.0) = 11.0$$

In general, for any hydronium ion concentration of 1.0×10^{-x} the pH is x. The pH scale is shown in Table 16.5. Acidic solutions have pH values smaller than 7, whereas basic solutions have pH values greater than 7.

The pH scale in Table 16.5 involves only changes by a factor of 10 in the hydronium ion concentration. Thus, a change of a few pH units actually involves large changes in hydronium ion concentration. A pH change of three units corresponds to a factor of 10^3 or 1000 in the hydronium ion concentration.

Dealing with pH for solutions with hydronium ion concentrations that can be written as 1.0×10^{-x} is fairly simple. Now let's consider a blood sample with a hydronium ion concentration equal to $4.5 \times 10^{-8}\,M$. What is the pH? In order to calculate the pH, you must know the following mathematical identity of logarithms.

$$\log(n \times p) = \log n + \log p$$

For a solution with $[H_3O^+] = 4.5 \times 10^{-8}$ we need to know the logarithms of 4.5 and 10^{-8}.

$$pH = -\log [H_3O^+]$$
$$pH = -\log(4.5 \times 10^{-8})$$
$$pH = -\log 4.5 - \log 10^{-8}$$

Hand calculators that have a key for base 10 logarithms can be used to calculate pH. Consult your instruction booklet for directions on how to calculate a logarithm on your calculator. The logarithm of 4.5 is 0.65321251 as displayed on a calculator. However, using two significant figures, only 0.65 is used in the calculation.

$$pH = -\log 4.5 - \log 10^{-8}$$
$$pH = -0.65 - (-8) = 7.35$$

The pH of blood is slightly on the basic side, as expected from the hydronium ion concentration.

Example 16.6

The hydronium ion concentration of a sample of urine is $2.0 \times 10^{-6}\,M$. What is its pH?

Solution The sample is more acidic than a solution with a hydronium ion concentration of $1 \times 10^{-6}\,M$. Therefore, the pH must be less than 6. Now substitute into the pH formula, and determine to two significant figures the logarithm of 2.0 on your calculator.

$$pH = -\log[H_3O^+]$$
$$pH = -\log(2.0 \times 10^{-6})$$
$$pH = -\log 2.0 - \log 10^{-6}$$
$$pH = -0.30 - (-6) = 5.70$$

The value of the pH is somewhat less than 6, as expected.

Example 16.7

A $0.0050\,M$ morphine solution has a $[OH^-] = 8.8 \times 10^{-5}\,M$. What is the pH of the solution?

Solution The solution is basic, since the hydroxide ion concentration is greater than $1 \times 10^{-7}\,M$. Therefore the pH must be greater than 7. First the hydronium ion concentration must be calculated. Substitution into the K_w expression shown previously in Example 16.5 gives $[H_3O^+] = 1.1 \times 10^{-10}\,M$. The pH is obtained by substitution into the expression for pH.

$$pH = -\log[H_3O^+]$$
$$pH = -\log(1.1 \times 10^{-10})$$
$$pH = -\log 1.1 - \log 10^{-10}$$
$$pH = -0.04 - (-10) = 9.96$$

As expected the pH is greater than 7 for this basic solution.

[*Additional examples may be found in 16.51–16.54 at the end of the chapter.*]

The pH values of several of our body fluids are given in Table 16.6. Except for gastric juices in which the main acid is hydrochloric acid, the majority of body fluids have pH values near 7. No body fluid is very basic.

Table 16.6   pH Values of Body Fluids

blood	7.35–7.45
gastric juice	1.6–1.8
bile	7.8–8.6
urine	5.5–7.0
saliva	6.2–7.4
interstitial fluid	7.4
muscle intracellular fluid	6.1
liver intracellular fluid	6.9
pancreatic juice	7.8–8.0

Measuring pH

The pH meter is an instrument that presents the pH either by a needle reading on a scale or a digital display. The instrument measures the voltage when an electric current passes between electrodes immersed in the solution. Such meters can measure pH to the nearest 0.01 pH unit.

A somewhat older and less accurate method of determining pH uses chemical indicators. Chemical indicators are certain weak acids that can be represented by HInd. The indicator can exist as an acid and its conjugate base Ind$^-$ when the indicator ionizes in water.

$$HInd + H_2O \rightleftharpoons H_3O^+ + Ind^-$$

$$K_a = \frac{[H_3O^+][Ind^-]}{[HInd]}$$

Each member of the conjugate pair is a different color, so that an indicator changes color as the concentrations of the two substances forming the conjugate pair change. In acidic solution the indicator will exist as HInd, and the color will be that of the acid form of the indicator. In basic solution the hydroxide ion will remove a proton from the indicator, which will then exist as its conjugate base. The ratio of the concentrations of the conjugate base and the acid form depends on the K_a of the indicator and the hydronium ion concentration.

Table 16.7   Colors of Indicators

Indicator	Color in the more acid range	pH range	Color in the more basic range
methyl violet	yellow	0–2	violet
thymol blue	pink	1.2–2.8	yellow
bromophenol blue	yellow	3.0–4.7	violet
methyl orange	pink	3.1–4.4	yellow
methyl red	red	4.8–6.0	yellow
bromocresol purple	yellow	5.2–6.8	purple
litmus	red	4.7–8.2	blue
phenolphthalein	colorless	8.3–10.0	pink
thymolphthalein	colorless	9.3–10.5	blue
alizarin yellow G	colorless	10.1–12.1	yellow
trinitrobenzene	colorless	12.0–14.3	orange

$$\frac{[\text{Ind}^-]}{[\text{HInd}]} = \frac{K_a}{[\text{H}_3\text{O}^+]}$$

The use of an indicator is limited only by the eye of the observer. However, when the ratio of the concentrations of the two colored species is about 0.1, the color is viewed as that of the acid form of the indicator. When the ratio of the concentrations is 10, the color is viewed as that of the conjugate base of the indicator. Therefore the range of the sensitivity of the indicators is about a hundred-fold range of H_3O^+, or two pH units. Phenolphthalein is colorless in solutions of pH less than 8.3, where it exists in its acidic form; it is pink in solutions of pH greater than 10.0, where it exists in its basic form. The colors of a number of other indicators are listed in Table 16.7 along with the approximate pH at which the color change occurs.

Use of the pH Scale

There are many practical uses of the pH scale. Because some crops flourish under acidic conditions and others require more basic conditions, agriculture depends on the proper soil pH. Soil is often tested to determine whether acidic or basic fertilizers are required for a particular crop. Soils in humid areas are acidic, whereas soils in arid areas tend to be basic or neutral. If the soil is too acidic, it can be "limed" by adding calcium carbonate.

$$2\,\text{H}^+ + \text{CaCO}_3 \longrightarrow \text{Ca}^{2+} + \text{H}_2\text{O} + \text{CO}_2$$

In Table 16.8 are listed the pH values of a number of common foods. All of those listed are acidic. The pH of canned goods is important. If green beans are not canned under sterile conditions and if the pH is greater than 4.5, *Clostridium botulinum*, which causes botulism, can grow.

Table 16.8 pH of Food

Food	pH
apples	2.9–3.3
cabbage	5.2–5.4
corn	6.0–6.5
grapes	3.5–4.5
lemons	2.2–2.4
milk	6.3–6.6
oranges	3.0–4.0
peas	5.8–6.4
potatoes	5.6–6.0
tomatoes	4.0–4.4

16.7 Hydrolysis of Salts

The pH of the solutions of some salts is not 7.0. For a solution of sodium acetate, the pH is greater than 7, indicating that the hydroxide ion concentration is greater than that of pure water. In a solution of ammonium chloride, the pH is less than 7, indicating that the hydronium ion concentration is greater than that of pure water. In both instances an ion of the salt reacts with water to change the balance of hydronium and hydroxide ions. The reaction of ions with water is called **hydrolysis.**

Because acetic acid is a weak acid, the acetate ion is a moderately strong base. When acetate ions dissolve in water, some protons are abstracted from water.

$$\text{CH}_3\text{CO}_2^- + \text{H}_2\text{O} \rightleftharpoons \text{CH}_3\text{CO}_2\text{H} + \text{OH}^-$$

As a result, the hydroxide ions formed cause the solution to have a pH greater than 7. The equilibrium constant expression for the hydrolysis reaction is

$$K_h = \frac{[\text{CH}_3\text{CO}_2\text{H}][\text{OH}^-]}{[\text{CH}_3\text{CO}_2^-]}$$

where K_h is the **hydrolysis constant** that contains the constant term for the concentration of water. Hydrolysis constants are not tabulated because they are equal to the quotient K_w/K_a. The weaker the acid, the smaller is the acid dissociation constant. As a consequence the hydrolysis constant is larger, and the hydroxide ion concentration of the solution is larger.

Table 16.9   **Acid–Base Properties of Some Ions**

Neutral	Basic	Acidic
Cl^-	CO_3^{2-}	HSO_4^-
Br^-	HCO_3^-	$H_2PO_4^-$
I^-	F^-	
NO_3^-	S^{2-}	
ClO_4^-	HS^-	
	PO_4^{3-}	
	HPO_4^{2-}	
	CN^-	
	$CH_3CO_2^-$	

$$K_h = \frac{K_w}{K_a} = \frac{[H_3O^+][OH^-]}{[H_3O^+][CH_3CO_2^-]/[CH_3CO_2H]} = \frac{[CH_3CO_2H][OH^-]}{[CH_3CO_2^-]}$$

The ammonium ions of ammonium chloride can donate protons to water to yield hydronium ions and ammonia. The resultant solution of ammonium ions has a pH less than 7.

$$NH_4^+ + H_2O \rightleftharpoons NH_3 + H_3O^+$$

Of the two types of hydrolysis reactions illustrated, the hydrolysis of the salt of the conjugate base of a weak acid is the more common. A list of the acid–base properties of some anions is given in Table 16.9. Note that the conjugate bases of strong acids give neutral solutions because the conjugate bases are too weak to remove protons from water.

Example 16.8

Which solution is the more basic, a 0.1 M solution of sodium cyanide or a 0.1 M solution of sodium acetate?

Solution From Table 16.4 it can be seen that the ionization constant of HCN is less than that of acetic acid. Thus HCN is a weaker acid than CH_3CO_2H, and we can conclude that CN^- is a stronger base than $CH_3CO_2^-$. The cyanide ion will abstract more protons from water than will the acetate ion. The cyanide ion solution will be the more basic.

[*Additional examples may be found in 16.55 and 16.56 at the end of the chapter.*]

Example 16.9

Determine the hydroxide ion concentration of a 0.10 M solution of sodium acetate in water. The K_a of acetic acid is 1.8×10^{-5}.

Solution First calculate the hydrolysis constant of the acetate ion.

$$K_h = \frac{K_w}{K_a} = \frac{1.0 \times 10^{-14}}{1.8 \times 10^{-5}} = 5.555556 \times 10^{-10}$$

To two significant figures the hydrolysis constant is 5.6×10^{-10}. The concentration of hydroxide ion produced by the hydrolysis of acetate ion will be equal to the concentration of the acetic acid produced.

$$CH_3CO_2^- + H_2O \rightleftharpoons CH_3CO_2H + OH^-$$

The concentration of the acetate ion will be essentially $0.10\ M$ because the hydrolysis constant is small. The concentration of the hydroxide ion may be represented by x. Since the concentration of acetic acid is equal to that of the hydroxide ion, its concentration is also x. Substituting into the equilibrium constant expression, we have

$$K_h = \frac{[CH_3CO_2H][OH^-]}{[CH_3CO_2^-]} = \frac{(x)(x)}{0.10} = 5.6 \times 10^{-10}$$
$$x^2 = (5.6 \times 10^{-10})(0.10)$$
$$x = 7.483315 \times 10^{-6}$$

The correct answer to the two required significant figures is $7.5 \times 10^{-6}\ M$.

[*Additional examples may be found in 16.57–16.60 at the end of the chapter.*]

16.8 Buffers

The term to buffer means to prevent changes or to lessen the shock of changes. In chemistry, a **buffer** is a solution that will prevent a drastic pH change. A chemical buffer is prepared by dissolving both a weak acid and a salt of its conjugate base in water, for example, acetic acid and sodium acetate. The buffer will react with any strong acid or base that may be added. Hydronium ions will react with the acetate ions. Hydroxide ions will react with the acetic acid.

$$H_3O^+ + CH_3CO_2^- \longrightarrow H_2O + CH_3CO_2H$$
$$OH^- + CH_3CO_2H \longrightarrow H_2O + CH_3CO_2^-$$

Thus, either H_3O^+ or OH^-, which could drastically change the pH, is neutralized by reacting with one of the components of the buffer.

Buffers may be prepared from any ratio of concentrations of weak acids and the salt of the weak acid. A $1:1$ ratio of acid to salt is the most efficient for handling the addition of either a base or an acid. If the buffer contains much more acid than the conjugate base, it will be less efficient in handling an acid. Alternatively, a buffer with much more of the conjugate base than of acid cannot efficiently counteract the addition of a base. For example, the H_2CO_3/HCO_3^- buffer in blood has a $1:20$ ratio of acid to conjugate base. Blood is then more effective in counteracting acids, which are the products of metabolism.

A variety of hydronium ion concentrations may be obtained for various ratios of a weak acid and the salt of the weak acid. A relationship that gives the $[H_3O^+]$ in terms of the acid ionization constant and the two components of the buffer can be derived from the acid ionization constant.

$$K_a = \frac{[H_3O^+][A^-]}{[HA]}$$
$$[H_3O^+] = K_a \frac{[HA]}{[A^-]}$$

Example 16.10

Determine the hydronium ion concentration of a buffer consisting of 0.10 M sodium acetate and 0.20 M acetic acid in water. The K_a of acetic acid is 1.8×10^{-5}.

Solution   The concentration of the hydronium ion in equilibrium with both acetic acid and acetate ion is controlled by the following equilibrium.

$$CH_3CO_2H + H_2O \rightleftharpoons CH_3CO_2^- + H_3O^+$$

The concentration of the acetate ion will be essentially 0.10 M because the hydrolysis constant is small. Similarly the concentration of acetic acid will be essentially 0.20 M because the ionization of acetic acid is small. Using x as the concentration of the hydronium ion and substituting into the equilibrium constant expression, we have

$$K_a = \frac{[CH_3CO_2^-][H_3O^+]}{[CH_3CO_2H]} = \frac{(0.10)(x)}{0.20} = 1.8 \times 10^{-5}$$

$$x = \frac{(1.8 \times 10^{-5})(0.20)}{0.10}$$

$$x = 3.6 \times 10^{-5} M = [H_3O^+]$$

[*Additional examples may be found in 16.69–16.72 at the end of the chapter.*]

16.9   Buffers in the Body

The normal metabolic reactions that take place in your muscles as a result of exercise continuously produce a variety of acids. On the average, 10 moles of a variety of acids are produced in your body each day.

There are two ways to decrease the concentration of an acid that results from these metabolic reactions. One is by breathing and expelling the carbon dioxide that is formed from the carbonic acid. The second is by excreting the acid in urine (Table 16.6). Both mechanisms depend on buffers. The carbonate buffer, H_2CO_3/HCO_3^-, is present in the blood, whereas the phosphate buffer, $H_2PO_4^-/HPO_4^{2-}$, is involved in kidney functions.

Blood has a normal pH range of 7.35–7.45, which corresponds to hydronium ion concentrations of 4.5×10^{-8} to $3.6 \times 10^{-8} M$. Any acids formed in the blood react with bicarbonate ions to give carbonic acid, which then decomposes to give carbon dioxide.

$$HCO_3^- + H_3O^+ \rightleftharpoons H_2CO_3 + H_2O$$
$$H_2CO_3 \rightleftharpoons H_2O + CO_2$$

The carbon dioxide is removed from the blood by the lungs and exhaled.

Although it is a less common occurrence, the H_2CO_3/HCO_3^- buffer can prevent an increase in base concentration. The carbonic acid can react to neutralize a base.

$$H_2CO_3 + OH^- \rightleftharpoons HCO_3^- + H_2O$$

However, the capacity of the buffer to handle a base is much more limited than its capacity to neutralize acid because the $[H_2CO_3]/[HCO_3^-]$ ratio is about 0.05.

The $H_2PO_4^-/HPO_4^{2-}$ buffer is important within cells. Many reactions in cells involve complex compounds with covalently bonded phosphate groups. The proper pH in cellular fluids is maintained by the reaction of a base with $H_2PO_4^-$ and the reaction of an acid with HPO_4^{2-}.

$$H_2PO_4^- + OH^- \rightleftharpoons HPO_4^{2-} + H_2O$$
$$HPO_4^{2-} + H_3O^+ \rightleftharpoons H_2PO_4^- + H_2O$$

However, the formation of an acid is the more common result of metabolism, and HPO_4^{2-} is the more important component of the buffer. The normal $H_2PO_4^-/HPO_4^{2-}$ ratio in the cell is about 1:4, which allows the buffer to neutralize acid more efficiently than base. The $H_2PO_4^-$ formed from the reaction of HPO_4^{2-} with acid is then eliminated by excretion in the urine.

16.10 Normality

Normality (N) is the number of equivalents of acid or base per liter of solution.

$$N = \frac{\text{equivalents}}{\text{liter of solution}}$$

An **equivalent** of an acid is equal to the weight of an acid containing 1 mole of protons. An equivalent of a base is that weight which will react with 1 mole of protons supplied by an acid. The equivalent weight of an acid is calculated by dividing the molecular weight of the acid by the number of moles of ionizable hydrogen atoms contained in 1 mole of the acid. Similarly, the equivalent weight of a base is calculated by dividing the molecular weight of the base by the number of moles of hydrogen atoms that will react with 1 mole of the base. A listing of equivalent weights of several acids and bases is given in Table 16.10.

The advantage of the normality unit is that one equivalent of any acid will react with one

Table 16.10 **Equivalent Weights of Selected Acids and Bases**

Compound	Molecular weight	Equivalent weight
HCl	36.5	$\dfrac{36.5}{1} = 36.5$
H_2SO_4	98.0	$\dfrac{98.0}{2} = 49.0$
H_3PO_4	98.0	$\dfrac{98.0}{3} = 32.7$
NaOH	40.0	$\dfrac{40.0}{1} = 40.0$
$Ca(OH)_2$	74.0	$\dfrac{74.0}{2} = 37.0$
$Al(OH)_3$	78.0	$\dfrac{78.0}{3} = 26.0$

equivalent of any base. If molarity is used, a mole ratio factor derived from the balanced equation must always be applied before determining whether equal numbers of equivalents of acid or base are present for neutralization. For example, whereas 1 L of a 1 M solution of HCl will neutralize 1 L of a 1 M solution of NaOH, a 1 M solution of H_2SO_4 contains twice as much hydrogen and will require 2 L of a 1 M NaOH solution. However, 1 L of a 1 N solution of any acid will neutralize 1 L of a 1 N solution of any base.

Example 16.11

What is the normality of a solution of H_3PO_4 obtained by dissolving 49.0 g of the acid in sufficient water to produce 3.00 L of solution?

Solution First the equivalent weight of H_3PO_4 must be used to determine how many equivalents are present in 49.0 g.

molecular weight of H_3PO_4 = 98.0 g

equivalent weight of H_3PO_4 = 98.0 g/3 = 32.6̲6 g

$$49.0 \text{ g} \times \frac{1 \text{ equivalent}}{32.6̲6 \text{ g}} = 1.50 \text{ equivalents}$$

Now the normality may be calculated by dividing the number of equivalents by the volume.

$$\frac{1.50 \text{ equivalents}}{3.00 \text{ L}} = 0.500 \ N$$

[*Additional examples may be found in 16.75 and 16.76 at the end of the chapter.*]

Example 16.12

Calculate the number of grams of H_2SO_4 required to make 100 mL of a 0.20 N H_2SO_4 solution.

Solution First the number of equivalents of H_2SO_4 in the solution must be calculated.

$$\frac{0.20 \text{ equivalent}}{1 \text{ L}} \times 0.100 \text{ L} = 0.020 \text{ equivalent}$$

The equivalent weight of H_2SO_4 is one-half the molecular weight.

$$\frac{98.0 \text{ g}}{\text{mole}} \times \frac{1 \text{ mole}}{2 \text{ equivalents}} = \frac{49.0 \text{ g}}{\text{equivalent}}$$

Now the number of grams required to make the solution can be calculated.

$$0.020 \text{ equivalent} \times \frac{49.0 \text{ g}}{\text{equivalent}} = 0.98 \text{ g}$$

[*Additional examples may be found in 16.77 and 16.78 at the end of the chapter.*]

Example 16.13

What is the molarity of a $0.020\,N$ solution of $Ca(OH)_2$?

Solution In a mole of $Ca(OH)_2$ there are two equivalents.

$$\frac{1 \text{ mole } Ca(OH)_2}{2 \text{ equivalents } Ca(OH)_2}$$

Therefore, the definition of normality can be related to molarity by the use of this factor.

$$\frac{0.020 \text{ equivalent}}{1 \text{ L}} \times \frac{1 \text{ mole}}{2 \text{ equivalents}} = \frac{0.010 \text{ mole}}{1 \text{ L}}$$
$$= 0.010\,M$$

[*Additional examples may be found in 16.79 and 16.80 at the end of the chapter.*]

16.11 Titration

An acid–base **titration** is a procedure in which the concentration or amount of an acid or base is determined by reacting a sample of it with a known amount of a base or acid to achieve neutralization. The point of neutralization is known as the **endpoint** or **neutralization point.** The endpoint is detected by using an indicator.

In a titration, a measured volume or mass of an acid or base, whose concentration is the unknown, is placed in a flask, and a small amount of an indicator is added. The indicator chosen must undergo a change in color at a point where the pH of the solution is equal to that of the products of the titration. A solution of a base or acid of known concentration is then added dropwise from a buret. When the indicator changes color, the addition is terminated and the volume of solution added from the buret is recorded (Figure 16.3).

The number of equivalents of base or acid added from the buret is calculated by multiplying the volume used by its normality. This number of equivalents must be equal to the number of equivalents of the unknown because one equivalent of an acid is required to neutralize one equivalent of a base.

$$V_{acid} \times N_{acid} = V_{base} \times N_{base}$$

equivalents of acid = equivalents of base

Frequently, in a laboratory, the volumes of solutions used are not large. Therefore, the units of equivalents and liters are not convenient. One thousandth of a liter is a milliliter, and one thousandth of an equivalent is a milliequivalent (meq). Thus the ratio of the number of milliequivalents to the number of milliliters is also equal to the normality.

$$\frac{\text{milliequivalents}}{\text{milliliter}} = \frac{\text{meq}}{\text{mL}} = N$$

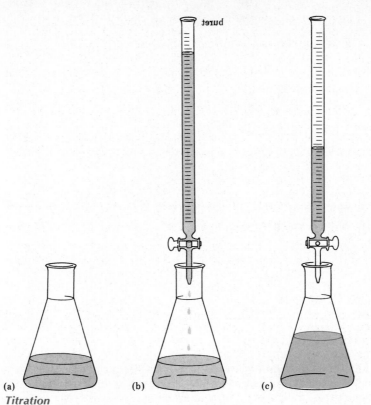

Figure 16.3

Titration
(a) A solution of an unknown concentration of an acid (or a base) is placed in a flask. (b) A titrating solution of a known concentration of a base (or an acid) is added dropwise from a buret. (c) The titration is stopped when the amounts of acid and base are equal as shown by a change in the color of the indicator.

Example 16.14

A 5.00 mL sample of household ammonia is titrated with 48.0 mL of a 0.200 N solution of hydrochloric acid to a methyl red endpoint. What is the molarity of the household ammonia?

Solution The acid solution contains 0.200 meq/mL of hydrochloric acid. Thus the number of milliequivalents of acid used is

$$\frac{0.200 \text{ meq}}{\text{mL}} \times 48.0 \text{ mL} = 9.60 \text{ meq}$$

The number of milliequivalents of ammonia must also be equal to 9.60 meq. This quantity is contained in 5.00 mL of the ammonia solution, and the normality must be

$$\frac{9.60 \text{ meq}}{5.00 \text{ mL}} = 1.92 \, N$$

For ammonia, which only reacts with one equivalent of acid per mole, the equivalent weight is the same as the molecular weight. Therefore, the molarity is the same as the normality, or 1.92 M.

Example 16.15

Baking soda, which is sodium bicarbonate, may be used to establish the normality of an acid solution such as hydrochloric acid by the following reaction.

$$NaHCO_3 + HCl \longrightarrow NaCl + CO_2 + H_2O$$

A 0.0420 g sample of sodium bicarbonate is dissolved in water in a flask, and 20.0 mL of a hydrochloric acid solution is required to neutralize it. What is the normality of the HCl solution?

Solution The equivalent weight of $NaHCO_3$ is 84.0. A milliequivalent will have a mass of 0.0840 g. Thus, the number of milliequivalents in the sample is

$$\frac{0.0420\ g}{0.0840\ g/meq} = 0.500\ meq$$

The number of milliequivalents in the 20.0 mL of HCl must also be 0.500 meq. Thus, the normality is

$$\frac{0.500\ meq}{20.0\ mL} = 0.0250\ N$$

[*Additional examples may be found in 16.81–16.84 at the end of the chapter.*]

Summary

The Brønsted–Lowry concept of acid–base reactions focuses on proton transfer from an acid to a base. The tendency of acids and bases to lose and gain protons, respectively, determines the direction of an acid–base reaction.

Acid strength refers to the degree of ionization of an acid. Strong acids are completely ionized. An equilibrium exists between weak acids and their conjugate bases. The acid ionization constant quantitatively describes the degree of ionization of an acid.

Water undergoes self-ionization to produce small concentrations of hydronium and hydroxide ions. The addition of acids or bases to water affects the concentrations of both hydronium ions and hydroxide ions in a manner predicted by Le Châtelier's principle. The concentration of hydronium ions in water may be expressed in pH units.

The hydrolysis of some salts results in solutions with a pH other than 7. The hydrolysis constant of the salt of a weak acid may be calculated by dividing K_w by the K_a of the conjugate acid.

Buffers have components, a weak acid and a salt of its conjugate base, that will react with added amounts of either a base or an acid. The effectiveness of a buffer against an added acid or base depends on the ratio of the concentrations of the two components of the buffer. Buffers play a critical role in controlling the pH of fluids in living organisms.

Concentrations of acids or bases may be expressed in normality units. This concentration indicates the number of equivalents of an acid or base in a liter of solution.

Titration is a procedure in which an acid and a base reach a point of neutralization. An acid–base indicator is used in a titration.

New Terms

An **acid** (Brønsted–Lowry) is a proton donor.
An **acidic solution** has a higher concentration of hydronium ions than pure water.

The **acid ionization constant** (K_a) is the equilibrium constant for the ionization of a weak acid.
An **amphoteric** substance can act as either an acid or a base.

A **base** (Brønsted–Lowry) is a proton acceptor.

A **basic solution** has a lower concentration of hydronium ions than pure water.

A **buffer** is a solution that resists a change in pH and consists of a weak acid and a salt of its conjugate base.

A **conjugate acid** is the acid formed when a base gains a proton.

A **conjugate base** is the base formed when an acid donates a proton.

A **diprotic acid** can transfer two protons.

The **endpoint** is the point of neutralization in an acid–base titration.

An **equivalent** of an acid contains 1 mole of protons.

The **hydrolysis** of a salt involves the reaction of the salt with water, which sometimes results in a solution with a pH other than 7.

The **hydronium ion** (H_3O^+) is the principal form in which protons are found in aqueous solution.

Ionization of an acid involves the transfer of a proton to water.

A **milliequivalent** is one thousandth of an equivalent.

A **monoprotic acid** can transfer only one proton.

The **normality** of a solution indicates the number of equivalents of acid or base in a liter of solution.

The **pH** is a logarithmic expression of the hydronium ion concentration.

A **strong acid** is a substance that essentially completely transfers its proton to water.

A **strong base** essentially completely removes the proton of an acid.

Titration is a procedure to determine the amount of an acid or base by neutralization.

A **triprotic acid** can transfer three protons.

A **weak acid** is only partially ionized in water.

Exercises

Terminology

16.1 Distinguish between each pair of terms.
(a) acid and base
(b) hydroxide ion and hydronium ion

16.3 Describe the relationship between an acid and its conjugate base.

16.2 Distinguish between each pair of terms.
(a) strong acid and weak acid
(b) molarity and normality

16.4 Describe the relationship between a base and its conjugate acid.

Properties of Acids and Bases

16.5 Explain why acids such as hydrochloric acid are electrolytes in aqueous solution.

16.7 What precautions should you take when working with acids and bases in the laboratory?

16.6 Explain why bases such as sodium hydroxide are electrolytes in aqueous solution.

16.8 Why should you wear eye protection when working with drain cleaner or oven cleaner?

Common Acids and Bases

16.9 What is the formula for each of the following?
(a) phosphoric acid   (b) acetic acid
(c) sulfuric acid   (d) calcium hydroxide
(e) potassium hydroxide (f) aluminum hydroxide

16.11 Glucose ($C_6H_{12}O_6$) contains hydrogen but is not an acid. What test could you use to confirm this fact?

16.13 Name one acid and one base that have industrial uses.

16.10 What is the formula for each of the following?
(a) hydrochloric acid   (b) sodium hydroxide
(c) nitric acid   (d) magnesium hydroxide
(e) perchloric acid   (f) barium hydroxide

16.12 Sodium bicarbonate ($NaHCO_3$) is often kept in laboratories to place on acid or base spills. What reaction occurs when an acid reacts with $NaHCO_3$? What reaction occurs when a base reacts with $NaHCO_3$?

16.14 How is ammonium hydroxide used in the home?

Brønsted–Lowry Theory

16.15 Classify each of the following as a monoprotic, diprotic, or triprotic acid.
(a) $HClO_4$ (b) H_3PO_4 (c) HNO_3
(d) H_2SO_4 (e) H_2CO_3 (f) CH_3CO_2H

16.17 Can a substance act as either an acid or a base? Explain, giving examples.

16.19 Write equations for the stepwise ionization of H_2CO_3.

16.16 Classify each of the following as a monoprotic, diprotic, or triprotic acid.
(a) HCl   (b) $HBrO_3$ (c) HNO_2
(d) H_2SO_3 (e) $HClO_3$ (f) HBr

16.18 What is meant by the term amphoterism?

16.20 Write equations for the stepwise ionization of H_3PO_4.

Conjugate Acids and Bases

16.21 What is the conjugate acid of each of the following?
(a) ClO_4^- (b) Cl^- (c) NO_3^-
(d) CO_3^{2-} (e) $CH_3CO_2^{2-}$ (f) HSO_4^-

16.22 What is the conjugate acid of each of the following?
(a) NH_3 (b) H_2O (c) HCO_3^-
(d) ClO_4^- (e) $H_2PO_4^-$ (f) SO_4^{2-}

16.23 What is the conjugate base of each of the following?
(a) H_2O (b) HCl (c) HCO_3^-
(d) H_2CO_3 (e) HNO_3 (f) H_2SO_4

16.24 What is the conjugate base of each of the following?
(a) HSO_4^- (b) H_3PO_4 (c) $H_2PO_4^-$
(d) HPO_4^{2-} (e) CH_3CO_2H (f) HBr

16.25 Pyruvic acid, an acid produced in metabolic reactions, has the following structure. What is the molecular formula of the conjugate base? What is the charge of this ion?

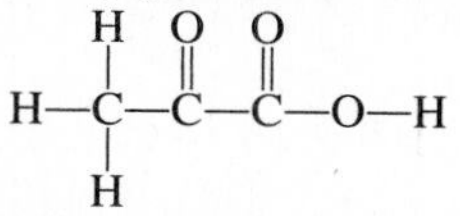

16.26 Methylamine is an organic base. From its structure write the conjugate acid. What is the charge of the ion?

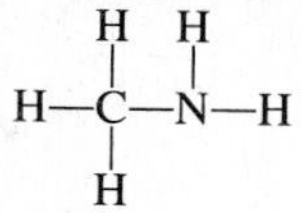

16.27 Codeine ($C_{18}H_{21}NO_3$) is a cough suppressant that is a base. What is the formula of the conjugate acid that results from the gain of a proton?

16.28 Aspartic acid is a diprotic amino acid with the molecular formula $C_4H_7NO_4$. What is the molecular formula and charge of the ion that results from the loss of two protons?

Acid–Base Strength

16.29 Ammonia is very soluble in water. Why is a 27% solution of ammonia still considered to be a solution of a weak base?

16.30 A 0.1 M solution of HCl is a solution of a strong acid. Why is the term strong used for a solution of such low concentration?

16.31 Fill in the blanks in each of the following statements.
(a) A strong acid has a _____ conjugate base.
(b) A strong base has a _____ conjugate acid.

16.32 Fill in the blanks in each of the following statements.
(a) A weak acid has a _____ conjugate base.
(b) A weak base has a _____ conjugate acid.

16.33 Classify each of the following as a weak or strong acid.
(a) CH_3CO_2H (b) HCl (c) HNO_3
(d) HCN (e) H_2SO_4 (f) H_2O

16.34 Classify each of the following as a weak or strong acid.
(a) HF (b) HNO_2 (c) H_2SO_3
(d) HBr (e) $HClO_4$ (f) H_3PO_4

Acid Ionization Constant

16.35 Write the expression for the acid ionization constant of the weak acid HCN.

16.36 Oxalic acid has the following structure. Write equilibrium reactions and acid ionization constant expressions for this diprotic acid.

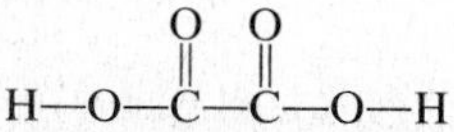

16.37 Write an acid ionization constant expression for each ionization step of the triprotic acid H_3PO_4.

16.38 Write an acid ionization constant expression for each ionization step of the diprotic acid H_2SO_3.

16.39 A 0.040 M solution of HF is 12% ionized. Calculate K_a.

16.40 A 0.010 M solution of an acid is 15% ionized. Calculate K_a.

16.41 What is the hydronium ion concentration of a 0.280 M acid solution whose ionization constant is 3.10×10^{-9}?

16.42 What is the hydronium ion concentration of a 0.510 M acid solution whose ionization constant is 1.62×10^{-7}?

16.43 What is the ionization constant of an acid if the hydronium ion concentration of a 0.100 M solution is $7.66 \times 10^{-4} M$?

16.44 What is the ionization constant of an acid if the hydronium ion concentration of a 0.400 M solution is $1.40 \times 10^{-4} M$?

16.45 A 0.10 M ammonia solution has an equilibrium NH_4^+ concentration equal to 0.0013 M. Calculate K for the following equilibrium.

$$NH_3 + H_2O \rightleftharpoons NH_4^+ + OH^-$$

16.46 Urea, a metabolic product found in urine, is a weak base. The concentrations of urea and its conjugate base are $4.0 \times 10^{-1} M$ and $4.0 \times 10^{-8} M$, respectively, in a solution whose hydronium ion concentration is $7.6 \times 10^{-8} M$. Calculate the equilibrium constant for the base.

Ionization of Water

16.47 What are the hydronium ion and hydroxide ion concentrations in each of the following solutions?
(a) 0.1 M HCl   (b) 10^{-2} M HNO$_3$
(c) 10^{-3} M KOH   (d) 0.05 M H$_2$SO$_4$
(e) 10^{-4} M NaOH (f) 0.005 M Mg(OH)$_2$

16.49 Explain what occurs to the concentration of hydronium ions in water when solid sodium hydroxide is added.

16.48 What are the hydronium ion and hydroxide ion concentrations in each of the following solutions?
(a) 0.02 M HBr   (b) 2×10^{-2} M HNO$_3$
(c) 3×10^{-4} M LiOH (d) 0.04 M H$_2$SO$_4$
(e) 5×10^{-4} M KOH (f) 0.002 M Sr(OH)$_2$

16.50 Explain what occurs to the concentration of hydroxide ions when gaseous hydrogen chloride is bubbled into water.

pH Values

16.51 Calculate the pH of the following solutions.
(a) 10^{-3} M HCl   (b) 2×10^{-3} M HNO$_3$
(c) 10^{-2} M NaOH (d) 3×10^{-3} M NaOH

16.53 Calculate the pH of the following items.
(a) a soft drink with [H$_3$O$^+$] = 2×10^{-4} M
(b) milk with [H$_3$O$^+$] = 2×10^{-7} M
(c) ammonia with [H$_3$O$^+$] = 2×10^{-12} M
(d) gastric juice with [H$_3$O$^+$] = 0.1 M

16.52 Calculate the pH of the following solutions.
(a) 10^{-2} M HBr   (b) 2×10^{-4} M HClO$_4$
(c) 10^{-1} M KOH (d) 4×10^{-2} M LiOH

16.54 Calculate the pH of the following items.
(a) vinegar with [H$_3$O$^+$] = 7.5×10^{-3} M
(b) orange juice with [H$_3$O$^+$] = 2.9×10^{-4} M
(c) lye with [OH$^-$] = 0.05 M
(d) arterial blood with [OH$^-$] = 2.5×10^{-7} M

Hydrolysis of Salts

16.55 Given the following K_a values, determine which conjugate base is the strongest base.
(a) HF, 6.8×10^{-4} (b) HNO$_2$, 7.2×10^{-4}
(c) HCNO, 2.2×10^{-4}

16.57 What is the hydrolysis constant of the OI$^-$ ion? The ionization constant of HOI is 2.0×10^{-11}.

16.59 What is the hydroxide ion concentration of 0.810 M OBr$^-$? The ionization constant of HOBr is 2.0×10^{-9}.

16.56 Given the following K_a values, determine which conjugate base is the strongest base.
(a) H$_2$SO$_3$, 1.2×10^{-2} (b) HNO$_2$, 7.2×10^{-4}
(c) HCN, 4.4×10^{-10}

16.58 Sodium propionate (NaC$_3$H$_5$O$_2$) is used as a food preservative. What is the hydrolysis constant of the propionate ion? The ionization constant of propionic acid is 1.34×10^{-5}.

16.60 What is the hydronium ion concentration in a 0.680 M CN$^-$ solution? (K_a HCN = 4.4×10^{-10})

Indicators

16.61 Some natural fruit juices change color when a base is added. Explain why.

16.63 What will be the color of methyl red at pH 3? What will be the color at pH 7?

16.62 How does an indicator work?

16.64 What will be the color of phenolphthalein at pH 7? What will be the color at pH 11?

Buffers

16.65 Explain how a buffer works.

16.67 How does the acid/conjugate base ratio affect the pH of the buffer?

16.69 What is the hydronium ion concentration of a solution that contains 0.260 M acid and 0.640 M of its conjugate base if the ionization constant is 2.06×10^{-6}?

16.71 What is the pH of a solution that contains 0.170 M acid and 0.500 M of its conjugate base if the ionization constant is 2.87×10^{-9}?

16.66 Could a solution of HCl and NaCl be a buffer?

16.68 How does the acid/conjugate base ratio affect the effectiveness of the buffer toward acid or base?

16.70 What is the hydronium ion concentration of a solution that contains 0.700 M acid and 0.440 M of its conjugate base if the ionization constant is 3.59×10^{-8}?

16.72 What is the pH of a solution that contains 0.810 M acid and 0.650 M of its conjugate base if the ionization constant is 4.92×10^{-7}?

Normality

16.73 How many milliequivalents of acid are present in each of the following solutions?
   (a) 50 mL of 0.01 N HCl
   (b) 5 mL of 0.1 N HNO_3
   (c) 1 L of 1 N HCl

16.75 What is the normality of each solution?
   (a) 4.0 g NaOH in 500 mL of solution
   (b) 3.65 g HCl in 1 L of solution
   (c) 4.9 g H_2SO_4 in 100 mL of solution
   (d) 0.327 g H_3PO_4 in 10 mL of solution
   (e) 0.0037 g $Ca(OH)_2$ in 500 mL of solution

16.77 Lactic acid ($C_3H_6O_3$) is a monoprotic acid. How many grams of lactic acid are required to produce 250 mL of a 0.01 N solution?

16.79 What is the normality of each of the following solutions?
   (a) 0.15 M H_3PO_4 (b) 0.20 M H_2SO_4
   (c) 0.25 M HBr   (d) 0.05 M H_2SeO_3
   (e) 0.40 M $HClO_4$ (f) 0.30 M HNO_3

16.74 How many milliequivalents of acid are present in each of the following solutions?
   (a) 50 mL of 0.01 N HCl
   (b) 5 mL of 0.1 N HNO_3
   (c) 1 L of 1 N HCl

16.76 What is the normality of each solution?
   (a) 0.56 g KOH in 250 mL of solution
   (b) 4.05 g HBr in 500 L of solution
   (c) 0.98 g H_2SO_4 in 10 mL of solution
   (d) 3.27 g H_3PO_4 in 100 mL of solution
   (e) 0.00185 g $Ca(OH)_2$ in 750 mL of solution

16.78 Oxalic acid ($C_2H_2O_4$) is a diprotic acid. How many grams of oxalic acid are required to produce 500 mL of a 0.02 N solution?

16.80 What is the normality of each of the following solutions?
   (a) 0.25 M lactic acid (monoprotic)
   (b) 0.30 M oxalic acid (diprotic)
   (c) 0.10 M citric acid (triprotic)
   (d) 0.010 M pyrophosphoric acid (tetraprotic)
   (e) 0.010 M triphosphoric acid (pentaprotic)

Titration

16.81 What quantity of 0.10 N NaOH is required to titrate 25.0 mL of a 0.060 N HCl solution?

16.83 A 0.1060 g sample of Na_2CO_3 is neutralized with 48.0 mL of a $HClO_4$ solution. What is the normality of the $HClO_4$ solution?

16.82 A 50.0 mL solution of $Ca(OH)_2$ is titrated with 36.0 mL of a 0.05 N HCl solution to an endpoint. What is the molarity of the $Ca(OH)_2$ solution?

16.84 A 0.100 g sample of NaOH is tritrated to an endpoint with 12.5 mL of HCl. What is the normality of the HCl?

Additional Exercises

16.85 What is the difference between hydrogen chloride and hydrochloric acid?

16.87 Write equations for the stepwise ionization of H_2SO_3.

16.89 What is the conjugate acid of each of the following?
   (a) Br^- (b) HCO_3^- (c) OH^-
   (d) HPO_4^{2-} (e) SO_3^{2-} (f) PO_4^{3-}

16.91 Veronal ($C_8H_{12}N_2O_3$), a barbiturate drug, is an acid. What is its conjugate base?

16.93 A sample of orange juice has a hydrogen ion concentration of 2.9×10^{-4} M. What is the pH of the orange juice?

16.95 What is the pH of a 0.012 M solution of $Ca(OH)_2$?

16.97 What is the hydronium ion concentration of a 0.110 M acid solution whose ionization constant is 5.45×10^{-7}?

16.99 What is the ionization constant of an acid if the hydronium ion concentration of a 0.500 M solution is 1.70×10^{-4} M?

16.101 What is the hydrolysis constant of the CN^- ion? The ionization constant of HCN is 4.4×10^{-10}.

16.86 What is the difference between ammonia and ammonium hydroxide?

16.88 Write equations for the stepwise ionization of the diprotic acid, $C_2H_2O_4$.

16.90 What is the conjugate base of each of the following?
   (a) $HClO_4$ (b) H_2SO_3 (c) H_3O^+
   (d) HI   (e) HClO   (f) $HBrO_2$

16.92 Niacin ($C_6H_5NO_2$), also known as nicotinic acid, is a vitamin. What is the conjugate base?

16.94 What is the pH of a sample of gastric juice whose hydronium ion concentration is 4.5×10^{-2} M?

16.96 What is the pH of a 0.005 M solution of $Sr(OH)_2$?

16.98 What is the hydronium ion concentration of a 0.160 M acid solution whose ionization constant is 1.62×10^{-9}?

16.100 What is the ionization constant of an acid if the hydronium ion concentration of a 0.500 M solution is 1.59×10^{-3} M?

16.102 What is the hydrolysis constant of the OCl^- ion? The ionization constant of HOCl is 3.0×10^{-8}.

16.103 What is the hydronium ion concentration of a solution of 0.200 M acid and 0.600 M of its conjugate base if the ionization constant is 2.88×10^{-7}?

16.105 What mass (g) of calcium carbonate, $CaCO_3$, is required to react with the sulfuric acid in 38.5 mL of a 0.973 M solution of H_2SO_4?

16.107 What volume of 0.0250 M HNO_3 solution is needed to titrate 125 mL of a 0.0100 N $Ca(OH)_2$ solution?

16.104 What is the hydronium ion concentration of a solution of 0.480 M acid and 0.680 M of its conjugate base if the ionization constant is 5.78×10^{-7}?

16.106 A 10.0 mL sample of sulfuric acid solution from an automobile battery reacts exactly with 35.0 mL of a 1.87 M solution of sodium hydroxide (NaOH). What is the molar concentration of the battery acid?

16.108 Phosphoric acid is a triprotic acid. How many grams of phosphoric acid are required to produce 100 mL of a 0.25 N solution?

17 Oxidation–Reduction Reactions

Learning Objectives

As a result of studying Chapter 17 you should be able to

1. Identify which substances are oxidized, and which are reduced, in a chemical reaction.

2. Identify which substances are oxidizing agents, and which are reducing agents, in a chemical reaction.

3. Balance redox equations by the oxidation number method.

4. Balance redox equations by the half-reaction method.

5. Use a table of standard electrode potentials to calculate the potential of a galvanic cell and to predict whether a reaction will proceed as written.

6. Illustrate the utility of electrolysis.

17.1 Oxidation and Reduction

At one time one definition of oxidation was the reaction of an element, such as a metal, with oxygen. The reaction of magnesium in a flashbulb and the rusting of iron are examples of the oxidation of metals.

$$2 \, Mg + O_2 \longrightarrow 2 \, MgO$$
$$4 \, Fe + 3 \, O_2 \longrightarrow 2 \, Fe_2O_3$$

The burning of natural gas (methane) involves the combination of carbon with oxygen and is also considered an oxidation reaction.

$$CH_4 + 2 \, O_2 \longrightarrow CO_2 + 2 \, H_2O$$

Note that in this case the carbon atom also loses its hydrogen atoms. Thus, oxidation has also been viewed as the loss of hydrogen.

The opposite of oxidation is known as reduction. Reduction at one time was described as the removal of oxygen from a compound. For example, the removal of oxygen from rust by carbon involves the reduction of iron.

$$2 \, Fe_2O_3 + 3 \, C \longrightarrow 4 \, Fe + 3 \, CO_2$$

Reduction has also been viewed as the combination of hydrogen with the oxygen of a compound, as in the reaction with copper(II) oxide, or the direct combination of hydrogen with a substance, as in the reaction with nitrogen to form ammonia.

$$CuO + H_2 \longrightarrow Cu + H_2O$$
$$N_2 + 3 \, H_2 \longrightarrow 2 \, NH_3$$

Today chemists have consolidated all oxidation–reduction reactions (also called **redox reactions**) with new definitions that have wider application than the older definitions. These definitions deal with the transfer of electrons and the changes in oxidation number.

Currently, **oxidation** is defined as the loss of electrons by a substance or an increase in its oxidation number. In the reaction of magnesium with oxygen, the oxidation number of magnesium has changed from 0 to $+2$ and the magnesium has been oxidized. An oxidation reaction does not necessarily involve oxygen. Thus in the reaction of sodium with chlorine, the sodium is oxidized because its oxidation number increases from 0 to $+1$.

$$2 \, Na + Cl_2 \longrightarrow 2 \, NaCl$$

Reduction is defined as the gain of electrons by a substance or a decrease in its oxidation number. In the case of the reaction of oxygen with magnesium, the oxidation number of oxygen changes from 0 in the element to -2 in the magnesium oxide. The oxygen has become reduced. In the reaction of sodium with chlorine, the oxidation number of chlorine changes from 0 in the element to -1 in sodium chloride. The chlorine is reduced.

When a substance is oxidized, it loses electrons to another substance, which then is reduced. For example, the electrons lost by magnesium in its reaction with oxygen are gained by oxygen. From a slightly different point of view, it follows that when a substance becomes reduced, the electrons that it gains are obtained from a substance that becomes oxidized. This is why reactions in which an electron transfer occurs are called oxidation–reduction reactions.

The close and necessary relationship between oxidation and reduction is emphasized further by the use of terms oxidizing agent and reducing agent. In an oxidation–reduction reaction, the substance that is reduced is called the **oxidizing agent** because, by gaining electrons, it causes

Table 17.1 **Oxidation–Reduction Terminology**

Term	Electron change	Oxidation number change
oxidation	loss of electrons	increase
reduction	gain of electrons	decrease
oxidizing agent	accepts electrons	decrease
reducing agent	donates electrons	increase

oxidation in another substance. The substance that is oxidized is called the **reducing agent** because, by losing its electrons, it causes the reduction of another substance. In terms of the changes in oxidation number, an oxidizing agent is a substance whose oxidation number decreases; a reducing agent is a substance whose oxidation number increases. A summary of the terms used in this section is given in Table 17.1.

Example 17.1

What is oxidized and what is reduced in the following reaction?

$$3 P + 5 HNO_3 + 2 H_2O \longrightarrow 5 NO + 3 H_3PO_4$$

Solution The oxidation numbers of hydrogen and oxygen are +1 and −2, respectively, and are not changed in the reaction. The oxidation number of nitrogen in HNO_3 is +5, whereas it is +2 in NO. Thus nitrogen is reduced. The oxidation number of phosphorus changes from 0 in elemental phosphorus to +5 in H_3PO_4. The phosphorus is oxidized.

[Additional examples may be found in 17.9–17.12 at the end of the chapter.]

Example 17.2

What is the oxidizing agent and what is the reducing agent in the reaction given by the following equation?

$$4 NH_3 + 5 O_2 \longrightarrow 4 NO + 6 H_2O$$

Solution Oxygen is present as a free element as a reactant but is combined in compounds in the products. Its oxidation number changes from 0 to −2. Oxygen becomes reduced and serves as an oxidizing agent. The oxidation number of nitrogen changes from −3 in NH_3 to +2 in NO. Nitrogen in NH_3 is oxidized and serves as the reducing agent.

[Additional examples may be found in 17.13–17.16 at the end of the chapter.]

17.2 Balancing Redox Equations Using Oxidation Numbers

Some oxidation–reduction equations are simple enough to balance by just looking at them. However, for more complex equations, the oxidation number method provides a convenient way to balance equations. There is no hard and fast procedure or set of rules using oxidation numbers that works in all circumstances. However, some guidelines can be given that, if used with some common sense, will produce a balanced equation.

1. Determine the oxidation number of each element.
2. Establish which elements undergo changes in their oxidation numbers as they are converted from reactant to product.
3. Write the oxidation number of each element oxidized or reduced above the elemental symbol.
4. Draw an arrow from the element in the reactant to that element in the product for each change in oxidation number.
5. Calculate the change in the oxidation number for each element undergoing oxidation or reduction. Record this change as an increase or a decrease over the proper arrow.
6. Place a factor in front of the change in oxidation numbers so that the total increase is balanced by the total decrease.
7. Put this factor as a coefficient in front of the corresponding formula for the reactant and product if the element is present as a single atom per molecular or formula unit.
8. If more than one atom is contained per molecular or formula unit, divide the factor by the number of atoms to obtain the proper coefficient. If a fraction results, multiply by a number to produce an integer coefficient. Then multiply all other coefficients by the same number. This must be done to keep the total increase in oxidation number equal to the total decrease in oxidation number.
9. Balance the remaining parts of the equation by inspection. Do not change the coefficients for substances undergoing changes in oxidation number.

The best way to learn these guidelines is to use them in balancing equations. Several examples follow, and additional equations are given at the end of this chapter.

Example 17.3

Balance the following equation.

$$Zn + AgNO_3 \longrightarrow Zn(NO_3)_2 + Ag$$

Solution According to guideline 1, the oxidation number of each element should be determined. However, note that both nitrogen and oxygen are present only as the nitrate ion in both reactants and products. Clearly these elements do not undergo changes in oxidation number (guideline 2). Thus only zinc and silver change oxidation number. The oxidation numbers are written above the elemental symbol (guideline 3).

$$\overset{(0)}{Zn} + \overset{(+1)}{Ag}\overset{}{NO_3} \longrightarrow \overset{(+2)}{Zn}(NO_3)_2 + \overset{(0)}{Ag}$$

Draw an arrow between elements as reactants and products (guideline 4) and indicate the change in oxidation number (guideline 5).

$$\overset{(0)}{Zn} + \overset{(+1)}{Ag}NO_3 \longrightarrow \overset{(+2)}{Zn}(NO_3)_2 + \overset{(0)}{Ag}$$

(increase by two)

(decrease by one)

The factors necessary to balance the total increase by the total decrease in oxidation number are 1 and 2, respectively.

$$\overset{\displaystyle\overbrace{\qquad}^{1\,\times\,(\text{increase by two})\,=\,+2}}{\underset{(0)\qquad(+1)\qquad\qquad(+2)\qquad\qquad(0)}{\text{Zn} + \text{AgNO}_3 \longrightarrow \text{Zn(NO}_3)_2 + \text{Ag}}}$$

$$\underbrace{\qquad\qquad\qquad}_{2\,\times\,(\text{decrease by one})\,=\,-2}$$

The coefficients are therefore 1 and 2 (guideline 7) because each element is present as a single atom per formula unit.

$$\text{Zn} + 2\,\text{AgNO}_3 \longrightarrow \text{Zn(NO}_3)_2 + 2\,\text{Ag}$$

Example 17.4

Balance the following oxidation–reduction reaction.

$$\text{HNO}_3 + \text{I}_2 \longrightarrow \text{NO}_2 + \text{HIO}_3 + \text{H}_2\text{O}$$

Solution In this reaction, hydrogen and oxygen appear only in compounds and their oxidation numbers are $+1$ and -2, respectively. The elements nitrogen and iodine change oxidation numbers and these values are written above the elemental symbol.

$$\underset{\text{HNO}_3 + \text{I}_2 \longrightarrow \text{NO}_2 + \text{HIO}_3 + \text{H}_2\text{O}}{(+5)\qquad(0)\qquad\quad(+4)\qquad(+5)}$$

Next an arrow is drawn between the elements in the reactants and products, and the change in oxidation number is indicated.

$$\overset{(\text{decrease by one})}{\underset{(+5)\qquad(0)\qquad\quad(+4)\qquad(+5)}{\text{HNO}_3 + \text{I}_2 \longrightarrow \text{NO}_2 + \text{HIO}_3 + \text{H}_2\text{O}}}$$
$$(\text{increase by five})$$

The factors necessary for the total increase in oxidation number of iodine and the total decrease in oxidation number of nitrogen are 1 and 5, respectively.

$$\overset{5\,\times\,(\text{decrease by one})\,=\,-5}{\underset{(+5)\qquad(0)\qquad\quad(+4)\qquad(+5)}{\text{HNO}_3 + \text{I}_2 \longrightarrow \text{NO}_2 + \text{HIO}_3 + \text{H}_2\text{O}}}$$
$$1\,\times\,(\text{increase by five})\,=\,+5$$

These factors may not be placed as coefficients in front of the formulas for the reactants and products because iodine is present as a diatomic molecule in one of the reactants. Division of the factor by the number of iodine atoms in I_2 gives $\frac{1}{2}$ as the coefficient. In order to obtain an integer coefficient, we multiply by 2. The factor of 5 for nitrogen compounds must then also be multiplied by 2. The coefficients may now be placed in the equation.

$$10\,\text{HNO}_3 + \text{I}_2 \longrightarrow 10\,\text{NO}_2 + 2\,\text{HIO}_3 + \text{H}_2\text{O}$$

The remaining coefficient for water is easily established as 4 by noting that there are 10 hydrogen atoms in the reactant but only 2 accounted for in the 2 HIO_3. Similarly, one could achieve a balance by noting that there are 30 oxygen atoms in the reactant but only 26 oxygen atoms in two of the products whose coefficients are fixed. Thus 4 oxygen atoms must be produced in the form of H_2O.

$$10 \, HNO_3 + I_2 \longrightarrow 10 \, NO_2 + 2 \, HIO_3 + 4 \, H_2O$$

Example 17.5

Balance the following oxidation–reduction equation.

$$Na_2CrO_4 + FeCl_2 + HCl \longrightarrow CrCl_3 + FeCl_3 + NaCl + H_2O$$

Solution   The elements changing oxidation numbers are chromium and iron. The oxidation numbers, arrows, and changes in oxidation number are written first.

decrease by 3

$$\underset{(+6)}{Na_2CrO_4} + \underset{(+2)}{FeCl_2} + HCl \longrightarrow \underset{(+3)}{CrCl_3} + \underset{(+3)}{FeCl_3} + NaCl + H_2O$$

(increase by 1)

To achieve a balance of the total increase in oxidation number with a total decrease in oxidation number, the factors for chromium and iron are 1 and 3, respectively. Since each element appears only as a single atom in the reactants and products, the factors may be used directly as coefficients.

$$Na_2CrO_4 + 3 \, FeCl_2 + HCl \longrightarrow CrCl_3 + 3 \, FeCl_3 + NaCl + H_2O$$

The coefficient for NaCl is next set as 2 because there are two sodium atoms in Na_2CrO_4. Similarly, the four oxygen atoms in Na_2CrO_4 are balanced by placing a coefficient of 4 in front of H_2O.

$$Na_2CrO_4 + 3 \, FeCl_2 + HCl \longrightarrow CrCl_3 + 3 \, FeCl_3 + 2 \, NaCl + 4 \, H_2O$$

The eight hydrogen atoms contained in 4 H_2O are balanced by placing a coefficient of 8 in front of HCl.

$$Na_2CrO_4 + 3 \, FeCl_2 + 8 \, HCl \longrightarrow CrCl_3 + 3 \, FeCl_3 + 2 \, NaCl + 4 \, H_2O$$

A final check reveals that there are 14 chlorine atoms in both the reactants and products.

[*Additional examples may be found in 17.17–17.22 at the end of the chapter.*]

17.3   Balancing Equations Using the Half-reaction Method

A second method of balancing oxidation–reduction equations involves considering the reaction as two half-reactions. A **half-reaction** is an oxidation or a reduction reaction that cannot occur alone. The two half-reactions can only occur simultaneously within the total reaction. The method of achieving the final balanced equation involves balancing each half-reaction and then adding them together.

In Chapter 15 you learned that reactions can be expressed in ionic equations in which only the ions or molecules that are directly involved are written. The spectator ions are not involved in a net ionic equation. Although half-reactions can be written using all of the components of the total equation, it is much simpler to use only the substances that undergo change. Thus, adding the half-reactions together yields a net ionic equation.

Again there is a set of guidelines that will help you obtain the net ionic equations and enable you to balance equations by the half-reaction method.

1. Examine the equation and rewrite it in net ionic form.
2. Write the two partial equations: an oxidation equation and a reduction equation.
3. With the exception of hydrogen and oxygen, balance every atom on each side of the partial equations by using coefficients.
4. If hydrogen or oxygen is present in either reactants or products but not represented on the other side of the equation, then add appropriate quantities of H^+, OH^-, or H_2O. In acidic solutions, H^+ is added as a source for every unit of hydrogen required. In acidic solution, water is added for every unit of oxygen required, with $2\,H^+$ for each added water placed on the opposite side of the equation. In basic solutions, H_2O is added for every hydrogen required and an OH^- is placed on the other side of the equation. For every oxygen atom required in basic solution, $2\,OH^-$ are added along with an H_2O on the opposite side of the equation.
5. Balance the charge of each half-reaction by adding electrons on the proper side of the equation.
6. Multiply each half reaction by a factor so that the total number of electrons gained in one half-reaction equals the total number lost in the other half-reaction.
7. Add the two half-reactions and cancel electrons, water molecules, or ions that are common to each side of the equation and are not needed for the net ionic equation.

Example 17.6

Balance the following oxidation–reduction equation using the half-reaction method.

$$Ag_2SO_4 + Zn \longrightarrow ZnSO_4 + Ag$$

Solution   The equation in net ionic form is obtained by eliminating the spectator ion, SO_4^{2-} (guideline 1). The two half-reactions are selected from the unbalanced equation (guideline 2).

$$Ag^+ + Zn \longrightarrow Zn^{2+} + Ag$$
$$Ag^+ \longrightarrow Ag \quad \text{(reduction)}$$
$$Zn \longrightarrow Zn^{2+} \quad \text{(oxidation)}$$

The equations as written are balanced with respect to the elements (guideline 3). Furthermore, no hydrogen or oxygen is needed (guideline 4). Thus only electrons are added to the equation (guideline 5).

$$Ag^+ + 1\,e^- \longrightarrow Ag$$
$$Zn \longrightarrow Zn^{2+} + 2\,e^-$$

Now the reduction equation is multiplied by 2 so that electrons gained and lost are balanced (guideline 6). Finally, the two equations are added (guideline 7).

$$2\,Ag^+ + 2\,e^- \longrightarrow 2\,Ag$$
$$\underline{\hspace{1.5cm} Zn \longrightarrow Zn^{2+} + 2\,e^-}$$
$$2\,Ag^+ + Zn + \cancel{2\,e^-} \longrightarrow 2\,Ag + Zn^{2+} + \cancel{2\,e^-}$$
$$2\,Ag^+ + Zn \longrightarrow 2\,Ag + Zn^{2+}$$

Example 17.7

Balance the following equation for a reaction that occurs in acidic solution.

$$H_2S + MnO_4^- + H^+ \longrightarrow Mn^{2+} + S + H_2O$$

Solution Guidelines 1 and 2: Write the equations.

$$H_2S \longrightarrow S \quad \text{(oxidation)}$$
$$MnO_4^- \longrightarrow Mn^{2+} \quad \text{(reduction)}$$

Guidelines 3 and 4: Add hydrogen and oxygen in the correct form.

$$H_2S \longrightarrow S + 2\,H^+$$
$$8\,H^+ + MnO_4^- \longrightarrow Mn^{2+} + 4\,H_2O$$

Guideline 5: Add electrons to balance charges.

$$H_2S \longrightarrow S + 2\,H^+ + 2\,e^-$$
$$5\,e^- + 8\,H^+ + MnO_4^- \longrightarrow Mn^{2+} + 4\,H_2O$$

Guideline 6: Multiply by a factor to equalize the electrons gained and lost.

$$5\,(H_2S \longrightarrow S + 2\,H^+ + 2\,e^-)$$
$$2\,(5\,e^- + 8\,H^+ + MnO_4^- \longrightarrow Mn^{2+} + 4\,H_2O)$$

Guideline 7: Add the two half-reactions and subtract or cancel any factors common to each side.

$$5\,H_2S + \overset{6}{\cancel{16}}\,H^+ + 2\,MnO_4^- + \cancel{10\,e^-} \longrightarrow 5\,S + \cancel{10\,H^+} + 2\,Mn^{2+} + 8\,H_2O + \cancel{10\,e^-}$$
$$5\,H_2S + 6\,H^+ + 2\,MnO_4^- \longrightarrow 5\,S + 2\,Mn^{2+} + 8\,H_2O$$

[*Additional examples may be found in 17.23–17.28 at the end of the chapter.*]

17.4 Galvanic Cells

In Section 6.7 an order of reactivity of metals called the activity series was presented. A metal in the series will displace any of those metals below it from an aqueous solution of the salt of the second metal. Thus zinc will react spontaneously with an aqueous solution of copper(II) sulfate to yield copper and zinc sulfate.

$$Zn + CuSO_4 \longrightarrow ZnSO_4 + Cu$$

This replacement reaction is an oxidation–reduction reaction. The net ionic equation is

$$Zn + Cu^{2+} \longrightarrow Zn^{2+} + Cu$$

The equation can be represented by two half-reactions.

$$Zn \longrightarrow Zn^{2+} + 2\,e^-$$
$$2\,e^- + Cu^{2+} \longrightarrow Cu$$

If this spontaneous reaction is carried out by immersing zinc in the copper(II) sulfate solution, electron transfer occurs but electrical energy has not been made available. However, this reaction, as well as most oxidation–reduction reactions, can be carried out with proper experimental modifications to produce electrical energy. It is necessary to allow the oxidation and reduction half-reactions to occur in separate half-cells connected by a conductor through which electrons may flow. The electrons lost from the substance being oxidized in one half-cell flow through the wire to the substance being reduced in the other half-cell. The transfer of electrons through the conductor allows the oxidation and reduction half-reactions to occur and the energy of the reaction is released as a useful electric current. The experimental device that produces electric current from spontaneous redox reactions is called a **galvanic cell** or a **voltaic cell.**

The oxidation of zinc metal by the copper(II) ion (Cu^{2+}) can be carried out in a cell constructed so that the zinc metal and Cu^{2+} are in separate half-cells (Figure 17.1). A container is partitioned into two compartments by a porous divider that allows solutions placed in the separate compartments to be in contact but retards the rate of mixing. In one compartment a bar

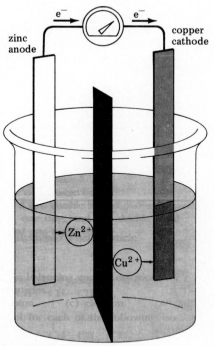

Figure 17.1

The zinc–copper galvanic cell

of zinc is immersed in a solution of zinc nitrate, $Zn(NO_3)_2$. In the other compartment a bar of copper is immersed in a solution of copper(II) nitrate, $Cu(NO_3)_2$. If the two metal bars are then connected with a conductor, an electric current results.

The two metal bars in the galvanic cell are called **electrodes.** At the zinc electrode, zinc is oxidized and enters the solution as zinc ions. The electrode at which oxidation occurs is called the **anode.** In the oxidation, the zinc enters solution as Zn^{2+} and leaves two electrons per atom on the anode. Electrons travel via the conductor to the copper bar where copper ions in solution accept them, are reduced, and are plated out on the copper bar. The electrode at which reduction occurs is called the **cathode.**

17.5 Standard Electrode Potentials

The activity series presented in Chapter 6 is based on experimental observations made using galvanic cells. Thus as shown in the preceding section, zinc is more active than copper. If a voltmeter is placed between the two electrodes for the reaction in which zinc replaces copper(II) ions in solution, a voltage of 1.10 volts (V) is recorded. This voltage is a quantitative measure of the replacement tendency or activity of one metal relative to another. The voltage indicates how strongly electrons are attracted from the anode of one half-cell to the cathode of the other half-cell.

Just as the redox reaction in a cell is divided into two half-cells where half-reactions occur, the cell potential as measured by voltage can be divided into two electrode potentials. An electrode potential is the potential of one half-reaction. However, the individual electrode potential cannot be measured directly. Only the potential of the cell, which is a sum of the two electrode potentials, can be measured. Therefore all electrode potentials are determined relative to a defined standard electrode.

Table 17.2   **Standard Electrode Potentials**

	$E°$ (V)					$E°$ (V)
$Li^+ + 1\,e^- \longrightarrow Li$	-3.04		$Cd^{2+} + 2\,e^- \longrightarrow Cd$			-0.40
$K^+ + 1\,e^- \longrightarrow K$	-2.92		$Ni^{2+} + 2\,e^- \longrightarrow Ni$			-0.25
$Cs^+ + 1\,e^- \longrightarrow Cs$	-2.92		$Sn^{2+} + 2\,e^- \longrightarrow Sn$			-0.14
$Ba^{2+} + 2\,e^- \longrightarrow Ba$	-2.90		$Pb^{2+} + 2\,e^- \longrightarrow Pb$			-0.13
$Sr^{2+} + 2\,e^- \longrightarrow Sr$	-2.89		$2\,H^+ + 2\,e^- \longrightarrow H_2$			0.00
$Ca^{2+} + 2\,e^- \longrightarrow Ca$	-2.76		$Cu^{2+} + 2\,e^- \longrightarrow Cu$			0.34
$Na^+ + 1\,e^- \longrightarrow Na$	-2.71		$I_2 + 2\,e^- \longrightarrow 2\,I^-$			0.54
$Mg^{2+} + 2\,e^- \longrightarrow Mg$	-2.37		$Ag^+ + 1\,e^- \longrightarrow Ag$			0.80
$Al^{3+} + 3\,e^- \longrightarrow Al$	-1.66		$Hg^{2+} + 2\,e^- \longrightarrow Hg$			0.85
$Mn^{2+} + 2\,e^- \longrightarrow Mn$	-1.18		$Br_2 + 2\,e^- \longrightarrow 2\,Br^-$			1.06
$Zn^{2+} + 2\,e^- \longrightarrow Zn$	-0.76		$Cl_2 + 2\,e^- \longrightarrow 2\,Cl^-$			1.36
$Cr^{3+} + 3\,e^- \longrightarrow Cr$	-0.74		$Au^+ + 1\,e^- \longrightarrow Au$			1.69
$Fe^{2+} + 2\,e^- \longrightarrow Fe$	-0.44		$F_2 + 2\,e^- \longrightarrow 2\,F^-$			2.87

In Table 17.2 are listed a series of standard electrode potentials ($E°$) for the reaction of the oxidized form of a metal. The **standard electrode potentials** are the voltages of a galvanic cell in which the metal is compared to a hydrogen electrode, whose potential is defined as 0.0 V.

$$2\,H^+(aq) + 2\,e^- \longrightarrow H_2(g)$$

The standard electrode potentials are assigned positive values if the ion of the metal can gain electrons from hydrogen molecules. Negative values indicate the metal will lose electrons to hydrogen ions in solution to form hydrogen gas.

For the measurement of standard electrode potentials, a hydrogen electrode consisting of a strip of platinum (which is resistant to oxidation) immersed in a solution of hydrogen ions is used. Hydrogen gas is bubbled over the surface of the platinum so that the gas and hydrogen ions come in contact. If hydrogen ions gain electrons to produce hydrogen, the electrode is acting as a cathode. Alternatively, if hydrogen molecules give up electrons and produce hydrogen ions, the hydrogen electrode is acting as an anode. The behavior of the hydrogen electrode depends upon the electrode coupled to it. In the presence of zinc and zinc ions, the hydrogen electrode functions as a cathode. The voltage of the hydrogen and zinc galvanic cell is 0.76 V (Figure 17.2).

$$
\begin{aligned}
Zn &\longrightarrow Zn^{2+} + 2\,e^- \\
2\,H^+ + 2\,e^- &\longrightarrow H_2 \\
\hline
Zn + 2\,H^+ &\longrightarrow H_2 + Zn^{2+} \qquad E° = 0.76\ V
\end{aligned}
$$

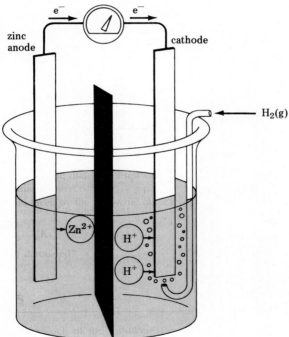

Figure 17.2

The zinc–hydrogen galvanic cell

Since zinc is oxidized in this half-reaction, the direction of the reaction must be reversed in order to list Zn^{2+} in a table of standard electrode potentials, which are written as reduction reactions. Therefore the electrode potential for the reduction of Zn^{2+} must be reversed to -0.76 V.

$$Zn \longrightarrow Zn^{2+} + 2\,e^- \qquad \text{(observed)}$$
$$2\,e^- + Zn^{2+} \longrightarrow Zn \qquad \text{(standard electrode potential)}$$

If copper metal, immersed in a solution containing copper(II) ions, is coupled to the hydrogen electrode, the copper ions will be reduced. In this case the hydrogen electrode acts as an anode, and the voltage for the cell is 0.34 V (Figure 17.3).

$$2\,e^- + Cu^{2+} \longrightarrow Cu$$
$$H_2 \longrightarrow 2\,H^+ + 2\,e^-$$
$$\overline{Cu^{2+} + H_2 \longrightarrow Cu + 2H^+ \qquad E° = 0.34\ V}$$

Thus the standard electrode potential for the reduction of Cu^{2+} is defined as $+0.34$ V.

All standard electrode equations are written as reductions. For oxidation reactions, the sign of the related standard electrode potential must be reversed. Therefore, the metals above hydrogen in the table of standard electrode potentials, which undergo oxidation compared to the hydrogen electrode, will have a positive oxidation potential, whereas those below hydrogen will have a negative oxidation potential.

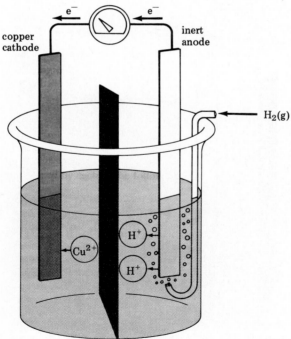

Figure 17.3

The copper–hydrogen galvanic cell

Some important generalizations can be made about the reactions listed in the table of standard electrode potentials. The reactants at the top of the table are not easily reduced and are the weakest oxidizing agents because they have the most negative potential. The reactants at the bottom of the table are the most easily reduced and are the strongest oxidizing agents. Based on the reverse of the equations in the list of standard electrode potentials, the products at the top of the table are the most easily oxidized and are the strongest reducing agents. The products at the bottom of the table are the least easily oxidized and are the weakest reducing agents.

Example 17.8

Will magnesium reduce Zn^{2+} in an aqueous solution?

Solution The electrode potentials to consider are

$$2\,e^- + Mg^{2+} \longrightarrow Mg \qquad E° = -2.37\ V$$
$$2\,e^- + Zn^{2+} \longrightarrow Zn \qquad E° = -0.76\ V$$

Thus Zn^{2+} has the larger tendency to accept electrons and become reduced. The electrons released by magnesium, which acts as the reducing agent, will reduce Zn^{2+}.

$$Mg + Zn^{2+} \longrightarrow Mg^{2+} + Zn$$

[*Additional examples may be found in 17.29–17.32 at the end of the chapter.*]

Example 17.9

Will the following reaction occur as written?

$$Cu^{2+} + Hg \longrightarrow Cu + Hg^{2+}$$

Solution The electrode potentials to consider are

$$2\,e^- + Cu^{2+} \longrightarrow Cu \qquad E° = +0.34\ V$$
$$2\,e^- + Hg^{2+} \longrightarrow Hg \qquad E° = +0.85\ V$$

Thus Hg^{2+} has a greater tendency to gain electrons and become reduced than does Cu^{2+}. The reaction given involves the reduction of Cu^{2+} and therefore will not occur. The reverse of the reaction can occur.

[*Additional examples may be found in 17.33 and 17.34 at the end of the chapter.*]

Calculating Cell Potentials

The potential of a redox reaction can be calculated from the electrode potentials of two half-cell reactions. Since the standard electrode potentials are given for half-reactions in which reduction occurs, one of the half-reactions must be reversed. The sign of $E°$ must be changed for the reverse reaction, which is now an oxidation reaction. The potentials for the reduction and oxidation half-reactions are added. For a positive sum, the redox reaction will be spontaneous in the direction written. If a negative sum results, the reverse of the reaction will be spontaneous.

The reaction of zinc and copper(II) in a galvanic cell has a voltage of 1.10 V.

$$Zn + Cu^{2+} \longrightarrow Zn^{2+} + Cu$$

This voltage is equal to the difference in the abilities of Cu^{2+} and Zn^{2+} to gain electrons in the reactions given as reductions. Copper(II) has a standard electrode potential that is 1.10 V more positive than zinc.

$$Cu^{2+} + 2\,e^- \longrightarrow Cu \qquad E° = +0.34 \text{ V}$$
$$Zn^{2+} + 2\,e^- \longrightarrow Zn \qquad E° = -0.76 \text{ V}$$

The voltage of the cell in which the reaction occurs can be obtained by summing the potential for the half-reactions. The potential of the oxidation half-reaction of zinc is obtained by reversing the sign of the standard potential.

$$
\begin{array}{lll}
Cu^{2+} + 2\,e^- \longrightarrow & Cu & E° = +0.34 \text{ V} \\
\underline{\phantom{Cu^{2+} + 2\,e^-} Zn \longrightarrow \quad Zn^{2+} + 2\,e^-} & & \underline{E° = +0.76 \text{ V}} \\
Cu^{2+} + Zn \longrightarrow & Cu + Zn^{2+} & E° = +1.10 \text{ V}
\end{array}
$$

Example 17.10

What is the voltage of a cell so constructed that Mg reduces Pb^{2+}?

Solution   The proposed reaction is first written.

$$Mg + Pb^{2+} \longrightarrow Mg^{2+} + Pb$$

The half-reactions for the reduction and the oxidation are

$$
\begin{array}{lll}
Mg \longrightarrow & Mg^{2+} + 2\,e^- & E° = +2.37 \text{ V} \\
\underline{2\,e^- + Pb^{2+} \longrightarrow \quad Pb} & & \underline{E° = -0.13 \text{ V}} \\
& & E° = +2.24 \text{ V}
\end{array}
$$

The voltage is +2.24 V. Remember that the potential for the oxidation half-reaction is obtained by reversing the sign of the standard electrode potential.

[*Additional examples may be found in 17.35 and 17.36 at the end of the chapter.*]

An activity series for the nonmetallic halogens was given in Chapter 6. Fluorine is more active than chlorine and will replace the chloride ion from an aqueous solution.

$$F_2 + 2\,Cl^- \longrightarrow 2\,F^- + Cl_2$$

Fluorine will also replace bromide and iodide. Chlorine will replace bromide and iodide, whereas bromine will replace only iodide. A quantitative measure of these tendencies is given by the standard electrode potentials listed in Table 17.2. Fluorine has the highest positive electrode potential followed by $Cl_2 > Br_2 > I_2$. Thus the tendency to gain electrons is $F_2 >$

$Cl_2 > Br_2 > I_2$. The voltage derived from the reaction of chlorine and bromide is calculated by summing the proper electrode potentials for the half reactions.

$$\begin{array}{llr}
Cl_2 + 2\,e^- & \longrightarrow & 2\,Cl^- & 1.36\text{ V} \\
2\,Br^- & \longrightarrow & Br_2 + 2\,e^- & -1.06\text{ V} \\
\hline
Cl_2 + 2\,Br^- & \longrightarrow & 2\,Cl^- + Br_2 & 0.30\text{ V}
\end{array}$$

17.6 Electrolysis

The application of voltage and electric power in a cell to cause a nonspontaneous reaction to occur is called **electrolysis.** The reaction would not occur without the electrical energy supplied to the cell. Such processes are extremely important in the industrial preparation of many metals and nonmetals.

Many ions of metals are difficult to reduce because there are no readily available chemical reducing agents. As an example, magnesium ions could be reduced by sodium or potassium metals, which have higher electrode potentials, but such a reduction process involves many difficulties in the handling of the very active metals. The commercial preparation of magnesium involves electrolytic reduction of Mg^{2+} at a cathode and electrolytic oxidation of chloride ion at the anode in the molten salt $MgCl_2$ (Figure 17.4).

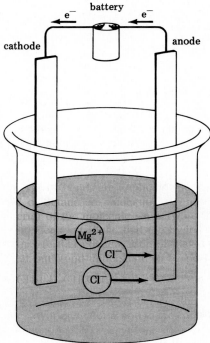

Figure 17.4

The electrolysis of molten magnesium chloride

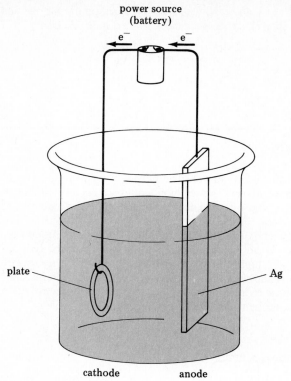

power source
(battery)

e^- e^-

plate

Ag

Figure 17.5

cathode   anode

Electroplating of silver

cathode: $Mg^2 + 2\,e^- \longrightarrow Mg$
anode: $2\,Cl^- \longrightarrow Cl_2 + 2\,e^-$

The net reaction decomposes the compound $MgCl_2$ into its constituent elements. The process of compound formation from magnesium and chlorine is a spontaneous reaction. Thus the decomposition reaction is not spontaneous and occurs only because a source of electrical energy forces it to occur.

Electrolytic processes are used to obtain pure metals from less pure samples. If two strips of copper are connected to the terminals of a voltage source and are placed in a solution of copper sulfate ($CuSO_4$), copper can be made to dissolve from one electrode and deposit on the other. The copper that is oxidized at the anode is changed into Cu^{2+} ions, which then are reduced at the cathode to elemental copper. If the anode is 99.0% "pure" copper, the copper deposited at the cathode will be approximately 99.98% "pure" copper.

anode: $Cu \longrightarrow Cu^{2+} + 2\,e^-$
cathode: $Cu^{2+} + 2\,e^- \longrightarrow Cu$

Electrolysis is also an important process in industry where metals are coated or plated with a thin layer of a second metal. The second metal may be plated onto the surface for either decorative purposes, as in the case of silver plated utensils, or for protective purposes, as in the case of nickel. In either case, the object to be plated is immersed in a solution containing the ions of the metal to be used in plating. Then a bar of the free metal to be used for plating is also immersed in the solution. An electric current is passed through the solution through a wire connecting the object and the pure metal (Figure 17.5). The direction of the current flow makes the object to be plated the cathode and the metal bar the anode. When plating an object with silver, the silver at the anode is oxizided and enters solution as Ag^+.

anode: $Ag \longrightarrow Ag^+ + e^-$

The Ag^+ in solution becomes deposited at the cathode as silver metal and the object is coated with silver.

cathode: $e^- + Ag^+ \longrightarrow Ag$

Summary

Oxidation is the loss of electrons by a substance. Reduction is the gain of electrons by a substance. The substance reduced is also called the oxidizing agent. The substance oxidized is called the reducing agent.

Oxidation–reduction reactions may be balanced by using oxidation numbers or half-reactions. A half-reaction is a reaction that cannot occur alone; it may be an oxidation half-reaction or a reduction half-reaction.

Galvanic cells produce electricity from redox reactions that are arranged to occur as separate half-reactions. Oxidation occurs at the anode while reduction occurs at the cathode.

The standard electrode potential of a metal is the voltage of a galvanic cell in which the metal is compared to a hydrogen electrode. It is positive if the metal ion is reduced to the metal by hydrogen gas. It is negative if the metal is oxidized by the hydrogen ion. Standard electrode potentials and their reverse may be added to obtain the voltage of a galvanic cell. Positive potentials are obtained for spontaneous reactions.

Electrolysis is the application of an external voltage to cause a nonspontaneous reaction to occur. Electrolysis may be used to reduce metal ions and to purify metals and electroplate metals.

New Terms

The **anode** is the electrode at which oxidation occurs.
The **cathode** is the electrode at which reduction occurs.
Electrodes are bars in a cell where oxidation and reduction reactions occur.
Electrolysis is a nonspontaneous reaction that occurs with the application of power in a cell.
A **galvanic cell** is a device that produces electric current from a spontaneous redox reaction.
A **half-reaction** is an oxidation or reduction reaction that cannot occur alone.

Oxidation is the loss of electrons by a substance.
An **oxidizing agent** is a substance that gains electrons and is reduced.
Redox reaction is a common short term used for oxidation–reduction reaction.
A **reducing agent** is a substance that loses electrons and is oxidized.
Reduction is the gain of electrons by a substance.
The **standard electrode potential** is the voltage of a half-reaction compared to the hydrogen electrode.

Exercises

Definitions

17.1 Define each of the following terms.:
(a) oxidation   (b) reduction
(c) oxidizing agent   (d) reducing agent

17.3 Why is the standard electrode potential of the hydrogen electrode zero?

17.2 Define each of the following terms.
(a) anode   (b) cathode
(c) half-reaction   (d) standard electrode potential
(e) galvanic cell   (f) electrolysis

17.4 What occurs at the anode and cathode in a galvanic cell?

Oxidation Numbers

17.5 What is the oxidation number of the indicated element in each of the following substances?
(a) H in $HClO_4$
(b) O in $HClO_4$
(c) Cl in $HClO_4$
(d) S in SO_2
(e) S in H_2S
(f) S in H_2SO_4

17.7 What is the oxidation number of the indicated element in each of the following ions?
(a) Fe in $FeCl_2^+$
(b) Cr in $CrCl_4^-$
(c) S in HSO_4^-
(d) P in PO_4^{3-}
(e) I in IO_3^-
(f) Al in $Al(OH)_4^-$

17.6 What is the oxidation number of the indicated element in each of the following substances?
(a) N in NO_2
(b) N in HNO_3
(c) Mn in $KMnO_4$
(d) As in H_3AsO_4
(e) Cr in Na_2CrO_4
(f) Mn in MnO_2

17.8 What is the oxidation number of the indicated element in each of the following ions?
(a) S in SO_4^{2-}
(b) S in SO_3^{2-}
(c) Br in BrO^-
(d) Mn in MnO_4^-
(e) C in CO_3^{2-}
(f) Cr in CrO_4^{2-}

Oxidation and Reduction

17.9 Determine which substance is oxidized and which substance is reduced in each of the following equations.
(a) $2\,Al + 3\,CoCl_2 \longrightarrow 2\,AlCl_3 + 3\,Co$
(b) $Cu + Br_2 \longrightarrow CuBr_2$
(c) $3\,H_2SO_3 + 2\,HNO_3 \longrightarrow 2\,NO + H_2O + 3\,H_2SO_4$
(d) $2\,S + 3\,O_2 \longrightarrow 2\,SO_3$

17.11 Determine which substance is oxidized and which substance is reduced in each of the following equations.
(a) $2\,K + Br_2 \longrightarrow 2\,KBr$
(b) $Fe_2O_3 + 2\,Al \longrightarrow Al_2O_3 + 2\,Fe$
(c) $H_2O + CO \longrightarrow H_2 + CO_2$
(d) $2\,ZnS + 3\,O_2 \longrightarrow 2\,ZnO + 2\,SO_2$

17.10 Determine which substance is oxidized and which substance is reduced in each of the following equations.
(a) $N_2 + 3\,H_2 \longrightarrow 2\,NH_3$
(b) $Cl_2 + 2\,NaBr \longrightarrow 2\,NaCl + Br_2$
(c) $WO_3 + 3\,H_2 \longrightarrow W + 3\,H_2O$
(d) $SnO_2 + 2\,C \longrightarrow Sn + 2\,CO$

17.12 Determine which substance is oxidized and which substance is reduced in each of the following equations.
(a) $TiCl_4 + 2\,Mg \longrightarrow 2\,MgCl_2 + Ti$
(b) $SiO_2 + 3\,C \longrightarrow SiC + 2\,CO$
(c) $B_2O_3 + 3\,Mg \longrightarrow 3\,MgO + 2\,B$
(d) $3\,Ca + N_2 \longrightarrow Ca_3N_2$

Oxidizing Agents and Reducing Agents

17.13 Identify the oxidizing agent and reducing agent in each of the following equations.
(a) $HgS + O_2 \longrightarrow Hg + SO_2$
(b) $SnO_2 + 2\,C \longrightarrow Sn + 2\,CO$
(c) $WO_3 + 3\,H_2 \longrightarrow W + 3\,H_2O$
(d) $2\,NiS + 3\,O_2 \longrightarrow 2\,NiO + 2\,SO_2$

17.15 Identify the oxidizing agent and reducing agent in each of the following equations.
(a) $Sb_2O_3 + 3\,Fe \longrightarrow 2\,Sb + 3\,FeO$
(b) $Sb_2O_4 + 4\,C \longrightarrow 4\,CO + 2\,Sb$
(c) $XeF_4 + 2\,H_2O \longrightarrow Xe + 4\,HF + O_2$
(d) $Cl_2 + F_2 \longrightarrow 2\,ClF$

17.14 Identify the oxidizing agent and reducing agent in each of the following equations.
(a) $N_2 + 3\,H_2 \longrightarrow 2\,NH_3$
(b) $Br_2 + 2\,NaI \longrightarrow 2\,NaBr + I_2$
(c) $Fe_2O_3 + 3\,CO \longrightarrow 2\,Fe + 3\,CO_2$
(d) $TiO_2 + 2\,Cl_2 + 2\,C \longrightarrow TiCl_4 + 2\,CO$

17.16 Identify the oxidizing agent and reducing agent in each of the following equations.
(a) $NaBrO_3 + F_2 + H_2O \longrightarrow NaBrO_4 + 2\,HF$
(b) $Te + 4\,HNO_3 \longrightarrow TeO_2 + 2\,H_2O + 4\,NO_2$
(c) $As_4 + 6\,Cl_2 \longrightarrow 4\,AsCl_3$
(d) $CS_2 + 3\,Cl_2 \longrightarrow CCl_4 + S_2Cl_2$

Balancing Equations by Oxidation Numbers

17.17 Balance each of the following reactions using the oxidation number method.
(a) $S + O_2 \longrightarrow SO_3$
(b) $Fe + Cl_2 \longrightarrow FeCl_3$
(c) $ZnS + O_2 \longrightarrow ZnO + SO_2$
(d) $CuO + NH_3 \longrightarrow Cu + N_2 + H_2O$

17.19 Balance each of the following reactions using the oxidation number method.
(a) $PbO_2 + Sb + NaOH \longrightarrow PbO + NaSbO_2 + H_2O$
(b) $MnO_2 + HBr \longrightarrow Br_2 + MnBr_2 + H_2O$
(c) $SnCl_2 + PbCl_4 \longrightarrow SnCl_4 + PbCl_2$
(d) $H_2SO_4 + HBr \longrightarrow SO_2 + Br_2 + H_2O$

17.18 Balance each of the following reactions using the oxidation number method.
(a) $H_2S + Br_2 + H_2O \longrightarrow H_2SO_4 + HBr$
(b) $HNO_3 + I_2 \longrightarrow NO_2 + H_2O + HIO_3$
(c) $NF_3 + AlCl_3 \longrightarrow N_2 + Cl_2 + AlF_3$
(d) $As_4O_6 + Cl_2 + H_2O \longrightarrow H_3AsO_4 + HCl$

17.20 Balance each of the following reactions using the oxidation number method.
(a) $CO + NO \longrightarrow CO_2 + N_2$
(b) $Fe(OH)_2 + O_2 + H_2O \longrightarrow Fe(OH)_3$
(c) $IF_5 + Fe \longrightarrow FeF_3 + IF_3$
(d) $K + KNO_3 \longrightarrow N_2 + K_2O$

17.21 Balance each of the following reactions using the oxidation number method.
(a) $NaIO_4 + NaI + HCl \longrightarrow NaCl + I_2 + H_2O$
(b) $KClO_3 \longrightarrow KCl + O_2$
(c) $Cl_2 + KOH \longrightarrow KClO_3 + KCl + H_2O$
(d) $KI + H_2SO_4 \longrightarrow K_2SO_4 + I_2 + H_2S + H_2O$

17.22 Balance each of the following reactions using the oxidation number method.
(a) $Cu + HNO_3 \longrightarrow Cu(NO_3)_2 + NO + H_2O$
(b) $NaOCl + Na_3AsO_3 \longrightarrow NaCl + Na_3AsO_4$
(c) $H_2O + SO_2 + I_2 \longrightarrow H_2SO_4 + HI$
(d) $C + HNO_3 \longrightarrow NO_2 + H_2O + CO_2$

Balancing Equations Using Half-Reactions

17.23 Balance each of the following using the half-reaction method.
(a) $Fe^{3+} + Sn^{2+} \longrightarrow Fe^{2+} + Sn^{4+}$
(b) $Cl_2 + I^- \longrightarrow I_2 + Cl^-$
(c) $Cu + H^+ + NO_3^- \longrightarrow Cu^{2+} + NO + H_2O$
(d) $Cr_2O_7^{2-} + Fe^{2+} + H^+ \longrightarrow Cr^{3+} + Fe^{3+} + H_2O$

17.24 Balance each of the following using the half-reaction method.
(a) $I^- + MnO_4^- + H^+ \longrightarrow Mn^{2+} + I_2 + H_2O$
(b) $H^+ + Cr_2O_7^{2-} + I_2 \longrightarrow Cr^{3+} + IO_3^- + H_2O$
(c) $ClO_3^- + H^+ + Fe^{2+} \longrightarrow Fe^{3+} + Cl^- + H_2O$
(d) $I^- + MnO_2 + H^+ \longrightarrow I_2 + Mn^{2+} + H_2O$

17.25 Balance each of the following using the half-reaction method.
(a) $I^- + HNO_2 + H^+ \longrightarrow I_2 + NO + H_2O$
(b) $Cu + NO_3^- + H^+ \longrightarrow Cu^{2+} + NO_2 + H_2O$
(c) $HNO_2 + Cr_2O_7^{2-} + H^+ \longrightarrow Cr^{3+} + NO_3^- + H_2O$
(d) $SnO_2^{2-} + Bi^{3+} + OH^- \longrightarrow SnO_3^{2-} + Bi + H_2O$

17.26 Balance each of the following using the half-reaction method.
(a) $SO_3^{2-} + CrO_4^{2-} + H_2O \longrightarrow SO_4^{2-} + CrO_2^- + OH^-$
(b) $HSnO_2^- + CrO_4^{2-} + H_2O \longrightarrow HSnO_3^- + CrO_2^- + OH^-$
(c) $I^- + MnO_4^- + OH^- \longrightarrow IO_4^- + MnO_4^{2-} + H_2O$
(d) $NO_2^- + MnO_4^- + H_2O \longrightarrow NO_3^- + MnO_2 + OH^-$

17.27 Balance each of the following using the half-reaction method.
(a) $Zn + MnO_4^- + H_2O \longrightarrow Zn(OH)_2 + MnO_2 + OH^-$
(b) $ClO_2^- + MnO_4^- + H_2O \longrightarrow ClO_4^- + MnO_2 + OH^-$
(c) $I_2 + ClO_3^- + H_2O \longrightarrow IO_3^- + Cl^- + H^+$
(d) $Br_2 + SO_2 + H_2O \longrightarrow H^+ + Br^- + SO_4^{2-}$

17.28 Balance each of the following using the half-reaction method.
(a) $BiO_3^- + Mn^{2+} + H^+ \longrightarrow MnO_4^- + Bi^{3+} + H_2O$
(b) $MnO_4^- + H_2S + H^+ \longrightarrow Mn^{2+} + S + H_2O$
(c) $Cu + H^+ + NO_3^- \longrightarrow Cu^{2+} + NO + H_2O$
(d) $MnO_4^{2-} + H_2O \longrightarrow MnO_4^- + OH^- + MnO_2$

Standard Electrode Potentials

17.29 Using the list of standard electrode potentials, indicate which of each of the following has the greater tendency to be reduced.
(a) Zn^{2+} or Cu^{2+} (b) Mg^{2+} or Zn^{2+}
(c) Cd^{2+} or Hg^{2+} (d) Ag^+ or Hg^{2+}
(e) Br_2 or I_2 (f) Pb^{2+} or Sn^{2+}

17.30 Using the list of standard electrode potentials, indicate which of each of the following has the greater tendency to be reduced.
(a) Mn^{2+} or Cr^{3+} (b) Mg^{2+} or Al^{3+}
(c) Fe^{2+} or Hg^{2+} (d) Ag^+ or Mg^{2+}
(e) Cl_2 or I_2 (f) Ni^{2+} or Sn^{2+}

17.31 Using the list of standard electrode potentials, indicate which of each of the following has the greater tendency to be oxidized.
(a) Li or K   (b) Na or Zn
(c) Zn or Cu   (d) Ni or Pb
(e) Cu or Hg   (f) I^- or Br^-

17.32 Using the list of standard electrode potentials, indicate which of each of the following has the greater tendency to be oxidized.
(a) Ba or K   (b) Ca or Ba
(c) Mn or Fe   (d) Sn or Cr
(e) Ag or Cu   (f) Cl^- or Br^-

Galvanic Cells

17.33 Using the list of standard electrode potentials, predict which of the following reactions will proceed as written.
(a) $Ni + 2 Ag^+ \longrightarrow Ni^{2+} + 2 Ag$
(b) $3 Ba + 2 Al^{3+} \longrightarrow 3 Ba^{2+} + 2 Al$
(c) $Pb + Hg^{2+} \longrightarrow Pb^{2+} + Hg$
(d) $Zn^{2+} + Fe \longrightarrow Zn + Fe^{2+}$

17.34 Using the list of standard electrode potentials, predict which of the following reactions will proceed as written.
(a) $Cd + 2 Ag^+ \longrightarrow Cd^{2+} + 2 Ag$
(b) $3 Ba + 2 Cr^{3+} \longrightarrow 3 Ba^{2+} + 2 Cr$
(c) $Pb + Cu^{2+} \longrightarrow Pb^{2+} + Cu$
(d) $Mn^{2+} + Fe \longrightarrow Mn + Fe^{2+}$

17.35 Calculate the voltage that can be produced for each of the following reactions.
(a) $Mg + Zn^{2+} \longrightarrow Mg^{2+} + Zn$
(b) $Mg + Fe^{2+} \longrightarrow Mg^{2+} + Fe$
(c) $Zn + Pb^{2+} \longrightarrow Zn^{2+} + Pb$
(d) $Cu + 2 Ag^+ \longrightarrow Cu^{2+} + 2 Ag$

17.36 Calculate the voltage that can be produced for each of the following reactions.
(a) $Br_2 + 2 I^- \longrightarrow 2 Br^- + I_2$
(b) $Mg + Pb^{2+} \longrightarrow Mg^{2+} + Pb$
(c) $Mg + Cd^{2+} \longrightarrow Mg^{2+} + Cd$
(d) $Zn + Sn^{2+} \longrightarrow Zn^{2+} + Sn$

Electrolysis Cell

17.37 In your own words, describe how metals may be purified by electrolysis.

17.39 What is the difference between a galvanic cell and an electrolysis cell?

17.38 Describe the use of an electrolysis cell in electroplating.

17.40 Copper and silver are two examples of metals that can be used in electroplating. Magnesium and aluminum cannot be used in electroplating. What reason can you suggest for this difference?

Additional Exercises

17.41 What is the oxidation number of the indicated element in each of the following substances?
(a) V in Ca_2VO_4 (b) Ta in Na_3TaF_8
(c) Ta in $TaOCl_3$

17.43 What is the oxidation number of the indicated element in each of the following substances?
(a) W in $K_2W_4O_{13}$ (b) Zr in $K_2Zr_2O_5$
(c) P in $Na_3P_3O_9$

17.45 Determine which substance is oxidized and which substance is reduced in each of the following equations.
(a) $2\,ReCl_5 + SbCl_3 \longrightarrow 2\,ReCl_4 + SbCl_5$
(b) $3\,IF_5 + 2\,Fe \longrightarrow 3\,IF_3 + 2\,FeF_3$
(c) $I_2O_5 + 5\,CO \longrightarrow I_2 + 5\,CO_2$

17.47 Determine which substance is the oxidizing agent and which substance is the reducing agent in each of the following equations.
(a) $2\,KMnO_4 + 16\,HCl \longrightarrow 2\,MnCl_2 + 5\,Cl_2 + 2\,KCl + 8\,H_2O$
(b) $2\,KMnO_4 + KI + H_2O \longrightarrow 2\,MnO_2 + KIO_3 + 2\,KOH$
(c) $3\,Cu + 8\,HNO_3 \longrightarrow 3\,Cu(NO_3)_2 + 2\,NO + 4\,H_2O$

17.49 Balance each of the following by the oxidation number method.
(a) $ReCl_5 + SbCl_3 \longrightarrow ReCl_4 + SbCl_5$
(b) $K_2S_2O_3 + I_2 \longrightarrow K_2S_4O_6 + KI$
(c) $Cu + HNO_3 \longrightarrow Cu(NO_3)_2 + NO + H_2O$

17.51 Balance each of the following by the half-reaction method.
(a) $NO_3^- + I_2 + H^+ \longrightarrow IO_3^- + NO_2 + H_2O$
(b) $Br_2 + SO_2 + H_2O \longrightarrow H^+ + Br^- + SO_4^{2-}$
(c) $MnO_4^{2-} + Se^{2-} + H_2O \longrightarrow MnO_2 + Se + OH^-$

17.53 Using the list of standard electrode potentials, indicate which of each of the following has the greater tendency to be reduced.
(a) Cd^{2+} or Cu^{2+} (b) Mg^{2+} or Hg^{2+}
(c) Zn^{2+} or Hg^{2+} (d) Ag^+ or Zn^{2+}
(e) Br_2 or Cl_2 (f) Pb^{2+} or Cd^{2+}

17.55 Using the list of standard electrode potentials, indicate which of each of the following has the greater tendency to be oxidized.
(a) Na or K   (b) Ni or Cd
(c) Hg or Cu   (d) K or Pb
(e) Zn or Hg   (f) Cl^- or Br^-

17.42 What is the oxidation number of the indicated element in each of the following substances?
(a) U in Mg_3UO_6 (b) Sn in Na_2SnO_3
(c) Nb in Li_3NbOF_6

17.44 What is the oxidation number of the indicated element in each of the following substances?
(a) U in $Na_2U_2O_7$ (b) V in $Na_6V_{10}O_{28}$
(c) Mo in $Na_4Mo_4Cl_8$

17.46 Determine which substance is oxidized and which substance is reduced in each of the following equations.
(a) $OF_2 + H_2O \longrightarrow O_2 + 2\,HF$
(b) $2\,K_2S_2O_3 + I_2 \longrightarrow K_2S_4O_6 + 2\,KI$
(c) $3\,S + 4\,HNO_3 \longrightarrow 3\,SO_2 + 4\,NO + 2\,H_2O$

17.48 Determine which substance is the oxidizing agent and which substance is the reducing agent in each of the following equations.
(a) $8\,NaI + 4\,H_2SO_4 \longrightarrow Na_2S + 3\,Na_2SO_4 + 4\,H_2O + 4\,I_2$
(b) $2\,NaBr + H_2SO_4 \longrightarrow Na_2SO_3 + H_2O + Br_2$
(c) $2\,Cr(OH)_3 + 3\,NaOCl + 4\,NaOH \longrightarrow 2\,Na_2CrO_4 + 3\,NaCl + 5\,H_2O$

17.50 Balance each of the following by the oxidation number method.
(a) $H_2SO_3 + HNO_3 \longrightarrow NO + H_2O + H_2SO_4$
(b) $TiO_2 + Cl_2 + C \longrightarrow TiCl_4 + CO$
(c) $XeF_4 + H_2O \longrightarrow Xe + HF + O_2$

17.52 Balance each of the following by the half-reaction method.
(a) $Cu + H^+ + NO_3^- \longrightarrow Cu^{2+} + NO_2 + H_2$
(b) $SO_3^{2-} + MnO_4^- + H^+ \longrightarrow SO_4^{2-} + Mn^{2+} + H_2O$
(c) $Cr_2O_7^{2-} + H^+ + Fe^{2+} \longrightarrow Cr^{3+} + Fe^{3+} + H_2O$

17.54 Using the list of standard electrode potentials, indicate which of each of the following has the greater tendency to be reduced.
(a) Fe^{2+} or Cr^{3+} (b) Hg^{2+} or Mg^{2+}
(c) Mn^{2+} or Hg^{2+} (d) Ag^+ or Al^{3+}
(e) Br_2 or I_2 (f) Ni^{2+} or Zn^{2+}

17.56 Using the list of standard electrode potentials, indicate which of each of the following has the greater tendency to be oxidized.
(a) Mg or Na   (b) K or Ba
(c) Mn or Cu   (d) Cd or Cr
(e) Ag or Fe   (f) I^- or Br^-

17.57 Using the list of standard electrode potentials, predict which of the following reactions will proceed as written.

(a) $Ni + Sn^{2+} \longrightarrow Ni^{2+} + Sn$

(b) $3 Sr + 2 Al^{3+} \longrightarrow 3 Sr^{2+} + 2 Al$

(c) $Mg + Hg^{2+} \longrightarrow Mg^{2+} + Hg$

(d) $Zn^{2+} + Cd \longrightarrow Zn + Cd^{2+}$

17.59 Calculate the voltage that can be produced for each of the following reactions.

(a) $Ni + 2 Ag^+ \longrightarrow Ni^{2+} + 2 Ag$

(b) $3 Ba + 2 Al^{3+} \longrightarrow 3 Ba^{2+} + 2 Al$

(c) $Zn + Fe^{2+} \longrightarrow Fe + Zn^{2+}$

(d) $Cd + 2 Ag^+ \longrightarrow Cd^{2+} + 2 Ag$

17.58 Using the list of standard electrode potentials, predict which of the following reactions will proceed as written.

(a) $Mn + 2 Ag^+ \longrightarrow Mn^{2+} + 2 Ag$

(b) $3 Mg + 2 Al^{3+} \longrightarrow 3 Mg^{2+} + 2 Al$

(c) $Cu + Hg^{2+} \longrightarrow Cu^{2+} + Hg$

(d) $Zn^{2+} + Ni \longrightarrow Zn + Ni^{2+}$

17.60 Calculate the voltage that can be produced for each of the following reactions.

(a) $Mn + Zn^{2+} \longrightarrow Mn^{2+} + Zn$

(b) $Mg + Pb^{2+} \longrightarrow Mg^{2+} + Pb$

(c) $Zn + Co^{2+} \longrightarrow Zn^{2+} + Co$

(d) $Cu + 2 Au^+ \longrightarrow Cu^{2+} + 2 Au$

18 Nuclear Chemistry

Learning Objectives

As a result of studying Chapter 18 you should be able to

1. Distinguish among the types of nuclear radiation.

2. Describe the units of measurement of radiation.

3. Balance nuclear equations.

4. Calculate the fraction of a radioisotope remaining in a reaction given its half-life and use both to calculate the age of an object.

5. Describe medical uses of radioisotopes.

6. Describe the methods used in nuclear transmutations.

7. Compare nuclear fusion and nuclear fission.

18.1 Nuclear Chemistry—A Perspective

Up to this point, the elements have been discussed in terms of the number and arrangement of electrons about the nucleus of an atom. In ordinary chemical reactions, changes in the arrangement of the electrons occur in the reactants to yield products. The energy changes in these chemical reactions are in the order of kilocalories.

Now we will examine reactions that involve the protons and neutrons in the nucleus of the atom. The energy required to cause a stable nucleus to react is extremely high compared to the energy required for ordinary chemical reactions. Thus, when unstable nuclei of radioactive isotopes react in high concentrations, a large amount of energy is released and a rapid chain reaction or even an atomic explosion may occur.

Today's generation of college students has grown up in a nuclear age in which the chemistry of the nucleus is part of modern life. On one hand, we live in fear of nuclear holocaust; on the other, we enjoy the many benefits of radioisotopes.

The atomic bombs that were used to devastate two Japanese cities and end World War II were small compared to the hydrogen bombs of today. This aspect of nuclear chemistry continues as the major political powers try to limit and control their respective arms and to prevent the further spread of nuclear devices to smaller countries.

Nuclear energy for peaceful uses, such as the generation of power, is a continuing and controversial story. The world is limited in its supply of fossil fuels, and nuclear energy is currently being used as an alternative for the generation of electricity. However, there is much concern about the radioactivity hazards and environmental side effects of nuclear power plants.

One major benefit from nuclear chemistry is the development of radioisotopes for use in medicine. In this chapter the principles of nuclear chemistry are presented and some practical aspects of nuclear processes are outlined.

18.2 Radioactivity

Recall from Chapter 4 that isotopes of an element have the same number of protons but different numbers of neutrons. In the general formula $^A_Z E$, Z is the atomic number or number of protons, and A is the mass number or the sum of the numbers of protons and neutrons. Thus, $^{24}_{11}Na$ represents an isotope of sodium that contains 11 protons and 13 neutrons. In this chapter we will also use the symbols ^{24}Na, Na-24, or sodium-24. The number following the elemental symbol or element name is the mass number of the isotope.

Some elements exist in nature as only one or two isotopes, whereas others have many more isotopes. Fluorine exists only as fluorine-19, whereas tin exists in ten isotopic forms. As a result of the existence of multiple isotopes of the elements, there are over 300 naturally occurring isotopes. Table 18.1 gives a few of the elements and their isotopes.

Some naturally occurring isotopes contain nuclei that are stable for indefinite periods of time, whereas other isotopes undergo nuclear decay reactions. The reactant nucleus or parent nucleus decays to form another nucleus known as the **daughter nucleus.** In this process some high-energy particles collectively known as radiation are formed. The daughter nucleus formed is more stable than the original nucleus. However, the daughter nucleus itself may undergo further decay reactions. Thus, a whole series of decay processes may occur before a very stable nucleus is ultimately formed.

Table 18.1   Naturally Occurring Isotopes of Some Selected Elements

Element	Isotope	Percent abundance	Element	Isotope	Percent abundance
neon	$^{20}_{10}Ne$	90.92	selenium	$^{74}_{34}Se$	0.87
	$^{21}_{10}Ne$	0.26		$^{76}_{34}Se$	9.02
	$^{22}_{10}Ne$	8.82		$^{77}_{34}Se$	7.58
silicon	$^{28}_{14}Si$	92.21		$^{78}_{34}Se$	25.52
	$^{29}_{14}Si$	4.70		$^{80}_{34}Se$	49.82
	$^{30}_{14}Si$	3.09		$^{82}_{34}Se$	9.19
potassium	$^{39}_{19}K$	93.10	tin	$^{112}_{50}Sn$	0.96
	$^{40}_{19}K$	0.02		$^{114}_{50}Sn$	0.66
	$^{41}_{19}K$	6.89		$^{115}_{50}Sn$	0.35
zinc	$^{64}_{30}Zn$	48.89		$^{116}_{50}Sn$	14.30
	$^{66}_{30}Zn$	27.81		$^{117}_{50}Sn$	7.61
	$^{67}_{30}Zn$	4.11		$^{118}_{50}Sn$	24.03
	$^{68}_{30}Zn$	18.57		$^{119}_{50}Sn$	8.58
	$^{70}_{30}Zn$	0.62		$^{120}_{50}Sn$	32.85
				$^{122}_{50}Sn$	4.72
				$^{124}_{50}Sn$	5.94

Isotopes that undergo nuclear change and produce radiation are called **radioisotopes.** Some radioisotopes occur in nature and, in addition, many radioisotopes have been produced by nuclear reactions in nuclear laboratories. Regardless of the source of radioactive substances, radiation can cause extensive biological damage. However, under controlled conditions radioisotopes are used for medical diagnosis and therapy.

The three common types of nuclear radiation emitted from naturally radioactive isotopes are alpha (α), beta (β), and gamma (γ). A radioactive isotope does not emit all three types of radiation in a single process. However, gamma radiation is produced simultaneously with either alpha or beta radiation. A single sample of radioactive material may emit all three types of radiation if a succession of several radioactive processes is occurring.

Alpha Particles

An **alpha particle** is a helium nucleus, which consists of two protons and two neutrons, and may be represented as $^4_2He^{2+}$. However, the +2 charge is ordinarily omitted in writing balanced nuclear equations because the nuclear equation deals only with the content of the nucleus.

When alpha particles are emitted from a nucleus of an atom, they travel at approximately one tenth the speed of light, or 18,600 miles per second. These particles have little penetrating power and can be stopped by this sheet of paper (Figure 18.1). A 0.05 mm layer of dead cells on the surface of skin will stop alpha particles. However, an intense dose of alpha radiation can cause skin burns. If radioactive dust particles get inside the body, the alpha particles emitted can affect cells and biological damage will occur.

Beta Particles

Beta particles are high-energy electrons produced within a radioactive nucleus. Although the electrons do not exist as such in the nucleus, they are produced during a process in which a neutron is transformed into a proton and a beta particle.

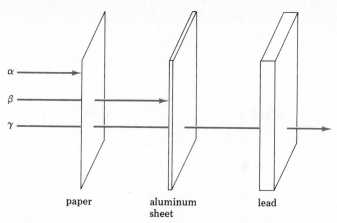

Figure 18.1

An illustration of the penetration power of radiation

$$\text{neutron} \longrightarrow \text{proton} + \text{electron}$$
$$^1_0\text{n} \longrightarrow ^1_1\text{H} + ^{\ 0}_{-1}\text{e}$$

A beta particle is written $^{\ 0}_{-1}\text{e}$ in nuclear equations. The superscript corresponds to the position of the atomic mass in elements. Since the mass of an electron is very much smaller than that of any element, it is assigned a value of zero. The -1 subscript occupies the position of the atomic number of an element. Since the atomic number is equal to the charge of the nucleus, the -1 value for the electron emphasizes its charge.

Beta particles travel at $\frac{9}{10}$ the speed of light, or about 165,000 miles per second. Although beta particles have greater penetrating power than alpha particles, they are stopped by thin sheets of metal (Figure 18.1). Beta particles can penetrate about 4 mm into living tissue. Radiation burns result, but the vital internal organs are not affected unless a radioactive source is ingested.

Gamma Rays

Gamma rays are not particles but are high-energy radiation similar to x-rays. When gamma rays are emitted by a nucleus, they travel at the speed of light and have great penetration power. Even 50 cm of tissue will reduce the intensity of gamma radiation by only 10%. A 5 cm sheet of lead will not stop all gamma rays (Figure 18.1).

18.3 Balancing Nuclear Equations

Unlike ordinary chemical reactions, nuclear reactions lead to changes in the composition of the nucleus. Thus, both atomic numbers and mass numbers must be included with the elemental symbol in a balanced nuclear equation. In a balanced nuclear equation, the sum of the mass numbers of the reactants must equal the sum of the mass numbers of the products. Furthermore, the sum of the atomic numbers of the reactants and the sum of the atomic numbers of the products must also be equal.

The proper symbols for alpha and beta particles are also included in a balanced nuclear equation. Gamma rays do not have any mass or charge and therefore are not used to balance the equation.

Several types of problems involving balancing nuclear equations follow. The only tools necessary for balancing are arithmetic and a list of elements and their atomic numbers.

Example 18.1

The isotope polonium-212 undergoes alpha decay to yield one alpha particle per atom and a single element. What is the element?

Solution First, write the symbol for polonium-212 to the left of an arrow using the information from a list of elements that its atomic number is 84. Place ^4_2He on the right of the arrow along with ^A_ZE to represent the unknown element.

$$^{212}_{84}\text{Po} \longrightarrow {}^4_2\text{He} + {}^A_Z\text{E}$$

The sum of the atomic numbers of the products must equal that of the single reactant, polonium.

$$84 = 2 + Z$$
$$82 = Z$$

Similarly, the sum of the mass numbers of the products must equal the mass of polonium.

$$212 = 4 + A$$
$$208 = A$$

Now it is established that the unknown element has an atomic number of 82 and a mass number of 208. In the list of elements one finds that lead has an atomic number of 82; therefore, the element is lead-208.

$$^A_Z\text{E} = {}^{208}_{82}\text{Pb}$$

The balanced equation is

$$^{212}_{84}\text{Po} \longrightarrow {}^4_2\text{He} + {}^{208}_{82}\text{Pb}$$

[*Additional examples may be found in 18.13 and 18.14 at the end of the chapter.*]

Example 18.2

The isotope carbon-14 is unstable and emits one beta particle per atom. What single element is produced?

Solution First write the unbalanced equation.

$$^{14}_{6}\text{C} \longrightarrow {}^{0}_{-1}\text{e} + {}^A_Z\text{E}$$

The atomic number must be given by the equation

$$6 = -1 + Z$$
$$7 = Z$$

The mass number of the element is equal to the mass number of carbon, as the beta particle has a negligible mass.

$$14 = 0 + A$$
$$14 = A$$

The element is nitrogen, as its atomic number is 7.

$$^{14}_{7}E = {}^{14}_{7}N$$

The balanced equation is

$$^{14}_{6}C \longrightarrow {}^{14}_{7}N + {}^{0}_{-1}e$$

[Additional examples are given in 18.11 and 18.12 at the end of the chapter.]

18.4 Half-lives

All radioactive decay reactions occur at rates characteristic for the specific parent nucleus. The rate is described by the time necessary for the parent to decay to give a daughter nucleus. The time required for one-half of a given amount of radioisotope to decay is its **half-life.**

The half-life of uranium-238 is 4.5×10^9 years! This means that a 100 g sample of uranium-238 will require 4.5×10^9 years to be decreased to 50 g of the isotope. The mass that is "missing" is in the decay products. In another 4.5×10^9 years, the 50 g of the isotope will have decreased to one-half of that value or 25 g. A list of half-life values is given in Table 18.2.

A graph that illustrates the decay pattern of radioactive materials is given in Figure 18.2. The isotope strontium-90 has a half-life of 28.1 years. This isotope is one of many produced in an atomic bomb explosion. Nuclear weapons tests in the atmosphere in the 1950s produced strontium-90. As shown in Figure 18.2, for 80 mg of the isotope produced, the decay process leaves 40 mg after 28.1 years. In another 28.1 years 20 mg remains. Thus, in the year 2006, there will still be 25% of the isotope from a 1950 test.

Table 18.2 **Half-life Period of Selected Naturally Occurring Isotopes**

Element	Isotope	Half-life (yr)
potassium	$^{40}_{19}K$	1.3×10^9
vanadium	$^{50}_{23}V$	6.0×10^{14}
samarium	$^{147}_{62}Sm$	1.1×10^{11}
samarium	$^{148}_{62}Sm$	1.2×10^{13}
samarium	$^{149}_{62}Sm$	4.0×10^{14}
lead	$^{204}_{82}Pb$	1.4×10^{19}
radium	$^{226}_{88}Ra$	1.6×10^{19}

Time (yr)	Amount (mg)
0	80
28.1	40
56.2	20
84.3	10
112.4	5

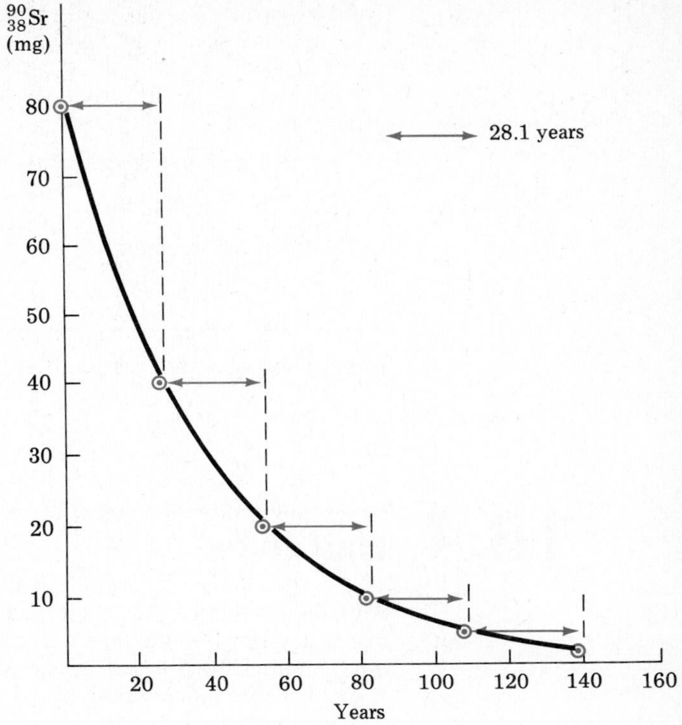

Figure 18.2

The half-life of strontium-90

Example 18.3

Sodium-24 is used in medicine to determine the effectiveness of blood circulation. Its half-life is 15 hours. What percent of the radioisotope will remain after 60 hours?

Solution After each half-life, the amount remaining will be one half of the amount remaining after the previous half-life. A table can be prepared to show this behavior.

Time (h)	Amount (%)
0	100
15	50
30	25
45	12.5
60	6.25

[*Additional examples are given in 18.17–18.24 at the end of the chapter.*]

One of the best-known applications of the radioisotope half-life is the carbon-14 dating method. The age of ancient materials made of plant matter can be established on the basis of their $^{14}_{6}C$ content. Carbon dioxide in the atmosphere consists of carbon-12 with trace amounts of carbon-14, which is radioactive and decays. However, the concentration of carbon-14 does not decrease because it is constantly being formed in the atmosphere from the action of cosmic rays on nitrogen, $^{14}_{7}N$. All plants absorb carbon dioxide from the atmosphere; as long as the plant is living, the amount of carbon-14 incorporated into the molecules it produces and uses will be a constant fraction of the amount of carbon present. When the plant dies, the carbon compounds in its cells are no longer interacting with the carbon dioxide of the atmosphere. The amount of carbon-14 in these plants therefore diminishes with time as the carbon-14 decays. Since the half-life of carbon-14 is 5730 years, it is possible to measure the age of an object by determining the amount of radioactive carbon remaining in it. A wooden dish that contains only 25% of the carbon-14 in trees living today is therefore approximately 11,000 years old. The carbon-14 dating technique cannot be used accurately for objects that are older than 50,000 years. After many half-lives have elapsed, such a small amount of carbon-14 remains that accurate measurement is not possible.

18.5 Ionizing Radiation and Its Biological Effects

As is generally known by the public, radiation is dangerous to humans and all other life forms. What is the reason for the effect that radiation has on cells?

All nuclear radiation travels with high speed. As a consequence of this high energy, chemical reactions occur when the radiation interacts with matter. Radiation that causes atoms and, more importantly, molecules to lose electrons and form ions is called **ionizing radiation.** The ionized material produced is then highly reactive in chemical reactions.

Consider the effect of radiation on the covalent bond of a hypothetical molecule A:B.

$$A:B \longrightarrow A \cdot B^+ + e-$$
$$A \cdot B^+ \longrightarrow A \cdot + B^+$$
$$A \cdot B^+ \longrightarrow A^+ + B \cdot$$

After an electron is "knocked out" of the covalent bond, the resultant cation does not have the necessary bonding electrons to remain together. Decomposition of the cation results in another cation and a species called a **radical,** which has an unpaired electron.

The disruption of the otherwise stable AB molecule by radiation produces reactive species. The subsequent reaction of cations and radicals with other molecules may result in abnormal processes that can seriously disrupt the normal functions of cells.

Radiation can affect any living cells that it reaches. Gamma radiation, with its great penetrating power, can literally plow through your entire body, and for this reason it is the most damaging type of radiation. Regardless of the type of radiation, the chemical effect on a cell is one of breaking molecules apart. The disruption of a stable organic molecule results in the formation of reactive materials that may recombine with themselves or react with neighboring molecules. Thus, chemical reactions can occur that are foreign to the cell and cause damage that may be serious. Cells are able to repair themselves up to a point, and low radiation exposure might not cause permanent biological damage. However, no exposure to radiation is ever 100% risk free.

The most serious effect of radiation is the mutation of a species. If radiation affects the nuclei of germ cells that produce eggs or sperm, then the offspring may be mutants. The seriousness of the mutation depends on the extent of the radiation.

The nucleus of any cell contains DNA (deoxyribonucleic acid) that is responsible for producing RNA (ribonucleic acid) and passing on genetic information for the formation of new cells. Excessive radiation could terminate DNA's ability to reproduce, and the death of the cell would result. If an alteration of the DNA occurred, then the RNA molecules that direct synthesis of enzymes and other proteins could be altered. Faulty enzymes could lead to serious health consequences and the death of the organism.

Tissues that reproduce at a rapid rate, such as bone marrow, lymphatic system tissue, and embryonic tissue, are the most sensitive to radiation damage. Since bone marrow is the site of red cell formation, one of the early signs of a radiation overexposure is a drop in the red cell count. A pregnant woman must avoid radiation exposure to protect the genetic health of her unborn child.

18.6 Radiation Detection

The method of detecting radiation depends on both the amount and the source of radiation. Methods used include the Geiger counter, the film badge, and the scintillation counter.

A Geiger counter consists of a metal tube containing a gas at low pressure (Figure 18.3). The gas chamber is connected to a battery. In the absence of radiation, there is no current flow through the gas chamber. When the gas is exposed to radiation, ions and electrons are produced. As a consequence, a current flows through the chamber. The higher the intensity of radiation, the stronger is the current flow. Either an audible clicking sound or swing of a needle in the Geiger counter gives a measure of the current flow proportional to the intensity of the radiation.

A film badge is used by workers to measure the amount of radiation that the body receives. The film badge consists of photographic film in a holder sealed from light. Radiation can pass through the holder, and the film becomes exposed in a manner similar to taking pictures. As the

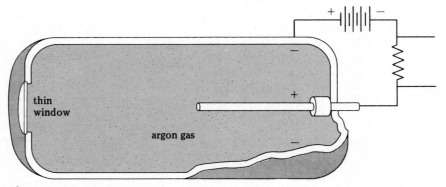

Figure 18.3

The Geiger counter
Radiation enters the Geiger tube through the thin window at the end of the tube. A voltage difference is maintained between the metal rod in the center of the tube and the lining of the tube. When ionizing radiation enters the tube, the argon atoms lose electrons and cations are formed. An electrical discharge results, and a current flows. The current is amplified and recorded as a clicking sound or on a meter.

amount of radiation increases, the developed film will be darker. Film badges are periodically processed to check the amount of exposure of the wearer to radiation.

Scintillation counters are important in detecting radiation in medical tests. A scintillation counter contains crystals of some chemicals that can produce scintillations or flashes of light when hit by radiation. An electronic device called a photomultiplier records and amplifies the scintillations. Scintillation counters are used to scan patients to measure the location and amount of radioisotopes present in the body.

18.7 Units of Radiation

Radiation has been measured quantitatively by two different methods. One method measures the intensity of the radiation emitted from the radioactive source. The other method is chemical or biochemical and measures the radiation absorbed by matter, such as air or tissue.

The Curie

The curie is a unit of radiation intensity. One **curie** (Ci) is equal to the amount of material that produces 3.7×10^{10} nuclear disintegrations per second. This number was chosen because it is the number of disintegrations that 1 g of radium undergoes in 1 s. The mass of other substances that produce 1 curie may be more or less than 1 g.

The Roentgen

The roentgen is a chemical measure of exposure to gamma radiation or x-rays. One **roentgen** (R) is the dose of radiation that produces 2.1×10^9 ions in 1 cm^2 of dry air at 0°C and 1 atm pressure. Geiger counters, which operate by detecting ionized air, are calibrated to read in milliroentgens per hour (mR/h).

The Rad

The rad, short for radiation absorbed dose, was selected to relate energy absorbed by tissue. The **rad** is defined as the absorption of 100 ergs (2.4×10^{-6} cal) of energy per gram of absorbing tissue. Although the roentgen and the rad are defined differently, they are numerically similar. One roentgen of gamma radiation causes 0.97 rad in muscle tissue. In bone, 1 roentgen delivers 0.92 rad. A dose of about 500 rads of gamma radiation would be lethal for a human.

RBE and the Rem

Even the dosage actually absorbed does not necessarily indicate the biological damage to an organ. In other words, a 10 rad dose of one type of radiation may result in different biological consequences than a 10 rad dose of another type of radiation. For this reason, factors known as dose equivalents are defined. These are factors that are multiplied by the rad dose.

The **relative biological effectiveness (RBE)** accounts for differences in biological damage caused by radiation. The standard for RBE is gamma radiation from cobalt-60. The RBE of a radiation source is the ratio of the dose of gamma radiation from cobalt-60 to the dose of the radiation in question required to cause the same biological effect. For example, 1 rad of alpha particles may have the same effect as 10 rads of gamma radiation from cobalt-60. The RBE of alpha particles is then 10. Highly charged and heavy particles such as alpha particles cause more ionization in matter than lighter and singly charged beta particles. The RBE of beta particles is about 1.

The **rem,** short for roentgen equivalent for man, is defined as the product of rads and RBE. For alpha particles, 1 rad equals 10 rem, whereas for beta particles 1 rad equals 1 rem.

$$rem = rad \times RBE$$

The radiation exposure of workers in radiation laboratories is monitored in millirem (mrem) since it accounts for the effect of radiation regardless of the source. The accumulated radiation then can be added in millirem units, and medical records are more easily maintained.

18.8 Radiation Sickness and Radiation Safety

Radiation sickness is the result of overexposure to ionizing radiation that creates ions and radicals. These particles cause cellular chemical changes that can be minor or can cause death.

Overexposure to radiation can be the result of a single large dose or many cumulative small doses. It is estimated than in 1 year the average American receives 200 mrem of radiation, half of which is from natural environmental background and cannot be decreased. The second major radiation source is x-rays used in medical diagnosis.

The exact effect of radiation can only be stated in terms of the probability of biological consequences. A listing is provided in Table 18.3.

Exposing the entire body to less than 25 rem results in no detectable clinical short-term effect. The consequences of increased exposure, 25–100 rem, are evident in the blood-forming tissue of bone marrow. These cells divide very rapidly and are very easily affected by the chemical changes caused by radiation. The medical result is a marked decrease in white blood cells, which lowers resistance to infection.

A dose of 100–200 rem results in the longer term reduction of some blood cells and will cause fatigue and nausea. A dose of 200–300 rem results in nausea and vomiting on the first day of exposure. By the third week, fever, diarrhea, loss of hair, and a 50% chance of death occur. Exposures greater than 300 rem result in the same symptoms. In a few hours, nausea, vomiting, and diarrhea result. The probability of death is about 50%. The central nervous system, blood formation system, and gastrointestinal system are all disrupted by radiation doses above 500 rem. Death results in essentially all persons receiving such doses.

Radiation exposure limits have been proposed by the National Council on Radiation Protection. Although the average American is not exposed to more than 200 mrem (0.2 rem) in a year,

Table 18.3 **Effect of Short-Term Radiation Doses**

Dose (rem)	Probable effect
<25	no detectable short-term effect
25–100	decrease in white blood cell count, which lowers resistance to infection
100–200	reduction in number of blood cells, fatigue, nausea
200–300	nausea and vomiting on first day of exposure; fever, diarrhea, loss of hair by the third week
300–500	nausea, vomiting, and diarrhea, in a few hours; probability of death about 50%
>500	vomiting, severe changes in blood and gastrointestinal system; death within 2 months in essentially 100% of the cases

the recommended maximum is 0.5 rem. Some individuals are exposed to radiation in their occupation, and a 5 rem limit per year is suggested for whole-body exposure. Those in regular contact with radiation wear protective clothing, gloves, and masks. In addition, a film badge is used to monitor the dosage, since radiation cannot be seen and you cannot tell if you have been exposed.

Shielding material can provide substantial protection from radiation. Such precautions are required for technicians who routinely use x-ray machines. Both the thickness and composition of the shielding material are important in decreasing the chances of radiation damage. As indicated earlier, alpha particles can be stopped by a sheet of paper. Beta particles can be stopped by a thin sheet of metal or about 1 cm of wood. Gamma rays require concrete or lead shielding. The thickness required depends on the sources of the gamma radiation. For example, one half of the radiation of technetium-99 will be stopped by 0.2 mm of lead, whereas iodine-131 requires 6 mm for the same protection, and 12 mm are needed to stop 75% of the radiation of iodine-131. Thus, a thick shielding material is required for good protection.

The level of exposure to radiation is related to the distance between the individual and the radiation source. The dosage decreases as the square of the distance. Thus, at 2 ft the dosage is $(\frac{1}{2})^2$ or $\frac{1}{4}$ the dosage at 1 ft. Every incremental increase in the distance from the radiation source provides increased protection. At 4 ft the dosage is $(\frac{1}{4})^2 = \frac{1}{16}$ the dosage at 1 ft. This relationship can be stated as an inverse square law with I for intensity and d for distance where the subscripts a and b refer to specific intensities at two selected distances.

$$\frac{I_a}{I_b} = \frac{d_b^2}{d_a^2}$$

Example 18.4

The dosage at 5 ft from a radiation source is 50 mrem. What is the dosage at 25 ft from the same source?

Solution The inverse square law for radiation can be used and the values substituted into the equation. Let $d_a = 5$ ft and $I_a = 50$ mrem. When $d_b = 25$ ft, the value $I_b = x$ mrem and can be calculated as follows:

$$\frac{I_a}{I_b} = \frac{d_b^2}{d_a^2}$$

$$\frac{50 \text{ mrem}}{x \text{ mrem}} = \frac{(25 \text{ ft})^2}{(5 \text{ ft})^2}$$

$$x \text{ mrem} = 50 \text{ mrem} \times \frac{25 \text{ ft}^2}{625 \text{ ft}^2} = 2 \text{ mrem}$$

[*Additional examples are given in 18.35 and 18.36 at the end of the chapter.*]

18.9 Use of Radioisotopes in Diagnosis

There are five criteria a radiologist uses in selecting a radioisotope for diagnosis. First, the radioisotope must be effective in diagnosis at as low a concentration as possible and yet still be reliably detected. Second, in order to minimize any radiation damage to the organism, the

Table 18.4   Radioisotope Half-lives and Uses in Diagnosis

Isotope	Half-life	Part of body	Use in diagnosis
barium-131	11.6 d	bone	detection of bone tumors
chromium-51	27.8 d	blood	determination of blood volume and red blood cell lifetime
		kidney	assessment of kidney activity
copper-64	12.8 h	liver	diagnosis of Wilson's disease
gold-198	64.8 h	kidney	assessment of kidney activity
iodine-125	60 d	blood	determination of hormone level in blood
iodine-131	8.05 d	brain	detection of fluid buildup in the brain
		kidney	location of cysts
		lung	location of blood clots
		thyroid	assessment of iodine uptake by thyroid
iron-59	45 d	blood	evaluation of iron metabolism in blood
krypton-79	34.5 h	blood	evaluation of cardiovascular system
mercury-197	65 h	spleen	evaluation of spleen function
		brain	brain scans
selenium-75	120 d	pancreas	determination of size and shape of pancreas
technetium-99	6.0 h	brain	detection of brain tumors, hemorrhages, or blood clots
		spleen	measurement of size and shape of spleen
		thyroid	measurement of size and shape of thyroid
		lung	location of blood clots
xenon-133	5.3 d	lung	determination of lung volume

radioisotope must have a short half-life. The third criterion is the ready elimination of the radioisotope from the body after the diagnosis is complete. The nature of the chemical species chosen and its function in the body form the fourth criterion. Ideally, the chemical will be selectively transmitted to the part of the body where diagnosis is desired. Finally, the fifth criterion is dependent on the type of radiation emitted by the radioisotope. In order to detect the radioisotope, the radiation must have sufficient penetrating power to reach the instruments placed outside the body. Both alpha and beta particles have low penetrating power. Thus, only radioisotopes that emit gamma rays are appropriate for diagnosis (Table 18.4).

Iodine-131 is useful in diagnosing thyroid activity. A patient drinks water containing the radioisotope in the form of sodium iodide. If the thyroid is functioning normally, about one sixth of the radioisotope accumulates in the thyroid gland within 24 hours. A radiation detector placed at the neck is used to measure the concentration of iodine in the thyroid. If a hypothyroid condition exists, the amount accumulated is less than normal. If a greater than average amount accumulates, then a hyperthyroid condition exists. A thyroid scan can show the distribution of the iodine-131 in the thyroid as a radiation ''picture.'' If any part of the thyroid is missing from the "picture," isotopic iodine was not absorbed and a hypothyroid condition exists. The absence of radiation in an area is called a "cold spot."

Example 18.5

Iodine-131 and a beta particle are produced by bombarding an element with a neutron. What element is used?

Solution   Write the nuclear equation using $^A_Z E$ as the symbol for the element.

$$^A_Z E + {}^1_0 n \longrightarrow {}^{131}_{53} I + {}^{\ 0}_{-1} e$$

The mass number (A) of the element plus 1 for the neutron must equal 131.

$$A + 1 = 131 + 0$$
$$A = 130$$

The atomic number (Z) of the element must be equal to 52.

$$Z + 0 = 53 + (-1)$$
$$Z = 52$$

The atomic number of tellurium is 52. The element is tellurium-130.

[Additional examples are given in 18.39 and 18.40 at the end of the chapter.]

Technetium-99 is a useful radioisotope to scan the brain for tumors. The ionic compound sodium pertechnetate ($NaTcO_4$) is administered intravenously in a sodium chloride solution. Normal brain chemistry prevents the ionic pertechnetate salts from entering the brain. However, in the case of a tumor, the pertechnetate enters and concentrates in the area of the tumor. As a result, a scan of the brain will reveal the presence, size, and location of the tumor as a "hot spot."

Body fluids can also be studied with radioisotopes. The radioisotope sodium-24 is used to determine the effectiveness of blood circulation. A dose of sodium-24 is injected as $^{24}NaCl$ directly into the blood, and its progress is monitored. If it takes a longer time than average to reach a certain part of the body, then impaired circulation is indicated. In a similar manner, the cerebrospinal fluid that surrounds the spinal cord and brain can be labeled by using iodine-131 in a spinal tap. Any blockage that could cause a buildup of fluid can then be detected.

18.10   Use of Radioisotopes in Therapy

The objectives in using radioisotopes for therapy are somewhat different than those for diagnosis. The radioisotope must be placed in the proper part of the body and must selectively destroy cells or tissue. There is no need to monitor the radioisotope by an external detector. Thus, gamma ray emitters are not needed. Indeed, such radiation would be destructive to healthy tissue and organs in the high doses used for therapy. This means that only alpha or beta emitters are used because they will destroy tissue or cells in a localized area.

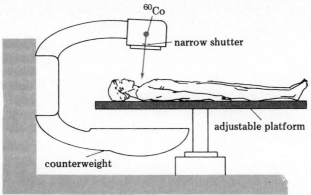

Figure 18.4

Use of cobalt-60 in radiation therapy
The cobalt source is heavily shielded with a metal alloy to prevent stray radiation from damaging healthy tissue.

Cells that are rapidly dividing are the most susceptible to radiation damage. Thus, cancer, which involves the rapid, uncontrolled division of abnormal cells, is affected by radiation. It is this principle that is applied in cancer therapy. Choice of a radioisotope that will do maximum damage to the cancerous cells while producing minimum damage to healthy cells is the goal of the radiologist.

Salts of the radioisotope $^{226}_{88}Ra$ were among the earliest used in cancer therapy; however, in addition to alpha particles, the radioactive decay of radium results in the emission of highly penetrating gamma rays.

$$^{226}_{88}Ra \longrightarrow \, ^{222}_{86}Rn + \, ^{4}_{2}He + \, ^{0}_{0}\gamma$$

Very high intensity treatment of cancers is now done with cobalt-60. This radioisotope emits beta particles and gamma rays. Its half-life is 5.2 years. Cobalt-60 in therapy involves using a beam of radiation that is focused on the small area of the body where the tumor is located (Figure 18.4). The radiation is carefully directed at the site of the cancer cells, but unfortunately some normal cells are always destroyed as well. In this method, the patient usually suffers the effects of radiation sickness while undergoing treatment.

Radiation therapy can also be accomplished by using sources of radiation located inside the body. For example, radium may be implanted near the tumor in very small hollow gold needles containing a radium salt. However, this therapeutic method has several disadvantages. Radium is very expensive and has a half-life of 1620 years. Therefore, the implant has to be removed at some later date to prevent continued radiation.

Yttrium-90 has considerable advantages over radium-226. It is a beta emitter with a half-life of 64 hours. An implant of an yttrium salt results in localized treatment of the tumor. Since no gamma rays are produced, healthy tissues a few centimeters away are unaffected. The short half-life is also useful because the radioisotope decays rapidly and is rendered harmless.

$$^{90}_{39}Y \longrightarrow \, ^{90}_{40}Zr + \, ^{0}_{-1}e$$

Some radioisotopes are injected into the body or ingested. The treatment is based on the fact that certain substances tend to concentrate at specific locations in the body. For example, iodine tends to be concentrated in the thyroid gland. An overreactive thyroid gland may be treated by administering orally a solution of sodium iodide containing some iodine-131. When this radioisotope is concentrated in the thyroid, it causes the destruction of some cells. Iodine-131 has a half-life of 8.1 days, which allows large doses to be used with safety. The radioactivity decreases rapidly after the treatment.

Phosphorus-32 has been used in the form of phosphate ions to treat some forms of leukemia. The phosphate ion can be transported to many parts of the body and can eventually be eliminated in the urine. However, phosphate ions are incorporated in bones, where they can affect the production of white blood cells in the bone marrow.

18.11 Nuclear Transmutations

In order to have radioisotopes available for medical and other uses, nuclear scientists must make them in the laboratory. High-energy nuclear particles are forced to collide with a target of some stable isotope. The collision of a high-energy particle with a nucleus to produce a different nucleus is called nuclear **transmutation.**

In nuclear transmutation the high-energy particle is called the **projectile.** Electrons ($_{-1}^{0}e$), neutrons ($_0^1n$), protons ($_1^1H$), deuterons ($_1^2H$), and alpha particles ($_2^4He$) have been used as projectiles.

The stable isotope used as a reactant in nuclear transmutation is called the **target nucleus.** Most stable isotopes available in nature have been used as targets. As a result, about 1000 artificial isotopes have been made in the laboratory.

Reaction of nitrogen-14 with high-energy neutrons produces carbon-14.

$$_7^{14}N + {}_0^1n \longrightarrow {}_6^{14}C + {}_1^1H$$

Carbon-14 is a beta emitter. One important use of carbon-14 is serving as a radioactive "tag" in organic molecules. The carbon-14 atoms are incorporated in an organic molecule, which then is allowed to proceed through either an organic or a biochemical reaction. As the compound reacts to form products, the location of the carbon-14 "tag" is followed. The process of photosynthesis was studied by such a method.

Radioactive isotopes needed for both medical diagnosis and therapy are also made by nuclear transmutation. Cobalt-60 is made from cobalt-59 by bombardment with neutrons.

$$_{27}^{59}Co + {}_0^1n \longrightarrow {}_{27}^{60}Co$$

By nuclear transmutation reactions, scientists have been able to produce elements of higher atomic number than exist in nature. Before the nuclear age, uranium was the element with the highest atomic number. Those elements with higher atomic number than uranium are called **transuranium elements.** A list of some of the isotopes of transuranium elements, with the transmutation reactions used to prepare them, is given in Table 18.5.

18.12 Nuclear Fission

The splitting of a nucleus of a heavy element into two or more lighter elements is called **nuclear fission.** One example of nuclear fission results from the bombardment of uranium-235 with neutrons. The unstable uranium-236 nucleus, which forms first, splits into nuclei of lower

Table 18.5 **Transmutation Reactions to Form Transuranium Elements**

Element	Atomic number	Reaction
neptunium, Np	93	$^{238}_{92}U + ^{1}_{0}n \longrightarrow ^{239}_{93}Np + ^{0}_{-1}e$
plutonium, Pu	94	$^{238}_{92}U + ^{2}_{1}H \longrightarrow ^{238}_{93}Np + 2^{1}_{0}n$
		$^{238}_{93}Np \longrightarrow ^{238}_{94}Pu + ^{0}_{-1}e$
americium, Am	95	$^{239}_{94}Pu + ^{1}_{0}n \longrightarrow ^{240}_{95}Am + ^{0}_{-1}e$
curium, Cm	96	$^{239}_{94}Pu + ^{4}_{2}He \longrightarrow ^{242}_{96}Cm + ^{1}_{0}n$
berkelium, Bk	97	$^{241}_{95}Am + ^{4}_{2}He \longrightarrow ^{243}_{97}Bk + 2^{1}_{0}n$
californium, Cf	98	$^{242}_{96}Cm + ^{4}_{2}He \longrightarrow ^{245}_{98}Cf + ^{1}_{0}n$
einsteinium, Es	99	$^{238}_{92}U + 15^{1}_{0}n \longrightarrow ^{253}_{99}Es + 7^{0}_{-1}e$
fermium, Fm	100	$^{238}_{92}U + 17^{1}_{0}n \longrightarrow ^{255}_{100}Fm + 8^{0}_{-1}e$
mendelevium, Md	101	$^{253}_{99}Es + ^{4}_{2}He \longrightarrow ^{256}_{101}Md + ^{1}_{0}n$
nobelium, No	102	$^{246}_{96}Cm + ^{12}_{6}C \longrightarrow ^{254}_{102}No + 4^{1}_{0}n$
lawrencium, Lr	103	$^{252}_{98}Cf + ^{10}_{5}B \longrightarrow ^{257}_{103}Lr + 5^{1}_{0}n$

atomic mass. Although many pairs of elements are produced, $^{89}_{37}Rb$ and $^{144}_{55}Cs$ are typical products.

$$^{235}_{92}U + ^{1}_{0}n \longrightarrow ^{236}_{92}U \longrightarrow ^{89}_{37}Rb + ^{144}_{55}Cs + 3^{1}_{0}n$$

The fission process can continue because the neutrons produced by the fission reaction can collide with other uranium atoms causing them to split apart. Because more neutrons are produced than are used in initiating the reaction, the fission process becomes a self-sustaining chain reaction that releases large amounts of energy. This principle is used in the construction of atomic bombs. A chain process with three neutrons per nuclear fission of an atom is illustrated in Figure 18.5. The chain process can also occur for reactions involving two neutrons per atom.

Although atomic energy in the arsenals of numerous countries is potentially destructive, controlled nuclear energy can be of service in a world of limited resources. The limited fossil-derived fuels (oil, gas, and coal) are being used at an ever-increasing rate, and this source will be exhausted in the future. However, there is also a limit to the availability of uranium-235. This isotope is only 0.71% of naturally occurring uranium. With the estimated world reserves and anticipated growth of nuclear power, it is possible that the supply of uranium-235 may be exhausted in two to three decades.

The reaction in a nuclear reactor is different from that which occurs in a nuclear bomb. The uranium in nuclear reactors is of lower purity than weapons-grade uranium. The rate of the chain reaction is also controlled by a moderator and control rods. Water is a moderator because it slows down the neutrons released by the uranium, which is suspended in water in nuclear reactors. Control rods, made of cadmium, absorb some of the neutrons generated. The control rods are positioned between the uranium fuel samples (Figure 18.6). When the reaction needs to be slowed, the rods are lowered. To speed up the reaction, the rods are raised.

One of the important considerations in using nuclear fission reactions for energy is the problem of the disposal of the spent material. The waste products are themselves radioactive and have quite long half-lives. The technological problem of separation and disposal of these waste products is complicated by the considerable political problem of where to store the material.

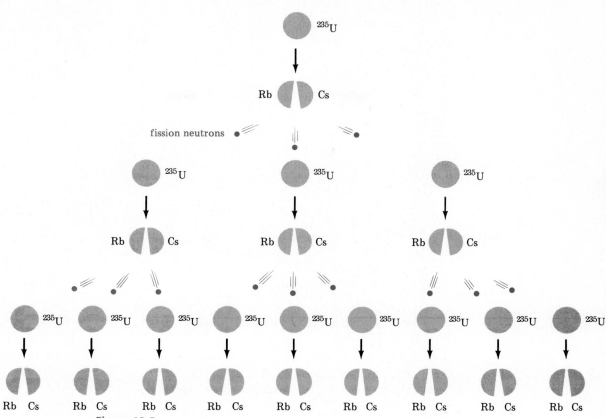

Figure 18.5
Nuclear chain reactions
Each nuclear event produces more neutrons than are used. The rate of the reaction then accelerates as more and more nuclei undergo fission.

18.13 Nuclear Fusion

The formation of a heavy element from two lower-mass elements is called **nuclear fusion.** Nuclear fusion is a source of even greater energy than nuclear fission. However, unlike nuclear fission, fusion requires extremely high temperature (40,000,000°C) for initiation. Fusion occurs in the sun, where the temperature is high enough to convert hydrogen into helium. The steps proposed to account for the reaction are

$$2\,{}^{1}_{1}\text{H} \longrightarrow {}^{2}_{1}\text{H} + {}^{0}_{1}\text{e}$$
$${}^{2}_{1}\text{H} + {}^{1}_{1}\text{H} \longrightarrow {}^{3}_{2}\text{He}$$
$$2\,{}^{3}_{2}\text{He} \longrightarrow {}^{4}_{2}\text{He} + 2\,{}^{1}_{1}\text{H}$$

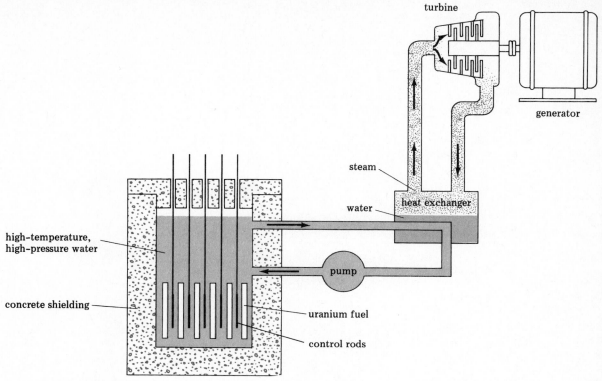

Figure 18.6
Generation of electricity from a nuclear power plant
The hot radioactive water is pumped to a heat exchanger that converts water in a second system into steam. The steam drives a turbine to produce electricity.

On earth, nuclear fusion has been achieved by using atomic bomb "triggers" to provide sufficient energy to initiate fusion. This is the principle upon which the hydrogen bomb is based. Since hydrogen is so readily available from water, controlled nuclear fusion for practical purposes would eliminate any concern about future sources of energy.

Obviously, atomic bombs cannot be used to produce fusion power in power plants. The fusion reaction will have to be initiated and controlled by technical means not yet devised. The United States has expended about a billion dollars in research to develop power from fusion reactors. Although progress is being made, some scientists estimate that fusion power may not become available until after the year 2000.

Summary

Many isotopes that occur naturally have existed since the earth was formed. Other isotopes are undergoing radioactive transformations that generate radiation. Radiation from natural sources includes alpha particles, beta particles, and gamma radiation.

Radioactive isotopes may be made in the nuclear chemistry laboratory by the bombardment of stable nuclei with subatomic particles. This method has been used to produce both isotopes for medical applications and new elements.

All radioisotopes decay at a rate that is given by their half-life. One of the applications of half-life data is the use of carbon-14 in dating ancient materials made from plant sources.

Radioactive substances cause surrounding material to ionize and undergo abnormal reactions. The effect of radiation on living tissue can be minor or so severe as to damage DNA to the extent that mutation of the species or death results.

Methods to detect radiation include the Geiger counter, film badges, and scintillation counters. The unit measuring the number of radioactive disintegrations is the curie. Roentgen, rad, and rem are used to measure dosage. The use of radioisotopes requires shielding, but the effect of radiation also depends on the distance between the worker and the radiation source.

Radioisotopes are used in medicine for both diagnosis and therapy. The choice of radioisotope is dictated by the purpose as well as the target organ. The radioisotopes are produced by nuclear transmutation reactions.

Two types of nuclear reaction, fusion and fission, are alternative fuel sources for the future. Fission processes are currently being used, but are limited by the availability of uranium-235 and the problem of waste disposal. Fusion has not been achieved in a self-sustaining reaction, but research continues to develop this promising source of energy.

New Terms

An **alpha particle** is a particle that consists of two protons and two neutrons and is identical to the nucleus of the helium atom.

A **beta particle** is an electron formed in the nucleus as the result of conversion of a neutron into a proton.

The **curie** is a quantity of radioactive material that produces 3.7×10^{10} nuclear disintegrations per second.

A **daughter nucleus** is the product of a nuclear disintegration reaction.

A **gamma ray** is a form of radiation, similar to x-rays, that is emitted from certain radioactive nuclei.

A **Geiger counter** is a device to detect and measure ionizing radiation.

The **half-life** of a radioisotope is the time required to decay to one half of its original amount.

Ionizing radiation causes molecules to lose electrons and form ions.

Nuclear fission is a radioactive decay process in which a heavy nucleus breaks apart into smaller nuclei and one or more neutrons.

Nuclear fusion is a nuclear process in which two small nuclei combine into a larger nucleus.

A **projectile** is a particle used to bombard a target nucleus.

One **rad** or radiation absorbed dose is the absorption of 100 ergs of energy per gram of absorbing tissue.

A **radical** is a species with an unpaired electron.

A **radioisotope** is an isotope that undergoes nuclear change and produces radiation.

The **RBE** or relative biological effectiveness is a multiplicative factor that takes into account the differing consequences of the source of radiation.

The **rem** or roentgen equivalent for man is a product of RBE and rad. The unit accounts for the varying effects of types of radiation on tissue.

The **roentgen** is a dose of radiation that produces 2.1×10^9 ions in 1 cm^3 or dry air at 0°C and 1 atm pressure.

A **target nucleus** is an isotope used as a reactant in nuclear transmutation.

Nuclear transmutation is the conversion of one element into another by the collision of high-energy particles.

A **transuranium element** is an element with an atomic number larger than that of uranium.

Exercises

Terminology

18.1 Distinguish between an alpha and a beta particle.

18.2 Distinguish between the terms rem and rad.

Nuclear Symbols

18.3 Write the symbols used in balancing nuclear equations for each of the following particles.
(a) neutron   (b) proton   (c) electron

18.4 Write the symbols used in balancing nuclear equations for each of the following particles.
(a) alpha   (b) beta

18.5 Write the nuclear symbol for each of the following isotopes.
(a) carbon-14   (b) boron-10   (c) uranium-238
(d) chlorine-37   (e) zinc-65   (f) sodium-24

18.6 Write the nuclear symbol for each of the following isotopes.
(a) cadmium-104   (b) krypton-81   (c) chlorine-35
(d) sulfur-35   (e) iron-55   (f) calcium-46

18.7 Indicate the number of protons and neutrons contained in each of the following.
(a) $^{206}_{82}Pb$ (b) $^{235}_{92}U$ (c) $^{127}_{53}I$
(d) $^{13}_{7}N$ (e) $^{25}_{11}Na$ (f) $^{3}_{1}H$

18.8 Indicate the number of protons and neutrons contained in each of the following.
(a) $^{104}_{47}Ag$ (b) $^{125}_{53}I$ (c) $^{116}_{51}Sb$
(d) $^{20}_{9}F$ (e) $^{47}_{20}Ca$ (f) $^{55}_{26}Fe$

Types of Radiation

18.9 Give the symbol, charge, and mass in amu of each of the following types of radiation.
(a) alpha (b) beta (c) gamma

18.10 Describe the penetrating power of alpha, beta, and gamma radiation.

Balancing Nuclear Equations

18.11 Write a balanced nuclear equation for beta emission from each of the following.
(a) fluorine-21 (b) silicon-31 (c) magnesium-28

18.12 Write a balanced nuclear equation for beta emission from each of the following.
(a) calcium-40 (b) potassium-40 (c) strontium-89

18.13 Write a balanced nuclear equation for alpha emission from each of the following.
(a) polonium-22 (b) curium-240 (c) einsteinium-252

18.14 Write a balanced nuclear equation for alpha emission from each of the following.
(a) thorium-229 (b) boron-11 (c) berkelium-245

18.15 Supply the correct symbol for the product of each of the following reactions.
(a) $^{243}_{96}Cm \longrightarrow ^{4}_{2}He + ?$ (b) $^{222}_{86}Rn \longrightarrow ^{4}_{2}He + ?$
(c) $^{234}_{94}Pu \longrightarrow ^{4}_{2}He + ?$ (d) $^{245}_{97}Bk \longrightarrow ^{4}_{2}He + ?$

18.16 Supply the correct symbol required to balance each of the following equations.
(a) $^{140}_{56}Ba \longrightarrow ? + ^{140}_{57}La$ (b) $^{245}_{96}Cm \longrightarrow ? + ^{241}_{94}Pu$
(c) $^{30}_{13}Al \longrightarrow ? + ^{30}_{14}Si$ (d) $^{11}_{5}B \longrightarrow ? + ^{7}_{3}Li$

Half-life

18.17 What percentage of an element will remain after each of the following number of half-lives?
(a) one (b) two (c) three

18.18 What percentage of an element will remain after each of the following number of half-lives?
(a) four (b) five (c) six

18.19 A wood object has 12.5% of the usual abundance of carbon-14 remaining. How old is the object?

18.20 A wooden object found in a cave in Greece has 6.2% of the normal abundance of carbon-14. How old is the object?

18.21 The half-life of mercury-203 is 47 days. Estimate how much will remain after 1 year.

18.22 The half-life of $^{32}_{15}P$ is 14 days. How many grams of the isotope in a 2.00 g sample will remain after 56 days?

18.23 The half-life of an isotope is 30 min. How many hours are required for a 10.00 g sample to decay to 0.62 g of this isotope?

18.24 The half-life of iodine-123 is 13 hours. If 10 ng is given to a patient, how much will remain after 3 days and 6 hours?

Biological Effects of Radiation

18.25 Explain the term "ionizing radiation."

18.26 Why is ionizing radiation harmful to living cells?

Detection of Radiation

18.27 Explain the operation of the Geiger counter.

18.28 How are film badges processed to detect radiation received by workers in nuclear laboratories?

Units of Radiation

18.29 The becquerel (Bq) is a new unit of radiation intensity. The equivalence is 1 Bq = 1 disintegration/s. How many Bq are in 1 Ci (curie)?

18.30 The gray (Gy) is a new unit of radiation absorbed dose. It is the absorption of 1 joule of energy per kilogram of tissue. A joule is equal to 10^7 ergs. Show that 1 Gy = 100 rad.

Radiation and Safety

18.31 Is there a minimum radiation that will have no biological effect? Explain how individuals working with radiation can minimize their exposure.

18.32 Which is more dangerous, a radioisotope that emits alpha particles or one that emits beta particles, providing they have the same half-life?

18.33 Which is more dangerous, a radioisotope with a short half-life or one with a long half-life, if they emit the same type of radiation?

18.34 A person receives 20 rem of radiation per year for 15 years and does not show any biological effect. However, a person who receives 300 rem in one dose may die. Explain the difference.

18.35 A radiologist measures 128 mrem at a distance of 2 m from a source. At what distance will the radiation measure 2 mrem?

18.36 A radiologist measures 25 mrem radiation at 10 m. What will the radiation be at 1 m?

Radiation in Diagnosis

18.37 Explain why radioisotopes used in diagnosis usually emit gamma radiation.

18.38 Explain why radioisotopes used in diagnosis should have a short half-life.

18.39 Phosphorus-32 for treatment of leukemia is made by bombarding an isotope with neutrons. What is the element?

$$? + {}^{1}_{0}n \longrightarrow {}^{32}_{15}P + {}^{1}_{1}H$$

18.40 A radioisotope required for a diagnostic procedure is produced by the following reaction. What is the radioisotope?

$$^{109}_{47}Ag + {}^{4}_{2}He \longrightarrow ? + 2\,{}^{1}_{0}n$$

Radiation and Therapy

18.41 How do isotopes used in therapy differ from isotopes used in diagnosis?

18.42 Radioisotopes can cause cancer and yet are used in therapy for cancer. Explain this apparent contradiction.

18.43 Explain how phosphorus-32 is used to treat leukemia.

18.44 Explain how iodine-131 is used to treat an overactive thyroid.

Transmutation of Elements

18.45 Nickel-58 produces an element and an alpha particle when bombarded with a proton. What is the elemental symbol of the product?

18.46 The bombardment of ${}^{7}_{3}Li$ by an alpha particle produces a neutron and an element. What is the element?

18.47 Bombardment of ${}^{238}_{92}U$ by ${}^{14}_{7}N$ yields five neutrons and an element. What is the element?

18.48 Gold-197 can yield mercury-197 and a neutron when bombarded with the proper particle. What is the particle?

18.49 Complete the following equations.
(a) ${}^{23}_{11}Na + ? \longrightarrow {}^{24}_{11}Na + {}^{1}_{1}H$
(b) ${}^{10}_{5}B + ? \longrightarrow {}^{13}_{7}N + {}^{1}_{0}n$
(c) ${}^{59}_{27}Co + ? \longrightarrow {}^{56}_{25}Mn + {}^{4}_{2}He$

18.50 Complete the following equations.
(a) ${}^{235}_{92}U + {}^{2}_{1}H \longrightarrow ? + {}^{0}_{-1}e$
(b) ${}^{27}_{13}Al + {}^{4}_{2}He \longrightarrow ? + {}^{1}_{0}n$
(c) ${}^{130}_{52}Te + {}^{1}_{0}n \longrightarrow ? + {}^{0}_{-1}e$

18.51 Complete the following equations.
(a) ${}^{27}_{13}Al + {}^{4}_{2}He \longrightarrow {}^{30}_{15}P + ?$
(b) ${}^{15}_{7}N + {}^{1}_{1}H \longrightarrow {}^{12}_{6}C + ?$
(c) ${}^{242}_{96}Cm + {}^{4}_{2}He \longrightarrow {}^{245}_{98}Cf + ?$

18.52 Complete the following equations.
(a) ${}^{209}_{83}Bi + ? \longrightarrow {}^{210}_{84}Po + {}^{1}_{0}n$
(b) ${}^{14}_{7}N + ? \longrightarrow {}^{17}_{8}O + {}^{1}_{1}H$
(c) ${}^{27}_{13}Al + ? \longrightarrow {}^{25}_{12}Mg + {}^{4}_{2}He$

18.53 Complete the following equations.
(a) ${}^{253}_{99}Es + {}^{4}_{2}He \longrightarrow ? + {}^{1}_{0}n$
(b) ${}^{249}_{98}Cf + {}^{18}_{8}O \longrightarrow ? + 4\,{}^{1}_{0}n$
(c) ${}^{238}_{92}U + {}^{14}_{7}N \longrightarrow ? + 5\,{}^{1}_{0}n$
(d) ${}^{238}_{92}U + {}^{16}_{8}O \longrightarrow ? + 5\,{}^{1}_{0}n$

18.54 Complete the following equations.
(a) ${}^{238}_{92}U + {}^{14}_{7}N \longrightarrow ? + 6\,{}^{1}_{0}n$
(b) ${}^{252}_{98}Cf + {}^{10}_{7}N \longrightarrow ? + 5\,{}^{1}_{0}n$
(c) ${}^{238}_{92}U + 15\,{}^{1}_{0}n \longrightarrow ? + 7\,{}^{0}_{-1}e$
(d) ${}^{245}_{96}Cm + {}^{13}_{6}C \longrightarrow ? + 5\,{}^{1}_{0}n$

Nuclear Fusion and Fission

18.55 What is the difference between the materials used for nuclear reactors and for atomic bombs?

18.56 How are control rods used in nuclear reactors?

18.57 How many neutrons are produced by the following nuclear fission reaction?

$$^{235}_{92}U + {}^{1}_{0}n \longrightarrow {}^{137}_{56}Ba + {}^{94}_{36}Kr + ?$$

18.58 How many neutrons are produced by the following nuclear fission reaction?

$$^{235}_{92}U + {}^{1}_{0}n \longrightarrow {}^{103}_{42}Mo + {}^{131}_{50}Sn + ?$$

18.59 How is a hydrogen bomb detonated?

18.60 Why haven't nuclear power plants employing nuclear fusion been built?

Additional Exercises

18.61 Write the nuclear symbol for each of the following isotopes.
(a) curium-243   (b) cobalt-60   (c) silver-104
(d) potassium-40   (e) oxygen-20   (f) mercury-184

18.62 Write the nuclear symbol for each of the following isotopes.
(a) bromine-82   (b) vanadium-51   (c) arsenic-73
(d) chromium-51   (e) sodium-24   (f) cobalt-59

18.63 Indicate the number of protons and neutrons contained in each of the following.
(a) $^{249}_{98}\text{Cf}$ (b) $^{242}_{96}\text{Cm}$ (c) $^{123}_{50}\text{Sn}$
(d) $^{35}_{16}\text{S}$ (e) $^{47}_{21}\text{Sc}$ (f) $^{245}_{97}\text{Bk}$

18.64 Indicate the number of protons and neutrons contained in each of the following.
(a) $^{186}_{73}\text{Ta}$ (b) $^{186}_{74}\text{W}$ (c) $^{234}_{90}\text{Th}$
(d) $^{27}_{13}\text{Al}$ (e) $^{113}_{48}\text{Cd}$ (f) $^{209}_{83}\text{Bi}$

18.65 In 1919 Earnest Rutherford proved the existence of a nuclear particle by the following experiment. What is the nuclear particle?

$$^{14}_{7}\text{N} + ^{4}_{2}\text{He} \longrightarrow ^{17}_{8}\text{O} + ?$$

18.66 In 1932 James Chadwick proved the existence of a nuclear particle by the following experiment. What is the nuclear particle?

$$^{9}_{4}\text{Be} + ^{4}_{2}\text{He} \longrightarrow ^{12}_{6}\text{C} + ?$$

18.67 What isotope was used to prepare the indicated transuranium element?
(a) $? + ^{2}_{1}\text{H} \longrightarrow ^{231}_{93}\text{Np} + 6\,^{1}_{0}\text{n}$
(b) $? + ^{2}_{1}\text{H} \longrightarrow ^{243}_{97}\text{Bk} + 3\,^{1}_{0}\text{n}$
(c) $? + ^{16}_{8}\text{O} \longrightarrow ^{249}_{100}\text{Fm} + 5\,^{1}_{0}\text{n}$

18.68 What isotope was used to prepare the indicated transuranium element?
(a) $? + ^{10}_{5}\text{B} \longrightarrow ^{257}_{103}\text{Lr} + 3\,^{1}_{0}\text{n}$
(b) $? + ^{4}_{2}\text{He} \longrightarrow ^{255}_{101}\text{Md} + 2\,^{1}_{0}\text{n}$
(c) $? + ^{4}_{2}\text{He} \longrightarrow ^{250}_{99}\text{Es} + ^{3}_{1}\text{H}$

18.69 What projectile is needed to accomplish each of the following transformations.
(a) $^{75}_{33}\text{As} + ? \longrightarrow ^{78}_{35}\text{Br} + ^{1}_{0}\text{n}$
(b) $^{31}_{15}\text{P} + ? \longrightarrow ^{32}_{15}\text{P} + ^{1}_{1}\text{H}$
(c) $^{59}_{27}\text{Co} + ? \longrightarrow ^{60}_{27}\text{Co}$
(d) $^{7}_{3}\text{Li} + ? \longrightarrow ^{7}_{4}\text{Be} + ^{1}_{0}\text{n}$

18.70 What projectile is needed to accomplish each of the following transformations.
(a) $^{209}_{83}\text{Bi} + ? \longrightarrow ^{210}_{84}\text{Po} + ^{1}_{0}\text{n}$
(b) $^{106}_{46}\text{Pd} + ? \longrightarrow ^{109}_{47}\text{Ag} + ^{1}_{1}\text{H}$
(c) $^{14}_{7}\text{N} + ? \longrightarrow ^{15}_{8}\text{O}$
(d) $^{45}_{21}\text{Sc} + ? \longrightarrow ^{45}_{20}\text{Ca} + ^{1}_{1}\text{H}$

18.71 What is the particle produced in each of the following transformations?
(a) $^{239}_{94}\text{Pu} + ^{4}_{2}\text{He} \longrightarrow ^{242}_{96}\text{Cm} + ?$
(b) $^{45}_{21}\text{Sc} + ^{1}_{0}\text{n} \longrightarrow ^{45}_{20}\text{Ca} + ?$
(c) $^{27}_{13}\text{Al} + ^{1}_{0}\text{n} \longrightarrow ^{24}_{11}\text{Na} + ?$

18.72 What is the particle produced in each of the following transformations?
(a) $^{7}_{3}\text{Li} + ^{2}_{1}\text{H} \longrightarrow ^{8}_{4}\text{Be} + ?$
(b) $^{75}_{33}\text{As} + ^{2}_{1}\text{H} \longrightarrow ^{76}_{33}\text{As} + ?$
(c) $^{10}_{5}\text{B} + ^{1}_{0}\text{n} \longrightarrow ^{7}_{3}\text{Li} + ?$

18.73 What element is produced in each of the following transformations?
(a) $^{20}_{8}\text{O} \longrightarrow ? + ^{0}_{-1}\text{e}$
(b) $^{234}_{92}\text{U} \longrightarrow ? + ^{4}_{2}\text{He}$
(c) $^{230}_{90}\text{Th} \longrightarrow ? + ^{4}_{2}\text{He}$
(d) $^{27}_{12}\text{Mg} \longrightarrow ? + ^{0}_{-1}\text{e}$

18.74 What element is produced in each of the following transformations?
(a) $^{218}_{84}\text{Po} \longrightarrow ? + ^{4}_{2}\text{He}$
(b) $^{186}_{73}\text{Ta} \longrightarrow ? + ^{0}_{-1}\text{e}$
(c) $^{82}_{35}\text{Br} \longrightarrow ? + ^{0}_{-1}\text{e}$
(d) $^{210}_{84}\text{Po} \longrightarrow ? + ^{4}_{2}\text{He}$

18.75 Complete each of the following equations.
(a) $^{45}_{21}\text{Sc} + ? \longrightarrow ^{45}_{20}\text{Ca} + ^{1}_{1}\text{H}$
(b) $^{27}_{13}\text{Al} + ^{1}_{0}\text{n} \longrightarrow ? + ^{4}_{2}\text{He}$
(c) $^{75}_{33}\text{As} + ^{2}_{1}\text{H} \longrightarrow ^{76}_{33}\text{As} + ?$

18.76 Complete each of the following equations.
(a) $^{51}_{23}\text{V} + ? \longrightarrow ^{51}_{24}\text{Cr} + 2\,^{1}_{0}\text{n}$
(b) $^{7}_{3}\text{Li} + ^{1}_{1}\text{H} \longrightarrow ? + ^{1}_{0}\text{n}$
(c) $^{9}_{4}\text{Be} + ^{1}_{1}\text{H} \longrightarrow ^{6}_{3}\text{Li} + ?$

18.77 Complete each of the following equations.
(a) $^{239}_{94}\text{Pu} + ? \longrightarrow ^{242}_{96}\text{Cm} + ^{1}_{0}\text{n}$
(b) $^{31}_{15}\text{P} + ^{2}_{1}\text{H} \longrightarrow ? + ^{1}_{1}\text{H}$
(c) $^{35}_{17}\text{Cl} + ^{1}_{0}\text{n} \longrightarrow ^{35}_{16}\text{S} + ?$

18.78 Complete each of the following equations.
(a) $^{10}_{5}\text{B} + ? \longrightarrow ^{7}_{3}\text{Li} + ^{4}_{2}\text{He}$
(b) $^{106}_{46}\text{Pd} + ^{4}_{2}\text{He} \longrightarrow ? + ^{1}_{1}\text{H}$
(c) $^{209}_{83}\text{Bi} + ^{2}_{1}\text{H} \longrightarrow ^{210}_{84}\text{Po} + ?$

18.79 How many neutrons are produced by each of the following reactions?
(a) $^{235}_{92}\text{U} + ^{2}_{1}\text{H} \longrightarrow ^{231}_{93}\text{Np} + ?\,^{1}_{0}\text{n}$
(b) $^{244}_{96}\text{Cm} + ^{2}_{1}\text{H} \longrightarrow ^{243}_{97}\text{Bk} + ?\,^{1}_{0}\text{n}$
(c) $^{240}_{95}\text{Am} + ^{4}_{2}\text{He} \longrightarrow ^{243}_{97}\text{Bk} + ?\,^{1}_{0}\text{n}$

18.80 How many neutrons are produced by each of the following nuclear fission reactions?
(a) $^{250}_{97}\text{Bk} + ^{10}_{5}\text{B} \longrightarrow ^{257}_{103}\text{Lr} + ?\,^{1}_{0}\text{n}$
(b) $^{253}_{99}\text{Es} + ^{4}_{2}\text{He} \longrightarrow ^{255}_{101}\text{Md} + ?\,^{1}_{0}\text{n}$
(c) $^{238}_{92}\text{U} + ^{16}_{8}\text{O} \longrightarrow ^{249}_{100}\text{Fm} + ?\,^{1}_{0}\text{n}$

19 Compounds of Carbon and Hydrogen

Learning Objectives

After studying Chapter 19 you should be able to

1. Write condensed structural formulas from complete structural formulas and vice versa.

2. Distinguish between conformers and isomers.

3. Recognize isomers from structural formulas.

4. Distinguish by chemical test between saturated and unsaturated hydrocarbons.

5. Give the IUPAC names for alkanes, alkenes, and alkynes.

6. Draw structures of alkanes, alkenes, and alkynes from IUPAC names.

7. Classify carbon and hydrogen atoms in alkanes.

8. Write the products of halogenation of an alkane.

9. Give the IUPAC names for cycloalkanes.

10. Calculate the molecular formula of a substance based on the number and type of bonds present.

11. Predict the product of addition of an unsymmetrical reagent to an alkene or alkyne.

12. Name substituted benzene compounds.

19.1 Carbon and Organic Compounds

Organic compounds contain carbon and only a few other elements such as hydrogen, oxygen, nitrogen, or sulfur. Approximately 5 million organic compounds are known, and approximately 5000 new compounds are being produced each week. With its four valence-shell electrons, carbon may form four covalent bonds to itself or to other atoms and achieve a Lewis octet of electrons. The other atoms then must contribute a total of four electrons. A single carbon atom can form

1. Four single bonds.
2. Two single bonds and a double bond.
3. One single bond and a triple bond.
4. Two double bonds.

Structures of carbon that include four single bonds are many and varied. Compounds with carbon bonded to hydrogen as well as to chlorine, oxygen, sulfur, and nitrogen are shown in Figure 19.1. The bonds for compounds of carbon containing four single bonds are drawn at right angles to each other. However, you should recall from Chapter 10 that the carbon bond angles in such compounds are 109.5°.

Carbon may bond to itself by sharing one, two, or three pairs of electrons.

$$\cdot \overset{\cdot\cdot}{C}-\overset{\cdot\cdot}{C}\cdot \qquad \overset{\cdot\cdot}{C}=\overset{\cdot\cdot}{C} \qquad \cdot C\equiv C\cdot$$

single bond   double bond   triple bond

The remaining unpaired electrons in each case can then be shared with hydrogen to give ethane, ethylene, or acetylene (Figure 19.1).

Since carbon can bond to itself, the units pictured with unpaired electrons can be joined to form longer chains of carbon atoms. Propane, used as a fuel, and butane, used in some cigarette lighters, contain chains of three and four carbon atoms, respectively (Figure 19.1). In this chapter, compounds containing only single bonds, called alkanes and cycloalkanes, will first be discussed. Subsequent sections will deal with compounds containing double and triple bonds, known as alkenes and alkynes, respectively.

19.2 Structural Formulas

The molecular formula of butane, C_4H_{10}, and that of ethyl alcohol, C_2H_6O, give their atomic composition. However, in order to understand the chemistry of organic compounds, it is necessary to know their structure. The structure of a molecule is represented by a **structural formula** that shows the arrangement of atoms and bonds. The structural formulas of both butane and ethyl alcohol are given in Figure 19.1.

Chemists draw shorthand or condensed versions of structural formulas. **Condensed structural formulas** show only specific bonds; other bonds are implied. The degree of condensation depends on which bonds are shown and which are implied. For example, since hydrogen forms only a single bond to carbon, such bonds need not be shown in a compound such as butane.

$$CH_3-CH_2-CH_2-CH_3$$

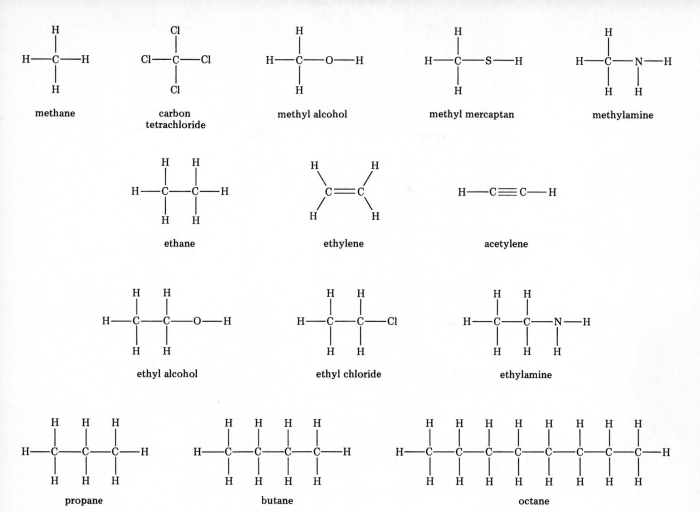

Figure 19.1
Structural formulas of some carbon compounds

The C—H bonds are implied. Since carbon always forms four bonds, the carbon atoms at each end of butane are understood to have three single bonds to hydrogen. Carbon atoms in the interior of the molecule show two carbon–carbon bonds, but the two carbon–hydrogen bonds are understood. Note that the hydrogen atoms are written after the carbon atom.

In a further condensation of the structural formula, both the C—H and C—C bonds can be "left out" and understood to be present.

$$CH_3CH_2CH_2CH_3$$

In this representation of butane, the carbon atom on the left is understood to be bonded to the three hydrogen atoms and to the carbon atom to the right. The second carbon atom from the left is bonded to the two hydrogen atoms to the right and to the carbon atoms to the immediate right and left, since carbon must have four bonds.

In the most condensed formula of butane, similar structural units are grouped together within

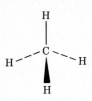

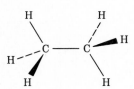
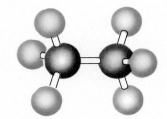

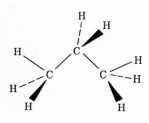
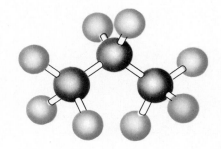

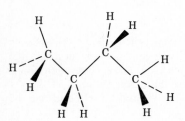

Figure 19.2
Perspective structures
and molecular models

parentheses. The —CH_2— unit is called a methylene group. The number of times the methylene unit is repeated is shown by a subscript following the closing parenthesis. The methylene group occurs twice in butane.

$$CH_3(CH_2)_2CH_3$$

Example 19.1

The structural formula of octane is given in Figure 19.1. Write three condensed structural formulas for octane.

Solution   With the C—H bonds understood, we may write

$$CH_3—CH_2—CH_2—CH_2—CH_2—CH_2—CH_2—CH_3$$

With both the C—H and C—C bonds understood, we may write

$$CH_3CH_2CH_2CH_2CH_2CH_2CH_2CH_3$$

In the most condensed version, the six methylene units are represented within parentheses.

$$CH_3(CH_2)_6CH_3$$

[*Additional examples may be found in 19.9–19.12 at the end of the chapter.*]

A structural formula of a molecule shown in two dimensions usually does not represent the actual shape of the molecule. For example, methane is not a planar molecule with carbon at the center of a square. The three dimensional shape of methane can be shown on paper by using wedge and dashed lines (Figure 19.2). The wedge is viewed as a bond extending out of the plane of the page toward the reader. The dashed line represents a bond directed behind the plane of the page. The other two lines are bonds in the plane of the page. Representations using wedge and dashed lines are **perspective structural formulas.**

Chemists use two types of molecular models, ball and stick and space filling (Figure 19.2). Each has certain advantages and disadvantages. The ball-and-stick models show the molecular framework and bond angles clearly: the balls represent the nuclei of the atoms, and the sticks represent the bonds. Space-filling models show the space occupied by the electrons surrounding each atom. As a consequence, the carbon skeleton and its bond angles are obscured.

When you build a molecular model, you may arrive at representations that appear to be different. Consider ethane as shown in Figure 19.3. The two representations differ in how the hydrogen atoms of one carbon are positioned relative to the hydrogen atoms of the other carbon atom. Which of the two forms represents ethane? The answer is that both do to some degree.

The bonding pair of electrons of the two carbon atoms are shared along the axis between the two nuclei. Each carbon atom and its three bonded hydrogen atoms can rotate about the C—C bond, but the two carbon atoms are still bonded (Figure 19.3). Only the positions of the hydrogen atoms on the adjacent carbon atoms are altered.

Molecules like ethane are in constant rotational and vibrational motion. The motion could be compared to the twisting and turning of your body while dancing. You may look different, but all parts of your body are still attached in a specific sequence. Only the orientation of your limbs is changing. A molecule can exist in different orientations or **conformations** by rotation about single bonds. The individual molecule in a given conformation is called a **conformer.**

When writing structural formulas of molecules you may write two or more representations of

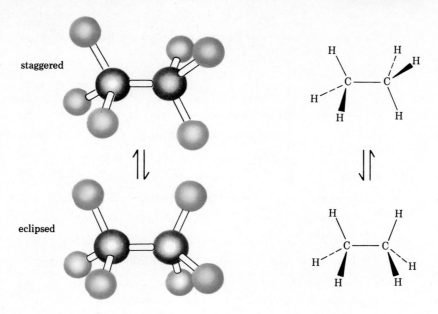

Figure 19.3

ball and stick models   perspective structures

Conformations of ethane
Rotation of the carbon atom on the right by 60° converts the top
conformation into the bottom conformation. The rotation occurs while the
bonding electrons remain between the carbon atoms.

the same molecule. Consider butane as shown in Figure 19.4. Each structure consists of four carbon atoms connected in a chain. Only the orientation of the chain differs in these structures, which represent different conformations. Butane is a three-dimensional molecule in which rotation can occur about each C—C bond. Just as there are numerous conformers possible, we can write many structural formulas for butane. By convention, chemists choose to write two-dimensional representations of carbon chains in a straight line, rather than as a bent chain.

19.3 Isomers

Compounds that have the same molecular formula but differ in structure are called **isomers.** The phenomenon of isomeric substances is one that contributes to the large number of organic compounds. Although there is only one substance represented by CH_4, C_2H_6, or C_3H_8, there are two substances with the molecular formula C_4H_{10}. There are 75 compounds or isomers with a molecular formula $C_{10}H_{22}$. Calculations give 62,491,178,805,831 possible isomers of $C_{40}H_{82}$!

Isomers have the same composition, but they differ in structure, that is, in the sequence in which the atoms are bonded. In order to understand this phenomenon, let's consider the two isomers of C_4H_{10}. One isomer is butane, which consists of an uninterrupted chain of carbon atoms (Figure 19.5). In the second isomer, called isobutane, only three carbon atoms are connected in sequence whereas the fourth forms a branch. Note that butane and isobutane are not conformers. One cannot be converted into the other by rotation about C—C bonds.

Structural formula	Perspective formula	Ball and stick model

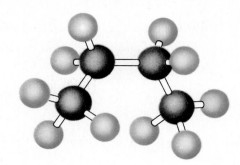

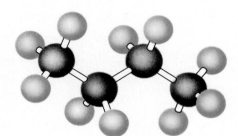

Figure 19.4
Structural formulas and conformations of butane

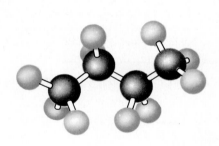

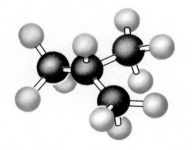

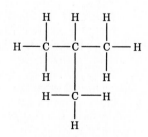

Figure 19.5
Isomers of C₄H₁₀

butane

isobutane

Butane and isobutane have different physical properties and can be separated if mixed together. Butane melts at $-138°C$ and boils at $-1°C$; isobutane melts at $-160°C$ and boils at $-12°C$.

Example 19.2

Consider the following structural formulas. Do they represent isomers?

$$Br—\underset{\underset{H}{|}}{\overset{\overset{H}{|}}{C}}—\underset{\underset{H}{|}}{\overset{\overset{H}{|}}{C}}—Cl \qquad Cl—\underset{\underset{H}{|}}{\overset{\overset{H}{|}}{C}}—\underset{\underset{H}{|}}{\overset{\overset{H}{|}}{C}}—Br$$

Solution The atomic compositions given in these structural formulas are identical. The molecular formula is C_2H_4ClBr. The condensed structural formula of the first structure is $BrCH_2CH_2Cl$ and the second is $ClCH_2CH_2Br$. However, the two structural formulas do not represent isomers. The order of bonds between atoms is Br—C—C—Cl in both cases. The second structural formula is merely the first structural formula written "in reverse."

Example 19.3

Consider the following structural formulas. Do they represent isomers?

$$Br—\underset{\underset{Cl}{|}}{\overset{\overset{H}{|}}{C}}—\underset{\underset{H}{|}}{\overset{\overset{H}{|}}{C}}—H \qquad H—\underset{\underset{Br}{|}}{\overset{\overset{H}{|}}{C}}—\underset{\underset{H}{|}}{\overset{\overset{Cl}{|}}{C}}—H$$

Solution The atomic compositions given in these structural formulas are identical, and therefore the molecular formula is C_2H_4ClBr. The condensed structural formula of the first structure is $CHClBrCH_3$ and the second is CH_2BrCH_2Cl. The two structural formulas represent isomers. The order of the bonds between bonded atoms is different in each compound.

[Additional examples may be found in 19.17 and 19.18 at the end of the chapter.]

19.4 Types of Hydrocarbons

A **hydrocarbon** is a compound containing only carbon and hydrogen atoms. A hydrocarbon is **saturated** if it has only carbon–carbon single bonds and thus has the maximum possible number of hydrogen atoms bonded to its carbon atoms. Saturated hydrocarbons are of two types: alkanes and cycloalkanes. **Alkanes** contain carbon atoms bonded in chains of atoms. In **cycloalkanes,** some of the carbon atoms are bonded to form a ring of atoms.

Hydrocarbons with carbon–carbon multiple bonds contain fewer than the largest number of hydrogen atoms possible based on the number of carbon atoms. For this reason we say that compounds that contain carbon–carbon multiple bonds are **unsaturated.** You may have heard the term unsaturated on television commercials as part of the word polyunsaturated. Polyunsaturated refers to the presence of several multiple bonds in a compound. Oils contain several double bonds and are polyunsaturated. There are three main classes of unsaturated compounds. These are based on the type of multiple bonds present in the molecules.

1. **Alkenes** are compounds that contain a carbon–carbon double bond.
2. **Alkynes** have a carbon–carbon triple bond.
3. **Aromatic hydrocarbons** contain a benzene ring or structural units like a benzene ring.

19.5 Alkanes

The names and condensed structural formulas of ten alkanes are given in Table 19.1. These compounds, called normal alkanes, consist of carbon atoms bonded one to another like links in a chain.

The general molecular formula for **straight-chain alkanes** is C_nH_{2n+2}. With the exception of the two carbon atoms at the ends of the chain, each carbon atom has two hydrogen atoms bonded to it, accounting for the $2n$ in the general formula. Each of the two end carbon atoms has an additional hydrogen atom bonded to it. This accounts for the $+2$ in the subscript on hydrogen. Each compound in Table 19.1 differs from the one above or below it by one —CH_2— unit. A series of compounds that differs from adjacent members by a repeating unit is called a **homologous series.**

Alkanes also exist as branched-chain alkanes. In **branched-chain alkanes,** one or more carbon atoms is bonded to more than two other carbon atoms. The carbon atom that is bonded to three or four other carbon atoms is the branching point. The group of carbon atoms attached to the main chain of carbon atoms at the branching point is called an **alkyl group.**

Isobutane is an example of a branched-chain alkane. There are three atoms in the main chain, and the alkyl group consists of a single carbon atom with three bonded hydrogen atoms. You may find it useful to regard the main carbon chain as a main street in a town. The branch then is a side street.

Table 19.1 **Structure and Nomenclature of Normal Alkanes**

Molecular formula	IUPAC prefix	IUPAC name	Structural formula
CH_4	meth-	methane	CH_4
C_2H_6	eth-	ethane	CH_3CH_3
C_3H_8	prop-	propane	$CH_3CH_2CH_3$
C_4H_{10}	but-	butane	$CH_3CH_2CH_2CH_3$
C_5H_{12}	pent-	pentane	$CH_3CH_2CH_2CH_2CH_3$
C_6H_{14}	hex-	hexane	$CH_3CH_2CH_2CH_2CH_2CH_3$
C_7H_{16}	hept-	heptane	$CH_3CH_2CH_2CH_2CH_2CH_2CH_3$
C_8H_{18}	oct-	octane	$CH_3CH_2CH_2CH_2CH_2CH_2CH_2CH_3$
C_9H_{20}	non-	nonane	$CH_3CH_2CH_2CH_2CH_2CH_2CH_2CH_2CH_3$
$C_{10}H_{22}$	dec-	decane	$CH_3CH_2CH_2CH_2CH_2CH_2CH_2CH_2CH_2CH_3$

Table 19.2   **Number of Isomers of the Alkanes**

Molecular formula	Number of isomers
CH_4	1
C_2H_6	1
C_3H_8	1
C_4H_{10}	2
C_5H_{12}	3
C_6H_{14}	5
C_7H_{16}	9
C_8H_{18}	18
C_9H_{20}	35
$C_{10}H_{22}$	75
$C_{20}H_{42}$	336,319
$C_{30}H_{62}$	4,111,846,763
$C_{40}H_{82}$	62,491,178,805,831

The general formula for a branched-chain alkane is also C_nH_{2n+2}. Each time a branch is present, the number of hydrogen atoms at the branching point, which is $2n$ in straight-chain hydrocarbons, is decreased by one. However, the branch now provides another "end" to the molecule, which means there is an additional hydrogen in the branch above the $2n$ value.

As the number of carbon atoms in an alkane increases, the number of isomers also increases. A list of the number of isomers is given in Table 19.2. Many of these isomers are not found in petroleum. However, any of them could be made in a chemistry laboratory.

Example 19.4

What is the molecular formula of an alkane containing 25 carbon atoms?

Solution   The value of n is 25. There must be $2(25) + 2$ hydrogen atoms. The molecular formula is $C_{25}H_{52}$. Any alkane, normal or branched, with 25 carbon atoms would have 52 hydrogen atoms.

[*Additional examples may be found in 19.21–19.24 at the end of the chapter.*]

Example 19.5

Are the following representations of the same substance or isomers?

Solution The longest chain in each representation contains six carbon atoms. Both have two branches located at the same points. The structures are both equivalent to

$$CH_3-\underset{\underset{H}{|}}{\overset{\overset{CH_3}{|}}{C}}-CH_2-\underset{\underset{H}{|}}{\overset{\overset{CH_3}{|}}{C}}-CH_2-CH_3$$

19.6 Classification and Nomenclature of Alkanes

It is convenient to designate parts of a structure according to the number of carbon and hydrogen atoms that are attached to a specific carbon atom. A **primary carbon atom** is directly bonded to only one other carbon atom.

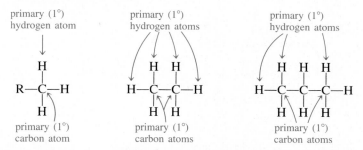

Ethane contains two primary carbon atoms. Propane contains two primary carbon atoms; the internal carbon atom is not primary because it is bonded to two other carbon atoms. The hydrogen atoms directly bonded to the primary carbon atoms are **primary hydrogen atoms.**

A **secondary carbon atom** is bonded to two other carbon atoms, as in the case of the internal carbon atom of propane. The hydrogen atoms bonded to the secondary carbon atom are **secondary hydrogen atoms.**

A **tertiary carbon atom** is bonded to three other carbon atoms. One of the four carbon atoms in isobutane is a tertiary carbon atom. The other three carbon atoms are primary. A hydrogen atom directly bonded to a tertiary carbon atom is a **tertiary hydrogen atom.**

Table 19.3 **The Structure of Alkyl Groups**

Parent alkane	Alkyl group	Name
CH_4	CH_3-	methyl
CH_3CH_3	CH_3CH_2-	ethyl
$CH_3CH_2CH_3$	$CH_3CH_2CH_2-$	propyl
	$\begin{matrix} H_3C \\ \diagdown \\ CH- \\ \diagup \\ H_3C \end{matrix}$	isopropyl
$CH_3CH_2CH_2CH_3$	$CH_3CH_2CH_2CH_2-$	butyl
	$\underset{\underset{CH_3}{\vert}}{CH_3CH_2CH-}$	s-butyl (secondary butyl)
$\underset{\underset{CH_3}{\vert}}{CH_3CHCH_3}$	$\underset{\underset{CH_3}{\vert}}{CH_3CHCH_2-}$	isobutyl
	$CH_3-\overset{\overset{CH_3}{\vert}}{\underset{\underset{CH_3}{\vert}}{C}}-$	t-butyl (tertiary butyl)

The IUPAC method for naming alkanes depends on identifying and numbering the longest carbon chain in a molecule. Groups such as halogens or alkyl groups, which are attached to this chain, are then designated by name and location. The system is then analogous to naming a street, numbering the houses on the street, and listing who lives at each address.

The names of the simpler alkyl groups are given in Table 19.3. A general shorthand representation of an alkyl group is R—. The alkyl group resembles an alkane that is missing one hydrogen atom. Alkyl groups are named by replacing the -ane ending of an alkane by -yl. Thus, CH_3- is the methyl group. In the case of halogens, the groups are called halo groups. Thus, Cl— and Br— are chloro- and bromo-, respectively.

The IUPAC rules for naming alkanes are as follows.

1. The longest continuous chain of carbon atoms is selected as the base. This chain is named according to the stem name listed in Table 19.1 plus the suffix -ane.
2. The carbon atoms in the chain are numbered consecutively from that end of the chain nearest a branch.
3. Each branch is located by the number of the atom to which it is attached on the chain.

$$\overset{4}{C}H_3-\overset{3}{C}H_2-\overset{2}{\underset{\underset{H}{\vert}}{\overset{\overset{CH_3}{\vert}}{C}}}-\overset{1}{C}H_3 \quad \text{is 2-methylbutane}$$

4. If two or more of the same types of branches occur, the number of them is indicated by

the prefixes di-, tri-, and so on, and the location of each on the main chain is indicated by a number.

$$
\overset{1}{C}H_3-\overset{2}{C}H-\overset{3}{\underset{\underset{CH_3}{|}}{C}H}-\overset{4}{C}H_3 \quad \text{is 2,3-dimethylbutane}
$$

with CH_3 above position 3 and CH_3 below position 2.

5. The numbers for the positions of the alkyl groups are placed immediately before the names of the groups, and hyphens are placed before and after the numbers. If two or more numbers occur together, commas are placed between them.

6. When several different alkyl groups are present, they are placed in alphabetical order and prefixed onto the stem name of the longest continuous chain. The whole name is written as a single word.

Be careful to locate the longest possible chain and also to number the chain correctly. The correct name and an incorrect one are given for two examples. Only if the rules are carefully applied can each structure correspond to a unique name and each name correspond to a unique structure.

$$
\overset{5}{C}H_3-\overset{4}{C}H_2-\overset{3}{C}H_2-\overset{2}{\underset{\underset{CH_3}{|}}{\overset{\overset{CH_3}{|}}{C}}}-\overset{1}{C}H_3
$$

2,2-dimethylpentane
(not 4,4-dimethylpentane)

$$
\overset{}{C}H_3-\overset{3}{C}H-\overset{4}{\underset{\underset{\underset{\underset{^1CH_3}{|}}{^2CH_2}}{|}}{C}H}-\overset{5}{C}H_2-\overset{6}{C}H_3
$$

3,4-dimethylhexane
(not 2-ethyl-3-methylpentane)

Example 19.6

Name the following compound.

$$
CH_3CH_2-\overset{\overset{CH_3}{|}}{\underset{\underset{CH_3}{|}}{C}}-\overset{}{\underset{\underset{CH_2CH_3}{|}}{C}H}-CH_3
$$

Solution The longest continuous chain has six carbon atoms. The chain is numbered so that the three methyl groups are located at positions 3, 3, and 4. The compound is 3,3,4-trimethylhexane. If the longest continuous chain is not identified, an incorrect name 2-ethyl-3,3-dimethylpentane is obtained.

$$
\overset{1}{C}H_3-\overset{2}{C}H_2-\overset{3}{\underset{\underset{CH_3}{|}}{\overset{\overset{CH_3}{|}}{C}}}-\overset{4}{\underset{\underset{\overset{5}{C}H_2-\overset{6}{C}H_3}{|}}{C}H}-CH_3
$$

[*Additional examples may be found in 19.29–19.32 at the end of the chapter.*]

19.7 Reactions of Alkanes

Carbon–carbon single bonds are very stable, and carbon–hydrogen bonds in alkanes are not very reactive. However at high temperatures, the carbon–hydrogen bond can undergo a substitution reaction by a halogen atom. When methane and chlorine are heated to a high temperature or are exposed to ultraviolet light, a chlorination reaction occurs.

$$CH_4 + Cl_2 \longrightarrow \quad CH_3Cl \quad + HCl$$
chloromethane
(methyl chloride)

The reaction can continue to produce more extensively chlorinated materials.

$$CH_3Cl + Cl_2 \longrightarrow \quad CH_2Cl_2 \quad + HCl$$
dichloromethane
(methylene chloride)

$$CH_2Cl_2 + Cl_2 \longrightarrow \quad CHCl_3 \quad + HCl$$
trichloromethane
(chloroform)

$$CHCl_3 + Cl_2 \longrightarrow \quad CCl_4 \quad + HCl$$
tetrachloromethane
(carbon tetrachloride)

Halogenated hydrocarbons are used for many industrial purposes. Unfortunately such compounds can cause severe biological disturbances such as liver damage and cancer. Chloroform was used as an anesthetic, and carbon tetrachloride was used as a drycleaning solvent; however, safer substances have replaced these compounds today.

Chlorination of higher-molecular-weight alkanes results in mixtures of chlorinated alkanes. For example, propane will yield two isomers.

$$CH_3CH_2CH_3 + Cl_2 \longrightarrow CH_3\overset{\overset{\displaystyle Cl}{|}}{C}HCH_3 + HCl$$
2-chloropropane

$$CH_3CH_2CH_3 + Cl_2 \longrightarrow CH_3CH_2CH_2Cl + HCl$$
1-chloropropane

Example 19.7

How many mono- and dichlorinated compounds can result from the chlorination of ethane?

Solution Both carbon atoms in ethane are equivalent, and only one monochlorinated compound can result.

$$CH_3CH_2Cl$$
chloroethane

After the first chlorine is substituted, the two carbon atoms are not equivalent. The second chlorine may substitute on the same carbon atom or on the other carbon atom.

$$CH_3CHCl_2 \qquad\qquad ClCH_2CH_2Cl$$
1,1-dichloroethane   1,2-dichloroethane

[Additional examples may be found in 19.41 and 19.42 at the end of the chapter.]

19.8 Cycloalkanes

Cyclic hydrocarbons, called **cycloalkanes,** have the general formula C_nH_{2n}. The reason for the smaller number of hydrogen atoms in cycloalkanes than in alkanes is that an additional carbon–carbon bond is needed to form the ring. The cycloalkanes are considered saturated hydrocarbons, since they contain only carbon–carbon single bonds.

The structural formulas of four cycloalkanes are shown in Table 19.4. The fully condensed formulas are simple polygons. It is understood that each corner is a carbon atom that has the correct number of carbon–hydrogen bonds.

Cyclopropane is a sweet-smelling, colorless gas used as an inhalation anesthetic. It produces unconsciousness in a few seconds. Cyclohexane is a widely used industrial solvent.

Cycloalkanes are named by the IUPAC system according to the number of carbon atoms in the ring, with the prefix cyclo-. When there is only one position containing an alkyl group or a halogen atom, there is only one possible compound and therefore no number is necessary. Thus, bromocyclopropane (Figure 19.6) defines the molecule completely.

When there is more than one group attached to the ring, the ring is numbered. One substituent is always given the number 1 and the others are given the next lowest possible number. The numbering must be done in a clockwise or counterclockwise direction so as to give the lowest combination of numbers for the groups attached to the ring.

Table 19.4   **Cycloalkanes**

Molecular formula	Name	Structural formula	Abbreviated notation
C_3H_6	cyclopropane	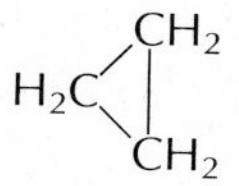	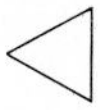
C_4H_8	cyclobutane	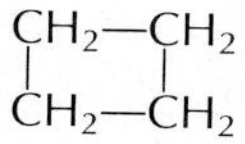	
C_5H_{10}	cyclopentane	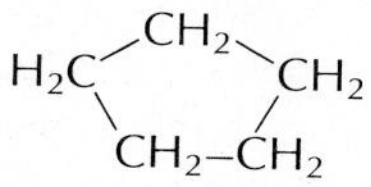	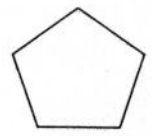
C_6H_{12}	cyclohexane	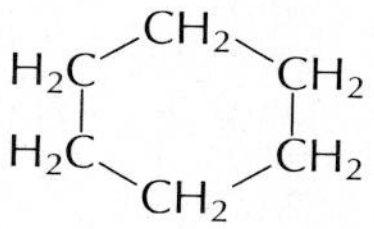	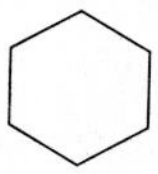

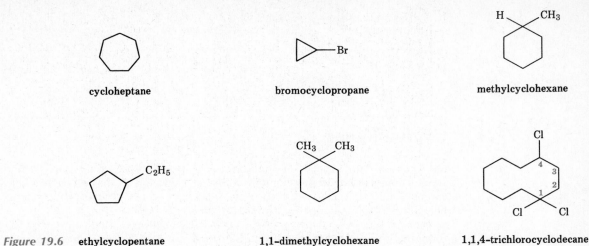

cycloheptane

bromocyclopropane

methylcyclohexane

ethylcyclopentane

1,1-dimethylcyclohexane

1,1,4-trichlorocyclodecane

Figure 19.6 *IUPAC nomenclature of cycloalkanes*

19.9 Alkenes

The double bond in an alkene decreases the number of hydrogen atoms in a molecule by two compared to the number in alkanes. As a result, the general formula for an alkene is C_nH_{2n}. Note that this general formula is the same as for cycloalkanes, which are thus isomeric with alkenes.

The structure of the simplest alkene, ethylene, is shown in Figure 19.7. All six atoms, two carbon atoms and four hydrogen atoms, are located in the same plane. The plane may be chosen to be that of the printed page or one perpendicular to it. If the plane is perpendicular to the printed page, the carbon–hydrogen bonds must project in front of and in back of the page. Wedge-shaped lines are used for bonds in front of the page and dotted lines for those behind the page. The bond angles H—C—H and H—C=C of ethylene are 118° and 121°, respectively. These values are close to the ideal value of 120° expected based on the VSEPR theory.

Example 19.8

What is the molecular formula for a hydrocarbon with six carbon atoms that contains a ring and a double bond?

Solution The presence of a ring decreases the number of hydrogen atoms by 2 below that of an alkane to give C_nH_{2n}. If a double bond is present, another two hydrogen atoms must be subtracted to give C_nH_{2n-2}. For six carbon atoms the molecular formula must be C_6H_{10}.

[*Additional examples may be found in 19.25 and 19.26 at the end of the chapter.*]

Geometric Isomerism

In an alkane there is free rotation about the carbon–carbon bonds, and the molecule can then twist and adopt many conformations. Thus, any of the following structures can represent 1,2-dichloroethane. The positions of the chlorine atoms in space vary constantly, and it doesn't matter which way the molecule is drawn.

$$
\begin{array}{ccc}
\overset{\displaystyle H\ \ H}{Cl-\underset{\displaystyle H\ \ H}{C-C}-Cl}
&
\overset{\displaystyle Cl\ \ Cl}{H-\underset{\displaystyle H\ \ H}{C-C}-H}
&
\overset{\displaystyle Cl\ \ H}{H-\underset{\displaystyle H\ \ Cl}{C-C}-H}
\end{array}
$$

The double bond does not permit free rotation. As a result of the restriction of rotation, alkenes have a rigid geometry for the groups attached to the carbon atoms of the double bond. The four atoms directly attached to the two double bonded carbon atoms lie in the same plane. These atoms can exist in different spatial or geometric arrangements, and isomerism about the double bond occurs. Isomers that differ from each other in the geometry of the molecules and not in the bonding order of the atoms are **geometric isomers.** They are also known as *cis-trans* isomers.

There are two geometric isomers of 1,2-dichloroethene. The structure with the chlorine atoms on the same "side" of the molecule is called the *cis* isomer. The structure with the chlorine atoms on opposite "sides" is called the *trans* isomer.

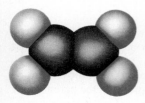

Figure 19.7
Structure of ethylene

cis-1,2-dichloroethene   *trans*-1,2-dichloroethene

The physical properties of the geometric isomers differ. For example, the boiling points of the *cis* and *trans* isomers of 1,2-dichloroethene are 60 and 47°C, respectively.

Geometric isomers do not occur for all alkenes. Only alkenes that have two different groups attached to each of the doubly bonded carbon atoms can exist as *cis-trans* isomers. 1,2-Dichloroethene contains a chlorine atom and a hydrogen atom on one unsaturated carbon atom. These groups are different. Similarly the groups on the other unsaturated carbon atom are different—a hydrogen atom and a chlorine atom.

different groups   H   H   different groups
C=C
Cl   Cl

Neither chloroethene nor 1,1-dichloroethene can exist as a *cis-trans* pair of geometric isomers.

different groups   H   H   identical groups
C=C
Cl   H

identical groups   Cl   H   identical groups
C=C
Cl   H

Nomenclature of Alkenes

The rules for naming alkenes are similar to those for alkanes but include an indication of the position of the double bond and the geometric arrangement about the double bond.

1. The longest continuous chain containing the double bond is used as the parent.

$$CH_3-\underset{\underset{\underset{CH_3}{|}}{\underset{CH_2}{|}}{C}}{\overset{\overset{CH_3}{|}}{C}}=C-CH_3$$

2. The longest chain is given the same stem name as an alkane but with -ene replacing -ane. The longest chain in the preceding structure contains five carbon atoms and the parent name is pentene.
3. The carbon atoms are numbered consecutively from the end nearest the double bond. The number of the first carbon atom of the double bond is used as a prefix to the parent name separated by a hyphen. The compound is, therefore, a substituted 2-pentene.

$$CH_3-\overset{2}{C}=\overset{3}{C}-CH_3$$

4. Alkyl groups and other substituents are named and their positions on the chain are determined by the numbering established by rule 3. Names and numbers are prefixed to the parent name.

$$CH_3-C=C-CH_3$$

2,3-dimethyl-2-pentene

5. If the compound can exist as a *cis* or *trans* isomer, the appropriate prefix followed by a hyphen is placed in front of the name. There are two identical CH_3 groups attached to the carbon at position 2 in 2,3-dimethyl-2-pentene so there is no possibility of *cis* or *trans* isomers.
6. In naming cycloalkenes, number the ring to give the double-bonded carbon atoms the numbers 1 and 2. Choose the direction of numbering so that the substituents receive the lowest numbers. The position of the double bond is not given because it is known to be between the carbon-1 and carbon-2.

Some examples of alkene nomenclature are given in Figure 19.8.

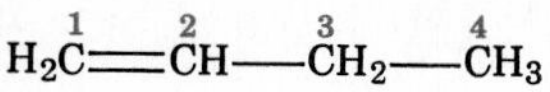

1-butene 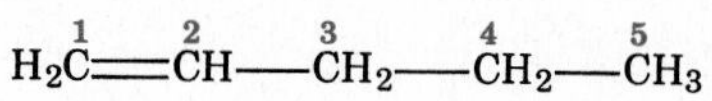   1-pentene 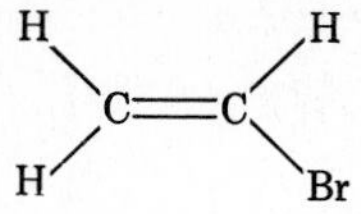   bromoethene 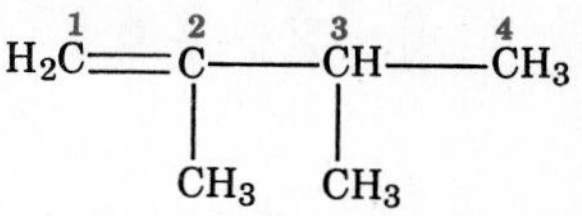   2,3-dimethyl-1-butene

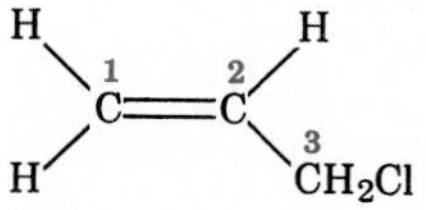

3-chloropropene 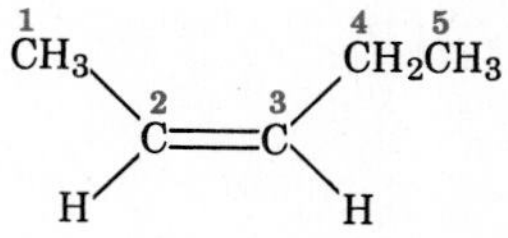   *cis*-2-pentene 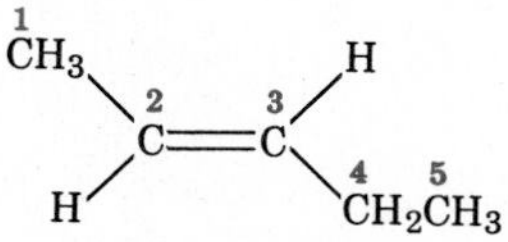   *trans*-2-pentene

cyclohexene 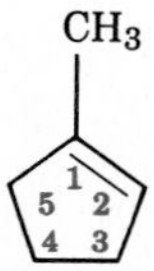   1-methylcyclopentene 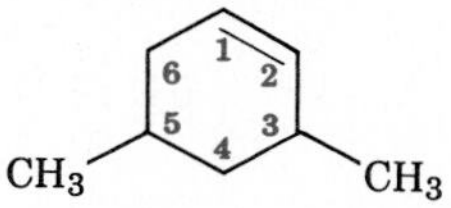   3,5-dimethylcyclohexene

Figure 19.8
Nomenclature of
alkenes

Example 19.9

Name the following compound.

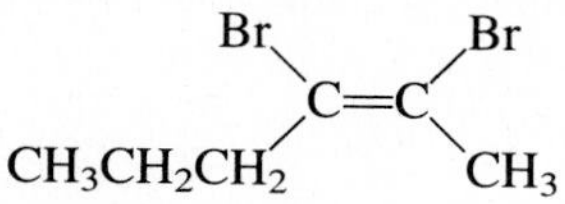

Solution There are six carbon atoms in the longest chain. It is numbered from right to left so that the double bond is at carbon-2. The parent is then 2-hexene. The bromine atoms are bonded to carbon-2 and carbon-3.

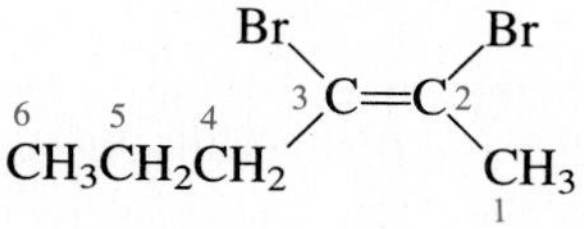

2,3-dibromo-2-hexene

Since there are two different groups of atoms on each unsaturated carbon atom, *cis* and *trans* isomers are possible. Inspection of the molecule indicates that this is *cis*-2,3-dibromo-2-hexene.

[*Additional examples may be found in 19.49 and 19.50 at the end of the chapter.*]

19.10 Reactions of Alkenes

Oxidation and Reduction Reactions

Alkenes can be attacked by oxidizing agents. For example, potassium permanganate oxidizes an alkene to a dialcohol or diol (two —OH groups).

$$3\ CH_3\text{—}CH\!\!=\!\!CH\text{—}CH_3 + 2\ KMnO_4 + 4\ H_2O \longrightarrow 3\ CH_3\text{—}\underset{\overset{|}{OH}}{CH}\text{—}\underset{\overset{|}{OH}}{CH}\text{—}CH_3 + 2\ MnO_2 + 2\ KOH$$

Potassium permanganate is a soluble purple compound; the MnO_2 is a brown solid that precipitates from solution. Since there is a color change, the oxidation of alkenes by potassium permanganate is a reaction that can be used to test visually for the presence of a double bond. Alkanes and cycloalkanes are not oxidized by potassium permanganate.

The reaction of hydrogen with an unsaturated molecule results in a saturated molecule. Although the process is one of reduction, the reaction is also called hydrogenation. The hydrogenation of 1-octene yields octane.

$$CH_3(CH_2)_5\text{—}CH\!\!=\!\!CH_2 + H\text{—}H \xrightarrow{\ Pt\ } CH_3(CH_2)_5\text{—}CH_2\text{—}CH_3$$
$$\quad\text{1-octene}\qquad\qquad\qquad\qquad\qquad\qquad\text{octane}$$

In order for the reaction to occur, finely divided platinum is used as a catalyst. Other catalysts that can be used include nickel and palladium, which are also members of Group VIIIB of the periodic table.

Hydrogenation is used commercially to convert liquid oils into solid fats. Oils tend to be unsaturated, whereas fats are more saturated. As the degree of saturation increases, the melting point increases.

Example 19.10

How many moles of hydrogen gas will react with the following compound? What is the molecular formula of the product?

Solution   There are two double bonds in the compound, so 1 mole of the compound will react with 2 moles of hydrogen gas. The product is a cycloalkane containing ten carbon atoms. Since the general molecular formula for a cycloalkane is C_nH_{2n}, the molecular formula of the product is $C_{10}H_{20}$.

Addition Reactions of Alkenes

An **addition reaction** occurs when two substances join together to form a compound containing all atoms present in the original substances. The carbon–carbon double bond provides the site for addition reactions of alkenes with other substances. The reactions of ethene with some common reagents are

$$CH_2{=}CH_2 + Br_2 \longrightarrow BrCH_2CH_2Br$$

$$CH_2{=}CH_2 + HCl \longrightarrow CH_3CH_2Cl$$

$$CH_2{=}CH_2 + H_2O \xrightarrow{H^+} CH_3CH_2OH$$

Note that one atom of the reagent becomes bonded to one carbon atom, and the second atom is bonded to the second carbon atom. As a result, the double bond is converted to a single bond.

In the case of the addition of bromine, the reaction is easily seen. Bromine is reddish brown; reaction leads to a colorless organic compound. The disappearance of the reddish color of bromine is then a useful test in determining whether a compound is unsaturated. Drops of bromine dissolved in carbon tetrachloride can be added to a compound. If the color disappears, the compound is unsaturated.

Reagents that add to alkenes can be classified into two types. **Symmetrical reagents** consist of two identical groups, each of which can become attached to the one carbon atom of the double bond. Bromine is an example of a symmetrical reagent. An **unsymmetrical reagent** consists of different groups, each of which can become attached to the one carbon atom of the double bond. Both HCl and H_2O are examples of unsymmetrical reagents.

There is only one possible product of the addition of a symmetrical reagent to an alkene. For example, bromine will react with propene to yield one specific compound.

$$CH_3CH{=}CH_2 + Br_2 \longrightarrow CH_3\overset{\overset{\displaystyle Br}{|}}{C}HCH_2Br$$

It doesn't make any difference which bromine atom becomes attached to which carbon atom. Bromine is a symmetrical molecule, and there is no way to distinguish the bromine atoms in the compound from each other.

When an unsymmetrical reagent such as HCl, is added to an unsymmetrical alkene, there are two possible products that can result. For a symmetrical alkene, such as ethene, only one product is possible because the two carbon atoms are identical.

$$\underset{\text{identical carbon atoms}}{CH_2{=}CH_2} + HCl \longrightarrow \underset{\text{identical structures}}{H{-}CH_2{-}CH_2{-}Cl \quad \text{or} \quad Cl{-}CH_2{-}CH_2{-}H}$$

Although two possible products could result from the addition of HCl to an unsymmetrical alkene, only one is actually formed. It seems as if addition of HCl to propene could yield either 1-chloropropane or 2-chloropropane. However, only 2-chloropropane is formed. The X written through one reaction arrow indicates that the reaction does not occur.

$$\underset{\substack{\text{not identical}\\ \text{carbon atoms}}}{CH_3CH{=}CH_2} + HCl \overset{\overset{\displaystyle \times}{\nearrow}}{\underset{\displaystyle \searrow}{}}$$

$$\underset{\text{(not formed)}}{CH_3\overset{\overset{\displaystyle H}{|}}{C}H{-}\overset{\overset{\displaystyle Cl}{|}}{C}H_2}$$

$$\underset{\text{only product}}{CH_3\overset{}{C}H{-}CH_2 \atop \overset{\displaystyle Cl \quad H}{}}$$

The Russian chemist Markovnikov observed that unsymmetrical reagents add to unsymmetrical double bonds in a specific way. **Markovnikov's rule** is that when a molecule of the general formula HX adds to a double bond, the hydrogen atom bonds to the unsaturated carbon atom that has the greater number of directly bonded hydrogen atoms. In the case of propene, one unsaturated carbon atom has two hydrogen atoms attached, and the other has only one hydrogen atom. Thus HCl is added so that the hydrogen atom is added to the terminal carbon atom, which already contains two hydrogen atoms, and the chlorine adds to the interior carbon atom, which has only one hydrogen atom.

Example 19.11

Predict the product formed when HCl adds to 1-methylcyclopentene.

Solution One unsaturated carbon atom has one attached hydrogen atom; the other carbon atom has no attached hydrogen atom. The product is then 1-chloro-1-methylcyclopentane.

1-methylcyclopentene   + H—Cl ⟶   1-chloro-1-methylcyclopentane

[*Additional examples may be found in 19.59 and 19.60 at the end of the chapter.*]

19.11 Alkynes

The triple bond in an alkyne decreases the number of hydrogen atoms in the molecule by four compared to alkanes. As a result, the general molecular formula for alkynes is C_nH_{2n-2}.

The simplest alkyne, C_2H_2, is commonly called acetylene. Unfortunately, the common name ends in -ene, which then might suggest that the compound contains a double bond. Such confusion is one reason why IUPAC names are so important for clear communication in chemistry. The IUPAC name for C_2H_2 is ethyne.

The structure of ethyne is shown in Figure 19.9. All four atoms lie in a straight line. Each H—C≡C bond angle is 180°. In other alkynes, the two triple-bonded carbon atoms and the two atoms directly attached to them all lie in a straight line.

Nomenclature of Alkynes

The rules for naming alkynes by the IUPAC system are analogous to those for alkanes and alkenes.

1. The longest continuous chain containing the triple bond is used as the parent.
2. The stem name of the alkane is changed by dropping the -ane suffix and adding -yne.
3. The carbon atoms are numbered consecutively from the end nearest the triple bond. The number of the first carbon atom with the triple bond is used as a prefix separated by a hyphen from the parent name.
4. Alkyl groups are named and their positions on the chain are determined by the numbering established by rule 3.

Figure 19.9

$$H - C \equiv C - H$$

The structure of ethyne

Examples of the correct nomenclature follow.

$$CH_3 - CH_2 - C \equiv C - CH_3$$

2-pentyne
(not 3-pentyne)

$$CH_3 - C \equiv C - CH_2 - \overset{\overset{\displaystyle CH_3}{|}}{CH} - CH_3$$

5-methyl-2-hexyne
(not 2-methyl-4-hexyne)

Reactions of Alkynes

The reactivity of the triple bond is similar to a double bond in an alkene. Thus, triple bonds undergo addition reactions just like those of the double bonds. However, a triple bond can add two molecules of a specific reagent. For example, propyne reacts with hydrogen to give propane.

$$CH_3 - C \equiv C - H \xrightarrow[\text{Pt}]{H_2} CH_3CH = CH_2 \xrightarrow[\text{Pt}]{H_2} CH_3CH_2CH_3$$

Although the reaction can be controlled with special catalysts to stop at the alkene, complete hydrogenation of an alkyne produces an alkane.

Addition of two molecules of halogens also occurs to produce a compound containing four halogen atoms.

$$CH_3 - C \equiv C - H \xrightarrow{Br_2} CH_3CBr = CHBr \xrightarrow{Br_2} CH_3CBr_2CHBr_2$$

Example 19.12

What is the molecular formula for the following compound? How many moles of hydrogen will react with the compound?

$$C \equiv C - H$$

Solution The compound contains eight carbon atoms. An eight-carbon cycloalkane without double or triple bonds would have the molecular formula C_8H_{16} based on the general molecular formula C_nH_{2n}. A double bond creates a deficiency of two hydrogen atoms, and a triple bond creates a deficiency of four hydrogen atoms. Thus, the molecular formula is C_8H_{10}.

One mole of hydrogen per mole of compound is required for the double bond; 2 moles of hydrogen per mole of compound are required for the triple bond. A total of 3 moles of hydrogen per mole of compound will react.

[*Additional examples may be found in 19.62 at the end of the chapter.*]

The product of the addition of hydrogen halides is that predicted by Markovnikov's rule.

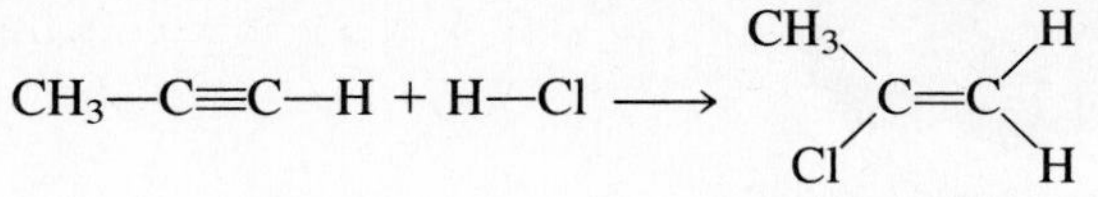

$$CH_3-C{\equiv}C-H + H-Cl \longrightarrow$$

(structure) 2-chloropropene

(structure) + H—Cl ⟶ $CH_3-\overset{Cl}{\underset{Cl}{C}}-\overset{H}{\underset{H}{C}}-H$

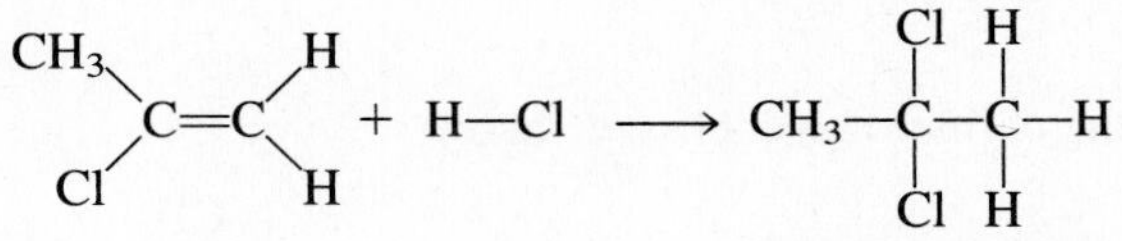

2,2-dichloropropane

19.12 Aromatic Hydrocarbons

Aromatic hydrocarbons are a class of unsaturated compounds that do not easily undergo addition reactions. Benzene (C_6H_6) does not add bromine as alkenes and alkynes do. In fact, benzene will react with bromine only in the presence of ferric bromide (or iron, which reacts with bromine to form ferric bromide), and then substitution rather than addition occurs.

$$C_6H_6 + Br_2 \xrightarrow{FeBr_3} C_6H_5Br + HBr$$

Benzene is a planar molecule in which all carbon–carbon bonds are of equal length. The bond angles of the ring are all 120°. Based on this data, benzene cannot be pictured with a conventional Lewis structure. An alternating series of single and double bonds satisfies the Lewis octet rule but not the fact that all bonds are equivalent. The bonding in benzene is rather special. Some of the electrons are shared between bonded atoms in the ordinary way. However, six electrons are shared over all carbon atoms of the entire molecule. The sharing of electrons over many atoms is called **delocalization.** It is this delocalization of electrons that accounts for the unique chemical stability of benzene. In order to write the structure for benzene, some chemists prefer to use a circle drawn within a hexagon; the circle indicates the six electrons that are spread out over the entire ring. Each corner of the hexagon represents a carbon atom and one attached hydrogen atom.

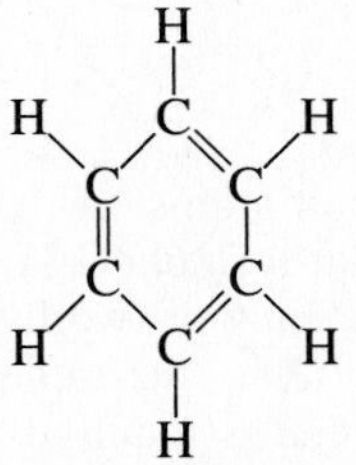

conventional Lewis structure 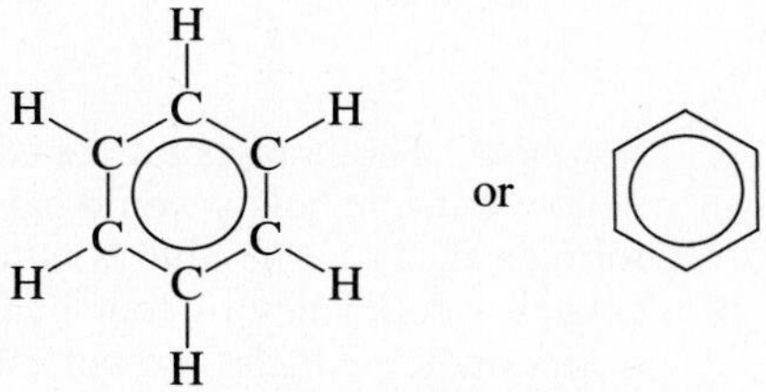   symbolic representations

All 12 atoms of benzene are in a plane. When a substituent group replaces a hydrogen atom, the atom of the group bonded to the ring is also in the plane of the ring.

Nomenclature of Aromatic Hydrocarbons

The IUPAC system of naming substituted aromatic hydrocarbons uses the name of the substituents as a prefix to benzene. Examples include the halogen-substituted benzenes.

fluorobenzene   chlorobenzene   bromobenzene

All of the compounds shown have the substituent at a "12 o'clock" position. However, all six positions on the benzene ring are equivalent, and you should recognize a compound such as bromobenzene no matter where the bromine atom is written.

is the same as   is the same as

Disubstituted compounds result when two hydrogen atoms are replaced by other groups. Three disubstituted isomers are possible. The three isomers of dichlorobenzene are designated *ortho, meta,* and *para.* These terms are abbreviated as *o, m,* and *p.* Alternatively a numbering system also may be used to give the lowest possible numbers to the carbon atoms bearing chlorine atoms.

o-dichlorobenzene   *m*-dichlorobenzene   *p*-dichlorobenzene
1,2-dichlorobenzene   1,3-dichlorobenzene   1,4-dichlorobenzene

When two groups are adjacent on the ring, they are in **ortho** positions. When the groups are separated by one carbon atom, they are in **meta** positions. When the groups are separated by two carbon atoms, they are in **para** positions. It is important to realize that any of these isomers could be written in a different orientation on the page without changing their identity or name.

is the same as   is the same as

When three or more substituents are attached on the benzene ring, the positions are numbered. Note in the following examples that the numbers are chosen to assign the lowest set of numbers for the substituents.

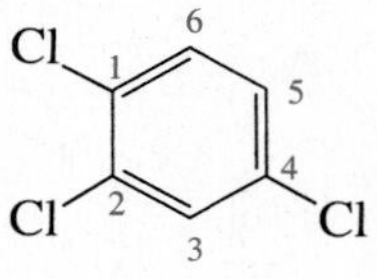

1,2,4-trichlorobenzene

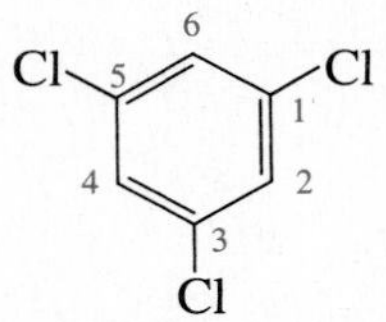

1,3,5-trichlorobenzene

Many compounds are better known by their common names than by their systematic names. These include toluene, phenol, and aniline.

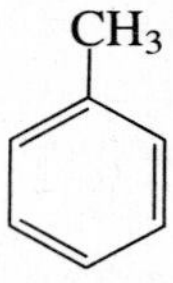

toluene
methylbenzene

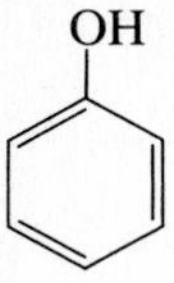

phenol
hydroxybenzene

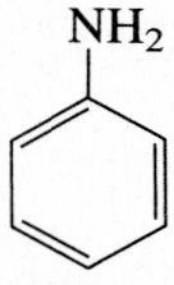

aniline
aminobenzene

Sometimes compounds are named with the common name of the monosubstituted aromatic compound as the parent. Three such cases are given below.

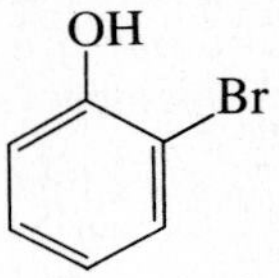

o-bromophenol

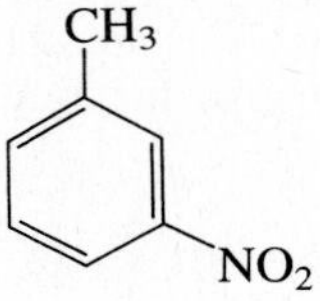

m-nitrotoluene

p-chloroaniline

In complex molecules, the benzene ring is often named as a substituent on a parent chain of carbon atoms. The aromatic group C_6H_5— derived from benzene is called a **phenyl** group. The general name for aromatic groups that are to be treated as substituents is **aryl,** symbolized Ar.

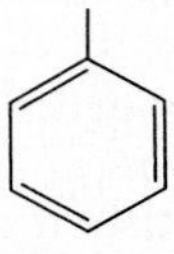

$$CH_3CH_2CH_2CH_2CHCH_2CH_3$$

phenyl
(an aryl group)

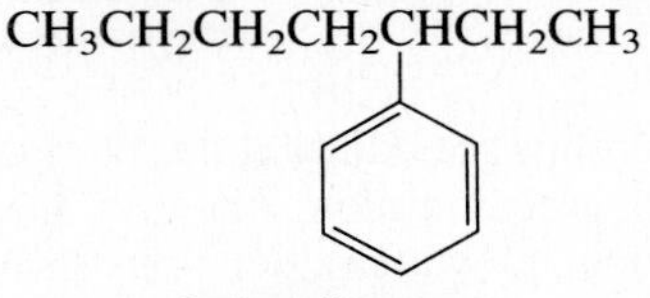

3-phenylheptane

Summary

Organic compounds have physical and chemical properties that are consequences of the carbon–carbon bond. The variety of organic compounds is the result of stable single, double, and triple bonds between carbon atoms as well as the incorporation of other atoms into bonds with carbon.

Isomerism is common among organic compounds. Isomers have

identical molecular formulas but different structures. Two or more models or written representations of the same organic compound that represent different orientations of atoms in space are called conformers. Conformers can be interconverted by rotation about carbon–carbon single bonds.

Organic compounds have specific names assigned by the International Union of Pure and Applied Chemistry. There is a one-to-one correspondence between a structure and its name.

Saturated hydrocarbons contain only carbon–carbon single bonds and carbon–hydrogen bonds. The atoms bonded to each carbon atom form a tetrahedral shape. There are two classes of saturated hydrocarbons: alkanes and cycloalkanes. The general molecular formula for alkanes is C_nH_{2n+2}. Each ring of a cycloalkane results in two fewer hydrogen atoms. The general formula for a cycloalkane is C_nH_{2n}.

Alkanes are named by selecting the longest continuous chain of carbon atoms as the parent compound. The group of saturated carbon atoms attached to the chain is an alkyl group. Both position and the identity of the alkyl group are prefixed to the parent name.

Saturated hydrocarbons are quite unreactive, but will undergo substitution reactions to yield a mixture of isomers.

Alkenes and alkynes are classes of unsaturated compounds. Alkenes contain a carbon–carbon double bond and alkynes a carbon–carbon triple bond. Alkenes have restricted rotation about the double bond. Geometric isomers or *cis-trans* isomers of alkenes have the same order of atoms bonded to each other, but have different orientations in space. However, the isomers cannot be interconverted by rotation about the carbon–carbon double bond.

The IUPAC system of naming alkenes and alkynes uses the -ene and -yne suffix, respectively, and designates the position of the double bond or triple bond by a number. The number of the substituents is dictated by the number assigned to the double bond or, if there is no double bond, to the triple bond. The characteristic reaction of alkenes and alkynes is addition to the multiple bond.

Aromatic compounds are a third type of unsaturated compound. These compounds have a special stability as the result of sharing electrons over the entire ring.

New Terms

An **addition reaction** is a reaction that occurs when two substances join to form a compound containing all of the atoms present in the original substances.

An **alkane** contains only carbon and hydrogen atoms bonded only by single bonds.

An **alkene** has a carbon–carbon double bond in a hydrocarbon.

An **alkyl group** is a group containing carbon and hydrogen atoms that resembles an alkane but has one less hydrogen atom.

An **alkyne** has a carbon–carbon triple bond in a hydrocarbon.

An **aromatic compound** has a benzene ring or structural units that resemble a benzene ring.

An **aryl** group is a substituent group derived from an aromatic compound.

A **branched-chain alkane** has carbon atoms bonded to more than two other carbon atoms.

A *cis* isomer is an isomer that has two groups of atoms oriented on the same side of a structural feature, such as an alkene.

A **condensed structural formula** shows only specific bonds; other bonds are implied.

A **conformation** is an orientation of a molecule that results from rotation about single bonds.

A **conformer** is a specific orientation of a molecule.

A **cycloalkane** is a hydrocarbon that contains a ring of carbon atoms.

Delocalization is the sharing of electrons over many atoms.

A **hydrocarbon** is a compound containing only carbon and hydrogen.

A **homologous series** is a series of compounds that differ from adjacent members by repeating units.

Isomers have the same molecular formula but differ in structure.

Markovnikov's rule states that when a compound HX adds to a double bond, the hydrogen atom bonds to the unsaturated carbon atom that has the greater number of directly bonded hydrogen atoms.

Perspective structural formulas are representations using wedge and dashed lines to show the geometry of a molecule.

The **phenyl** group derived from benzene is C_6H_5—.

A **primary carbon atom** is a carbon atom that is directly bonded to one other carbon atom.

Primary hydrogen atoms are bonded to a primary carbon atom.

A **saturated** hydrocarbon has only carbon–carbon single bonds and the maximum possible number of hydrogen atoms bonded to the carbon atoms.

A **secondary carbon atom** is a carbon atom that is directly bonded to two other carbon atoms.

Secondary hydrogen atoms are bonded to a secondary carbon atom.

Straight-chain alkanes have carbon atoms bonded to each other like links in a chain.

A **structural formula** shows the arrangement of atoms and bonds in a molecule.

A **substituent** is an atom or group of atoms attached to the skeleton of carbon atoms in a compound.

A **symmetrical reagent** consists of two identical groups, each of which can become attached to one carbon atom of a double bond in an addition reaction.

A **tertiary carbon atom** is a carbon atom that is directly bonded to three other carbon atoms.

A **tertiary hydrogen atom** is bonded to a tertiary carbon atom.

A *trans* isomer is an isomer that has two groups of atoms oriented on opposite sides of a structural feature, such as an alkene.

Unsaturated compounds contain carbon–carbon multiple bonds.

An **unsymmetrical reagent** consists of different groups, each of which can become attached to one carbon atom of a double bond in an addition reaction.

Exercises

Terminology

19.1 Explain the difference between a molecular formula and a structural formula.

19.2 Explain the difference between conformers and isomers.

19.3 Explain what is meant by the term saturated in organic chemistry.

19.4 What do the terms primary, secondary, and tertiary mean?

19.5 Distinguish between structural and geometric isomerism.

19.6 Describe the structural similarities and differences between alkenes and alkynes.

Molecular Formulas

19.7 Write the molecular formula for each of the following structures.
(a) $CH_3CH_2CH_2CH_2CH_3$
(b) $CH_3CH_2CH_2CH_3$
(c) $CH_3CH_2CH_2CH_2CH_2CH_3$
(d) $CH_3CH_2CH_2CH_2CH_2CH_2CH_2CH_2CH_3$

19.8 Write the molecular formula for each of the following structures.
(a) $CH_3CH_2C\equiv CH$
(b) $CH_3CH_2C\equiv CCH_3$
(c) $CH_3CH_2CH_2CH=CH_2$
(d) $CH_2=CHCH_2CH_3$

Condensed Structural Formulas

19.9 Write condensed structural formulas in which only the bonds to hydrogen are not shown.

(a) H—C—C—Br   (b) H—C—C—H

(c) Br—C—C—Br   (d) H—C—C—C—C—C—H

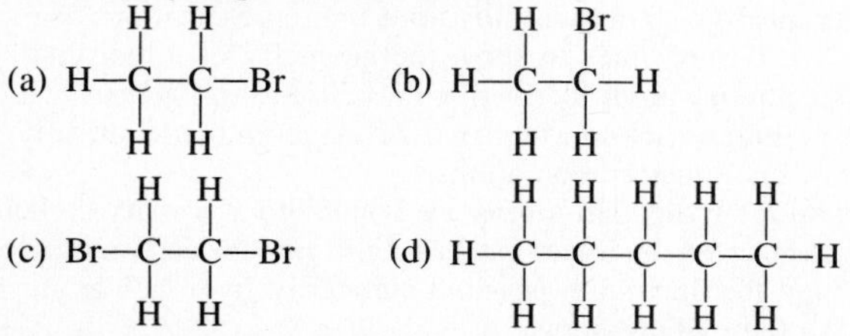

19.10 Write condensed structural formulas in which only the bonds to hydrogen are not shown.

(a) H—C—C—C—S—H   (b) H—C—C—C—N—H

(c) H—C—C—C—C—S—H   (d) H—C—C—C—C—Cl

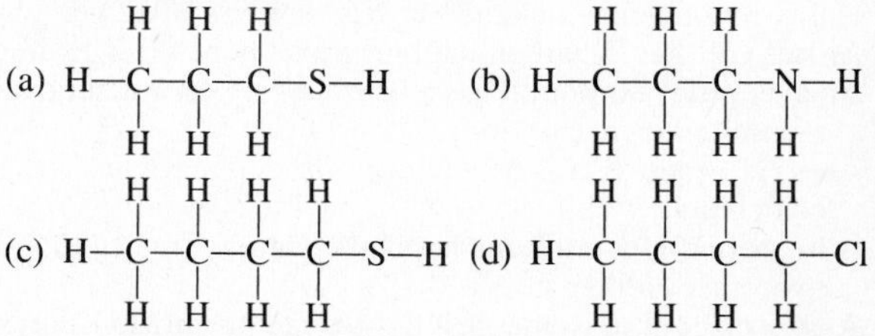

19.11 Write a condensed structural formula in which no bonds are shown for each substance in 19.9.

19.12 Write completely condensed formulas in which no bonds are shown for each substance in 19.10.

Structural Formulas

19.13 Write a complete structural formula showing all bonds for each of the following condensed formulas.
(a) $CH_3CH_2CH_2CH_3$ (b) $CH_3CH(CH_3)CH_2CH_3$
(c) $CH_3CH_2CH(CH_3)_2$ (d) $(CH_3)_3CCH_2CH_3$

19.14 Write a complete structural formula showing all bonds for each of the following condensed formulas.
(a) $CH_3CH_2CHBr_2$ (b) $CH_3CHBrCH_2CH_2CH_3$
(c) $CH_3CH_2CCl_2CH_2CH_3$ (d) $CH_2ClCH_2CHClCH_3$

19.15 Draw a three-dimensional structure for each of the following.
(a) CH_3Cl (b) CH_2Br_2 (c) CH_3F
(d) CHF_3 (e) CH_2ClBr (f) CCl_4

19.16 Draw a three-dimensional structure for each of the following.
(a) CH_3CH_3 (b) CH_3NH_2 (c) CH_3OH
(d) CH_3SH (e) CH_3CH_2Cl (f) CH_3CHCl_2

444

Isomers and Conformers

19.17 Indicate whether the following pairs of structures are isomers or conformers.

(a)
$$H-\overset{\overset{\displaystyle Br}{|}}{\underset{\underset{\displaystyle H}{|}}{C}}-\overset{\overset{\displaystyle H}{|}}{\underset{\underset{\displaystyle H}{|}}{C}}-Br \quad \text{and} \quad Br-\overset{\overset{\displaystyle H}{|}}{\underset{\underset{\displaystyle H}{|}}{C}}-\overset{\overset{\displaystyle H}{|}}{\underset{\underset{\displaystyle H}{|}}{C}}-Br$$

(b) CH_3-CH_2 and $CH_3-CH_2-CH_2-Cl$
 $\quad\quad\quad CH_2-Cl$

(c) $CH_3-\underset{\underset{\displaystyle CH_3}{|}}{CH}-Cl$ and $CH_3-CH_2-CH_2-Cl$

(d) $CH_3-\underset{\underset{\displaystyle CH_3}{|}}{CH}-CH_2-CH_3$ and $CH_3-CH_2-CH_2$
 $\quad\quad\quad\quad\quad\quad\quad\quad\quad\quad\quad\quad\quad\quad\quad CH_2-CH_3$

19.18 Which of the following represent the same compound drawn in different perspectives?

(a) $CH_3CH_2CH_2\overset{\overset{\displaystyle CH_3}{|}}{CH}CH_3$

(b) $CH_2\overset{\overset{\displaystyle CH_3}{|}}{CH}CH_2\overset{\overset{\displaystyle CH_3}{|}}{CH}CH_3$

(c) $CH_3\overset{\overset{\displaystyle CH_3}{|}}{CH}CH_2CH_2\overset{\overset{\displaystyle CH_3}{|}}{CH}CH_3$

(d)

(e)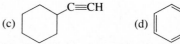

(f) $CH_3\overset{\overset{\displaystyle CH_3}{|}}{CH}CH_2\overset{\overset{\displaystyle CH_3}{|}}{CH}CH_2$
 $\quad\quad\quad CH_3 \quad\quad CH_3$

(g) $CH_3CHCH_2\overset{\overset{\displaystyle CH_3}{|}}{CH}CH_3$
 $\quad\quad\underset{\displaystyle CH_3}{|}$

Saturated and Unsaturated Hydrocarbons

19.19 Classify each of the following as saturated or unsaturated.

(a) $CH_3\overset{\overset{\displaystyle CH_3}{|}}{CH}CH_3$

(b) $CH_3-\overset{\overset{\displaystyle CH_3}{|}}{C}=CH_2$

(c) $CH_3-C\equiv C-CH_3$

(d) $CH_3-C\equiv C-CH=CH_2$

19.20 Classify each of the following as saturated or unsaturated.

(a)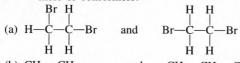

(b)

(c)

(d)

19.21 Which of the following molecular formulas for noncyclic hydrocarbons represent saturated compounds?

(a) C_6H_{12} (b) C_5H_{12} (c) C_8H_{18}
(d) $C_{10}H_{22}$ (e) $C_{18}H_{34}$ (f) $C_{20}H_{40}$

19.22 Which of the following molecular formulas for noncyclic hydrocarbons represent saturated compounds?

(a) C_6H_{10} (b) C_4H_{10} (c) C_7H_{16}
(d) $C_{10}H_{20}$ (e) $C_{17}H_{36}$ (f) $C_{20}H_{42}$

19.23 How many hydrogen atoms are contained in an alkane with the following number of carbon atoms?

(a) 10 (b) 8 (c) 5
(d) 15 (e) 6 (f) 4

19.24 How many hydrogen atoms are contained in an alkane with the following number of carbon atoms?

(a) 12 (b) 22 (c) 32
(d) 15 (e) 28 (f) 40

19.25 What is the molecular formula for each of the compounds with the following structural features?

(a) six carbon atoms and one double bond
(b) five carbon atoms and two double bonds
(c) seven carbon atoms, one ring, and one double bond
(d) four carbon atoms and one triple bond

19.26 What is the molecular formula for each of the compounds with the following structural features?

(a) four carbon atoms and two triple bonds
(b) four carbon atoms, one double bond, and one triple bond
(c) ten carbon atoms and two rings
(d) ten carbon atoms, two rings, and five double bonds

Nomenclature of Alkanes

19.27 Name each of the following alkyl groups.

(a) CH_3-

(b) $CH_3-\underset{\displaystyle |}{CH}-CH_3$

(c) $CH_3-\overset{\overset{\displaystyle CH_3}{|}}{\underset{\underset{\displaystyle CH_3}{|}}{C}}-$

(d) $CH_3-\underset{\underset{\displaystyle |}{|}}{CH}-CH_2-CH_3$

19.28 Name each of the following alkyl groups.

(a) CH_3-CH_2-

(b) $CH_3-\overset{\overset{\displaystyle |}{|}}{CH}-CH_2-$
 $\quad\quad\quad\quad\underset{\displaystyle CH_3}{|}$

(c) $CH_3-CH_2-CH_2-CH_2-$

(d) $CH_3-CH_2-CH_2-$

19.29 Give the IUPAC name for each of the following compounds.

(a) 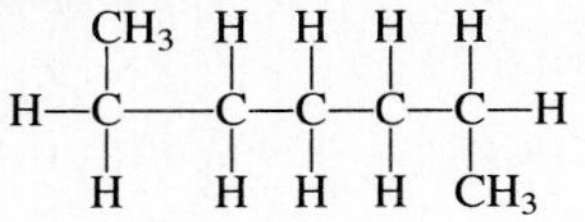
$$
\begin{array}{ccccc}
& CH_3 & H & H & H & H \\
H-&C&-&C&-&C&-&C&-&C&-H \\
& H & H & H & H & CH_3
\end{array}
$$

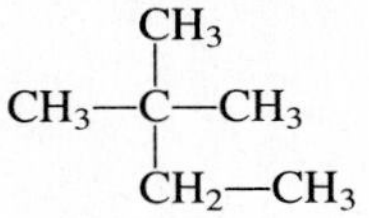

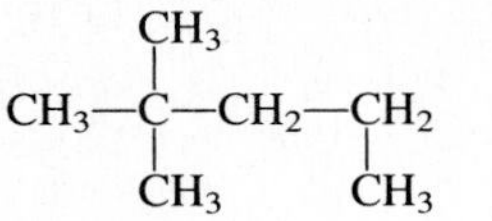

(b)
$$
CH_3-\overset{\displaystyle CH_3}{\underset{\displaystyle CH_2-CH_3}{C}}-CH_3
$$

(c)
$$
CH_3-\overset{\displaystyle CH_3}{\underset{\displaystyle CH_3}{C}}-CH_2-CH_2
$$
CH$_3$

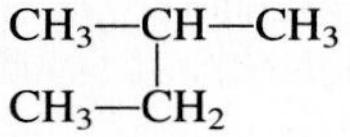

(d) 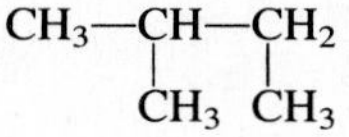
$$
CH_3-CH-CH_2-CH-\overset{\displaystyle CH_3}{CH}
$$
CH$_3$ CH$_3$—CH$_2$ CH$_3$

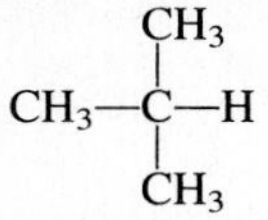

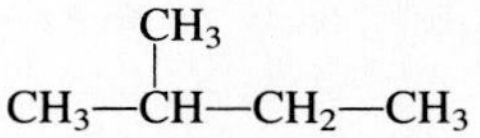

19.31 Write a structural formula for each of the following compounds.
(a) 3-methylpentane
(b) 3,4-dimethylhexane
(c) 2,2,3-trimethylpentane
(d) 4-ethylheptane
(e) 2,3,4,5-tetramethylhexane

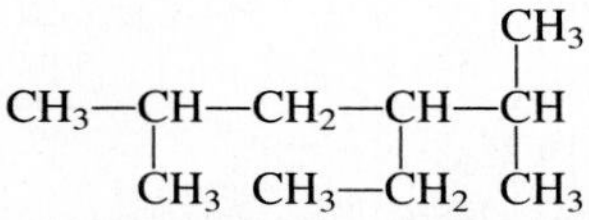

19.30 Give the IUPAC name for each of the following compounds.

(a)
$$
CH_3-\overset{\displaystyle CH_3}{\underset{\displaystyle CH_3}{C}}-H
$$

(b)
$$
CH_3-\overset{\displaystyle CH_3}{CH}-CH_2-CH_3
$$

(c)
$$
CH_3-CH-CH_3
$$
CH$_3$—CH$_2$

(d)
$$
CH_3-CH-CH_2
$$
CH$_3$ CH$_3$

19.32 Write the structural formula for each of the following compounds.
(a) 1,2,3-trichloropropane
(b) 2,2-dichloropentane
(c) 1,1,1-trichloropropane
(d) 1,1,2,3-tetrachlorobutane

Classification of Carbon Atoms

19.33 Determine the number of primary carbon atoms in each of the following.
(a) $CH_3CH_2CH_2CH_3$
(b) $CH_3CH_2CH_2CH_2CH_3$

(c) 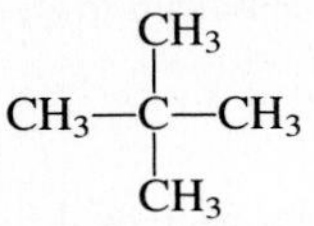
$$
CH_3-\overset{\displaystyle CH_3}{\underset{\displaystyle CH_3}{C}}-CH_3
$$

(d) 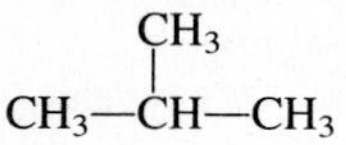
$$
CH_3-\overset{\displaystyle CH_3}{CH}-CH_3
$$

19.34 Determine the number of primary carbon atoms in each of the following.

(a) 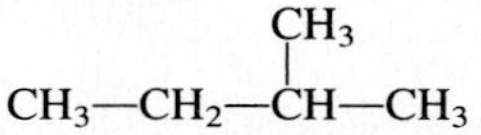
$$
CH_3-CH_2-\overset{\displaystyle CH_3}{CH}-CH_3
$$

(b) 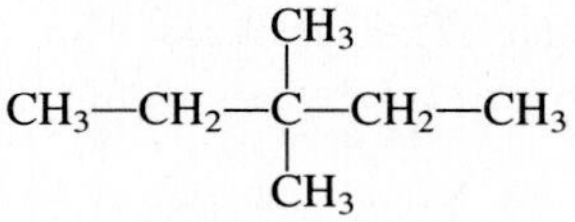
$$
CH_3-CH_2-\overset{\displaystyle CH_3}{\underset{\displaystyle CH_3}{C}}-CH_2-CH_3
$$

(c)
$$
CH_3-\overset{\displaystyle CH_3}{CH}-\overset{\displaystyle CH_3}{CH}-CH_3
$$

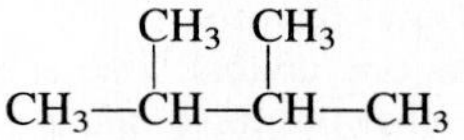

(d) 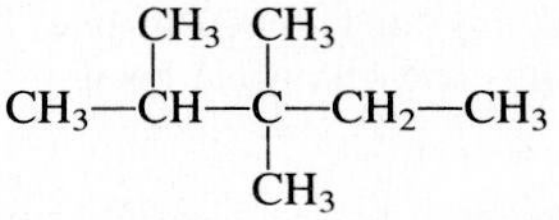
$$
CH_3-\overset{\displaystyle CH_3}{CH}-\overset{\displaystyle CH_3}{\underset{\displaystyle CH_3}{C}}-CH_2-CH_3
$$

19.35 Determine the number of secondary carbon atoms in each structure of 19.33.

19.37 Determine the number of tertiary carbon atoms in each structure of 19.33.

19.39 Determine the number of primary hydrogen atoms in each structure of 19.33.

19.36 Determine the number of secondary carbon atoms in each structure of 19.34.

19.38 Determine the number of tertiary carbon atoms in each structure of 19.34.

19.40 Determine the number of secondary hydrogen atoms in each structure of 19.34.

Halogenation of Alkanes

19.41 How many products can result from the substitution of one hydrogen atom by a chlorine atom in each of the following compounds?

(a) propane   (b) butane
(c) methylpropane   (d) pentane
(e) 2-methylbutane
(f) 2,2-dimethylbutane
(g) 2,3-dimethylbutane

19.42 How many products can result from the substitution of one hydrogen atom by a chlorine atom in each of the following compounds?

(a) 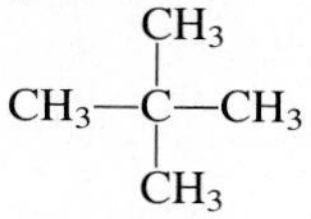 $CH_3\!-\!\overset{\displaystyle CH_3}{\underset{\displaystyle CH_3}{C}}\!-\!CH_3$

(b) $CH_3\!-\!\overset{\displaystyle CH_3}{CH}\!-\!CH_3$

(c) $CH_3\!-\!CH_2\!-\!\overset{\displaystyle CH_3}{\underset{\displaystyle CH_3}{C}}\!-\!CH_2\!-\!CH_3$

(d) 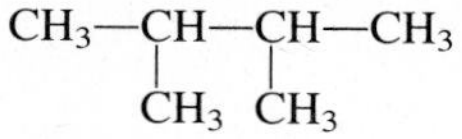 $CH_3\!-\!\overset{\displaystyle CH_3}{CH}\!-\!\overset{\displaystyle CH_3}{CH}\!-\!CH_3$

Cycloalkanes

19.43 Write fully condensed formulas for each of the following compounds.

(a) chlorocyclopropane
(b) 1,1-dimethylcyclobutane
(c) cycloheptane
(d) methylcyclopentane

19.44 Name each of the following compounds.

(a)

(b)

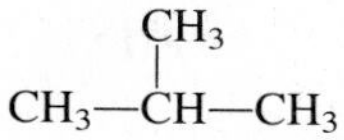

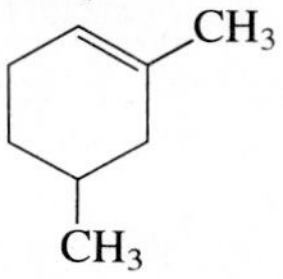

(c)

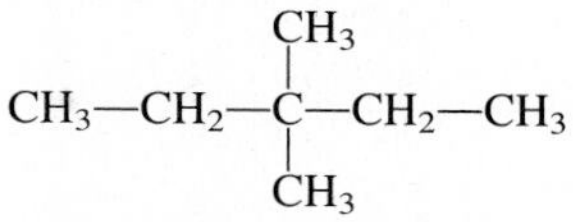

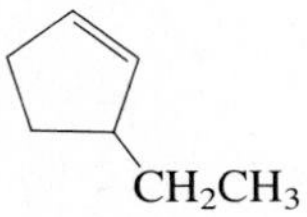

(d)

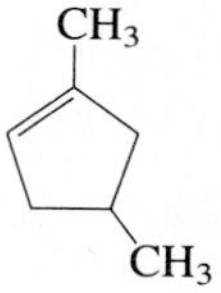

Isomers of Alkenes

19.45 There are four isomeric alkenes with the molecular formula C_4H_8. Draw structural formulas for the compounds, and name them.

19.46 There are six isomeric alkenes with the molecular formula C_5H_{10}. Write the structures, and name the compounds.

Geometric Isomers

19.47 Which of the following molecules can exist as *cis* and *trans* isomers?

(a) $CH_3CH\!=\!CHBr$ (b) $CH_2\!=\!CHCH_2Br$
(c) $CH_3CH\!=\!CHCH_2Cl$ (d) $(CH_3)_2C\!=\!CHCH_3$
(e) $CHBr\!=\!CHCl$ (f) $CBr_2\!=\!CCl_2$

19.48 Draw the structural formulas for the *cis* isomer of each of the following alkenes.

(a) $CH_3CH\!=\!CHCH_2CH_3$ (b) $CHCl\!=\!CHBr$
(c) $CH_2ClCH\!=\!CHCH_2Br$ (d) $CH_3CH\!=\!CHBr$

Nomenclature of Alkenes

19.49 Name each of the following compounds.

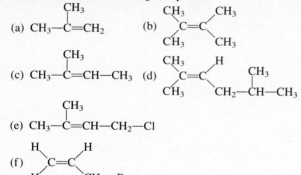

(a) $CH_3-\underset{\underset{CH_3}{|}}{C}=CH_2$

(b) $\underset{\underset{CH_3}{|}}{\overset{\overset{CH_3}{|}}{C}}=\underset{\underset{CH_3}{|}}{\overset{\overset{CH_3}{|}}{C}}$

(c) $CH_3-\underset{\underset{CH_3}{|}}{C}=CH-CH_3$

(d) $\underset{\underset{CH_3}{|}}{\overset{\overset{CH_3}{|}}{C}}=\underset{\underset{CH_2-CH-CH_3}{|}}{\overset{\overset{H}{|}}{C}}\ CH_3$

(e) $CH_3-\underset{\underset{CH_3}{|}}{C}=CH-CH_2-Cl$

(f) $\underset{\underset{H}{|}}{\overset{\overset{H}{|}}{C}}=\underset{\underset{CH_2-Br}{|}}{\overset{\overset{H}{|}}{C}}$

19.50 Name each of the following compounds.

(a) $\underset{\underset{H}{|}}{\overset{\overset{CH_3}{|}}{C}}=\underset{\underset{CH_2-CH_3}{|}}{\overset{\overset{H}{|}}{C}}$

(b) $\underset{\underset{CH_3}{|}}{\overset{\overset{H}{|}}{C}}=\underset{\underset{CH-CH_2-CH_3}{|}\atop\underset{CH_3}{|}}{\overset{\overset{H}{|}}{C}}$

(c) $\underset{\underset{H}{|}}{\overset{\overset{Cl}{|}}{C}}=\underset{\underset{CH_2-CH_3}{|}}{\overset{\overset{H}{|}}{C}}$

(d) $\underset{\underset{Cl}{|}}{\overset{\overset{H}{|}}{C}}=\underset{\underset{CH_2-CH_2-CH_3}{|}}{\overset{\overset{H}{|}}{C}}$

19.51 Write the structural formula for each of the following compounds.
(a) 2-methyl-2-pentene
(b) 1-hexene
(c) *cis*-2-methyl-3-hexene
(d) *trans*-5-methyl-2-hexene

19.52 Write the structural formula for each of the following compounds.
(a) cyclohexene
(b) 1-methylcyclopentene
(c) 1,2-dibromocyclohexene
(d) 4,4-dibromocyclohexene

Nomenclature of Alkynes

19.53 Name each of the following compounds.
(a) $CH_3CH_2CH_2C\equiv CH$
(b) $(CH_3)_3CC\equiv CCH_2CH_3$
(c) $Cl(CH_2)_2C\equiv C(CH_2)_3CH_3$
(d) $CH_3CHBrCHBrC\equiv CCH_3$
(e) $CH_3-C\equiv C-\underset{\underset{CH_2-CH_3}{|}}{CH}-CH_3$
(f) $CH_3-\underset{\underset{CH_2-CH_3}{|}}{CH}-CH_2-C\equiv C-\underset{\underset{Cl}{|}}{CH}-CH_3$

19.54 Write the structural formula for each of the following compounds.
(a) 2-hexyne
(b) 3-methyl-1-pentyne
(c) 5-ethyl-3-octyne
(d) 4,4-dimethyl-2-pentyne

Reactions of Alkenes

19.55 Describe the observation that should be made when 3-hexene reacts with Br_2. How can Br_2 be used to distinguish between 3-hexene and cyclohexane?

19.57 Complete the following equations.
(a) $CH_3CH_2CH=CH_2 + Br_2 \longrightarrow$
(b) $CH_3CH_2CH=CH_2 + HBr \longrightarrow$

19.59 Write the product of the reaction of each of the reagents in 19.58 with propene.

19.56 Describe the observation that should be made when 3-hexene reacts with potassium permanganate.

19.58 Classify the following addition reagents as symmetrical or unsymmetrical.
(a) Cl_2 (b) H_2O (c) HBr
(d) Br_2 (e) HCl

19.60 Write the product of the reaction of each of the reagents in 19.58 with methylpropene.

Reactions of Alkynes

19.61 Complete the following reactions.
(a) $CH_3C{\equiv}CH + 2 H_2 \longrightarrow$
(b) $CH_3C{\equiv}CH + 2 Br_2 \longrightarrow$
(c) $CH_3C{\equiv}CH + HBr \longrightarrow$
(d) $CH_3C{\equiv}CH + 2 HBr \longrightarrow$

19.62 How many moles of hydrogen will react with each of the following compounds?
(a) $CH_3C{\equiv}CC{\equiv}CH$ (b) $HC{\equiv}C{-}C{\equiv}CH$

(c)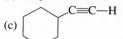

(d) $HC{\equiv}C{-}CH{=}CH{-}C{\equiv}CH$

Aromatic Compounds

19.63 Give the common names of each of the following compounds.

(a) 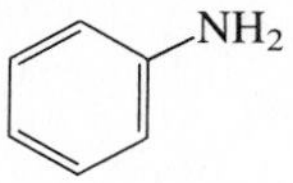 [benzene ring with NH₂]

(b) 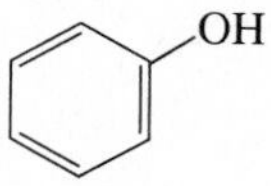 [benzene ring with OH]

(c) 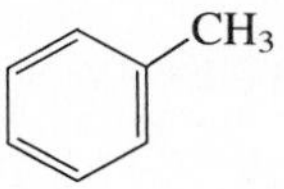 [benzene ring with CH₃]

(d) 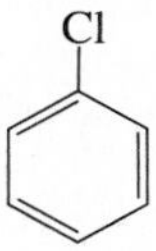 [benzene ring with Cl]

19.64 Name each of the following alkyl-substituted benzene compounds.

(a) 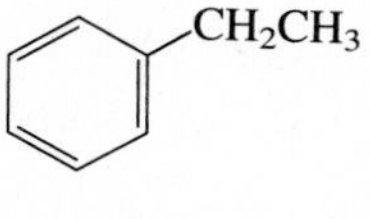 [benzene ring with CH₂CH₃]

(b) 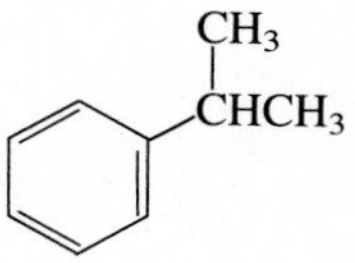 [benzene ring with CHCH₃ and CH₃ above]

(c) 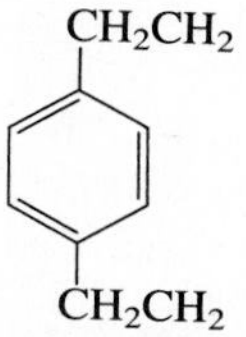 [benzene ring with CH₂CH₂ top and CH₂CH₂ bottom]

(d) 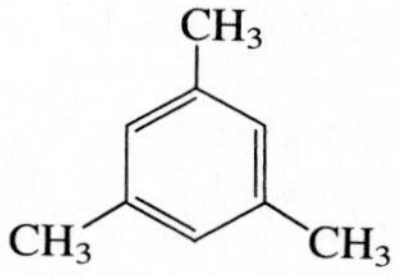 [benzene ring with CH₃, CH₃, CH₃]

19.65 Name each of the following compounds, using accepted common names where appropriate.

(a) 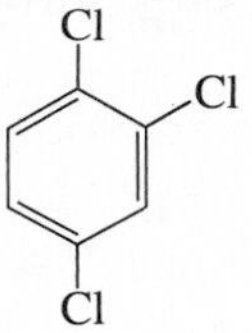 [benzene ring with Cl, Cl, Cl]

(b) 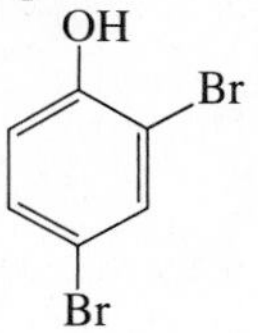 [benzene ring with OH, Br, Br]

(c) 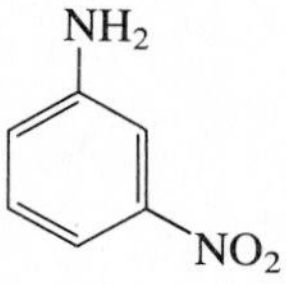 [benzene ring with NH₂ and NO₂]

(d) 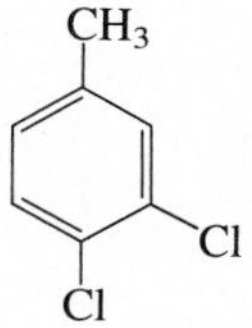 [benzene ring with CH₃, Cl, Cl]

19.66 Name the following compounds.

(a) 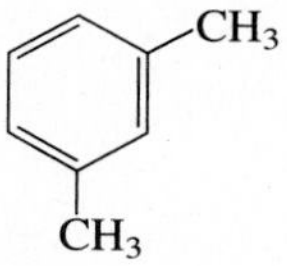 [benzene ring with CH₃ and CH₃]

(b)  [benzene ring with OH and Br]

(c) 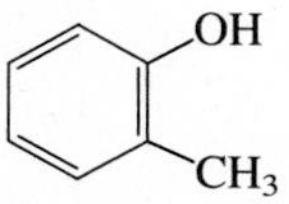 [benzene ring with OH and CH₃]

(d) 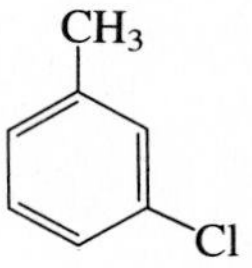 [benzene ring with CH₃ and Cl]

Additional Exercises

19.67 Write the molecular formula for each of the following structures.
(a) $CH_3CH_2CHCl_2$ (b) $CH_3CCl_2CH_3$
(c) $BrCH_2CH_2Br$ (d) $BrCH_2CH_2CH_2Cl$

19.68 Write the molecular formula for each of the following structures.
(a) $CH_3CH_2CH_2CHBr_2$ (b) $CH_3CF_2CH_2CH_3$
(c) $BrCH_2CH_2CH_2CH_2Br$ (d) $BrCH_2CH_2CHCl_2$

19.69 Draw each of the following so that the longest continuous chain is written horizontally.

(a) CH₃—CH₂
         |
        CH₂—CH₃

(b) CH₂—CH₂—CH—CH₂—CH₃
    | |
    CH₃   CH₂—CH₃

(c) CH₃—CH—CH₂—CH₃
        |
        CH₂—CH₃

(d)   CH₃
         |
  CH₃—CH—CH—CH₂—CH₃
        |
        CH₂—CH₃

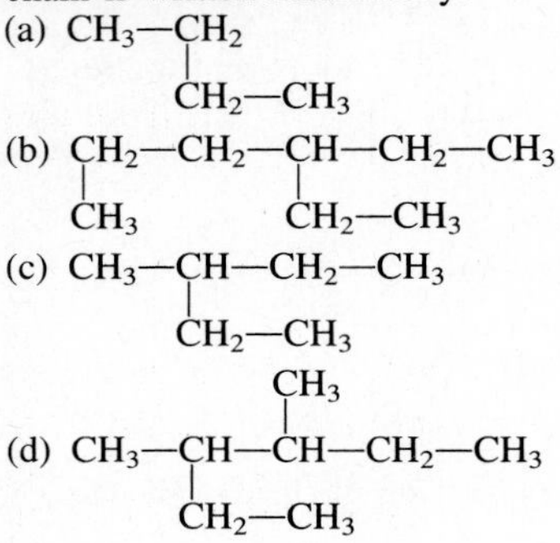

19.70 Draw each of the following so that the longest continuous chain is written horizontally.

(a) CH₃—CH—CH₂
        | |
        CH₃ CH₃

(b)   CH₃
          |
  CH₃—CH—CH—CH—CH₃
        | |
        CH₃   CH₂
                 |
                 CH₃

(c) CH₃—CH—CH₂—CH₂
        | |
        CH₂   CH₃
        |
        CH₃

(d)   CH₃
       |
  CH₃—CH—CH₂
           |
           CH₂
           |
  CH₃—CH₂

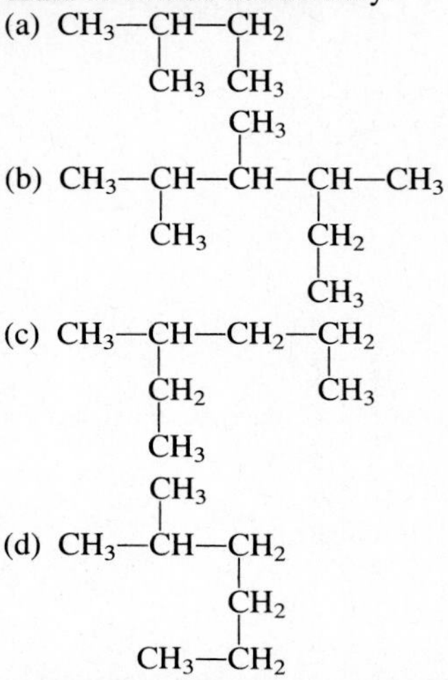

19.71 Which of the following cannot correspond to an actual molecule?

(a) C_6H_{14} (b) $C_{10}H_{23}$ (c) C_7H_{18}
(d) C_5H_{14} (e) $C_{10}H_{20}$ (f) C_5H_{10}

19.72 Which of the following cannot correspond to an actual molecule?

(a) C_6H_{12} (b) $C_{10}H_{22}$ (c) C_7H_{16}
(d) C_5H_{12} (e) $C_{10}H_{24}$ (f) C_5H_{11}

19.73 Classify the following compounds as saturated or unsaturated.

(a) $CH_3CH_2CH_3$ (b) $CH_3CH_2CH=CH_2$

(c)   (d) [cyclopentene with CH₃ substituent]

19.74 Classify the following compounds as saturated or unsaturated.

(a) $CH_3CH_2CH(CH_3)_2$ (b) $HC≡CCH_2CH_3$

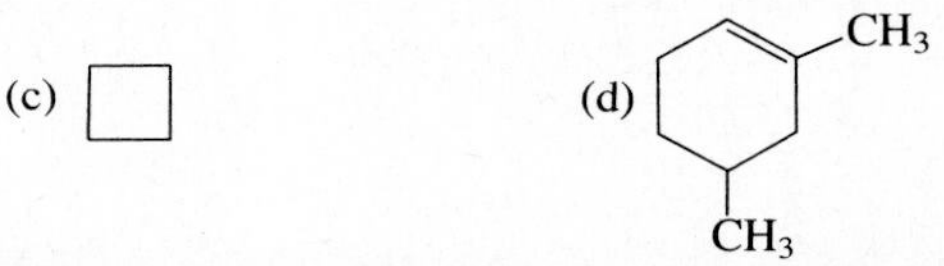

(c) [square/cyclobutane]   (d) [cyclohexene with two CH₃ substituents]

19.75 Write the molecular formula for each of the following compounds.

(a) [cyclohexadiene]   (b) [cyclohexane with $CH=CH_2$]

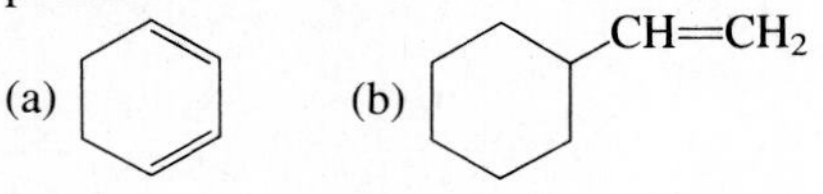

19.76 Write the molecular formula for each of the following compounds.

(a) [two cyclohexene rings connected by a C=C with H H]

(b) [structure with CH₃, H, C=C, H, C=C, H, $(CH_2)_6CH_2OH$]

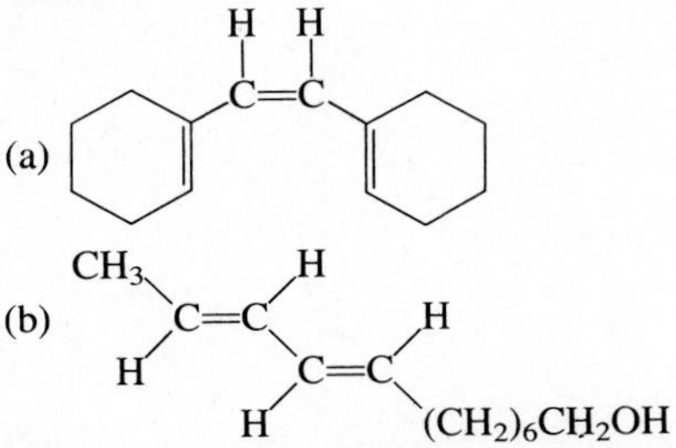

19.77 Name each of the following alkyl groups.

(a) CH₃—CH—   (b) CH₃—CH—
      | |
     CH₂—CH₃   CH₃

(c) CH₃—CH₂   (d) CH₃—CH₂
          | |
    CH₂—CH₂—   CH₂—

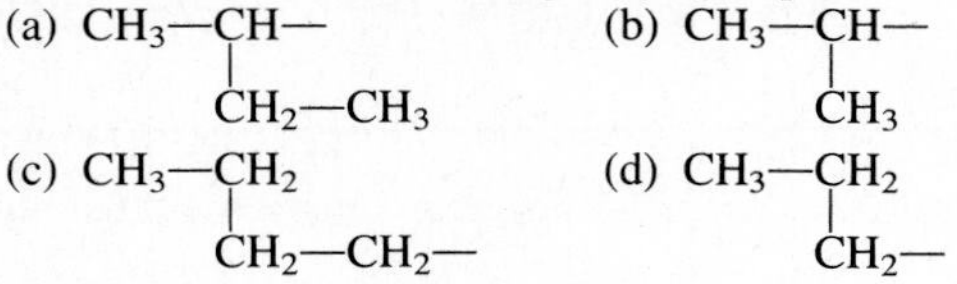

19.78 Name each of the following alkyl groups.

(a) CH₃—CH—   (b) CH₃—CH—CH₂—CH₃
      | |
     CH₃

(c) CH₃—CH—   (d) CH₃—CH—CH₂—
      | |
    CH₂—CH₃   CH₃

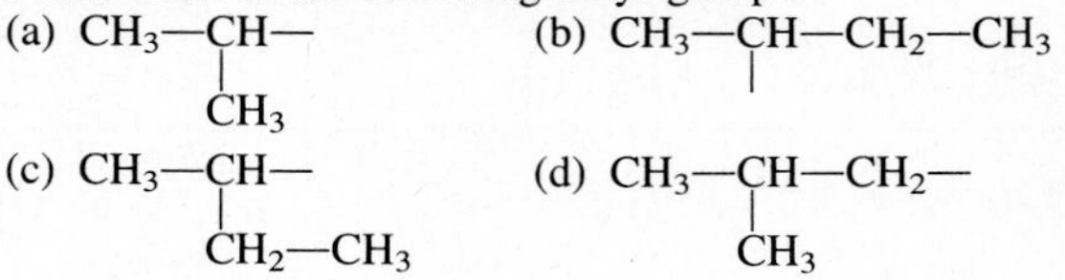

19.79 Give the IUPAC name for each of the following compounds.

(a)
$$H-\overset{\overset{\displaystyle H}{|}}{\underset{\displaystyle H}{C}}-\overset{\overset{\displaystyle H}{|}}{\underset{\displaystyle H}{C}}-\overset{\overset{\displaystyle H}{|}}{\underset{\displaystyle CH_3}{C}}-\overset{\overset{\displaystyle H}{|}}{\underset{\displaystyle H}{C}}-\overset{\overset{\displaystyle H}{|}}{\underset{\displaystyle H}{C}}-H$$

(b) $CH_3-\overset{\overset{\displaystyle CH_3}{|}}{\underset{\displaystyle CH_3}{C}}-CH_2-CH_3$

(c) $CH_3-\overset{\overset{\displaystyle CH_3}{|}}{\underset{\displaystyle CH_3}{C}}-CH_2-CH_2-CH_3$

(d) $CH_3-\overset{}{\underset{\displaystyle CH_3}{CH}}-CH_2-\overset{}{\underset{\displaystyle CH_3}{CH}}-\overset{}{\underset{\displaystyle CH_3}{CH_2}}$

19.81 Name each of the following compounds.
(a) CH_3CHCl_2 (b) $ClCH_2CH_2CH_2Cl$
(c) $CH_3CH_2CHCl_2$ (d) $CCl_3CH_2CH_2CH_2Cl$

19.83 Classify each carbon atom in the following compounds.
(a) $CH_3CH_2CH_2CH_2CH_3$

(b) $CH_3CH_2\overset{}{\underset{\displaystyle CH_3}{CH}}CH_2CH_3$

(c) $CH_3CH_2CH_2\overset{}{\underset{\displaystyle CH_3}{CH}}CH_3$

(d) $CH_3CH_2\overset{}{\underset{\displaystyle CH_3}{CH}}CH_2\overset{}{\underset{\displaystyle CH_3}{CH}}CH_2CH_3$

19.85 There are two isomeric cycloalkanes with the molecular formula C_4H_8. Write the structures of these compounds.

19.87 Which of the following compounds can exist as *cis* and *trans* isomers?
(a) 1-hexene   (b) 3-heptene
(c) 4-methyl-2-pentene   (d) 2-methyl-2-butene
(e) 3-methyl-3-hexene

19.89 Name each of the following compounds.

(a)   (b)

19.91 How many moles of hydrogen gas will react with 1 mole of each type of compound given in 19.25?

19.93 What product will be formed when Br_2 is added to each compound given in 19.49?

19.80 Name each of the following compounds.

(a)
$$H-\overset{\overset{\displaystyle CH_3}{|}}{\underset{\displaystyle H}{C}}-\overset{\overset{\displaystyle H}{|}}{\underset{\displaystyle H}{C}}-\overset{\overset{\displaystyle H}{|}}{\underset{\displaystyle H}{C}}-\overset{\overset{\displaystyle H}{|}}{\underset{\displaystyle H}{C}}-\overset{\overset{\displaystyle CH_3}{|}}{\underset{\displaystyle CH_2-CH_3}{C}}-H$$

(b) $CH_3-\overset{\overset{\displaystyle CH_3}{|}}{\underset{\displaystyle CH_2-CH_2-CH_3}{C}}-CH_2-CH_2-CH_3$

(c) $CH_3-\overset{}{\underset{\displaystyle CH_3-CH_2-CH_2}{CH}}-CH_2-\overset{\overset{\displaystyle CH_2-CH_3}{|}}{CH_2}$

(d) $\overset{\overset{\displaystyle CH_3}{|}}{CH_2}-\overset{}{\underset{\displaystyle CH_2-CH_3}{CH}}-CH_2-\overset{\overset{\displaystyle CH_3-CH_2}{|}}{\underset{\displaystyle CH_2-CH_3}{CH_2}}$

19.82 Name each of the following compounds.
(a) $CH_3CCl_2CH_2CH_3$ (b) $ClCH_2CH_2CHClCH_3$
(c) $CH_2FCH_2CH_2CHF_2$ (d) $CH_3CBr_2CH_3$

19.84 Classify each carbon atom in the following compounds.
(a) $CH_3CH_2CH_2CH_2CH_2CH_3$

(b) $CH_3CH_2\overset{}{\underset{\displaystyle CH_2CH_3}{CH}}CH_2CH_3$

(c) $CH_3CH_2CH_2\overset{}{\underset{\displaystyle CH_3}{CH}}CH_2CH_2CH_3$

(d) $CH_3\overset{}{\underset{\displaystyle CH_3}{CH}}CH_2\overset{}{\underset{\displaystyle CH_3}{CH}}CH_2\overset{}{\underset{\displaystyle CH_3}{CH}}CH_3$

19.86 There are three isomeric compounds with the molecular formula C_3H_4. Write the structures of these compounds.

19.88 Which of the following compounds can exist as *cis* and *trans* isomers?
(a) 2-pentene   (b) 1-heptene
(c) 3-methyl-2-pentene   (d) 2-methyl-2-pentene
(e) 3-ethyl-3-hexene

19.90 Name each of the following compounds.

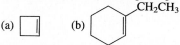

(a)   (b)

19.92 How many moles of hydrogen gas will react with 1 mole of each type of compound given in 19.26?

19.94 What product will be formed when HBr is added to each compound given in 19.49?

19.95 Indicate which of the following compounds is *ortho, meta,* or *para*.

(a) [structure: benzene ring with CH₃ groups at 1,3 positions]

(b) [structure: benzene ring with Br groups at adjacent positions]

(c) [structure: benzene ring with CH₂CH₂ groups at 1,4 positions]

19.96 Indicate which of the following compounds is *ortho, meta,* or *para*.

(a) [structure: benzene ring with Br groups at 1,4 positions]

(b) [structure: benzene ring with CH₃ groups at 1,3 positions]

(c) [structure: benzene ring with OH and CH₃ groups at adjacent positions]

19.97 Write the structure of each of the following compounds.
  (a) 3-phenylheptane
  (b) 1,1-diphenylethane
  (c) 1-chloro-2-phenylpentane
  (d) 1,2-diphenylethane

19.98 Write the structure of each of the following compounds.
  (a) 2-phenylpentane
  (b) 1,3-diphenylpropane
  (c) 2-bromo-3-phenylhexane
  (d) 1,2-diphenylpropane

20 Functional Groups in Organic Chemistry

Learning Objectives

After studying Chapter 20 you should be able to

1. Assign common names to alcohols and name them using the IUPAC system.

2. Classify alcohols as primary, secondary, and tertiary.

3. Predict the products of dehydration and oxidation of alcohols.

4. Assign IUPAC names to aldehydes and ketones.

5. Distinguish between aldehydes and ketones by oxidation reactions.

6. Assign common and IUPAC names to carboxylic acids and to their salts and esters.

7. Describe differences between and similarities of hydrolysis and saponification of esters.

8. Assign common and IUPAC names to and classify amines.

9. Assign IUPAC names to amides.

10. Relate the properties of amines and amides to their structure.

Table 20.1 Classes and Functional Groups of Organic Compounds

Class	Functional group	Example of expanded structural formula	Example of condensed structural formula
alkene	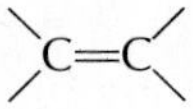 $\diagup\!\!\!C\!=\!C\diagdown$	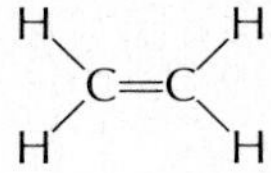	$CH_2\!=\!CH_2$
alkyne	$-C\!\equiv\!C-$	$H-C\!\equiv\!C-H$	$CH\!\equiv\!CH$
alcohol	$-O-H$	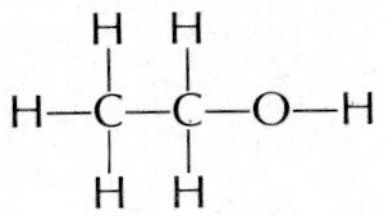	CH_3CH_2-OH
ether	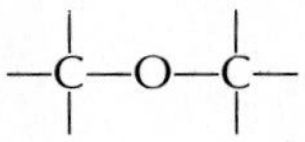 $-\overset{\mid}{\underset{\mid}{C}}-O-\overset{\mid}{\underset{\mid}{C}}-$	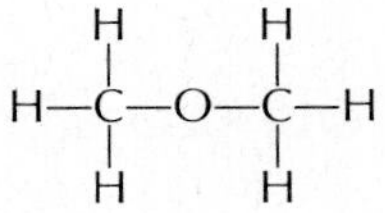	CH_3-O-CH_3
aldehyde	$-\overset{\displaystyle O}{\overset{\|}{C}}\diagdown_H$		$CH_3CH_2-\overset{\displaystyle O}{\overset{\|}{C}}H$
ketone	$-\overset{\displaystyle O}{\overset{\|}{C}}-$		$CH_3-\overset{\displaystyle O}{\overset{\|}{C}}-CH_3$
carboxylic acid	$-\overset{\displaystyle O}{\overset{\|}{C}}\diagdown_{O-H}$		$CH_3-\overset{\displaystyle O}{\overset{\|}{C}}-OH$
ester	$-\overset{\displaystyle O}{\overset{\|}{C}}\diagdown_{O-\overset{\mid}{\underset{\mid}{C}}-}$		$CH_3-\overset{\displaystyle O}{\overset{\|}{C}}-O-CH_3$
amine	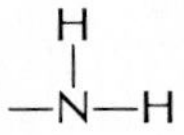 $\overset{\displaystyle H}{\overset{\mid}{-N}}-H$	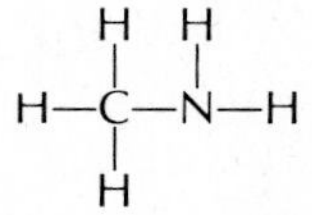	CH_3-NH_2
amide	$-\overset{\displaystyle O}{\overset{\|}{C}}-\overset{\displaystyle H}{\overset{\mid}{N}}-H$		$CH_3-\overset{\displaystyle O}{\overset{\|}{C}}-NH_2$
halide	$-X$ $(X\ =\ F,\ Cl,\ Br,\ I)$	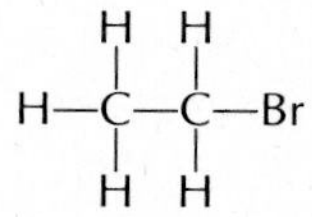	CH_3CH_2-Br
mercaptan or thiol	$-S-H$	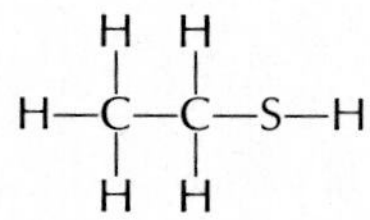	CH_3CH_2SH

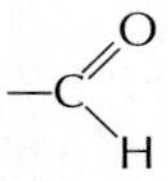

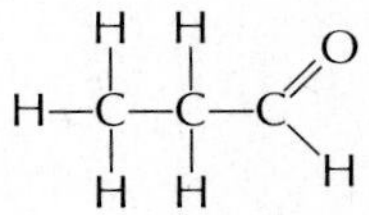

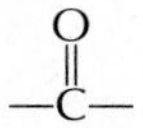

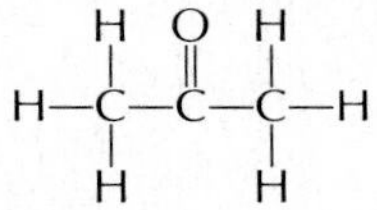

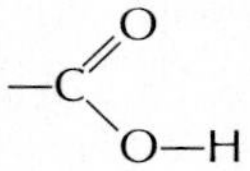

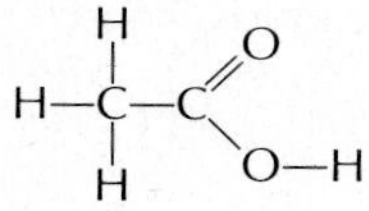

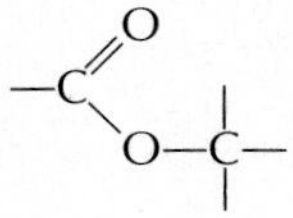

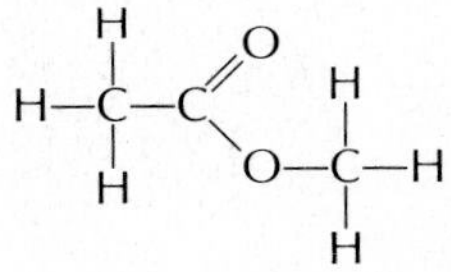

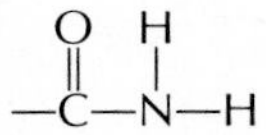

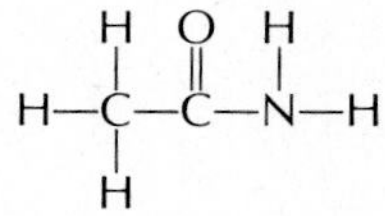

20.1 Functional Groups

Those atoms or groups of atoms and their bonds that confer on a molecule similar physical and chemical properties are called **functional groups** (Table 20.1). Thus, functional groups are used to classify molecules that have similar properties. Once you learn the properties and reactions of one functional group, you will know the properties and reactions of thousands of compounds in that class. The study of the millions of organic compounds is based on the functional groups contained in molecules.

A carbon–carbon double bond is a functional group. In ethylene, the double bond reacts in an addition reaction with hydrogen in the presence of a platinum catalyst. More complex compounds that contain a carbon–carbon double bond also react to add hydrogen. Functional groups also contain elements other than carbon; the most common elements are oxygen and nitrogen, although sulfur or the halogens may also be present. A dozen of the most common functional groups are listed in Table 20.1. These include the hydroxyl group of alcohols such as in ethyl alcohol, the aldehyde group as in formaldehyde, and the carboxylic acid group as in acetic acid.

20.2 Alcohols, Phenols, and Ethers

Alcohols contain the hydroxyl (—OH) functional group bonded to a saturated carbon atom. In methanol, the simplest alcohol, the hydroxyl group is bonded to a one-carbon alkyl group, the methyl group. The O—H bond in methanol is quite polar because of the difference in electronegativity between oxygen and hydrogen, and it resembles the O—H bonds in water. Alcohols can be viewed as the organic analogs of water, in which one hydrogen atom is replaced by an alkyl group (Figure 20.1).

Compounds in which the hydroxyl group is bonded to an aromatic ring are called **phenols.** The removal of one hydrogen atom from a carbon atom of an aromatic ring carbon results in an aryl group, symbolized by Ar. A phenol can be viewed as an organic analog of water in which one hydrogen atom is replaced by an aryl group (Figure 20.1). Ethers contain two groups, which may be alkyl or aryl groups, bonded to oxygen (Figure 20.1). Ethers are different from alcohols because they do not have an —OH group and have quite different chemical properties.

Nomenclature and Classification of Alcohols

The lower-molecular-weight alcohols have common names that frequently are used. The names consist of the alkyl group names plus the term alcohol (Figure 20.2) The IUPAC system of naming alcohols is as follows.

1. The longest continuous chain of carbon atoms that includes the hydroxyl group is chosen as the parent chain.
2. The parent name is obtained by substituting the ending -ol for the -e of the corresponding alkane.
3. The position of the hydroxyl group is indicated by the number of the carbon atom to

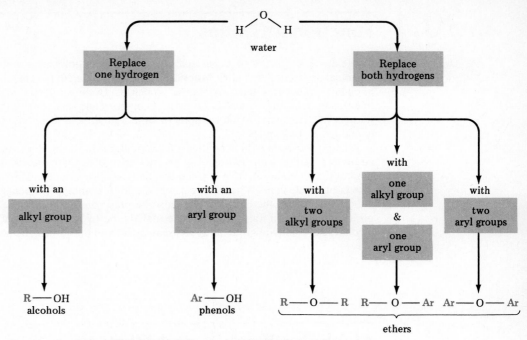

Figure 20.1

Structural relationship of organic oxygen compounds to water

which it is attached, with the numbering arranged so that the hydroxyl group receives the lowest possible number.

$$CH_3\underset{4}{-}\overset{\overset{\displaystyle CH_3}{|}}{\underset{\underset{\displaystyle H}{|}}{C}}_3\underset{2}{-}\overset{\overset{\displaystyle H}{|}}{\underset{\underset{\displaystyle OH}{|}}{C}}_2\underset{1}{-}CH_3$$

3-methyl-2-butanol (correct)
2-methyl-3-butanol (incorrect)

4. The ring of cyclic alcohols is numbered starting with the carbon atom bearing the hydroxyl group. Numbering continues to give the lowest numbers to carbon atoms bearing alkyl groups. The number 1 is not used in the name.

Example 20.1

Name the following compound.

$$CH_3-CH_2-CH_2-\overset{\displaystyle CH-}{\underset{\displaystyle CH_3-CH_2}{|}}\overset{\displaystyle CH-}{\underset{\displaystyle OH}{|}}CH_2-CH_3$$

Solution   The longest chain contains seven carbon atoms. The hydroxyl group is on the third carbon atom when the chain is numbered from right to left. The ethyl group is at position 4. The name is 4-ethyl-3-heptanol.

[*Additional examples may be found in 20.7–20.10 at the end of the chapter.*]

Alcohols are classified according to the structure of the alkyl group to which the hydroxyl group is attached. A **primary** (1°) alcohol is a compound in which the hydroxyl group is bonded to a primary carbon atom. In a **secondary** (2°) alcohol, the hydroxyl group is bonded to a secondary carbon atom. In a **tertiary** (3°) alcohol, the hydroxyl group is bonded to a tertiary carbon atom.

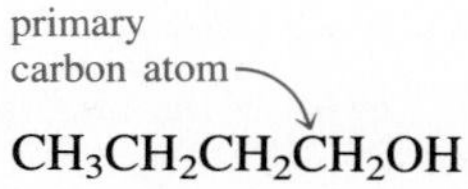

primary carbon atom

$CH_3CH_2CH_2CH_2OH$

a primary alcohol

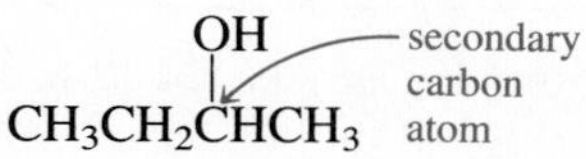

$CH_3CH_2CHCH_3$ — secondary carbon atom

a secondary alcohol

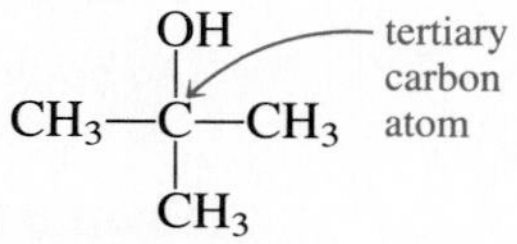

CH_3-C-CH_3 with OH and CH_3 — tertiary carbon atom

a tertiary alcohol

Figure 20.2
Nomenclature of alcohols

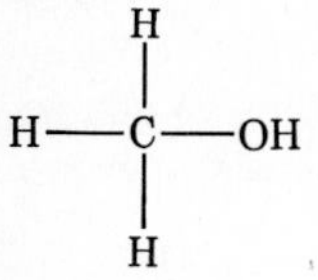

methanol
methyl alcohol

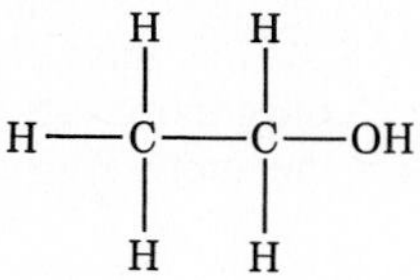

ethanol
ethyl alcohol

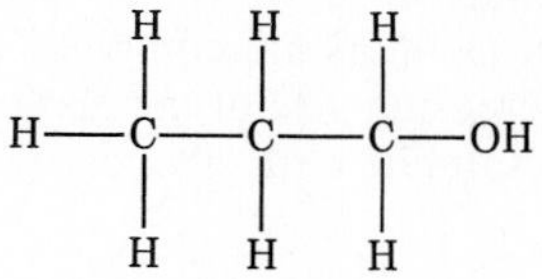

1–propanol
propyl alcohol

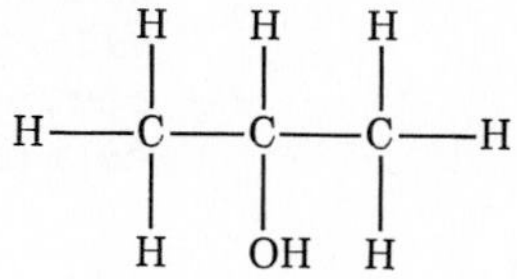

2–propanol
isopropyl alcohol

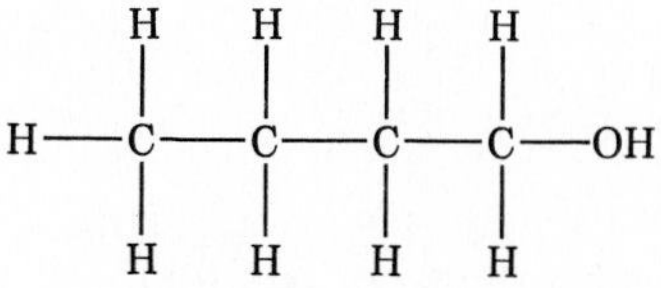

1–butanol
n–butyl alcohol

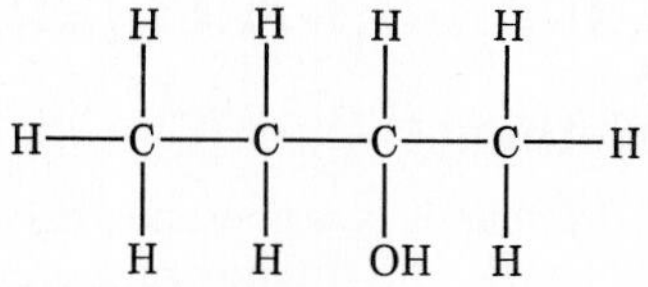

2–butanol
sec–butyl alcohol

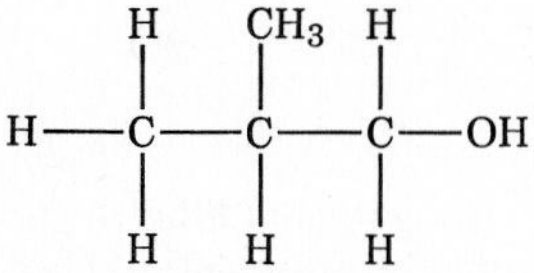

2–methyl–1–propanol
isobutyl alcohol

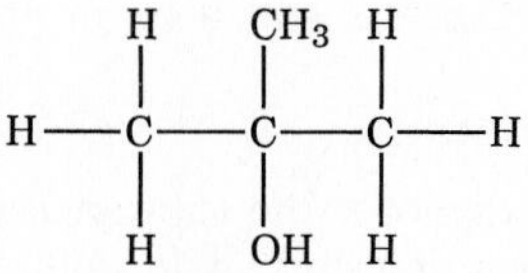

2–methyl–2–propanol
t–butyl alcohol

Example 20.2

Classify the following alcohol.

Solution   Although the compound is a cyclic alcohol, it is still classified on the basis of the carbon atom bearing the hydroxyl group. This carbon atom is bonded to a methyl group and to carbon-2 and carbon-6 of the ring. Thus, the carbon-1 is tertiary, and the compound is a tertiary alcohol.

[Additional examples may be found in 20.11–20.14 at the end of the chapter.]

Properties of Alcohols

Alcohols boil at higher temperatures than alkanes of comparable molecular weight. For example, ethanol, CH_3CH_2—OH, of molecular weight 46 boils at 78°C, whereas propane of molecular weight 44 boils at −42°C. This dramatic difference in boiling points is due to the hydrogen bonding of alcohols in the liquid state.

The solubility of alcohols in water is very different from the solubility of hydrocarbons. Table 20.2 lists the solubilities of some alcohols containing normal alkyl groups. The lower-molecular-weight alcohols are completely soluble in water, as they closely resemble it. However, with the increasing size of the alkyl group, the alcohol more closely resembles an alkane. As a result, the solubility of alcohols decreases with the increasing size of the alkyl group.

20.3   Common Alcohols

Methanol

Methanol is sometimes called wood alcohol because it was originally obtained by heating wood to a high temperature in the absence of air. Methanol is a toxic substance. Temporary blindness, permanent blindness, or death can result from the consumption of methanol that may be present as an impurity in illegal sources of ethanol. As little as 30 mL (1 fluid ounce) of pure methanol can cause death. The prolonged breathing of methanol vapors also is a serious health hazard. Methanol was used as a radiator antifreeze before it was replaced by ethylene glycol. Methanol is still used in windshield washer fluids.

Ethanol

Ethanol is the substance popularly known as alcohol. Fermentation of almost any substance containing sugar will produce ethanol but only in concentrations up to about 14%. Distillation of alcohol–water mixtures increases the concentration up to 95%. To increase the concentration to

Table 20.2   Boiling Points and Solubilities of Alcohols

Name	Formula	Boiling point (°C)	Solubility (g/100 mL in water)
methanol	CH_3OH	65	miscible
ethanol	CH_3CH_2OH	78	miscible
1-propanol	$CH_3CH_2CH_2OH$	97	miscible
1-butanol	$CH_3CH_2CH_2CH_2OH$	117	7.9
1-pentanol	$CH_3CH_2CH_2CH_2CH_2OH$	138	2.7
1-hexanol	$CH_3(CH_2)_4CH_2OH$	158	0.59
1-heptanol	$CH_3(CH_2)_5CH_2OH$	176	0.09
1-octanol	$CH_3(CH_2)_6CH_2OH$	194	insoluble
1-nonanol	$CH_3(CH_2)_7CH_2OH$	213	insoluble
1-decanol	$CH_3(CH_2)_8CH_2OH$	229	insoluble

100%, it is necessary to distill the mixture with a small amount of benzene. The resultant 100% alcohol is called **absolute alcohol,** but it contains a small amount of benzene. Ethanol is used as an industrial solvent. Substances are added in small amounts to render such alcohol unfit for drinking. Alcohol containing these adulterants, is called **denatured alcohol.**

Isopropyl Alcohol

Isopropyl alcohol is commonly known as rubbing alcohol. It is soluble in all proportions in water and is an excellent industrial solvent. Its low freezing point and solubility in water allow this compound to be used as a deicer for airplane wings and to prevent fuel-line freezeup in cars. In addition, it is used in inks, paints, and cosmetic preparations.

Polyhydroxy Alcohols

Ethylene glycol is a compound containing two hydroxyl groups, one each on the adjacent carbon atoms of ethane.

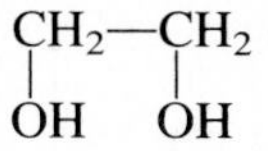

$$CH_2-CH_2$$
$$\;|\qquad\;|$$
$$OH\quad OH$$

It is used as an automobile antifreeze. One of the largest uses of ethylene glycol is in the production of Dacron and of Mylar film, which is used in tapes for recorders and computers.

Glycerol, also known as glycerin, contains three hydroxyl groups, one each on the three carbon atoms of propane.

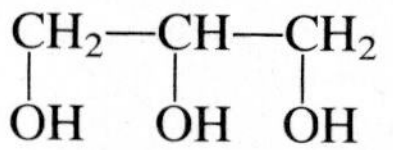

$$CH_2-CH-CH_2$$
$$\;|\qquad\;|\qquad\;|$$
$$OH\quad OH\quad OH$$

The moisture-retaining properties of glycerol make it useful in skin lotions, inks, and pharmaceuticals.

20.4 Reactions of Alcohols

Reaction with Active Metals

Alcohols react with alkali metals in the same way as water.

$$2\,Na + 2\,CH_3OH \longrightarrow H_2 + 2\,CH_3O^- + 2\,Na^+$$

$$2\,Na + 2\,H_2O \longrightarrow H_2 + 2\,HO^- + 2\,Na^+$$

Hydrogen gas is produced, and an alkoxide of the metal is formed. In the reaction with methanol, sodium methoxide is produced. The evolution of hydrogen gas is a good qualitative test to indicate the presence of an —OH group in an organic molecule.

Substitution Reactions

In a **substitution reaction** one atom or group of atoms is substituted for another atom or group of atoms. The hydroxyl group of alcohols can be substituted or replaced by halogens.

$$CH_3CH_2CH_2OH + HCl \xrightarrow{ZnCl_2} CH_3CH_2CH_2Cl + H_2O$$

$$3\,CH_3CH_2CH_2OH + PBr_3 \longrightarrow 3\,CH_3CH_2CH_2Br + H_3PO_3$$

In these substitution reactions the carbon–oxygen bond is broken, and the halogen replaces the hydroxyl group. The rate for the reaction with HCl, in the presence of a $ZnCl_2$ catalyst, decreases in the order of tertiary > secondary > primary alcohols. The reaction is used for classification purposes and is called the **Lucas test.**

Haloalkanes are insoluble in water, so they are easily detected in the Lucas test. If an alcohol is added to a Lucas reagent and turbidity (cloudiness) occurs immediately, the alcohol is probably tertiary. The turbidity results from the insoluble chloroalkane formed in the reaction. If the turbidity develops within 5–10 min, a secondary structure is indicated. Primary alcohols may never cause turbidity at room temperature.

Example 20.3

How could 1-pentanol and 2-pentanol be distinguished?

Solution   The reaction of 2-pentanol with HCl and $ZnCl_2$ should occur readily at room temperature. The reaction with 1-pentanol should be extremely slow or not occur at room temperature.

$$CH_3CH_2CH_2CH_2CH_2OH \qquad CH_3\underset{\underset{OH}{|}}{C}HCH_2CH_2CH_3$$

1-pentanol	2-pentanol
(a primary alcohol)	(a secondary alcohol)

[*Additional examples may be found in 20.23 and 20.24 at the end of the chapter.*]

Dehydration Reaction

The removal of water from an alcohol is a **dehydration reaction.** An acid catalyst such as sulfuric acid or phosphoric acid is used. The reaction is illustrated by the formation of ethene from ethyl alcohol. The reaction involves breaking a carbon–oxygen bond and the carbon–hydrogen bond of an adjacent carbon atom. For more complex alcohols, such as 2-butanol, the dehydration produces a mixture of products because there are two neighboring carbon atoms adjacent to the carbon atom bearing the hydroxyl group. In such cases, the isomers that contain the greatest number of alkyl groups attached to the unsaturated carbon atoms predominate in the mixture.

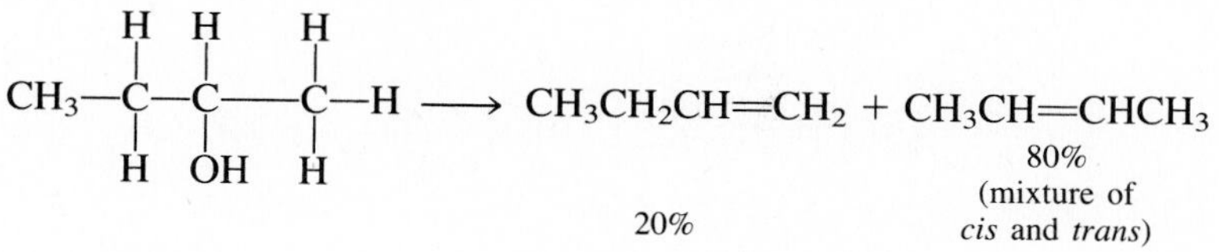

$$CH_3-\overset{\overset{\displaystyle H}{|}}{\underset{\underset{\displaystyle H}{|}}{C}}-\overset{\overset{\displaystyle H}{|}}{\underset{\underset{\displaystyle OH}{|}}{C}}-\overset{\overset{\displaystyle H}{|}}{\underset{\underset{\displaystyle H}{|}}{C}}-H \longrightarrow CH_3CH_2CH{=}CH_2 + CH_3CH{=}CHCH_3$$

20%   80%
(mixture of
cis and trans)

Example 20.4

What is the product of the dehydration of 1-methylcyclohexanol?

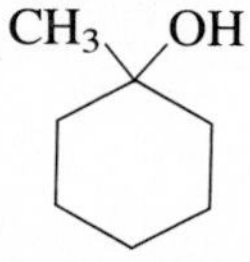

Solution This tertiary alcohol has three adjacent carbon atoms, which may lose a hydrogen atom in the dehydration. However, the loss of a hydrogen atom from either carbon-2 or carbon-6 will result in the same product. Thus, only two isomers result.

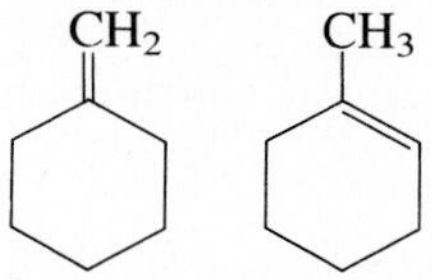

The second structure with the double bond within the six-membered ring predominates. This compound has three carbon atoms attached to the two double-bonded carbon atoms.

[*Additional examples may be found in 20.25 and 20.26 at the end of the chapter.*]

Oxidation of Alcohols

Primary alcohols, given by the general formula RCH_2OH, are oxidized to compounds called aldehydes.

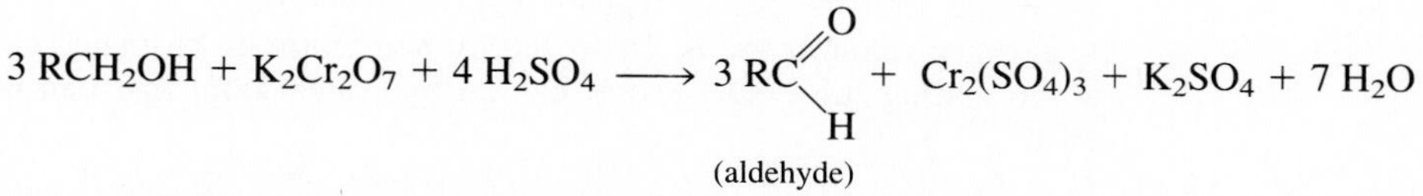

$$3\ RCH_2OH + K_2Cr_2O_7 + 4\ H_2SO_4 \longrightarrow 3\ RC\overset{\displaystyle O}{\underset{\displaystyle H}{\diagup}} + Cr_2(SO_4)_3 + K_2SO_4 + 7\ H_2O$$

(aldehyde)

Although the preceding balanced equation gives the stoichiometry of the reaction, equations for organic reactions are seldom balanced. The primary interest is usually in the changes of the organic compounds. Thus, the oxidation of a primary alcohol is simply written with the inorganic reagent displayed above the reaction arrow. Note that both the hydrogen atom of the —OH group and a hydrogen from the carbon bearing the —OH group are removed.

this hydrogen atom is removed

one of these two hydrogen atoms is removed

$$RC-O \xrightarrow[H_2SO_4]{K_2Cr_2O_7} RC\begin{smallmatrix}O\\\\H\end{smallmatrix}$$

primary alcohol   aldehyde

$$CH_3CH_2CH_2-\underset{\underset{H}{|}}{\overset{\overset{H}{|}}{C}}-OH \xrightarrow[H_2SO_4]{K_2Cr_2O_7} CH_3CH_2CH_2-C\begin{smallmatrix}O\\\\H\end{smallmatrix}$$

1-butanol   butanal

Aldehydes are easily oxidized further to produce acids. In this step, the oxidation is the result of a gain of an atom of oxygen.

$$RC\begin{smallmatrix}O\\\\H\end{smallmatrix} \xrightarrow[H_2SO_4]{K_2Cr_2O_7} R-C\begin{smallmatrix}O\\\\OH\end{smallmatrix}$$

$$CH_3CH_2CH_2C\begin{smallmatrix}O\\\\H\end{smallmatrix} \xrightarrow[H_2SO_4]{K_2Cr_2O_7} CH_3CH_2CH_2C\begin{smallmatrix}O\\\\OH\end{smallmatrix}$$

butanal   butanoic acid

Secondary alcohols are oxidized to form ketones, which are not further oxidized.

these hydrogen atoms are removed

$$R-\underset{\underset{H}{|}}{\overset{\overset{H-O}{|}}{C}}-R \xrightarrow[H_2SO_4]{K_2Cr_2O_7} R-\overset{\overset{O}{\|}}{C}-R \xrightarrow[H_2SO_4]{K_2Cr_2O_7} \text{ no reaction}$$

$$CH_3CH_2-\underset{\underset{H}{|}}{\overset{\overset{O-H}{|}}{C}}-CH_2CH_3 \xrightarrow[H_2SO_4]{K_2Cr_2O_7} CH_3CH_2-\overset{\overset{O}{\|}}{C}-CH_2CH_3$$

3-pentanol   3-pentanone

The reason for the stability of the ketones as compared to the aldehydes is the absence of a hydrogen atom on the oxygen-bearing carbon atom.

Tertiary alcohols are not oxidized by potassium dichromate. In tertiary alcohols, the carbon atom bearing the —OH group has no hydrogen atom that can be removed by the oxidizing agent.

$$R-\underset{\underset{R}{|}}{\overset{\overset{OH}{|}}{C}}-R \xrightarrow[H_2SO_4]{K_2Cr_2O_7} \text{ no oxidation products}$$

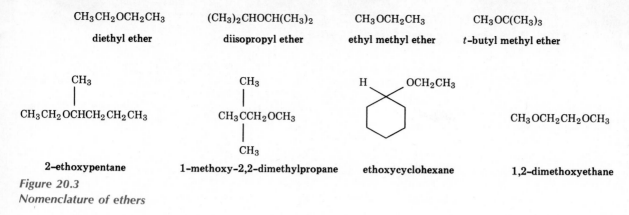

Figure 20.3
Nomenclature of ethers

Example 20.5

Which of the isomeric $C_4H_{10}O$ alcohols will react with potassium dichromate to produce a ketone, C_4H_8O?

Solution Inspection of Figure 20.2 reveals that two of the $C_4H_{10}O$ alcohols are primary, one is secondary, and one is tertiary. Only 2-butanol (*s*-butyl alcohol) can give a ketone.

[*Additional examples may be found in 20.27 and 20.28 at the end of the chapter.*]

20.5 Ethers

Simple ethers are named according to the alkyl groups attached to the oxygen atom. For higher-molecular-weight ethers, the smaller alkyl group and the oxygen are called an **alkoxy group,** and this group is regarded as a substituent on the larger alkane chain. Examples of nomenclature are given in Figure 20.3.

larger group is parent chain

$$CH_3CH_2CH_2CHCH_3$$
$$OCH_3$$

smaller group is substituent

2-methoxypentane

Diethyl ether is called simply ethyl ether or ether. It is the ether most familiar as a general anesthetic. Diethyl ether is a colorless liquid that boils at 35°C and thus can be administered as a vapor. Ether vapors depress the central nervous system and cause unconsciousness. Because its high flammability and volatility are hazards in the operating room, it must be administered by highly trained personnel.

20.6 Aldehydes and Ketones

A **carbonyl group** consists of a carbon atom and an oxygen atom bonded by a double bond. In the carbonyl group, oxygen shares two of its six valence electrons with carbon. The remaining four electrons of oxygen are present as two nonbonded pairs of electrons. Carbon shares two of its four valence electrons with oxygen, and the remaining two electrons are shared to form two bonds to other atoms.

$$\overset{..}{\underset{.}{C}} : : \overset{..}{\underset{..}{O}} : \qquad or \qquad \overset{..}{\underset{.}{C}} = \overset{..}{\underset{..}{O}} :$$

Aldehydes are compounds in which the carbonyl carbon atom is bonded to at least one hydrogen atom. The other group can be a hydrogen atom, an alkyl group (R), or an aromatic group (Ar). In formaldehyde, the simplest aldehyde, the second group is a hydrogen atom.

$$\begin{array}{c} H \\ \diagdown \\ \diagup \\ R \end{array} C = \overset{..}{\underset{..}{O}} : \qquad or \qquad \begin{array}{c} H \\ \diagdown \\ \diagup \\ Ar \end{array} C = \overset{..}{\underset{..}{O}} :$$

The condensed formula of an aldehyde is RCHO or ArCHO. Both the H and O following the carbon are understood to be bonded to the carbon.

Ketones are compounds in which the carbonyl carbon atom is bonded to two other carbon atoms. The bonded groups may be any combination of alkyl or aromatic groups.

$$\begin{array}{c} R \\ \diagdown \\ \diagup \\ R \end{array} C = \overset{..}{\underset{..}{O}} : \qquad \begin{array}{c} R \\ \diagdown \\ \diagup \\ Ar \end{array} C = \overset{..}{\underset{..}{O}} : \qquad \begin{array}{c} Ar \\ \diagdown \\ \diagup \\ Ar \end{array} C = \overset{..}{\underset{..}{O}} :$$

$$or \qquad or \qquad or$$

$$\begin{array}{c} :\overset{..}{O} \\ \parallel \\ R - C - R \end{array} \qquad \begin{array}{c} :\overset{..}{O} \\ \parallel \\ R - C - Ar \end{array} \qquad \begin{array}{c} :\overset{..}{O} \\ \parallel \\ Ar - C - Ar \end{array}$$

There are two ways to represent the structural formula of a ketone. The structural formula with 120° bond angles represents the actual shape of the molecule. However, for convenience in writing or printing, a linear arrangement of the atoms may be used.

The condensed formula of a ketone is RCOR. The oxygen atom is understood to be double bonded to the carbonyl carbon atom at the left of the oxygen atom. The alkyl group on the right is also understood to be bonded to the carbon atom of the carbonyl group and not to the oxygen.

Nomenclature of Aldehydes and Ketones

In the IUPAC system, aldehydes are named by the following rules.

1. Select the longest continuous carbon chain with the aldehyde carbon atom.
2. Replace the final -e of the parent hydrocarbon by the ending -al.
3. Number the chain by assigning the number 1 to the aldehyde carbon and numbering the other carbon atoms consecutively. The number 1 is not used in the name to indicate the position of the carbonyl carbon atom since it is understood to be located at the end of the chain.
4. Determine the identity and location of any branches and indicate this information by prefixing it to the parent name.

Examples of the nomenclature of aldehydes are given in Figure 20.4. An earlier nomenclature

Aldehydes

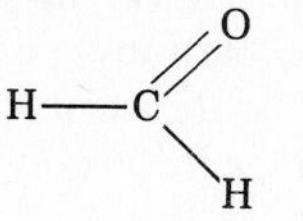

methanal
formaldehyde

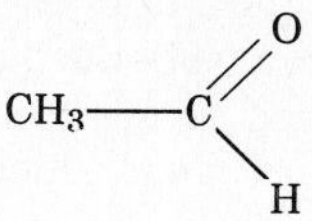

ethanal
acetaldehyde

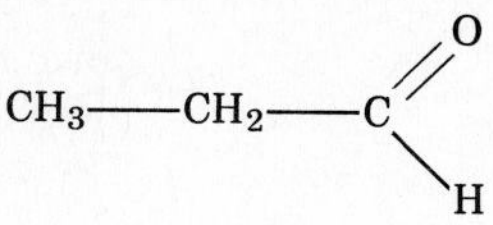

propanal
propionaldehyde

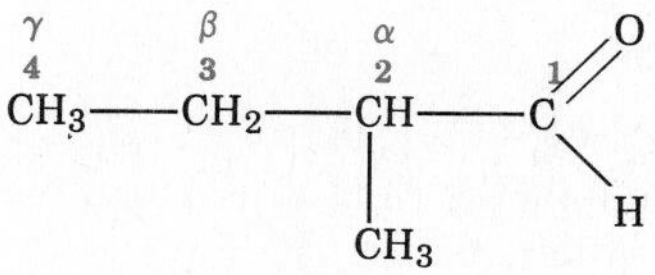

2-methylbutanal
α-methylbutyraldehyde

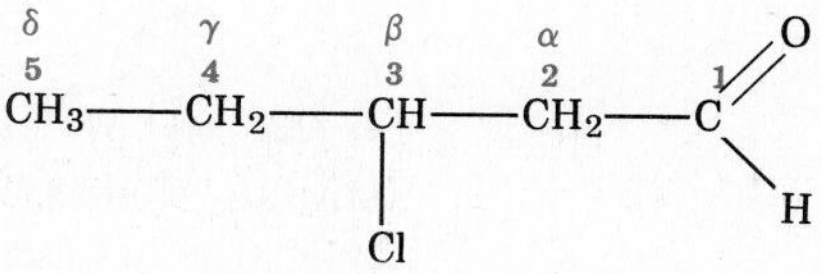

3-chloropentanal
β-chlorovaleraldehyde

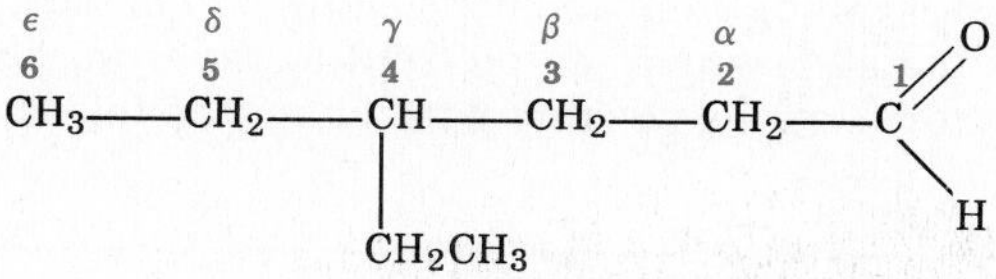

4-ethylhexanal
γ-ethylcaproaldehyde

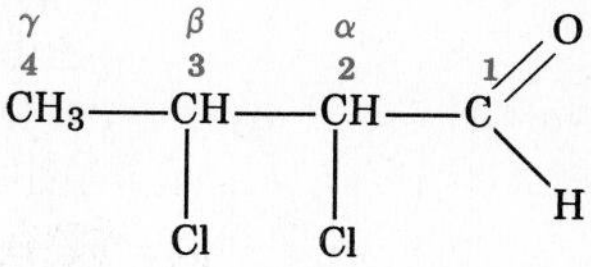

2,3-dichlorobutanal
α,β-dichlorobutyraldehyde

Ketones

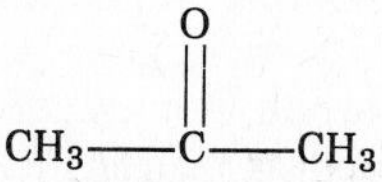

propanone
acetone
(dimethyl ketone)

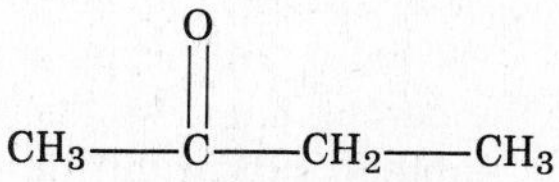

butanone
ethyl methyl ketone

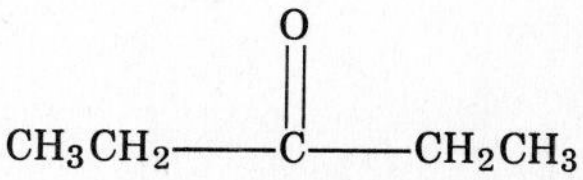

3-pentanone
diethyl ketone

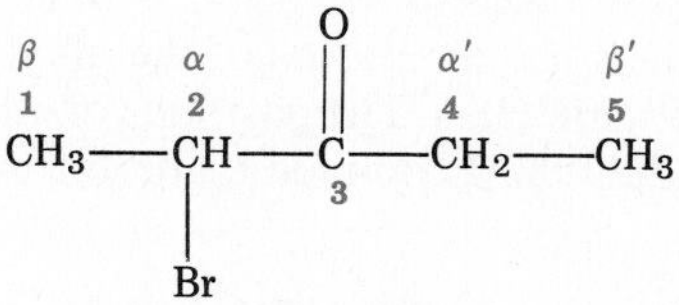

2-bromo-3-pentanone
α-bromodiethyl ketone

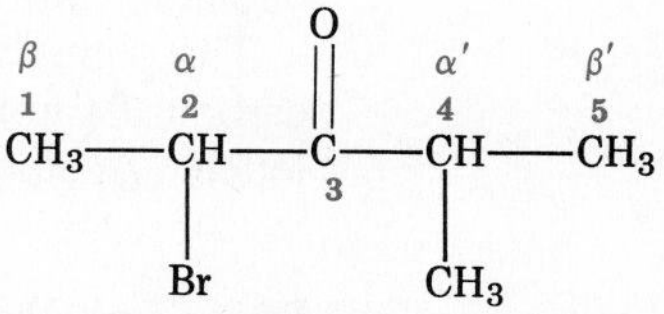

2-bromo-4-methyl-3-pentanone
α-bromo-α′-methyldiethyl ketone

Figure 20.4   *Nomenclature of aldehydes and ketones*

for aldehydes designates the positions on the carbon chain by the Greek letters α, β, γ, and so on. The α position is the carbon atom next to the carbonyl carbon. This convention was chosen since no substituent can be bonded directly to the carbonyl carbon of an aldehyde. The names of the aldehydes are based on the common names of the structurally related acids (Section 20.8). The common names of some aldehydes are given in Figure 20.4.

Example 20.6

Give the IUPAC name for the following compound.

$$CH_3-\underset{\underset{CH_3}{|}}{CH}-CH_2-\underset{\underset{Br}{|}}{CH}-CHO$$

Solution The aldehyde carbon atom on the right is assigned the number 1, and the five-carbon pentanal is numbered from right to left.

$$\underset{5}{CH_3}-\underset{4}{\underset{\underset{CH_3}{|}}{CH}}-\underset{3}{CH_2}-\underset{2}{\underset{\underset{Br}{|}}{CH}}-\underset{1}{CHO}$$

A methyl group is located at carbon atom number 4 and is part of the hydrocarbon skeleton. The position and identity of the alkyl group are prefixed to the word pentanal to give 4-methyl-pentanal. The bromine atom with its position number is then indicated in 2-bromo-4-methyl-pentanal.

[*Additional examples may be found in 20.33–20.40 at the end of the chapter.*]

Ketones are named by the following IUPAC rules.

1. Select the longest continuous carbon chain containing the carbonyl carbon atom.
2. Replace the final -e of the parent hydrocarbon by the ending -one.
3. Number the carbon chain so that the carbonyl carbon atom receives the lowest number.
4. Place the number indicating the carbonyl location as a prefix before the parent name.
5. Indicate the identity and location of substituents as a prefix on the parent ketone.
6. Cyclic ketones are named by assigning the carbonyl carbon atom as number 1. The ring is then numbered to give the lowest number(s) to the atom(s) bearing substitutents.

Several examples of IUPAC nomenclature are given in Figure 20.4. In some cases, such as butanone, no number is necessary because the name is unambiguous.

The common names for ketones use the names of the alkyl groups attached to the carbonyl carbon. Examples are given in Figure 20.4. The positions of attached groups are given by Greek letters. Acetone is the most widely used common name for the simplest ketone, CH_3COCH_3.

Physical Properties of Aldehydes and Ketones

The physical properties of aldehydes and ketones are distinctly different from those of either the alkanes or alcohols of similar molecular weight (Table 20.3). Since oxygen is more electronegative than carbon, the shared electrons are pulled toward oxygen, and the carbonyl group is polar. The polarity of the group is shown by use of $\leftrightarrow$ or $\delta+$ and $\delta-$. The physical properties of aldehydes and ketones are determined by this polarity.

Table 20.3 **Comparative Boiling Points of Carbonyl Compounds**

Compound	Structure	Molecular weight	Boiling point (°C)
ethane	CH_3CH_3	30	−89
methanol	CH_3OH	32	64.6
formaldehyde	HCHO	30	−21
propane	$CH_3CH_2CH_3$	44	−42
ethanol	CH_3CH_2OH	46	78.3
acetaldehyde	CH_3CHO	44	20
isobutane	$CH_3CH(CH_3)_2$	58	−12
2-propanol	$CH_3CH(OH)CH_3$	60	82.5
acetone	CH_3COCH_3	58	56.1

Aldehydes and ketones have higher boiling points than the alkanes because there are attractive forces between neighboring polar molecules. However, these attractive forces are smaller than those of hydrogen bonding. Therefore, alcohols have higher boiling points than aldehydes and ketones of similar molecular weight.

Aldehydes and ketones can form hydrogen bonds with water. As a consequence, the lower-molecular-weight compounds formaldehyde, acetaldehyde, and acetone are soluble in water in all proportions.

acetone hydrogen-bonded to water

As the molecular weights of the carbonyl compounds increase, the differences in physical properties compared with alkanes and alcohols are lessened. The dipole–dipole attractive forces are less important as the chain length increases. As a result, aldehydes and ketones as well as alcohols become more like hydrocarbons.

20.7 Reactions of Carbonyl Compounds

In Section 20.4 you learned that primary alcohols are oxidized to aldehydes, which then are easily oxidized to acids. Under the same conditions, secondary alcohols are oxidized to ketones, which are not oxidized further.

$$R-CH_2OH \xrightarrow[H^+]{K_2Cr_2O_7} R-\overset{\displaystyle O}{\underset{\displaystyle H}{C}} \xrightarrow[H^+]{K_2Cr_2O_7} R-\overset{\displaystyle O}{\underset{\displaystyle OH}{C}}$$

$$R-\overset{\displaystyle OH}{\underset{\displaystyle |}{C}}H-R \xrightarrow[H^+]{K_2Cr_2O_7} R-\overset{\displaystyle O}{\overset{\displaystyle \|}{C}}-R \xrightarrow[H^+]{K_2Cr_2O_7} \text{no reaction}$$

The difference in the reactivities of aldehydes and ketones toward oxidizing agents is useful to distinguish between these two classes of compounds. Fehling's solution contains cupric ion (Cu^{2+}) as a complex ion in a basic solution. The Cu^{2+} will oxidize aldehydes, but not ketones. The aldehyde is oxidized to an acid, and Cu^{2+} is reduced to Cu^+, which precipitates as Cu_2O. The characteristic blue color of Cu^{2+} in solution is diminished as a red precipitate of Cu_2O is formed. Since the solution is basic, it is the conjugate base of the carboxylic acid that is formed rather than the carboxylic acid itself. Unbalanced equations for this reaction are given below.

$$R-\overset{\displaystyle O}{\underset{\displaystyle H}{C}} + Cu^{2+} \longrightarrow R-\overset{\displaystyle O}{\underset{\displaystyle O^-}{C}} + Cu_2O\downarrow$$

$$\underset{\text{(blue solution)}}{} \qquad \underset{\text{(red precipitate)}}{}$$

$$CH_3CH_2CH_2CHO + Cu^{2+} \longrightarrow CH_3CH_2CH_2\overset{\displaystyle O}{\underset{\displaystyle O^-}{C}} + Cu_2O\downarrow$$

Tollens' reagent also reacts with aldehydes but not with ketones. The reagent is a basic solution of a silver ammonia complex ion. When an aldehyde is oxidized by $[Ag(NH_3)_2]^+$, metallic silver is deposited as a mirror on the walls of the test tube. Unbalanced equations are given below.

$$R-\overset{\displaystyle O}{\underset{\displaystyle H}{C}} + [Ag(NH_3)_2]^+ \longrightarrow R-\overset{\displaystyle O}{\underset{\displaystyle O^-}{C}} + Ag\downarrow$$

$$CH_3\overset{\displaystyle CH_3}{\underset{\displaystyle |}{C}}HCHO + [Ag(NH_3)_2]^+ \longrightarrow CH_3\overset{\displaystyle CH_3}{\underset{\displaystyle |}{C}}H\overset{\displaystyle O}{\underset{\displaystyle O^-}{C}} + Ag\downarrow$$

Example 20.7

How can the isomeric carbonyl-containing compounds of the molecular formula C_4H_8O be distinguished from each other by a simple chemical test?

Solution There are three isomeric carbonyl-containing compounds, two aldehydes and one ketone.

$$CH_3\overset{\displaystyle CH_3}{\underset{\displaystyle |}{C}}HCHO \qquad CH_3CH_2CH_2CHO \qquad CH_3CH_2\overset{\displaystyle O}{\overset{\displaystyle \|}{C}}CH_3$$

Both aldehydes will react with Tollens' reagent to produce a silver mirror. Therefore these two compounds cannot be distinguished from each other. However, the ketone will not react with

Tollens' reagent, and therefore the compound that does not yield a silver mirror is butanone.

[*Additional examples may be found in 20.43 and 20.44 at the end of the chapter.*]

20.8 Carboxylic Acids

Carboxylic acids are compounds containing the carboxyl group. The carboxyl group is represented in several ways.

$$-C\underset{OH}{\overset{O}{\lVert}} \qquad \text{or} \qquad -COOH \qquad \text{or} \qquad -CO_2H$$

Hydrogen, an alkyl group, or an aromatic group may be bonded to the carboxyl group. The simplest examples of such carboxylic acids are formic acid, acetic acid, and benzoic acid.

formic acid   acetic acid   benzoic acid

Carboxylic acids are abundant in nature and were long ago given common names based on their natural sources (Table 20.4). In the IUPAC system, the longest continuous chain of carbon atoms containing the carboxyl functional group is selected. The -e ending of the parent alkane is dropped and replaced by the ending -oic acid. The numbering of the chain starts at the carboxyl carbon, but since the carboxyl group must be at the end of the chain, the number 1 is not used in the name. Examples of this nomenclature are given in Figure 20.5.

When common names are used, the positions of groups attached to the parent chain are designated α, β, γ, and so on. The α-position is the carbon atom attached to the carboxyl carbon.

Table 20.4   Nomenclature of Unbranched Carboxylic Acids

Formula	Common name	IUPAC name
HCO_2H	formic acid	methanoic acid
CH_3CO_2H	acetic acid	ethanoic acid
$CH_3CH_2CO_2H$	propionic acid	propanoic acid
$CH_3(CH_2)_2CO_2H$	butyric acid	butanoic acid
$CH_3(CH_2)_3CO_2H$	valeric acid	pentanoic acid
$CH_3(CH_2)_4CO_2H$	caproic acid	hexanoic acid
$CH_3(CH_2)_6CO_2H$	caprylic acid	octanoic acid
$CH_3(CH_2)_8CO_2H$	capric acid	decanoic acid
$CH_3(CH_2)_{10}CO_2H$	lauric acid	dodecanoic acid
$CH_3(CH_2)_{12}CO_2H$	myristic acid	tetradecanoic acid
$CH_3(CH_2)_{14}CO_2H$	palmitic acid	hexadecanoic acid
$CH_3(CH_2)_{16}CO_2H$	stearic acid	octadecanoic acid

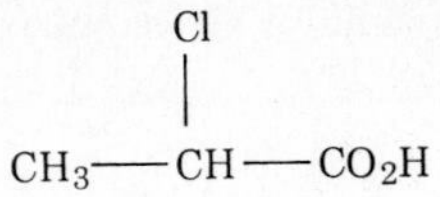

2-chloropropanoic acid
α-chloropropionic acid

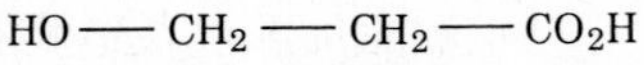

3-hydroxypropanoic acid
β-hydroxypropionic acid

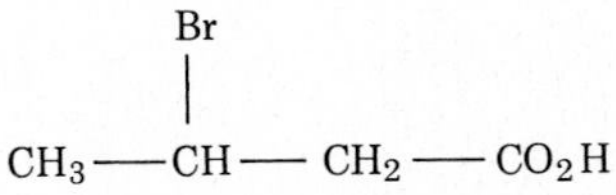

3-bromobutanoic acid
β-bromobutyric acid

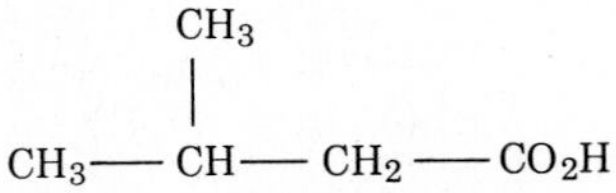

3-methylbutanoic acid
β-methylbutyric acid

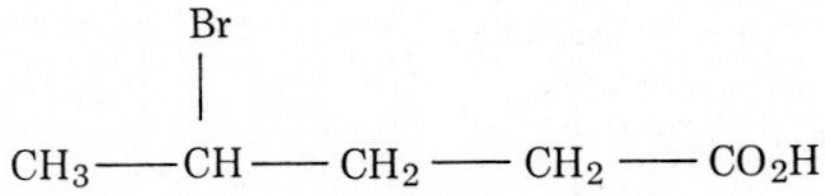

4-bromopentanoic acid
γ-bromovaleric acid

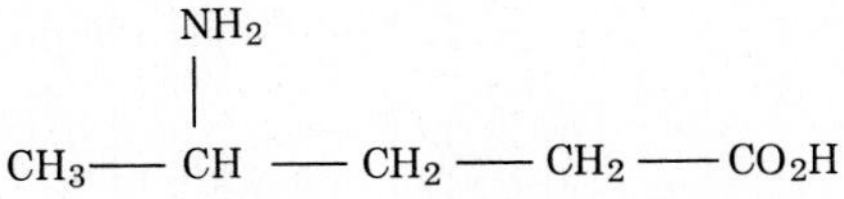

4-aminopentanoic acid
γ-aminovaleric acid

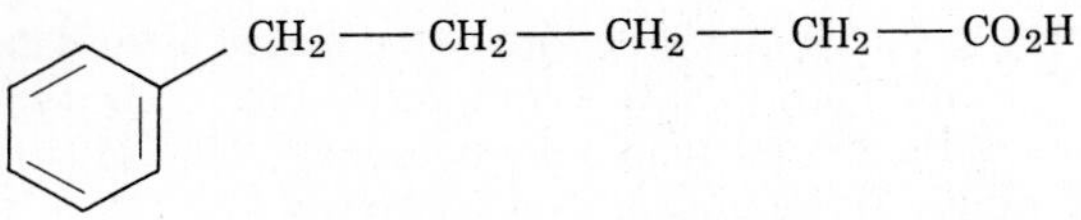

5-phenylpentanoic acid
δ-phenylvaleric acid

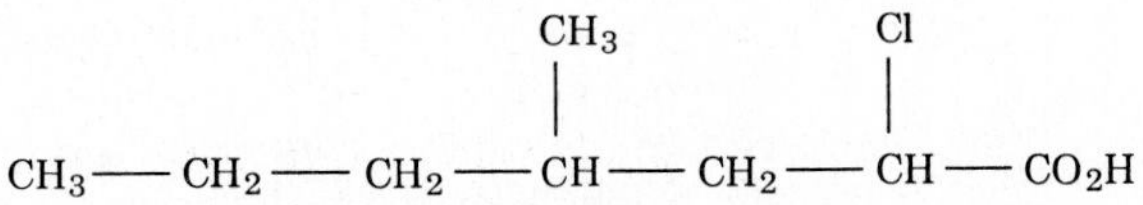

2-chloro-4-methylheptanoic acid
α-chloro-γ-methylenanthic acid

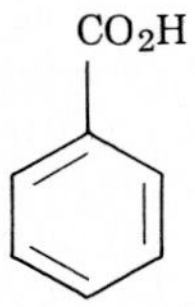

benzoic acid

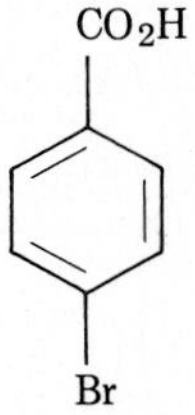

4-bromobenzoic acid
p-bromobenzoic acid

Figure 20.5
Nomenclature of carboxylic acids

Example 20.8

Give both the IUPAC and common names for the following carboxylic acid.

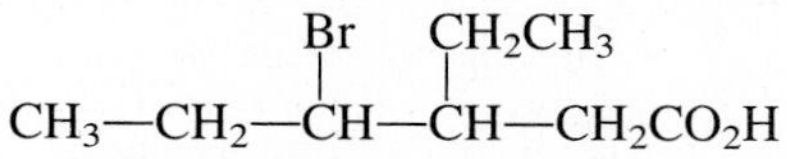

$$\begin{array}{ccc} & Br & CH_2CH_3 \\ & | & | \\ CH_3-CH_2-CH-CH-CH_2CO_2H & & \end{array}$$

Solution First determine that the longest continuous chain of the acid contains six carbon atoms. The parent name in the IUPAC system is hexanoic; the common name of the parent acid is caproic acid (Table 20.4).

The substituents are an ethyl group and a bromo group. These groups are on the numbers 3 and 4 carbon atoms, respectively, according to IUPAC rules, but the β- and γ-carbon atoms according to the older common name system.

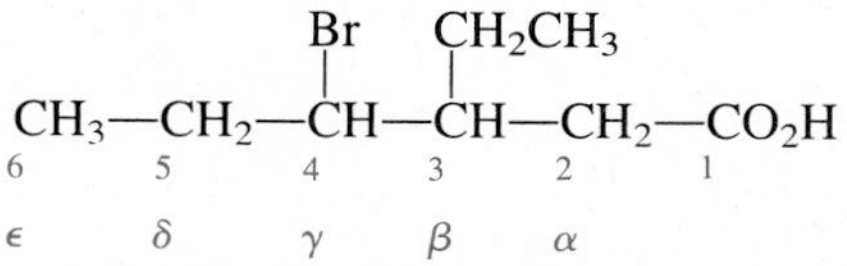

$$\begin{array}{cccccc} & & Br & CH_2CH_3 & & \\ & & | & | & & \\ CH_3-&CH_2-&CH-&CH-&CH_2-&CO_2H \\ 6 & 5 & 4 & 3 & 2 & 1 \\ \epsilon & \delta & \gamma & \beta & \alpha & \end{array}$$

The ethyl group is part of the hydrocarbon structure and is placed closest to the parent name. The bromo group is prefixed to this name. The names are 4-bromo-3-ethylhexanoic acid and γ-bromo-β-ethylcaproic acid.

[*Additional examples may be found in 20.45–20.48 at the end of the chapter.*]

The low-molecular-weight carboxylic acids are liquids at room temperature (Table 20.5). The higher-molecular-weight compounds are wax-like solids. Low-molecular-weight acids are soluble in water, but their solubility decreases with increasing chain length. The solubility of these compounds in water is due to hydrogen bonding between water and the carboxyl group. The boiling points of carboxylic acids are abnormally high. This phenomenon is due to very strong hydrogen bonding.

Liquid acids have sharp and unpleasant odors. Butyric acid occurs in rancid butter and aged cheese. Caproic, caprylic, and capric acids have the smell of goats. The Latin word for goat is *caper* and is the source of the common name of these acids.

Table 20.5 **Physical Properties of Carboxylic Acids**

IUPAC	Melting point (°C)	Boiling point (°C)	Solubility (g/100 g H_2O, at 20°C)
methanoic acid	8	101	miscible
ethanoic acid	17	118	miscible
propanoic acid	−21	141	miscible
butanoic acid	−5	164	miscible
pentanoic acid	−34	186	4.97
hexanoic acid	−3	205	0.97
octanoic acid	17	239	0.068
decanoic acid	32	270	0.015
dodecanoic acid	44	299	0.0055

Acids and Their Salts

You learned in Chapter 16 that acetic acid is a weak acid. The equilibrium for acetic acid and its conjugate base, the acetate ion, lies far to the left. A 1 M solution of acetic acid is only 0.4% ionized. The other carboxylic acids are also weak acids.

Although water is too weak a base to convert very much of a carboxylic acid into its conjugate base, the hydroxide ion is a sufficiently strong base to neutralize carboxylic acids. Carboxylic acids react with bases such as metal hydroxides to produce carboxylate salts. The names of salts are derived from the acid by changing the -ic ending to -ate and preceding this name by the name of the metal.

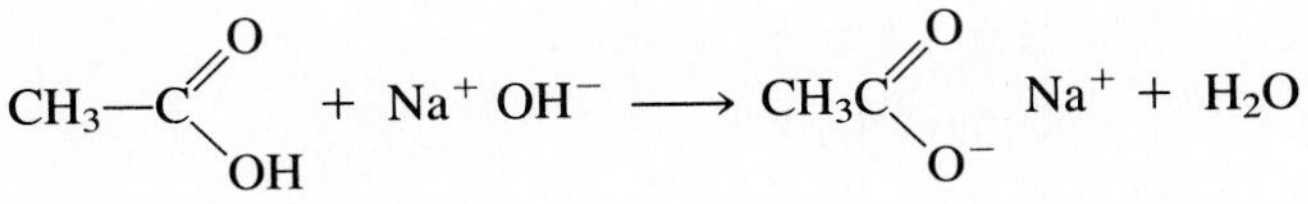

$$CH_3-C\overset{O}{\underset{OH}{}} + Na^+ OH^- \longrightarrow CH_3C\overset{O}{\underset{O^-}{}} Na^+ + H_2O$$

common: sodium acetate
IUPAC: sodium ethanoate

20.9 Esters

Esters are formed when a carboxylic acid reacts with an alcohol. The esterification reaction is catalyzed by inorganic acids.

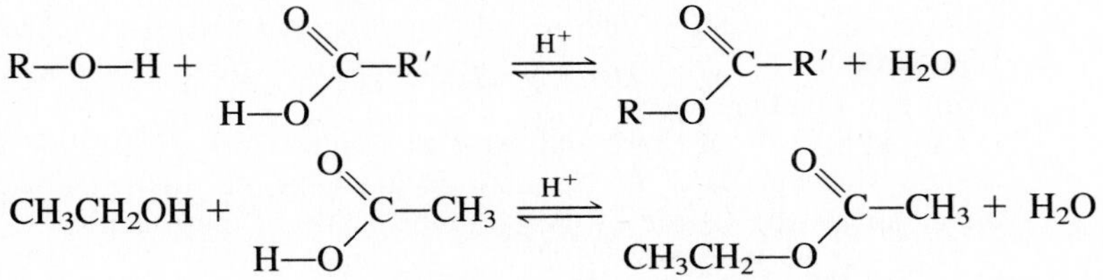

$$R-O-H + \underset{H-O}{\overset{O}{C}}-R' \underset{}{\overset{H^+}{\rightleftharpoons}} \underset{R-O}{\overset{O}{C}}-R' + H_2O$$

$$CH_3CH_2OH + \underset{H-O}{\overset{O}{C}}-CH_3 \underset{}{\overset{H^+}{\rightleftharpoons}} \underset{CH_3CH_2-O}{\overset{O}{C}}-CH_3 + H_2O$$

ethyl acetate

The equilibrium can be displaced to increase the amount of the ester by distilling the water out of the reaction mixture. The yield of ester can also be increased by using an excess of the cheaper reagent of the reaction. Ethyl esters of acids can be obtained by using ethanol as a solvent. Under such conditions, the high concentration of ethanol favors a high conversion of the acid to the ester. Both the removal of water and the addition of alcohol result in shifts in the equilibrium predicted by Le Châtelier's principle.

alcohol + acid $\rightleftharpoons$ ester + water

adding alcohol
"pushes" reaction
to the right

removal of water "pulls"
reaction to the right

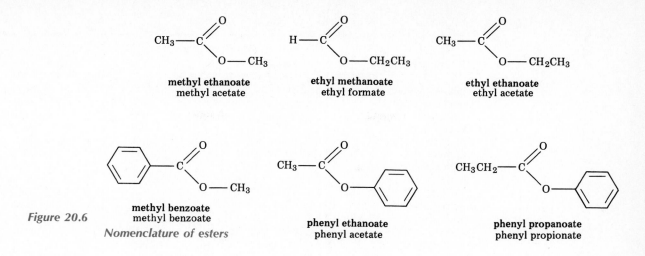

Figure 20.6
Nomenclature of esters

methyl ethanoate
methyl acetate

ethyl methanoate
ethyl formate

ethyl ethanoate
ethyl acetate

methyl benzoate
methyl benzoate

phenyl ethanoate
phenyl acetate

phenyl propanoate
phenyl propionate

Nomenclature of Esters

The esters are named using the name of the alkyl group of the alcohol followed by the name of the acid with the suffix -ate replacing -ic. Because there are two systems of naming carboxylic acids, there are two names for the esters. In the examples given in Figure 20.6, the alcohol portion of each molecule is on the right side of the structure. However, regardless of how the structure is written, the name of the alcohol is written first in the name of the ester.

The structure $R-C\overset{O}{\diagdown}$ that occurs in esters and other derivatives of carboxylic acids is called an **acyl group.** An acyl group is named by replacing the -ic acid of a carboxylic acid by -yl.

$$CH_3-C\overset{O}{\diagdown}$$

acetyl group
or
ethanoyl group

$$CH_3CH_2CH_2C\overset{O}{\diagdown}$$

butyryl group
or
butanoyl group

Properties of Esters

Esters are polar molecules, but have lower boiling points than carboxylic acids and alcohols of similar molecular weight. This lower boiling point reflects the absence of hydrogen bonds in esters.

$$CH_3CH_2CH_2CH_2CH_2OH$$

1-pentanol
b.p. 138°C

$$CH_3CH_2CH_2C\overset{O}{\underset{OH}{\diagdown}}$$

butanoic acid
b.p. 164°C

$$CH_3C\overset{O}{\underset{O-CH_2CH_3}{\diagup}}$$

ethyl acetate
b.p. 77°C

Table 20.6   **Properties of Esters of Carboxylic Acids**

Common name	Structure	Boiling point (°C)	Solubility (g/100 g H_2O at 20°C)
methyl formate	HCO_2CH_3	32	miscible
methyl acetate	$CH_3CO_2CH_3$	57	24.4
methyl propionate	$CH_3CH_2CO_2CH_3$	80	1.8
methyl butyrate	$CH_3(CH_2)_2CO_2CH_3$	102	0.5
methyl valerate	$CH_3(CH_2)_3CO_2CH_3$	126	0.2
methyl caproate	$CH_3(CH_2)_4CO_2CH_3$	151	0.06
methyl enanthate	$CH_3(CH_2)_5CO_2CH_3$	172	0.03
methyl caprylate	$CH_3(CH_2)_6CO_2CH_3$	208	0.007
methyl acetate	$CH_3CO_2CH_3$	57	24.4
ethyl acetate	$CH_3CO_2CH_2CH_3$	77	7.4
propyl acetate	$CH_3CO_2CH_2CH_2CH_3$	102	1.9
butyl acetate	$CH_3CO_2CH_2CH_2CH_2CH_3$	125	0.9

Esters can form hydrogen bonds with water molecules via their oxygen atoms and thus have some degree of water solubility. However, since esters do not themselves have a hydrogen atom with which to bond to an oxygen atom of water, they are less soluble than carboxylic acids. The solubilities and boiling points of esters are listed in Table 20.6.

Many of the odors of fruits are due to esters. For example, ethyl acetate is found in pineapples and octyl acetate in oranges. In recent years there has been an increased demand for substances that enhance the flavor and aroma of processed fruits and fruit products such as jams and jellies. However, the esters used in these products are not necessarily the same as those in the natural fruits; they may be other esters that produce the same odor or taste.

Reactions of Esters

Ester hydrolysis involves the breaking of the ester bond to produce an acid and an alcohol. The reaction is catalyzed by strong acids. Hydrolysis of an ester, then, is just the reverse of the esterification reaction.

$$R-\overset{\overset{\displaystyle O}{\|}}{C}-O-R' + H_2O \xrightarrow{\ H^+\ } R-\overset{\overset{\displaystyle O}{\|}}{C}-OH + R'OH$$

$$CH_3-\overset{\overset{\displaystyle O}{\|}}{C}-O-CH_3 + H_2O \xrightarrow{\ H^+\ } CH_3-\overset{\overset{\displaystyle O}{\|}}{C}-OH + CH_3OH$$

In order to favor the hydrolysis reaction a large excess of water is used. Note that when the ester bond is broken, the water is incorporated into the products.

ester + water ⇌ acid + alcohol

adding water "pushes" the reaction to the right

Saponification is the hydrolysis of a ester bond by a strong base. An example shows the saponification of methyl acetate.

$$R-\overset{\overset{\displaystyle O}{\|}}{C}-OR' + OH^- \quad\rightleftharpoons\quad R-\overset{\overset{\displaystyle O}{\|}}{C}-O^- + R'OH$$

$$CH_3-\overset{\overset{\displaystyle O}{\|}}{C}-O-CH_3 + OH^- \rightleftharpoons CH_3-\overset{\overset{\displaystyle O}{\|}}{C}-O^- + CH_3OH$$

Note that instead of the acetic acid that is produced in the hydrolysis reaction, the acetate ion is formed in basic solution. There is another difference between hydrolysis and the related saponification reaction. Hydrolysis is catalyzed by acid, whereas in the saponification reaction the hydroxide ion is a reagent because equal numbers of moles of hydroxide ion and ester are required. The hydroxide ion is consumed in the reaction, and the point of equilibrium is overwhelmingly to the right.

Example 20.9

What is the IUPAC name of the following ester?

$$CH_3CH_2-O-\overset{\overset{\displaystyle O}{\|}}{C}-CH_2\overset{\overset{\displaystyle CH_3}{|}}{C}HCH_3$$

Solution First, identify the alcohol portion of the ester, at the left of the molecule; it contains two carbon atoms. This is an ethyl ester.

$$\underset{\substack{\text{alcohol}\\\text{portion}}}{CH_3CH_2}-O-\underset{\substack{\text{acyl portion}\\\text{from acid}}}{\overset{\overset{\displaystyle O}{\|}}{C}-CH_2\overset{\overset{\displaystyle CH_3}{|}}{C}HCH_3}$$

The acyl portion contains a four-carbon chain with a methyl branch. This acid is called 3-methylbutanoic acid. Now change the -ic ending of the acid to -ate and write the name of the alkyl group of the alcohol as a separate word in front of the modified acid name to give ethyl 3-methylbutanoate.

[*Additional examples may be found in 20.53–20.56 at the end of the chapter.*]

Example 20.10

Isobutyl formate has the odor of raspberries. Draw its structural formula.

Solution The acid portion of the ester is formic acid. The alcohol portion is isobutyl alcohol.

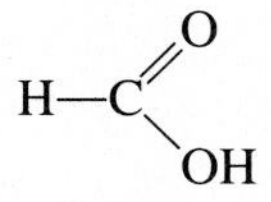

$$H-\overset{\overset{\displaystyle O}{\diagup\!\!\diagdown}}{\underset{\diagdown OH}{C}} \qquad CH_3-\overset{\overset{\displaystyle CH_3}{|}}{C}H-CH_2OH$$

The ester can be represented in two ways depending on which component is drawn on the left of the structure. Both structures are identical.

$$H-\overset{\overset{\displaystyle O}{\|}}{C}-O-CH_2-\overset{\overset{\displaystyle CH_3}{|}}{C}H-CH_3 \quad\text{or}\quad CH_3-\overset{\overset{\displaystyle CH_3}{|}}{C}H-CH_2-O-\overset{\overset{\displaystyle O}{\|}}{C}-H$$

20.10 Amines

Amines are organic derivatives of ammonia in which one or more of the hydrogen atoms is replaced by an alkyl or aromatic group. Thus, amines are organic bases. Amines are called primary (1°), secondary (2°), or tertiary (3°) depending on the number of groups attached to nitrogen.

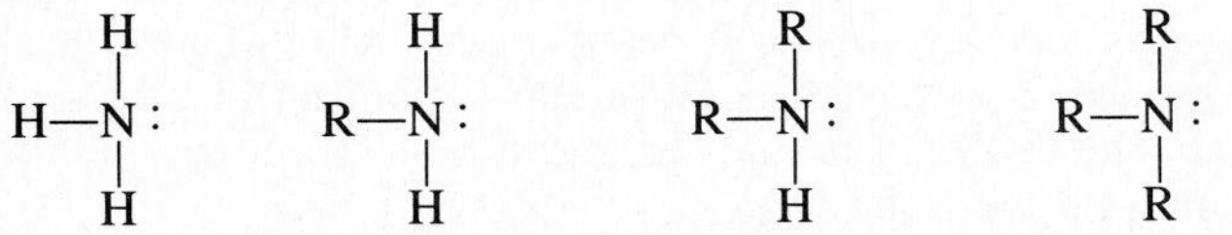

ammonia   primary amine   secondary amine   tertiary amine

This system of classifying amines is different than that previously used for alcohols. In alcohols, the classification is based on the groups attached to the carbon atom bearing the hydroxyl group. In amines, the focus is on the nitrogen atom and the number of groups attached to it. For example, *t*-butylamine has a *t*-butyl group, but the amine is primary since it has only one alkyl group bonded to the nitrogen atom. In *t*-butyl alcohol, the alcohol is classified as tertiary because the hydroxyl-bearing carbon atom has three alkyl groups.

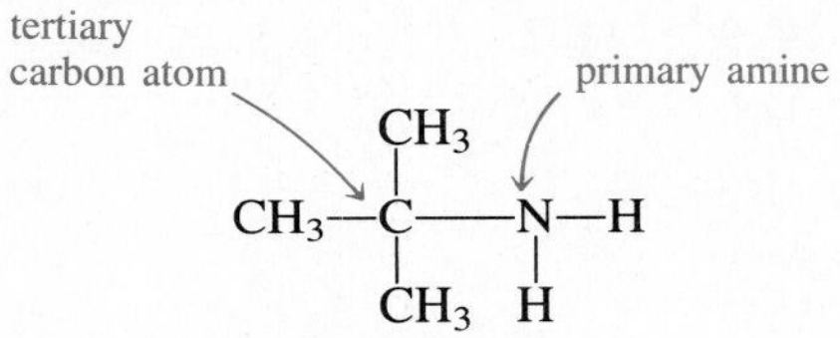

t-butylamine
(a primary amine)

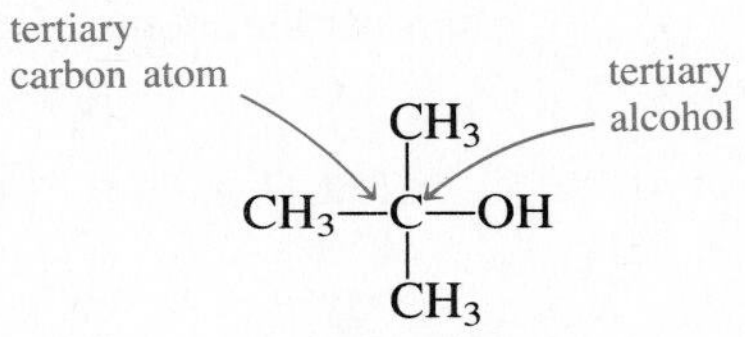

t-butyl alcohol
(a tertiary alcohol)

Example 20.11

The synthetic narcotic analgesic Demerol has the following structure. Is this a primary, secondary, or tertiary amine?

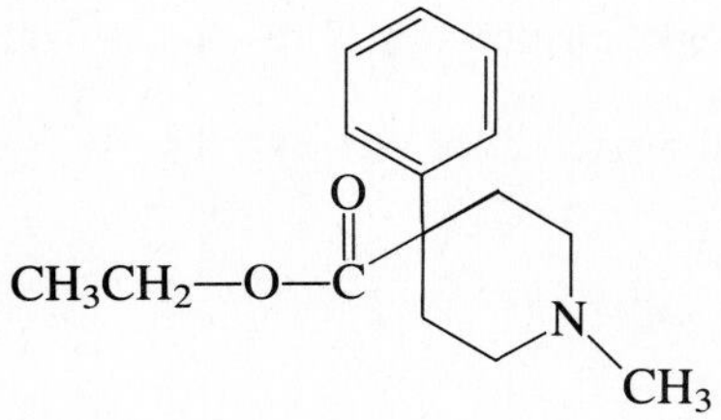

Solution   There are three carbon atoms bonded to nitrogen. Two of the carbon atoms are in the ring. The third carbon atom is the methyl group. The compound is a tertiary amine.

[*Additional examples may be found in 20.61–20.64 at the end of the chapter.*]

In the common system of nomenclature, amines are named by prefixing the names of the attached alkyl groups to the word amine. The entire name is written as one word. When two or more identical alkyl groups are present, the prefixes di- and tri- are used. Some examples illustrate this method.

$$CH_3$$

$$CH_3CH_2NH_2 \qquad (CH_3)_2NH \qquad CH_3CH_2CH_2NCH_2CH_3$$

ethylamine   dimethylamine   ethylmethylpropylamine

In the IUPAC system, amines are named as hydrocarbons and the location of the functional group is indicated. The —NH_2 group, called the amino group appears in 1° amines (Figure 20.7). In 2° and 3° amines, the largest alkyl group is considered the parent structure. The name of the N-alkylamino (—NHR) or the N,N-dialkylamino group (—NRR′) is then prefixed to the name of the hydrocarbon. The capital N indicates that the alkyl group is attached to nitrogen and not to the parent hydrocarbon. Some examples are given in Figure 20.7.

Figure 20.7
Nomenclature of amines

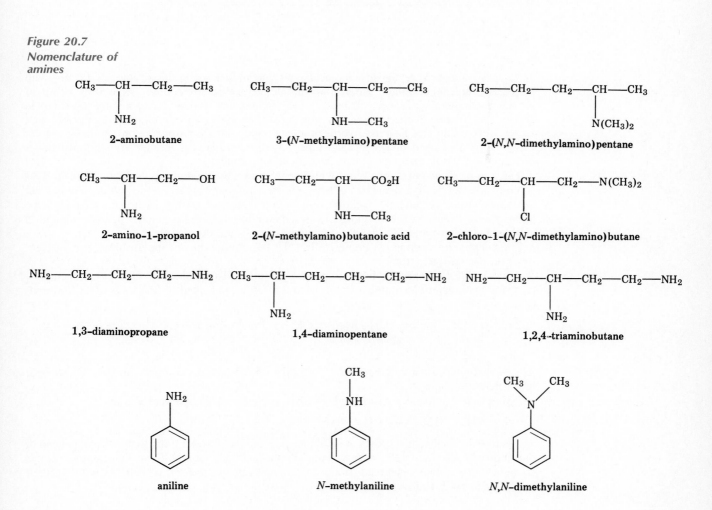

Table 20.7 **Boiling Points of Representative Amines**

Name	Structure	Boiling point (°C)
ammonia	NH_3	-33
methylamine	CH_3NH_2	-6
ethylamine	$C_2H_5NH_2$	17
propylamine	$n\text{-}C_3H_7NH_2$	48
butylamine	$n\text{-}C_4H_9NH_2$	78
t-butylamine	$t\text{-}C_4H_9NH_2$	44
dimethylamine	$(CH_3)_2NH$	7
trimethylamine	$(CH_3)_3N$	3
aniline	⬡—NH_2	184

At room temperature the low-molecular-weight amines are gases (Table 20.7), but with increasing molecular weight, amines are liquids and eventually solids. The boiling points of amines are higher than those of alkanes of similar molecular weight. This phenomenon is due in part to the polarity of the carbon–nitrogen bond, but hydrogen bonding is a more important factor. Tertiary amines have no hydrogen atoms bonded to the nitrogen atom, so these amines cannot form hydrogen bonds to each other. As a consequence, tertiary amines have lower boiling points than isomeric primary and secondary amines.

The low-molecular-weight amines with five or fewer carbon atoms are soluble in water. This solubility is due to hydrogen bonds between the amine and water. Even tertiary amines are soluble in water because the electron pair of nitrogen may form a hydrogen bond with the hydrogen atoms of water. Low-molecular-weight amines have sharp penetrating odors similar to ammonia. Higher-molecular-weight compounds smell like decaying fish.

The chemical properties of amines reflect the basic properties of the unshared pair of electrons on the nitrogen atom.

Amines are of comparable base strength to ammonia. The position of the following equilibrium for methylamine acting as a base is to the left as indicated by the base ionization constant, K_b.

$$CH_3NH_2 + H_2O \rightleftharpoons CH_3NH_3^+ + OH^- \qquad K_b = 4.4 \times 10^{-4}$$
$$\text{methylammonium ion}$$

Alkyl-substituted amines are stronger bases than ammonia. In contrast, anilines are weaker bases than ammonia.

Amines are completely protonated by strong acids such as hydrochloric acid.

$$CH_3-NH_2 + HCl \longrightarrow CH_3-NH_3^+ + Cl^-$$

The ammonium salts thus formed are ionic and are soluble in water if the hydrocarbon portion of the amine is not too large. Because ammonium salts are more soluble than amines, drugs containing the amine functional group are often given in the form of salts to improve their solubility in body fluids.

20.11 Amides

Amides are compounds in which an amino group or substituted amino group is bonded to a carbonyl carbon atom. The other two bonds of the nitrogen atom may be to hydrogen atoms, alkyl groups, or aryl groups. The amides are classified on the basis of the combined number of alkyl or aryl groups bonded to the nitrogen atom.

$$CH_3-CH_2-CH_2-\overset{\overset{\displaystyle O}{\|}}{C}-\underset{\underset{\displaystyle H}{|}}{N}-H \qquad CH_3-CH_2-\overset{\overset{\displaystyle O}{\|}}{C}-\underset{\underset{\displaystyle H}{|}}{N}-CH_3 \qquad CH_3-\overset{\overset{\displaystyle O}{\|}}{C}-\underset{\underset{\displaystyle CH_3}{|}}{N}-CH_3$$

an unsubstituted amide   a monosubstituted amide   a disubstituted amide

Nomenclature of Amides

In the common system amides are named by dropping the suffix -ic of the related acid and adding the suffix -amide. When there is a substituent on the nitrogen, the prefix *N*- followed by the name of the group bonded to nitrogen is prefixed to the name. Substituents on the acyl group are designated by α, β, γ, and so on, as in the case for the common names of acids. In the IUPAC system, the final -e of the alkane is replaced by -amide. The substituents on nitrogen are indicated by the same method as in the common system. Numbers are used for substituents on the parent chain in the same way as in acids. Some examples are given below.

$$CH_3CH_2-\overset{\overset{\displaystyle O}{\|}}{C}-\underset{\underset{\displaystyle H}{|}}{N}-CH_2CH_3 \qquad CH_3\overset{\overset{\displaystyle CH_3}{|}}{C}HCH_2\overset{\overset{\displaystyle O}{\|}}{C}-\underset{\underset{\displaystyle CH_2CH_3}{|}}{N}-CH_2CH_3$$

common: *N*-ethylpropionamide   *N,N*-diethyl-β-methylbutyramide
IUPAC: *N*-ethylpropanamide   *N,N*-diethyl-3-methylbutanamide

Properties of Amides

With the exception of formamide, the unsubstituted amides are solids at room temperature. Intermolecular hydrogen bonding between the amide hydrogen atom and the carbonyl oxygen atom is strong and accounts for the high melting and boiling points of unsubstituted amides. Substitution of the hydrogen atoms on the nitrogen atom by alkyl or aryl groups reduces the number of possible intermolecular hydrogen bonds and lowers the melting and boiling points.

Low-molecular-weight amides are soluble in water. The solubility is due to intermolecular hydrogen bonds of the amide with water.

Even low-molecular-weight disubstituted amides are water soluble because the carbonyl oxygen atom can hydrogen bond to the hydrogen atoms of water.

Amides can be hydrolyzed with water to break the carbon–nitrogen bond and form an acid and ammonia or an amine. A similar relationship was presented for esters in Section 20.9. There are, however, important differences. The hydrolysis of esters occurs relatively easily, whereas the amides are quite resistant to hydrolysis.

Amides can be hydrolyzed only by using a strong acid or strong base. Under basic conditions, the salt of the carboxylic acid is formed, and a mole of base per mole of amide is required.

$$R-\overset{\overset{\displaystyle O}{\|}}{C}-NHR' + OH^- \longrightarrow R-\overset{\overset{\displaystyle O}{\|}}{C}-O^- + R'NH_2$$

Under acidic conditions, the ammonium salt of the amine is formed, and a mole of acid per mole of amide is required. Amide hydrolysis requires heating for hours with a strong acid or base, whereas ester hydrolysis usually requires less severe conditions.

$$R-\overset{\overset{\displaystyle O}{\|}}{C}-NHR' + HCl + H_2O \longrightarrow R-\overset{\overset{\displaystyle O}{\|}}{C}-OH + R'NH_3{}^+ Cl^-$$

Summary

In alcohols the hydroxyl group is bonded to a saturated carbon atom. There are three classes of alcohols: primary, secondary, and tertiary. Alcohols have higher boiling points than alkanes because of hydrogen bonding. Lower-molecular-weight alcohols are soluble in water.

The oxygen–hydrogen bond of alcohols reacts with active metals to form hydrogen gas. Dehydration of alcohols forms alkenes. Oxidation of primary alcohols first produces aldehydes and then acids. Secondary alcohols are oxidized to ketones.

Phenols have the general formula Ar—OH.

Ethers have two groups bonded to an oxygen atom; these may both be alkyl, aryl, or one of each.

Aldehydes and ketones contain a carbonyl group. Aldehydes have one hydrogen atom bonded to the carbonyl carbon atom, whereas ketones have both bonds of the carbonyl group to other carbon atoms.

The carbonyl group is polar; as a result, carbonyl compounds have higher boiling points than hydrocarbons but lower boiling points than alcohols. Lower-molecular-weight aldehydes and ketones are soluble in water.

In the IUPAC system of nomenclature the terminal -e of the alkane is replaced by -al for aldehydes and -one by ketones.

Aldehydes are oxidized by Fehling's solution and by Tollens' reagent, whereas ketones are not oxidized by these reagents.

The structural characteristic of carboxylic acids is the carboxyl group, —CO_2H. The carboxyl group is very polar and forms hydrogen bonds. As a result the carboxylic acids have high boiling points and are very soluble in water. Carboxylic acids react with strong bases to form carboxylate salts.

Esters are formed by the reaction of an acid with an alcohol. Esters are polar compounds but cannot form hydrogen bonds to themselves. The hydrolysis of esters in the presence of acid gives an equilibrium mixture. The reaction of an ester with base yields an alcohol and a carboxylate salt.

Amines are classified as primary (1°), secondary (2°), or tertiary (3°) based on the number of hydrocarbon groups bonded to the nitrogen atom. Primary and secondary amines form intermolecular hydrogen bonds, but tertiary amines do not. Amines of low molecular weight are soluble in water. Amines are weak bases but react with strong acids to form ammonium salts.

Amides are derived from a carboxylic acid by replacing the hydroxyl group with a —NH_2, —NHR, or —NRR′ group. Unsubstituted amides have high melting points because of their intermolecular hydrogen bonds. The substitution of hydrocarbon groups on the nitrogen atom reduces the number of hydrogen bonds and lowers the melting point and boiling point.

New Terms

Absolute alcohol is 100% ethanol.

An **acyl group** is represented by RCO— and is part of the structure of acids and esters.

An **alcohol** (ROH) is a compound containing a hydroxyl group bonded to a saturated carbon atom.

An **aldehyde** (RCHO) has a carbonyl group bonded to one hydrogen atom and one carbon group.

An **alkoxyl group** is represented by RO— and is used to name ethers.

An **amide** ($RCONH_2$) is the functional group containing a nitrogen atom bonded to a carbonyl carbon atom.

An **amine** has alkyl or aryl groups bonded to a nitrogen atom.

The **amino** group is —NH_2.

The **carbonyl group** consists of a carbon atom and an oxygen atom bonded by a double bond.

A **carboxyl group** is represented by —CO_2H and is the functional group of acids.

A **carboxylate group** is the anion formed by the loss of a proton from a carboxylic acid.

Dehydration is the removal of water from a molecule.

Denatured alcohol is alcohol containing adulterants.

An **ester** is the product formed by the reaction of an alcohol and an acid.

Esterification is the formation of an ester from an alcohol and an acid.

Fehling's solution is an alkaline solution of cupric ion that is used as a test reagent for aldehydes.

A **functional group** is a group of atoms that confers on a molecule certain physical and chemical properties.

A **ketone** (RCOR) is a carbonyl compound with both bonds from the carbonyl carbon atom to carbon groups.

The **Lucas test** is a reaction used to classify alcohols.

A **phenol** (ArOH) is a substance that has the hydroxyl group bonded to an aromatic ring.

A **primary alcohol** has a hydroxyl group bonded to a primary carbon atom.

A **primary amine** (RNH_2) is a substance that has one carbon group bonded to a nitrogen atom.

Saponification is the hydrolysis of an ester by a strong base.

In a **substitution reaction** one atom or group of atoms is substituted for another atom or group of atoms.

A **secondary alcohol** is a substance that has a hydroxyl group bonded to a secondary carbon atom.

A **secondary amine** (RR′NH) is a substance that has two carbon groups bonded to the nitrogen atom.

A **tertiary alcohol** is a substance that has a hydroxyl group bonded to a tertiary carbon atom.

A **tertiary amine** (RR′R″N) is a substance that has three carbon groups bonded to the nitrogen atom.

Tollen's reagent is an alkaline solution of $[Ag(NH_3)_2]^+$ that is used as a test reagent for aldehydes.

Exercises

Terminology

20.1 Explain how alcohols are classified.

20.3 Explain the structural differences among an aldehyde, a ketone, and a carboxylic acid.

20.2 Why aren't aldehydes and ketones considered a single class of compounds?

20.4 How do an acid and an ester differ structurally?

20.5 What are the structural differences between an amide and an amine?

20.6 Explain the classification system of amines, and contrast it with the classification system of alcohols.

Nomenclature of Alcohols

20.7 Write the structural formula for each of the following.
(a) 2-methyl-2-pentanol
(b) 2-methyl-1-butanol
(c) 2,3-dimethyl-1-butanol
(d) 2-methyl-3-pentanol
(e) 3-ethyl-3-pentanol
(f) 4-methyl-2-pentanol

20.8 Write the structural formula for each of the following.
(a) propyl alcohol   (b) t-butyl alcohol
(c) n-butyl alcohol   (d) isopropyl alcohol
(e) ethyl alcohol   (f) s-butyl alcohol
(g) methyl alcohol   (h) isobutyl alcohol

20.9 Name each of the following compounds.

(a) $CH_3-CH_2-\overset{\overset{\displaystyle CH_3}{|}}{\underset{\underset{\displaystyle OH}{|}}{C}}-CH_3$

(b) $CH_3-\overset{\overset{\displaystyle CH_3}{|}}{CH}-\underset{\underset{\displaystyle OH}{|}}{CH}-CH_3$

(c) $CH_3-CH_2-\underset{\underset{\displaystyle CH_3}{|}}{CH}-CH_2-OH$

(d) $CH_3-\overset{\overset{\displaystyle CH_3}{|}}{\underset{\underset{\displaystyle CH_2-CH_3}{|}}{C}}-OH$

20.10 Name each of the following compounds.

(a) $CH_3-\underset{\underset{\displaystyle OH}{|}}{CH}-CH_2-CH_2-CH_3$

(b) $CH_3-\overset{\overset{\displaystyle CH_3}{|}}{\underset{\underset{\displaystyle CH_3}{|}}{C}}-CH_2-OH$

(c) $CH_3-\underset{\underset{\displaystyle OH}{|}}{CH}-CH_2-\underset{\underset{\displaystyle CH_3}{|}}{CH}-CH_2-CH_3$

(d) $CH_3-\overset{\overset{\displaystyle CH_3}{|}}{\underset{\underset{\displaystyle CH_3}{|}}{C}}-CH_2-CH_2-OH$

Classification of Alcohols

20.11 Classify each of the compounds in 20.9 as primary, secondary, or tertiary alcohols.

20.13 Locate and classify the alcohol functional groups in cortisone.

20.12 Classify each of the compounds in 20.10 as primary, secondary, or tertiary alcohols.

20.14 Locate and classify the alcohol functional groups in the following prostaglandin, which is a smooth-muscle stimulant.

Physical Properties of Alcohols

20.15 Ethylene glycol boils at 198°C. Why does this compound have such a high boiling point?

20.16 Explain why 1-butanol is less soluble in water than 1-propanol.

Reactions of Alcohols

20.17 Write a balanced equation for the reaction of ethanol with potassium.

20.19 Describe a simple visual test of a chemical reaction to distinguish between a sample of octane and a sample of 1-octanol.

20.18 Write a balanced equation for the reaction of methanol with sodium.

20.20 Describe the visual change observed in the reaction of a tertiary alcohol with the Lucas reagent.

20.21 Write the structure of the expected product from each of the following reactions.

(a) 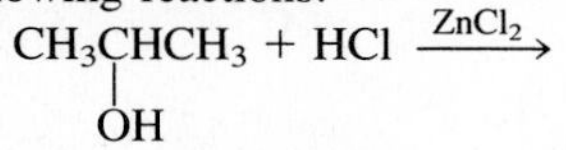 $CH_3CHCH_3 + HCl \xrightarrow{ZnCl_2}$
     |
     OH

(b) $CH_3CH_2CH_2CH_2OH + PBr_3 \longrightarrow$

20.22 Write the structure of the expected product from each of the following reactions.

(a) 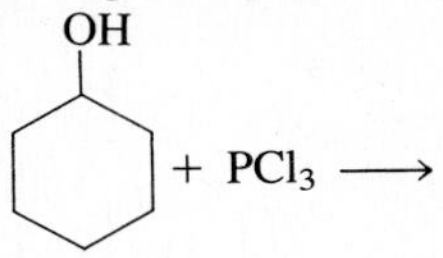 (cyclohexane with OH) $+ PCl_3 \longrightarrow$

(b) 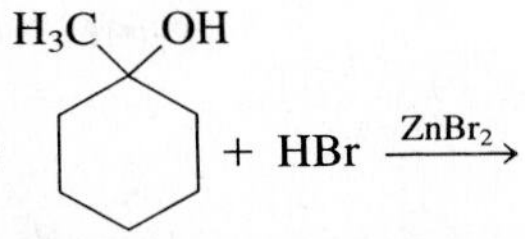 (cyclohexane with H_3C and OH) $+ HBr \xrightarrow{ZnBr_2}$

20.23 Describe how the following three compounds could be distinguished by using the Lucas reagent.

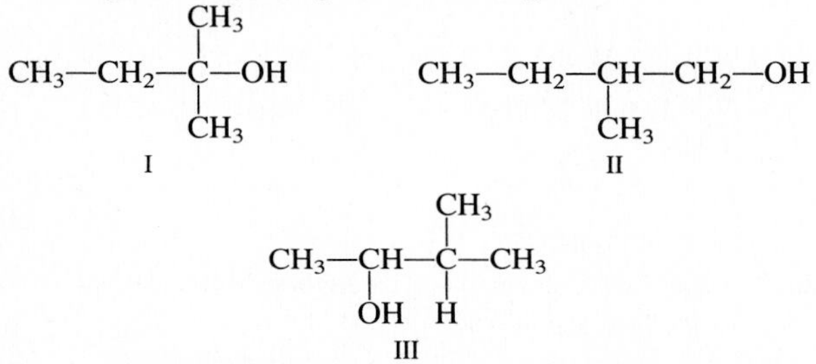

         CH_3
         |
$CH_3—CH_2—C—OH$
         |
         CH_3
        I

$CH_3—CH_2—CH—CH_2—OH$
              |
              CH_3
           II

          CH_3
          |
$CH_3—CH—C—CH_3$
      | |
      OH H
        III

20.24 Order the following compounds according to their reactivity with the Lucas reagent.

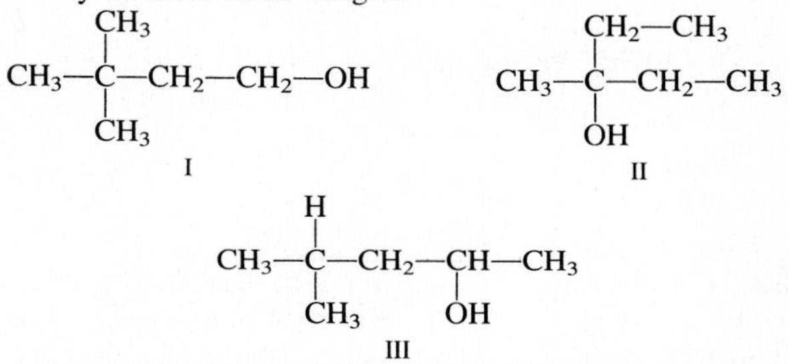

       CH_3
       |
$CH_3—C—CH_2—CH_2—OH$
       |
       CH_3
       I

      $CH_2—CH_3$
      |
$CH_3—C—CH_2—CH_3$
      |
      OH
      II

      H
      |
$CH_3—C—CH_2—CH—CH_3$
      | |
      CH_3 OH
        III

20.25 Draw the dehydration product(s) obtained by treating each of the following alcohols with H_2SO_4.

(a) 1-pentanol   (b) 2-pentanol
(c) 2-methyl-2-butanol

20.26 Draw the structures of the dehydration products when each of the following compounds reacts with sulfuric acid.

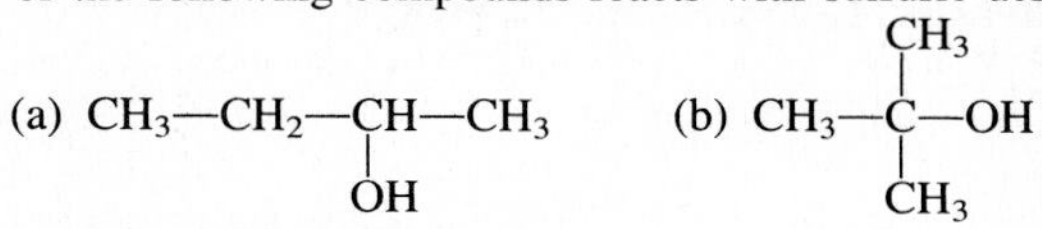

(a) $CH_3—CH_2—CH—CH_3$
                |
                OH

            CH_3
            |
(b) $CH_3—C—OH$
            |
            CH_3

(c) $CH_3—CH_2—CH_2—CH_2—OH$

(d) $CH_3—CH_2—CH_2—OH$

20.27 Write the structure of the product formed from the oxidation by $K_2Cr_2O_7$ of each of the compounds in 20.8.

20.28 Which of the compounds in 20.26 will give an acid as product when oxidized with potassium dichromate?

Ethers

20.29 Draw the three isomeric ethers with the molecular formula $C_4H_{10}O$.

20.30 Ethers are not soluble in water. Explain why.

20.31 Ethers do not react with sodium. Explain why no reaction occurs.

20.32 Explain how diethyl ether and 1-butanol can be distinguished by a visual chemical test.

Nomenclature of Aldehydes and Ketones

20.33 Write the structure of each of the following compounds.
(a) diethyl ketone
(b) methyl ethyl ketone
(c) acetaldehyde
(d) methyl phenyl ketone

20.34 Write the structure of each of the following compounds.
(a) butyraldehyde
(b) α-bromobutyraldehyde
(c) β-chloropropionaldehyde
(d) acetone

20.35 Write the structural formula for each of the following compounds.
(a) 2-methylbutanal
(b) 3-ethylpentanal
(c) 2-bromopentanal
(d) 3,4-dimethyloctanal

20.36 Write the structure of each of the following compounds.
(a) 3-bromo-2-pentanone
(b) 4-methyl-2-pentanone
(c) 2,4-dimethyl-3-pentanone
(d) 3,4-dimethyl-2-pentanone

20.37 Name each of the following compounds by IUPAC nomenclature.

    (a) $CH_3CH_2\overset{\underset{\displaystyle CH_3}{|}}{C}HCHO$

    (b) $CH_3\overset{\underset{\displaystyle CH_3}{|}}{C}HCHO$

    (c) $CH_3\overset{\underset{\displaystyle CH_3}{|}}{C}HCH_2CH_2CHO$

    (d) $CH_3CH_2\overset{\underset{\displaystyle CH_2CH_3}{|}}{\underset{|}{C}}CH_2CHO$ (with CH_3 above)

20.38 Give the IUPAC name for each of the following compounds.

    (a) $CH_3CH_2\overset{\overset{\displaystyle O}{||}}{C}CH_2CH_3$

    (b) $CH_3CH_2\overset{\overset{\displaystyle O}{||}}{C}CH_3$

    (c) $CH_3\overset{\overset{\displaystyle O}{||}}{C}-\overset{\underset{\displaystyle CH_3}{|}}{C}CH_3$ (with CH_3 above)

    (d) $CH_3CH_2\overset{\overset{\displaystyle O}{||}}{C}\overset{\underset{\displaystyle CH_3}{|}}{C}HCH_3$

20.39 Give the IUPAC name for each of the following compounds.

    (a) $CH_3\overset{\underset{\displaystyle Cl}{|}}{C}HCHO$

    (b) $BrCH_2CH_2CH_2CHO$

    (c) $ClCH_2\overset{\underset{\displaystyle CH_3}{|}}{C}HCH_2CHO$

    (d) ICH_2CH_2CHO

20.40 Give the IUPAC name for each of the following compounds.

    (a) $CH_3\overset{\underset{\displaystyle CH_3}{|}}{\overset{\overset{\displaystyle OH}{|}}{C}}CH_2\overset{\overset{\displaystyle O}{||}}{C}CH_3$

    (b) $CH_3\overset{\underset{\displaystyle CH_3}{|}}{\overset{\overset{\displaystyle OH}{|}}{C}}CH_2\overset{\overset{\displaystyle O}{||}}{C}CH_2CH_3$

    (c) $CH_3\overset{\overset{\displaystyle OH}{|}}{C}HCH_2CHO$

    (d) $CH_3\overset{\underset{\displaystyle CH_3}{|}}{\overset{\overset{\displaystyle OH}{|}}{C}}CH_2CHO$

Physical Properties of Aldehydes and Ketones

20.41 Aldehydes in general are less soluble in water than the corresponding alcohols of similar molecular weight. Explain why.

20.42 Which compound has the higher boiling point, cyclohexanol or cyclohexanone?

Reactions of Aldehydes and Ketones

20.43 What observation is made when an aldehyde reacts with Fehling's solution?

20.44 What observation is made when an aldehyde reacts with Tollens' reagent?

Nomenclature of Acids

20.45 Give the common name for each of the following acids.
    (a) $CH_3CH_2CO_2H$ (b) HCO_2H
    (c) $CH_3CH_2CH_2CO_2H$ (d) CH_3CO_2H

20.46 Give the IUPAC name for each of the acids in 20.45.

20.47 Give the common name for each of the following acids.
    (a) $CH_3(CH_2)_3CO_2H$ (b) $(CH_3)_2CHCO_2H$

    (c) $CH_3\overset{\underset{\displaystyle CH_2CH_3}{|}}{C}HCH_2CO_2H$

    (d) $CH_3\overset{\underset{\displaystyle CH_3}{|}}{C}HCH_2\overset{\underset{\displaystyle CH_3}{|}}{C}HCO_2H$

    (e) $CH_3\overset{\underset{\displaystyle CH_3}{|}}{C}HCH_2CO_2H$

    (f) $CH_3\overset{\underset{\displaystyle CH_3}{|}}{C}CH_2\overset{\underset{\displaystyle CH_3}{|}}{C}HCH_2CO_2H$ (with CH_3 above first C)

20.48 Give the IUPAC name for each of the acids in 20.47.

Properties of Acids

20.49 List each of the following compounds in order of increasing boiling point: butanal, 1-butanol, butanoic acid

20.50 Why is 1-butanol less soluble in water than butanoic acid?

Salts of Carboxylic Acids

20.51 Give the common and IUPAC names for each of the following.
    (a) $CH_3CH_2CH_2CO_2Na$ (b) $CH_3(CH_2)_{16}CO_2K$
    (c) $(CH_3CH_2CO_2)_2Ca$

20.52 Write the formula of each of the following compounds.
    (a) sodium stearate   (b) potassium propanoate
    (c) lithium butyrate   (d) calcium acetate

Nomenclature of Esters

20.53 Give the common name for each of the following compounds.

(a) $CH_3CH_2-O-\overset{\overset{\displaystyle O}{\|}}{C}-H$

(b) $CH_3-O-\overset{\overset{\displaystyle O}{\|}}{C}-CH_2CH_2CH_3$

(c) $CH_3(CH_2)_7-O-\overset{\overset{\displaystyle O}{\|}}{C}-CH_3$

20.55 Give the common name for each of the following.

(a) $CH_3-\overset{\overset{\displaystyle O}{\|}}{C}-O-$⬡

(b) $CH_3CH_2-\overset{\overset{\displaystyle O}{\|}}{C}-O-$⬡

20.54 Give the common name for each of the following compounds.

(a) $CH_3CH_2-O-\overset{\overset{\displaystyle O}{\|}}{C}-CH_2CH_2CH_3$

(b) $CH_3(CH_2)_4-O-\overset{\overset{\displaystyle O}{\|}}{C}-CH_2CH_2CH_3$

(c) $CH_3(CH_2)_4-O-\overset{\overset{\displaystyle O}{\|}}{C}-CH_3$

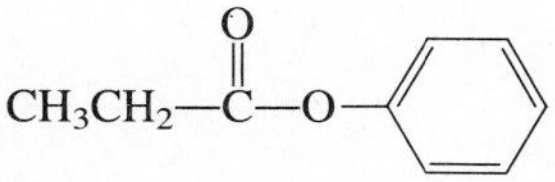

20.56 Give the common name for each of the following.
(a) $CH_3CH_2CH_2CO_2CH_2CH_2CH_2CH_3$
(b) $CH_3CO_2CH_2(CH_2)_3CH_3$
(c) $CH_3CO_2CH_2(CH_2)_6CH_3$
(d) $CH_3CH_2CH_2CO_2CH_3$

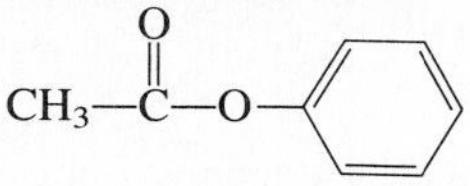

Reactions of Esters

20.57 Write the products of hydrolysis of each of the following.

(a) 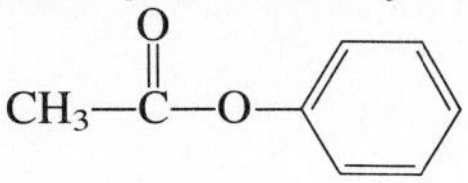 $CH_3-\overset{\overset{\displaystyle O}{\|}}{C}-O-$⬡

(b) ⬡$-\overset{\overset{\displaystyle O}{\|}}{C}-O-CH_2CH_3$

(c) $H-\overset{\overset{\displaystyle O}{\|}}{C}-O-CH_2CH_3$

(d) $CH_3-\overset{\overset{\displaystyle O}{\|}}{C}-O-CH_2CH_2CH_3$

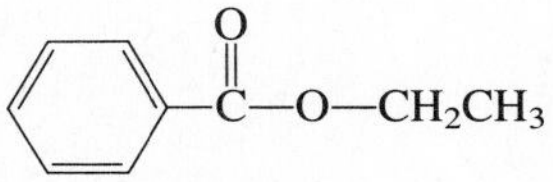

20.59 Write the products of the saponification with NaOH of each of the compounds in 20.57.

20.58 Write the products of hydrolysis of each of the following.

(a) $CH_3CH_2CH_2-\overset{\overset{\displaystyle O}{\|}}{C}-O-CH_2CH_2CH_2CH_3$

(b) $CH_3CH_2-\overset{\overset{\displaystyle O}{\|}}{C}-O-CH_2CH_2CH_3$

(c) $CH_3-\overset{\overset{\displaystyle O}{\|}}{C}-O-CH_2CH_3$

(d) $CH_3-\overset{\overset{\displaystyle O}{\|}}{C}-O-CH_2CH_2CH_3$

20.60 Write the products of the saponification with NaOH of each of the compounds in 20.58.

Classification of Amines

20.61 Coniine, the poison in hemlock, has the following structure. Classify this amine.

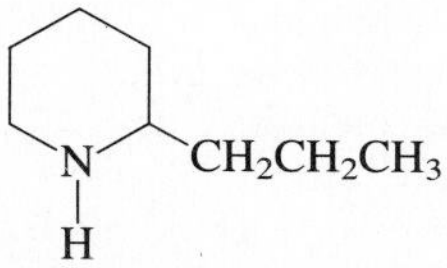

⬡ with N–H and $CH_2CH_2CH_3$

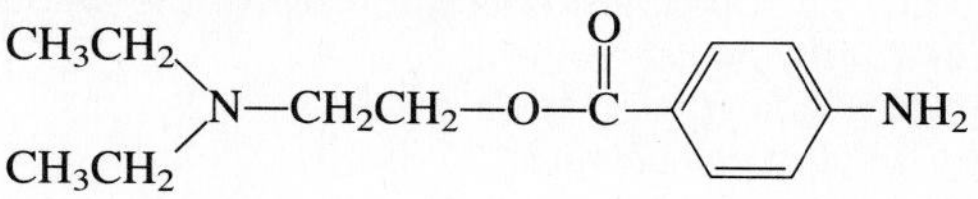

20.63 Classify each of the following amines.

(a) $CH_3-\overset{\overset{\displaystyle H}{|}}{N}-CH_2CH_3$

(b) $CH_3CH_2-\overset{\overset{\displaystyle CH_2CH_3}{|}}{N}-CH_2CH_3$

20.62 Classify the amine sites in novocaine.

$\begin{matrix} CH_3CH_2 \\ \\ CH_3CH_2 \end{matrix}$ $N-CH_2CH_2-O-\overset{\overset{\displaystyle O}{\|}}{C}-$⬡$-NH_2$

20.64 Classify each of the following amines.

(a)  NH_2–⬡–CH_3

(b) 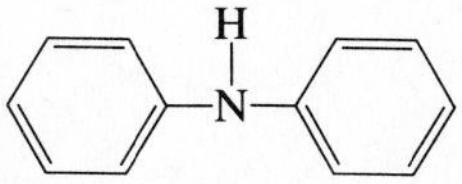 ⬡$-\overset{\overset{\displaystyle H}{|}}{N}-$⬡

(c) 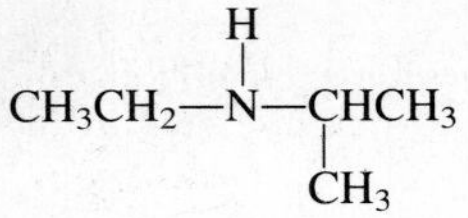 $CH_3CH_2-\underset{\underset{CH_3}{|}}{\overset{\overset{H}{|}}{N}}-CHCH_3$

(d) 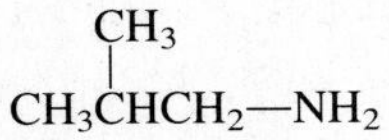 $CH_3\underset{\underset{CH_3}{|}}{CH}CH_2-NH_2$

(e) 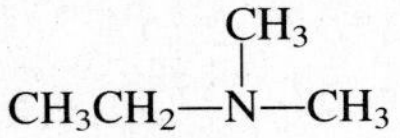 $CH_3CH_2-\underset{\underset{CH_3}{|}}{\overset{\overset{CH_3}{|}}{N}}-CH_3$

(f) 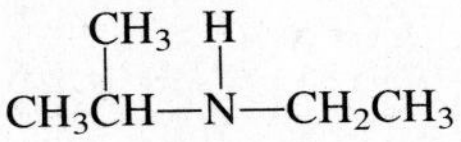 $CH_3\underset{\underset{CH_3}{|}}{CH}-\overset{\overset{H}{|}}{N}-CH_2CH_3$

(c) 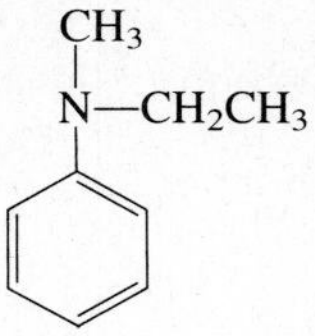 $\underset{\bigcirc}{}\overset{\overset{CH_3}{|}}{\underset{}{N}}-CH_2CH_3$

(d) 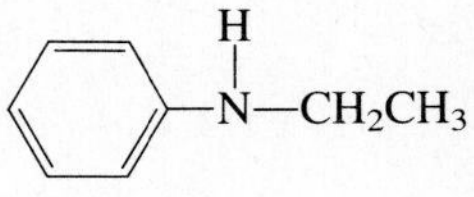 $\underset{\bigcirc}{}\overset{\overset{H}{|}}{N}-CH_2CH_3$

Nomenclature of Amines

20.65 Give the IUPAC name of each of the following compounds.

(a) 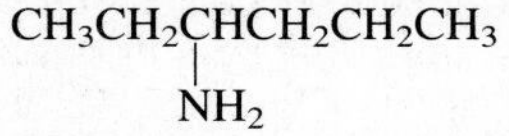 $CH_3CH_2\underset{\underset{NH_2}{|}}{CH}CH_2CH_2CH_3$

(b) 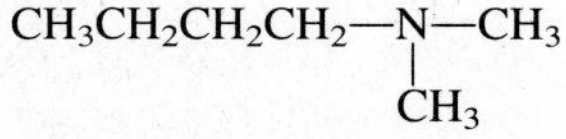 $CH_3CH_2CH_2CH_2-\underset{\underset{CH_3}{|}}{N}-CH_3$

(c) 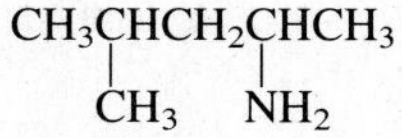 $CH_3\underset{\underset{CH_3}{|}}{CH}CH_2\underset{\underset{NH_2}{|}}{CH}CH_3$

20.66 Give the IUPAC name of each of the following compounds.

(a) 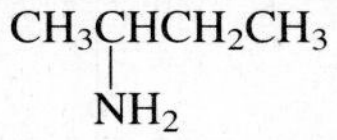 $CH_3\underset{\underset{NH_2}{|}}{CH}CH_2CH_3$

(b) 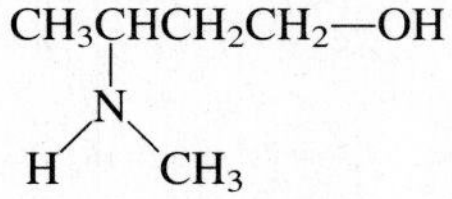 $CH_3\underset{\underset{\underset{H}{}\overset{}{N}\overset{}{CH_3}}{|}}{CH}CH_2CH_2-OH$

(c) 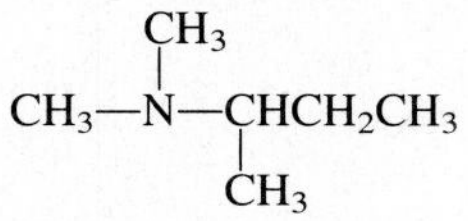 $CH_3-\underset{\underset{CH_3}{|}}{\overset{\overset{CH_3}{|}}{N}}-CHCH_2CH_3$

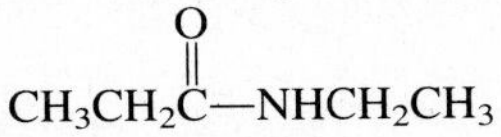

Nomenclature of Amides

20.67 Give the common name of each of the following amides.

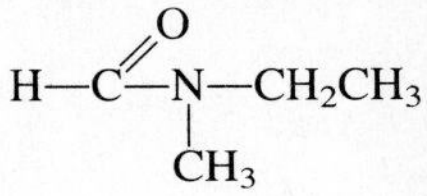

(a) 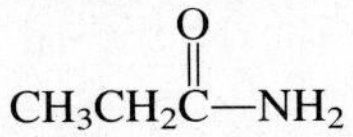 $CH_3CH_2\overset{\overset{O}{||}}{C}-NH_2$

(b) $CH_3CH_2\overset{\overset{O}{||}}{C}-NHCH_2CH_3$

(c) $CH_3CH_2\underset{\underset{CH_3}{|}}{CH}\overset{\overset{O}{||}}{C}-NH_2$

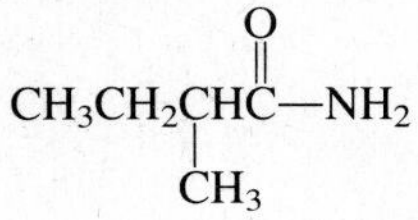

20.68 Give the IUPAC name of each of the following amides.

(a) 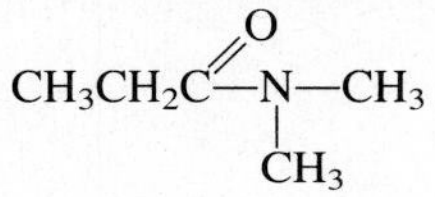 $CH_3CH_2\overset{\overset{O}{||}}{C}-\underset{\underset{CH_3}{|}}{N}-CH_3$

(b) $H-\overset{\overset{O}{||}}{C}-\underset{\underset{CH_3}{|}}{N}-CH_2CH_3$

(c) $CH_3\underset{\underset{CH_3}{|}}{CH}CH_2\overset{\overset{O}{||}}{C}-\underset{\underset{H}{|}}{N}-CH_3$

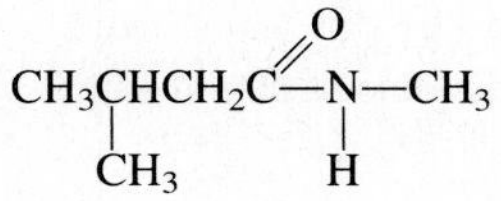

Reactions of Amides

20.69 Write the structural formula of each of the following.
(a) N-ethylpropanamide
(b) N,N-dimethylbutanamide
(c) N-phenylethanamide
(d) pentanamide

20.70 Write the structural formula of each of the following compounds.
(a) N-methylbutanamide
(b) N,N-dimethylpropanamide
(c) N,N-diphenylmethanamide
(d) ethanamide

20.71 Write the products of each of the following reactions.

(a) 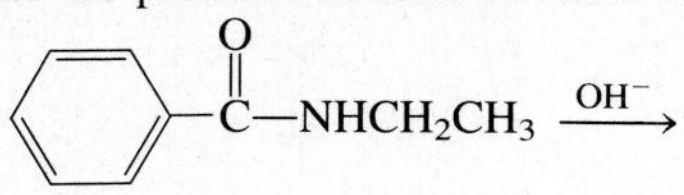 $\underset{\bigcirc}{}\overset{\overset{O}{||}}{C}-NHCH_2CH_3 \xrightarrow{OH^-}$

(b) 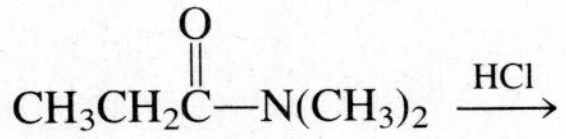 $CH_3CH_2\overset{\overset{O}{||}}{C}-N(CH_3)_2 \xrightarrow{HCl}$

20.72 Write the products of each of the following reactions.

(a) 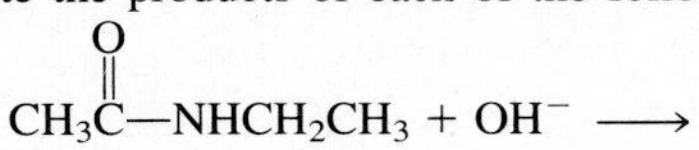 $CH_3\overset{\overset{O}{||}}{C}-NHCH_2CH_3 + OH^- \longrightarrow$

(b)  $CH_3CH_2\overset{\overset{O}{||}}{C}-NHCH_3 + H_3O^+ \longrightarrow$

Additional Exercises

20.73 Name each of the following compounds.
(a) $CH_3—CH_2—CH—CH_3$
     $|$
     OH
(b) $CH_3—CH_2—CH_2—CH_2—OH$
(c) $CH_3—CH_2—CH_2—CH—CH_3$
                   $|$
                   $CH_2—OH$
(d) $CH_3—CH—CH_2—CH—CH_3$
           $|$ $|$
           CH_3 CH_3

(In (d) the upper CH_3 is attached to the fourth carbon.)

20.75 Classify each of the following alcohols.

(a)
$—CH_2—CH—CH_2—OH$ with Br on the middle carbon
(b) a cyclohexane ring with CH_3 and OH

20.77 Describe how the Lucas reagent could be used to distinguish between samples of 3-pentanol and 2-methyl-2-butanol.

20.79 What products are formed from the oxidation by $K_2Cr_2O_7$ of each of the compounds in 20.23?

20.81 Write the structure of each of the following compounds.
(a) cyclopentanone
(b) 2-methylcyclohexanone
(c) 3-bromocyclobutanone
(d) 4,4-dimethylcyclohexanone

20.83 Give the IUPAC name for each of the following.

(a) cyclohexanone with CH_3

(b) cyclobutanone

(c) cyclohexanone with CH_3

(d) cyclopentanone with CH_3

20.85 The fishy odor of an aqueous solution of propylamine is eliminated when an equimolor amount of HCl is added. Explain why.

20.87 Write the products of hydrolysis of each of the following esters.

(a) $CH_3CHCH_2—O—\overset{\displaystyle O}{\overset{\displaystyle \|}{C}}—H$
         $|$
         CH_3

20.74 Name each of the following compounds.

(a) cyclohexane with $—OH$
(b) cyclopentane with $—OH$

(c) cyclopentane with CH_3 and OH

20.76 Classify each of the following alcohols.

(a) cyclopentane with $—CH_2—OH$
(b) benzene ring with $—CH_2CH_2OH$

20.78 Describe how the Lucas reagent could be used to distinguish between samples of 2-pentanol and 1,1-dimethyl-1-propanol.

20.80 Write the structure of the product formed when each of the compounds in 20.24 reacts with sodium dichromate.

20.82 Name each of the following compounds by IUPAC nomenclature.
(a) CH_3CHO
(b) $CH_3CH_2CH_2CH_2CHO$
(c) $CH_3CH_2CH_2CHO$
(d) $(CH_3)_3CCH_2CHO$

20.84 Give the IUPAC name for each of the following.

(a) benzene ring $—CH_2\overset{\displaystyle O}{\overset{\displaystyle \|}{C}}CH_2CH_3$

(b) benzene ring $—CH_2CH_2\overset{\displaystyle O}{\overset{\displaystyle \|}{C}}CH_3$

(c) $CH_3CH\overset{\displaystyle O}{\overset{\displaystyle \|}{C}}CH_2CH_3$
         $|$
         CH_3

(d) $CH_3CHCH_2\overset{\displaystyle O}{\overset{\displaystyle \|}{C}}CH_2CH_3$
         $|$
         CH_3

20.86 The boiling point of 1,2-diaminoethane is 116°C. Explain why this compound boils at a much higher temperature than propylamine (49°C), which has a similar molecular weight.

20.88 Write the structure of the ester formed in each reaction of acid and alcohol.
(a) $CH_3—CO_2H + CH_3—CH_2—OH$
(b) $CH_3—CH_2—CO_2H + CH_3—CH_2—CH_2—OH$

(b) $CH_3CH_2CH_2-O-\overset{\displaystyle O}{\overset{\displaystyle \|}{C}}-CH_3$

(c) 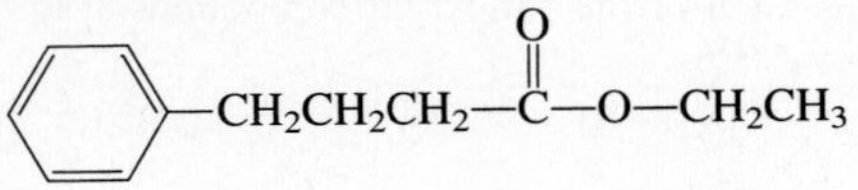 ⟨benzene ring⟩$-CH_2CH_2CH_2-\overset{\displaystyle O}{\overset{\displaystyle \|}{C}}-O-CH_2CH_3$

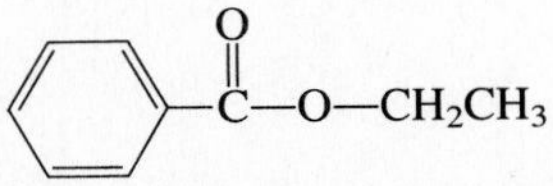

(d) ⟨benzene ring⟩$-\overset{\displaystyle O}{\overset{\displaystyle \|}{C}}-O-CH_2CH_3$

(c) ⟨benzene ring⟩$-CO_2H + CH_3-OH$

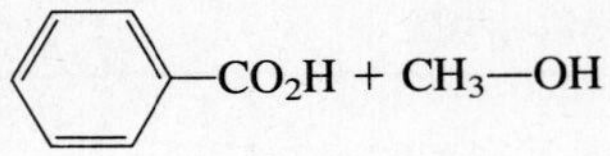

(d) $CH_3-CO_2H +$ ⟨benzene ring⟩$-OH$

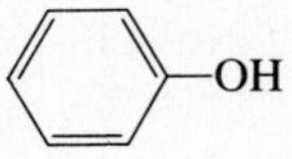

20.89 Classify arecoline as an amine. The compound is a narcotic found in the nut of the betel palm.

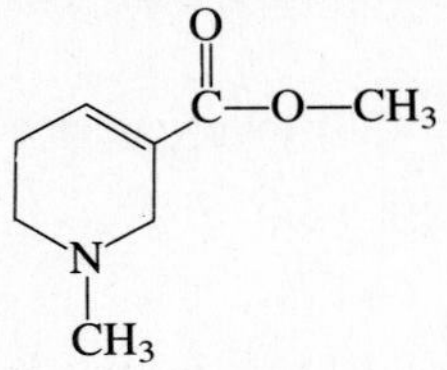

⟨ring structure with $\overset{\displaystyle O}{\overset{\displaystyle \|}{C}}-O-CH_3$ group and N–CH_3⟩

20.90 Classify each of the following amines.

(a) 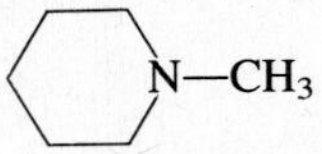 ⟨ring⟩N—CH_3

(b) 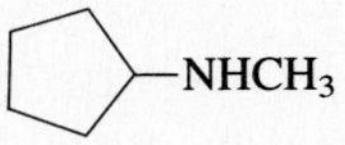 ⟨ring⟩$-NHCH_3$

(c) 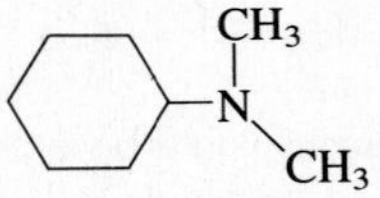 ⟨ring⟩$-N\overset{\displaystyle CH_3}{\underset{\displaystyle CH_3}{}}$

(d) 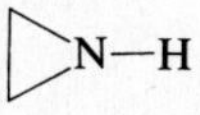 ⟨ring⟩N—H

Mathematical Review

A.1 Addition and Subtraction

The notation for addition is the plus sign, $+$, that is placed between two numbers to be added. The result of addition is the **sum** of the numbers.

The plus sign also is used to indicate that a number is positive. However, a number without a sign always is considered to be positive. Thus a number used as an exponent is considered to be positive if no sign is indicated.

The notation for subtraction is the minus sign, $-$, that is placed between two numbers to be subtracted. The result of subtraction is the **difference,** which indicates the change in going from one number to another.

A.2 Multiplication

The process of adding a given number a certain number of times is called multiplication. Therefore, 3 times 4 means 3 added four times or 4 added three times to give the **product** 12. The various notations used in expressing multiplication of numbers such as 4 and 3 are

$$4 \times 3 \qquad 4 \cdot 3 \qquad 4(3) \qquad (4)(3)$$

When a complex quantity contained within parentheses is multiplied, each term must be multiplied.

$$\tfrac{1}{2}(2x + y) = \tfrac{1}{2}(2x) + \tfrac{1}{2}(y) = x + y/2$$

If the quantities enclosed in parentheses can be simplified, do so before performing the multiplication. The conversion of a temperature in °F to °C involves multiplication of a difference contained within a set of parentheses.

$$°C = (5°C/9°F) \times (122°F - 32°F) = (5/9) \times (90) = 50°C$$

A.3 Division

The process of finding out how many times a number is contained in another number is called division. The notations used in expressing division are

$$8 \div 4 \qquad \frac{8}{4} \qquad 8/4$$

The number to the left of the divide symbol, above the horizontal line, or to the left of the slant line is called the **numerator;** the other number is the **denominator.** The **quotient** is the number obtained by dividing one number into another.

A.4 Fractions

The division of one number by another can be expressed as a fraction, which is a ratio of one quantity to another. Thus a fraction consist of a numerator above a line over a denominator. When a quantity is multiplied by a fraction, it is multiplied by the numerator and is divided by the denominator of the fraction. The same result is obtained regardless of the order of the two operations. The product of $\frac{9}{5}$ and 20 is 36.

$$\frac{9}{5} \times 20 = \frac{9 \times 20}{5} = \frac{180}{5} = 36$$

$$\frac{9}{5} \times 20 = 9 \times \frac{20}{5} = 9 \times 4 = 36$$

Proper fractions are those in which the numerator is smaller than the denominator; others are **improper fractions.** The quantity of $\frac{9}{5}$ is an improper fraction since it may be converted to the **mixed fraction** $1\frac{4}{5}$. However, in this chemistry text it is usually more convenient to convert to the equivalent decimal 1.8.

In order to add or subtract two fractions, the denominators must be changed to a common number. The numerators then can be added or subtracted and the resultant placed over the **common denominator.**

$$\frac{4}{7} + \frac{2}{5} = \frac{4(5)}{7(5)} + \frac{2(7)}{5(7)} = \frac{20}{35} + \frac{14}{35} = \frac{34}{35}$$

Note that conversion of fractions to a common denominator requires that both the numerator and denominator be multiplied by the same number. Identical operations of either multiplication or division on both the numerator and denominator of a fraction leave the fraction mathematically unchanged.

Fractions are multiplied by multiplying both the numerators and the denominators. The product of the numerators is placed over the product of the denominators. The fraction is then reduced to lowest terms by division of numerator and denominator by identical quantities.

$$\frac{4}{7} \times \frac{2}{5} = \frac{4 \times 2}{7 \times 5} = \frac{8}{35}$$

$$\frac{2}{5} \times \frac{3}{8} = \frac{2 \times 3}{5 \times 8} = \frac{6}{40} = \frac{6/2}{40/2} = \frac{3}{20}$$

Division of fractions can be regarded as a multiplication of the numerator by the inverse of the fraction in the denominator.

$$\frac{2/5}{3/7} = \frac{2}{5} \times \frac{7}{3} = \frac{14}{15}$$

The equivalence is achieved by multiplying both numerator and denominator by the inverse of the fraction in the denominator. This changes the denominator to 1.

$$\frac{\dfrac{2}{5}}{\dfrac{3}{7}} = \frac{\dfrac{2}{5} \times \dfrac{7}{3}}{\dfrac{3}{7} \times \dfrac{7}{3}} = \frac{\dfrac{14}{15}}{\dfrac{21}{21}} = \frac{\dfrac{14}{15}}{1} = \frac{14}{15}$$

A.5 Decimals

A decimal is a fraction in which the denominator is an unexpressed power of 10. Addition and subtraction of decimals are accomplished in the same manner as using whole numbers, but the decimals must be placed in the proper column.

add $5.46 + 130.21$

$$
\begin{array}{r}
5.46 \\
\underline{130.21} \\
135.67
\end{array}
$$

subtract $130.21 - 5.46$

$$
\begin{array}{r}
130.21 \\
\underline{5.46} \\
124.75
\end{array}
$$

Decimals are multiplied as if they were whole numbers. The location of the decimal point in the product is determined by adding the number of digits to the right of the decimal point in all of the numbers multiplied together. The product contains the same total number of digits to the right of the decimal point. However, it may be necessary to round off the product to express the result to the proper number of significant figures (Chapter 1).

$$
\begin{array}{r}
12.041 \\
\underline{\times\ 0.15} \\
60205 \\
\underline{12041} \\
1.80615
\end{array}
$$

Division of decimals involves relocation of the decimal points in both the numerator and denominator prior to actually carrying out the division. The relocation is accomplished by multiplying by a power of 10 such that the denominator becomes a whole number. Then the division is carried out and the decimal point is located immediately above its position in the dividend.

$$
\frac{86.52}{2.1} = \frac{86.52 \times 10}{2.1 \times 10} = \frac{865.2}{21}
$$

$$
\begin{array}{r}
41.2 \\
21\overline{)865.2} \\
\underline{84} \\
25 \\
\underline{21} \\
42 \\
\underline{42} \\
0
\end{array}
$$

A.6 Exponents

Many numbers encountered in science are best expressed in exponential form $a \times 10^b$. The a represents a **coefficient** and b is the **exponent** or the power of 10.

$$1.234 \times 10^3 = 1.234 \times (10 \times 10 \times 10) = 1234$$

For both large and small numbers exponents of the base 10 are used to make the number more compact and easier to handle. A number multiplied by 10^2 is equivalent to another number in which the decimal point is moved two places to the right.

$$3 \times 10^2 = 300$$

If a number is multiplied by 10^{-3}, the decimal is moved three places to the left.

$$3467 \times 10^{-3} = 3.467$$

In expressing a number in exponential form the decimal point is moved to a new position so that the number value is between 1 and 10. This new number is then multiplied by the proper power of 10 to maintain its original value. The exponent is determined by counting the number of places that the decimal point is moved. If the decimal point is moved to the left, the exponent is a positive number; if moved to the right, the exponent is a negative number.

$$5243 = 5.243 \times 10^3$$
$$0.0467 = 4.67 \times 10^{-2}$$

The movement of the decimal point and the introduction of a power of 10 actually involves simultaneous multiplication and division by the same power of 10.

$$5243 = 5243 \times \frac{10^3}{10^3} = \frac{5243}{10^3} \times 10^3 = 5.243 \times 10^3$$

$$0.0467 = 0.0467 \times \frac{10^{-2}}{10^{-2}} = \frac{0.0467}{10^{-2}} \times 10^{-2} = 4.67 \times 10^{-2}$$

To multiply exponential numbers, first multiply the numerical portion of the number and then algebraically add the exponents of the powers of 10.

$$(4 \times 10^2) \times (2 \times 10^3) = (4 \times 2) \times (10^2 \times 10^3) = 8 \times 10^5$$

Division is carried out in the usual manner with the numerical portion of the number. The power of 10 then is calculated by subtracting the exponent of the denominator from the exponent of the numerator.

$$\frac{4 \times 10^2}{2 \times 10^3} = \frac{4}{2} \times \frac{10^2}{10^3} = 2 \times \frac{10^2}{10^3} = 2 \times 10^{-1}$$

A.7 Roots

The notation for a square root and occasionally for cube and higher roots is $\sqrt[n]{}$, where n is the order of the root. For square roots the value of $n = 2$, but the 2 is usually omitted. The horizontal line should be extended over the entire number whose root is sought.

The square root of 49 is written $\sqrt{49}$ and is equal to 7. The square root of a quantity is the number that when squared (multiplied by itself) gives the original quantity. Thus $7^2 = 49$.

Since both $(+7)^2$ and $(-7)^2$ equal 49, both $+7$ and -7 are roots of 49. However, in problems based on chemical quantities the square root of 49 is considered to be $+7$. The negative root has no physical reality in describing quantities such as concentration.

To find the root of a number expressed in exponential notation with an even power of 10, find the root of the coefficient and then the root of the power of 10, which is done by dividing the exponent by 2.

$$\sqrt{4 \times 10^8} = \sqrt{4} \times \sqrt{10^8} = 2 \times 10^4$$

To find the root of a number expressed in exponential notation with an odd power of 10, change the coefficient in order to obtain an even power of 10 for the exponent. This is done by multiplying the coefficient by 10 and decreasing the exponent by 1. Then proceed as above for the modified exponential number.

$$\sqrt{2.5 \times 10^7} = \sqrt{25 \times 10^6} = \sqrt{25} \times \sqrt{10^6} = 5 \times 10^3$$

If you have a calculator, you may obtain the square root of a number expressed in exponential form regardless of whether the exponent is even or odd.

A.8 Logarithms

The power to which the number 10 must be raised to equal a desired number is its logarithm. The log of 100 is 2, because $10^2 = 100$. Similarly, the log of 0.0001 is -4. Since most numbers are not integral powers of 10, the logarithms may be nonintegers. A calculator will give you the logarithm of a number. If you do not use a calculator, Table A.1, which gives logarithms of numbers between 1 and 10 to the number of decimal places needed in this text, will be useful. In order to obtain the logarithm of 6.3, look for 6 in the first vertical column and then across to the column 0.3 to obtain the logarithm 0.799. For numbers greater than 10 or less than 1 write the number in exponential form so it is a number between 1 and 10 multiplied by a power of 10. The number 6300 is 6.3×10^3. Since the logarithm of a product $m \times n$ is equal to the log of m plus the log of n, then

$$\log(6.3 \times 10^3) = \log 6.3 + \log 10^3$$
$$= 0.799 + 3$$
$$= 3.799$$

The only use of logarithms in this text is in the calculation of pH, which is the negative logarithm of the hydronium ion concentration. For a hydronium ion concentration of 0.00063 the pH is calculated as follows:

$$pH = -\log(0.00063)$$
$$= -\log(6.3 \times 10^{-4})$$
$$= -(\log 6.3 + \log 10^{-4})$$
$$= -(0.799 - 4) = -(-3.201) = +3.201$$

Table A.1   **Table of logarithms**

	0.0	0.1	0.2	0.3	0.4	0.5	0.6	0.7	0.8	0.9
1	0.000	0.041	0.079	0.114	0.146	0.176	0.204	0.230	0.255	0.279
2	0.301	0.322	0.342	0.362	0.380	0.398	0.415	0.431	0.447	0.462
3	0.477	0.491	0.505	0.519	0.532	0.544	0.556	0.568	0.580	0.591
4	0.602	0.613	0.623	0.634	0.644	0.653	0.663	0.672	0.681	0.690
5	0.699	0.708	0.716	0.724	0.732	0.740	0.748	0.756	0.763	0.771
6	0.778	0.785	0.792	0.799	0.806	0.813	0.820	0.826	0.833	0.839
7	0.845	0.851	0.857	0.863	0.869	0.875	0.881	0.887	0.892	0.898
8	0.903	0.909	0.914	0.919	0.924	0.929	0.935	0.940	0.945	0.949
9	0.954	0.959	0.964	0.969	0.973	0.978	0.982	0.987	0.991	0.996

A.9   Proportionality

It is said that x is proportional to y if x is related to or depends on y. If a certain percentage change in x produces an equal percentage change in y, then x is **directly proportional** to y. For example, mass and volume are directly proportional to each other because the mass of a substance depends on the volume of the sample under consideration. Mathematically, the direct proportionality between mass m and volume V is expressed as

$$m \propto V \quad \text{or} \quad m = kV$$

The symbol $\propto$ means "is directly proportional to" and is the proportionality sign. In the second equation the k is a proportionality constant and allows the replacement of the proportionality sign by an equal sign. In the case of mass and volume the k is equal to the density of the substance, m/V.

The proportionality constant k in a direct proportion indicates the ratio of the two quantities. If one quantity is changed by a factor of 2, then the other quantity also must change by a factor of 2 in order to maintain the equality. For example, the mass and volume of a given substance always must be related such that if a larger mass is considered, the volume of the sample also must be larger. Thus for a given substance

$$\frac{m_1}{V_1} = \frac{m_2}{V_2} = \frac{m_3}{V_3} = k = \text{density}$$

where the subscripts refer to different quantities of the material.

An **inverse proportionality** indicates that as one variable, x, is increased by a certain percentage change the related variable y is decreased by the same percentage change. The inverse proportionality may be indicated as

$$x \propto \frac{1}{y} \quad \text{or} \quad x \propto y^{-1}$$

If a proportionality constant, k, is introduced, the following equations result.

$$x = k\left(\frac{1}{y}\right) \quad \text{or} \quad x = ky^{-1}$$

An alternate way of expressing the inverse relationship between the two variables results from rearranging the above equations.

$xy = k$

This expression indicates that whatever value of x is chosen, y must be such that the product of the two equals a constant, k.

At constant temperature the pressure P and the volume V of a gas are inversely related. Therefore, it follows that the equalities listed below are correct. The subscripts refer to different experimental conditions for the same sample of a gas.

$P_1V_1 = P_2V_2 = P_3V_3 = k$

A.10 Graphs

The relationship between two properties can be conveniently represented by a graph in which the values of one property are plotted along the horizontal axis (**abscissa**) and of the other along the vertical axis (**ordinate**). In Figure A.1 a graph of a direct proportion $y = kx$ is illustrated. The points define a straight line that goes through the origin, the point at which both y and x are zero. The **slope** of the line is equal to k and represents the ratio of the change in the quantity y per unit change in the quantity x.

In Figure A.2 the straight line does not pass through the origin. The equation of the line is $y = kx + c$. The value of c is a constant that is equal to the value of y when x is zero, that is, the **y intercept** or the point on the ordinate when x is zero. The value k is still the slope of the line. The equation is said to indicate a linear relationship between x and y.

In Figure A.3 the inverse proportionality $xy = k$ is graphically illustrated. Note that at no point does the curve intersect the x or y axis. As either quantity approaches zero, the other quantity approaches infinity. An inverse proportionality can be graphed to yield a straight line if one quantity is plotted against the reciprocal of the other quantity, as illustrated in Figure A.4. The labels of the points in Figures A.3 and A.4 are for identical values of x and y.

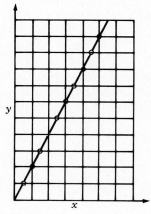

Figure A.1

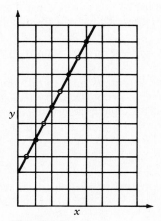

Figure A.2

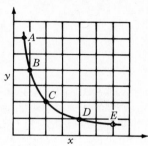

Figure A.3

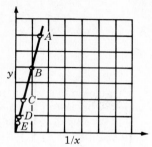

Figure A.4

A.11 Solving Equations

An equation relates quantities found on opposite sides of an equality sign. Although the two sides may appear to be different, they do represent the same thing. Therefore the equation may be reversed and the equality is still maintained. Thus $m = 2$ and $2 = m$ are equivalent.

In order to maintain an equality, a change made on one side of the equation must be made on the other side. For example, if $x = y$, a quantity such as 2 may be added or subtracted from both sides of the equation without changing the meaning of the equation.

addition of 2 $x + 2 = y + 2$
subtraction of 2 $x - 2 = y - 2$

Multiplication and division operations must be applied to the entire side and include all terms not just one term.

$$x + 4 = y + 3$$
multiplication by 3 $3(x + 4) = 3(y + 2)$
$$3x + 12 = 3y + 6$$

$$2x + 6 = 4y + 8$$
division by 2 $\dfrac{2x + 6}{2} = \dfrac{4y + 8}{2}$
$$x + 3 = 2y + 4$$

Raising equations to a power or obtaining a square root requires that all terms on each side of the equation must be used.

$$x = 3 + 2$$
squaring the equation $x^2 = (3 + 2)^2$
$$x^2 = (5)^2$$
$$x^2 = 25$$

$$x^2 = 4 + 5$$

obtaining the square root $\quad \sqrt{x^2} = (4 + 5)^{1/2}$

$$x = \sqrt{9}$$

$$x = 3$$

Solving an equation that contains an unknown is done by isolating the unknown on one side of the equation. This is accomplished by using an inverse operation to eliminate the undesired quantities from the side of the equation containing the unknown. If a number is contained in the form of addition, then it is removed by subtraction. That operation also must be performed on the other side of the equation.

$$x + 3 = 7$$

$$x + 3 - 3 = 7 - 3$$

$$x = 4$$

Similarly, if the quantity is contained in the form of subtraction, then it is removed by addition.

$$y - 4 = 9$$

$$y - 4 + 4 = 9 + 4$$

$$y = 13$$

If a number multiplies the unknown, it is removed by division; if a number divides the unknown, it is removed by multiplication.

$$2x = 6$$

division by 2 $\quad \dfrac{2x}{2} = \dfrac{6}{2}$

$$x = 3$$

$$\dfrac{y}{3} = 4$$

multiplication by 3 $\quad 3\left(\dfrac{y}{3}\right) = 4 \times 3$

$$y = 12$$

More than one step is required to solve many equations. Although there are several combinations of steps possible, the objective is to group all terms containing the unknown on one side of the equation and isolate all other terms on the other side. Then the necessary operations are performed to solve for the unknown.

$$\dfrac{3x}{4} + 5 = 17$$

subtraction of 5 $\quad \dfrac{3x}{4} + 5 - 5 = 17 - 5$

$$\dfrac{3x}{4} = 12$$

$$\text{multiplication by 4} \quad 4\left(\frac{3x}{4}\right) = 4(12)$$

$$3x = 48$$

$$\text{division by 3} \quad \frac{3x}{3} = \frac{48}{3}$$

$$x = 16$$

A.12 Calculators

Every student in a chemistry course should purchase a calculator. In addition to the normal function keys of addition, subtraction, division, and multiplication, the calculator should have other capabilities. These should include squares, square root, exponentials, and logarithm. The most common calculators operate on the Algebraic Operating System (AOS). A second system, the Reverse Polish Notation (RPN), is less common. Consult your instruction book concerning the individual characteristics of your calculator and the symbolism of the keys. Described here are the general procedures that should be followed in using a calculator.

You may find that your calculator displays a different number of figures from that of a friend. Do not be concerned about this because the number of figures displayed does not indicate anything about the worthiness of your calculator. Some calculators can display a variable number of figures. Rarely in this text will more than four figures be required; the remaining figures displayed will be rounded off and dropped. Make sure that you round off the answer on the calculator display to give a quantity that has the correct number of significant figures.

Your calculator should not be used carelessly and with a blind faith that what appears on display is the correct answer for the problem that you are trying to solve. The calculator doesn't make mistakes, but you do. There are a number of errors that you can make in keying the problem into the calculator. Therefore you should have an idea about the relative magnitude of the expected answer so that you may question whether the answer displayed is actually correct. You should learn how to estimate answers in your head or on paper. If the calculator answer seems correct, then write it down. However, be cautious and rekey the operations to see if the same result is obtained. If a different answer is obtained, then something is wrong, and you should check your procedure a third time very carefully.

In order to be sure that the calculator answer is reasonable, you should learn to estimate the answer. There is no specific procedure that you must use in estimating an answer. You will have to develop your own skills so that you can estimate an answer rapidly.

Estimation involves changing the given numbers very roughly to quantities that you can easily handle. For example, if you must divide 159.4 by 3.2245, change the numbers to 150 and 3. Why these two numbers? The answer is that they are close enough to the given and you can divide 150 by 3 in your head. Thus you should expect the calculator to display an answer close to 50. If the answer is 49.4340, then you have probably done the operation correctly. However, repeat the operation to be sure. If the displayed answer is significantly different, such as 513.9853, then you should suspect that the answer is in error. In this case the answer is one that would be obtained if you pressed the $\times$ key rather than the $\div$ key.

If the answer is 60.5985, you may suspect that something is wrong and should repeat the calculation carefully. If the next calculation gives 49.4340, then you know that one of the answers is incorrect. The 60.5989 is the result of dividing 195.4 by 3.2245 rather than 159.4 by

3.2245. In other words, two digits have been transposed in entering the quantity in the calculator.

If division of 159.4 by 3.2245 gives you 0.0049434, you should know that something is really wrong. In this case the decimal point in 3.2245 was not entered or the key did not respond to your touch. Always make it a habit to check the display to see that the correct digits and decimal have been entered.

Answers to Selected Exercises

Chapter 1

1.5	75%
1.7	9 lb
1.9	2 lb
1.11	13 ounces copper
1.13	43% carbon, 57% oxygen
1.15	82% carbon, 18% hydrogen

1.17 H—S
      |
      H

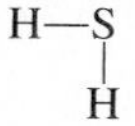

1.19
H
|
H—C—O—H
|
H

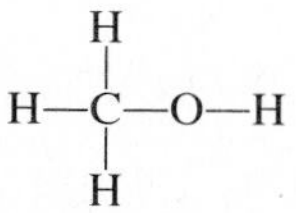

1.21 sulfur + oxygen $\longrightarrow$ sulfur dioxide

1.23 Hydrogen and nitrogen are the reactants, ammonia is the product.

1.29 no exceptions

Chapter 2

2.5 5 ft

2.7 4.2 kula

2.9 0.25 tun

2.15 (a) 5 (b) 3 (c) 5 (d) 3 (e) 6 (f) 2 (g) 4 (h) 6

2.17 (a) 3 (b) 4 (c) 3 (d) 2 (e) 4 (f) 3

2.19 (a) 2.4489×10^2 (b) 4.76×10^{-2} (c) 4.1438×10^3 (d) 4.55×10^{-4} (e) 4.16469×10^1 (f) 5.6×10^{-5} (g) 3.340×10^1 (h) 3.30540×10^1

2.21 (a) 50200 (b) 1256 (c) 0.000315 (d) 0.0000012 (e) 0.00002100 (f) 0.00201

2.23 (a) 344 (b) 160.6 (c) 125.50 (d) 104.8 (e) 24.5 (f) 3665

2.25 (a) 0.51 (b) 445.36 (c) 0.1243 (d) 0.31 (e) 451 (f) 149.0

2.27 (a) 14.9 (b) 28.3 (c) 0.002 (d) 0.47 (e) 0.9 (f) 2.0×10^1

2.29 (a) 4.70 (b) 497 (c) 6 (d) 1.7×10^2 (e) 1×10^1 (f) 16

2.31 (a) 147.9 (b) 0.038 (c) 2147 (d) 0.00041 (e) 21.65 (f) 0.000004 (g) 1.24 (h) 24.0 (i) 2.10×10^2

2.33 (a) 5.6×10^4 (b) 1.26×10^3 (c) 3.2×10^{-4} (d) 2×10^{-6} (e) 5.50×10^{-5} (f) 7.3×10^{-30}

2.35 (a) mL (b) pg (c) nm (d) dm (e) cL (f) kg (g) μg (h) μm (i) pL

2.37 (a) milligram (b) picometer (c) centiliter (d) decimeter (e) microgram (f) nanoliter (g) kilogram (h) milliliter (i) kilometer

2.39 (a) 16 in (b) 36.8 cm (c) 0.540 lb (d) 96.2 kg (e) 0.478 qt (f) 2.0 L (g) 11.1 m (h) 18 ft

2.41 25 mi/hr

2.43 44 lb

2.45 2.54 mm

2.47 11 kg

2.49 1×10^2 mg

2.51 (a) 1 dL (b) 1 nm (c) 1 kg (d) 1 μm (e) 1 cg (f) 1 dL (g) 1 cg (h) 1 mL (i) 1 km

2.53 (a) 5.9 cm (b) 1.53×10^{-3} km (c) 0.348 L (d) 5238 mL (e) 0.248 g (f) 56 g

2.55 (a) 11 mL (b) 0.243 μg (c) 0.3562 nm (d) $4.56 \ 10^{-3}$ dg (e) 1.2×10^3 pg (f) 3.4 nm

2.57 2.5×10^3 mL

2.59 2×10^{-3} g Vitamin C

2.61 1.5×10^6 nm

2.63 3.9×10^{-6} g

Chapter 3

3.3 19.3 g/cm^3
3.5 22.5 g/mL
3.7 7.8 g/mL
3.9 11.4 g/mL, not silver
3.11 78.0 g
3.13 5.0×10^2 mL
3.15 7.86 g/mL, iron
3.17 1.10 g/mL, 1.10
3.19 12 g
3.21 35°C
3.23 100°F
3.25 248°F
3.27 106°F
3.29 961°C
3.31 40°C
3.33 11 kg
3.35 3.8 Cal
3.37 72 g
3.39 0.052 cal/g°C
3.41 (a) gas (b) solid (c) liquid
(d) solid (e) solid (f) solid
3.43 liquid
3.45 (a) physical (b) physical
(c) chemical (d) physical
3.47 (a) chemical (b) chemical
(c) chemical (d) physical
3.49 (a) physical (b) chemical
(c) chemical (d) physical
3.51 physical, low density; chemical, it can burn
3.53 No, 0.160 g of a gas is formed
3.55 0.032 g sulfur
3.61 (a) oxygen (b) nitrogen (c) fluorine
(d) phosphorus (e) hydrogen (f) carbon
(g) iodine (h) uranium (i) sulfur
3.63 (a) sodium (b) potassium (c) silver
(d) gold (e) antimony (f) iron
(g) lead (h) copper (i) mercury
3.65 (a) Ag (b) He (c) Zn (d) Ba (e) U
(f) As (g) Fe (h) Pb (i) Li (j) Cr
3.67 mercury, neptunium, plutonium, uranium
3.71 Nd, first and fourth; Ne, first two; Np, first and third
3.73 hydrogen
3.75 iron and oxygen
3.77 C is a compound.
3.79 No statement can be made.
3.81 19.5 g lead required; 0.5 g of lead remains
3.83 3.12 g oxygen required; 23.1 g of lead oxide produced
3.87 metal

Chapter 4

4.5 (a) 24.0 amu magnesium
(b) 18.0 amu oxygen
(c) 235 amu uranium
4.7 6.03×10^{23} atoms Ar
4.9 1.21×10^{-8} cm
4.11 1.04 Å
4.15 Number of electrons and protons is the same.
4.17 nucleus is 10^{-13} cm; atom is 10^{-8} cm
4.19 atomic number is the number of protons
4.21 (a) 8 protons, 8 electrons, 8 neutrons
(b) 11 protons, 11 electrons, 12 neutrons
(c) 13 protons, 13 electrons, 14 neutrons
(d) 16 protons, 16 electrons, 16 neutrons
(e) 18 protons, 18 electrons, 22 neutrons
(f) 20 protons, 20 electrons, 20 neutrons
4.23 (a) $^{19}_{9}F$ (b) $^{30}_{14}Si$ (c) $^{28}_{14}Si$
(d) $^{31}_{15}P$ (e) $^{39}_{19}K$ (f) $^{48}_{22}Ti$
4.27 $^{131}_{53}I$ isotope has 78 neutrons; $^{127}_{53}I$ isotope has 74 neutrons.
4.29 $^{64}_{30}Zn$
4.33 neon-20
4.35 39.53%
4.37 (a) atoms
(b) molecules of 2 H atoms
(c) molecules of 4 P atoms
(d) atoms
(e) molecules of 8 S atoms
(f) molecules of 2 N atoms
4.39 Se_8
4.41 (a) one molecule contains 2 hydrogen atoms and 1 sulfur atom
(b) one molecule contains 1 hydrogen atom and 1 chlorine atom
(c) one molecule contains 2 nitrogen atoms and 4 hydrogen atoms
(d) one molecule contains 2 carbon atoms and 2 hydrogen atoms
(e) one molecule contains 2 carbon atoms and 6 hydrogen atoms
(f) one molecule contains 2 hydrogen atoms and 2 oxygen atoms
4.43 C_8H_{18}
4.45 20 carbon atoms, 30 hydrogen atoms, 1 oxygen atom, total 51 atoms
4.47 Se^{2-}; 36 electrons
4.49 Mn^{2+}; 23 electrons
4.51 (a) O^{2-} (b) S^{2-} (c) I^-
(d) Br^- (e) N^{3-} (f) F^-
4.53 (a) sulfide ion (b) iodide ion (c) oxide ion
(d) fluoride ion (e) nitride ion (f) bromide ion
4.55 (a) SO_4^{2-} (b) PO_4^{3-} (c) OH^-
(d) NH_4^+ (e) CN^- (f) CO_3^{2-}
4.57 (a) LiF (b) ZnO (c) $MgCO_3$ (d) KNO_3
(e) NaCN (f) Al_2S_3 (g) $Ca(ClO)_2$ (h) $Ba(NO_2)_2$
4.59 (a) FeF_3 (b) Cu_2O (c) $FeSO_4$
(d) $Cu(CN)_2$ (e) Fe_2O_3 (f) $Cu(NO_3)_2$
4.61 (a) $FeCl_3$ (b) NaOH (c) $Mg(OH)_2$
(d) CdS (e) MnF_2 (f) Cs_3N

4.63   (a) calcium hydroxide   (b) lithium perchlorate
      (c) sodium phosphate   (d) potassium sulfate
      (e) sodium nitrate   (f) ammonium nitrite
      (g) magnesium chloride   (h) aluminum bromide

4.65   (a) barium hydroxide   (b) potassium chlorite
      (c) cesium phosphate   (d) potassium sulfite
      (e) sodium nitrite   (f) ammonium nitrate
      (g) calcium bromide   (h) gallium chloride

Chapter 5

5.7 (a) 28.0 (b) 44.0 (c) 64.1
    (d) 80.1 (e) 30.0 (f) 46.0
    (g) 60.1 (h) 111.0 (i) 54.0

5.9 (a) 154.0 (b) 104.1 (c) 392.2
    (d) 270.7 (e) 208.5 (f) 71.0
    (g) 87.0 (h) 136.9 (i) 70.1

5.11 (a) 61.8 (b) 82.0 (c) 63.0
    (d) 227.9 (e) 229.6 (f) 107.9
    (g) 82.1 (h) 258.0 (i) 178.2

5.13 (a) 227.0 (b) 227.0

5.15 (a) 58.5 (b) 119.0 (c) 25.9
    (d) 56.4 (e) 119.0 (f) 103.6
    (g) 41.0 (h) 100.7 (i) 189.7

5.17 (a) 104.6 (b) 218.2 (c) 254.2
    (d) 231.9 (e) 86.9 (f) 270.0
    (g) 100.0 (h) 257 (i) 239.2

5.19 (a) 142.1 (b) 138.2 (c) 93.9
    (d) 164.1 (e) 223.3 (f) 259.3
    (g) 342.3 (h) 137.0 (i) 120.2

5.21 (a) 106.0 (b) 109.9 (c) 212.3
    (d) 151.9 (e) 187.5 (f) 158.0
    (g) 146.3 (h) 58.3

5.23 (a) 55.6% metal   (b) 46.7% element
    (c) 63.6% metal

5.25 (a) 65.6%, $FeCl_3$; 55.9%, $FeCl_2$
    (b) 25.8%, Na_2O; 17.0%, K_2O
    (c) 44.4%, CaS; 29.1%, K_2S

5.27 (a) %C = 42.9%; %O = 57.1%
    (b) %C = 27.2%; %O = 72.7
    (c) %S = 50.0%; %O = 50.0%
    (d) %S = 40.1%; %O = 59.9%
    (e) %N = 46.7%; %O = 53.3%
    (f) %N = 30.4%; %O = 69.6%
    (g) %P = 56.4%; %O = 43.6%
    (h) %P = 43.7%; %O = 56.3%
    (i) %Cl = 38.8%; %O = 61.2%
    (j) %I = 76.0%; %O = 24.0%
    (k) %Xe = 73.2%; %O = 26.8%
    (l) %Xe = 67.2%; %O = 32.8%

5.29 (a) %C = 7.8%; %Cl = 92.2%
    (b) %Si = 27.0%; %F = 73.0%
    (c) %Ge = 18.5%; %Br = 81.5%
    (d) %P = 11.5%; %Br = 88.5%

    (e) %P = 14.9%; %Cl = 85.1%
    (f) %N = 19.7%; %F = 80.3%
    (g) %S = 22.0%; %F = 78.0%
    (h) %P = 24.6%; %F = 75.4%
    (i) %Br = 45.7%; %F = 54.3%
    (j) %Se = 51.0%; %F = 49.0%
    (k) %Xe = 77.6%; %F = 22.4%
    (l) %Xe = 63.3%; %F = 36.7%

5.31 (a) %Na = 39.3%; %Cl = 60.7%
    (b) %K = 32.9%; %Br = 67.1%
    (c) %Li = 26.6%; %F = 73.4%
    (d) %Mg = 43.1%; %S = 56.9%
    (e) %Ca = 33.7%; %Se = 66.3%
    (f) %Sr = 84.6%; %O = 15.4%
    (g) %Na = 74.2%; %O = 25.8%
    (h) %K = 70.9%; %S = 29.1%
    (i) %Li = 14.9%; %Se = 85.1%
    (j) %Si = 46.8%; %O = 53.2%
    (k) %Ga = 39.6%; %Cl = 60.4%
    (l) %Sn = 78.8%; %O = 21.2%

5.33 (a) %Na = 32.4%; %S = 22.6%; %O = 45.0%
    (b) 56.6% K; 8.7% C; 34.7% O
    (c) 14.7% Li; 34.2% S; 51.1% O
    (d) 24.4% Ca; 17.1% N; 58.5% O
    (e) 10.9% Mg; 31.8% Cl; 57.3% O
    (f) 53.0% Ba; 0.8% H; 9.3% C; 37.0% O
    (g) 42.1% Na; 18.9% P; 39.9% O
    (h) 24.7% K; 34.7% Mn; 40.5% O
    (i) 12.6% Li; 29.2% S; 58.2% O
    (j) 27.9% Fe; 24.1% S; 48.0% O
    (k) 27.5% Co; 30.8% Cl; 41.6% O
    (l) 78.2% Ag; 4.4% C; 17.4% O

5.35 (a) 43.4% Na; 11.3% C; 45.3% O
    (b) 12.6% Li; 29.2% S; 58.2% O
    (c) 55.3% K; 14.6% P; 30.1% O
    (d) 36.7% Fe; 21.1% S; 42.1% O
    (e) 33.9% Cu; 14.9% N; 51.2% O
    (f) 24.7% K; 34.7% Mn; 40.5% O
    (g) 60.1% K; 18.4% C; 21.5% N
    (h) 23.1% Fe; 17.4% N; 59.6% O

5.37 A mole contains Avogadro's number of units of matter. A mole is 6.02×10^{23} units.

5.39 Divide the mass of the substance by the molar mass.

5.41 (a) 2.00 (b) 0.498 (c) 0.010
    (d) 1.00 (e) 0.010 (f) 0.020

5.43 (a) 10.0 (b) 0.010 (c) 2.0
    (d) 0.100 (e) 0.0053 (f) 5.0

5.45 (a) 0.249 (b) 9.99×10^{-4}
    (c) 0.0400 (d) 0.0167
    (e) 0.498 (f) 0.0667
    (g) 0.0400

5.47 (a) 0.00998 (b) 0.0500
    (c) 1.00 (d) 0.500
    (e) 0.0200 (f) 0.00500

5.49 (a) 0.00509 (b) 0.00250

(c) 0.00499   (d) 0.00200

(e) 0.00501   (f) 0.250

5.51   (a) 0.200   (b) 0.200

(c) 0.00999   (d) 0.0168

(e) 0.0200   (f) 0.0250

5.53   (a) 0.5 mole He, 0.25 mole Ar

(b) 0.5 mole CH_4, 0.75 mole CO_2

(c) 0.18 mole CaO, 0.14 mole CaS

5.55   (a) 0.2   (b) 0.4

(c) 0.3   (d) 0.50

(e) 0.6

5.57   (a) 23 g   (b) 0.69 g

(c) 80 g   (d) 6.1 g

(e) 1.1×10^2 g   (f) 13 g

5.59   (a) 18 g   (b) 18 g

(c) 23 g   (d) 90.0 g

(e) 6.8 g   (f) 13 g

5.61   (a) 29.2 g   (b) 2.4 g

(c) 51.8 g   (d) 85 g

(e) 48 g   (f) 10.4 g

5.63   360 moles

5.65 5.5×10^{-3} mole

5.67   (a) 6.00×10^{22} (b) 6.02×10^{24}

(c) 3.01×10^{23} (d) 6.02×10^{21}

(e) 3.00×10^{21} (f) 1.21×10^{22}

5.69   (a) 6.02×10^{22} (b) 3.0×10^{24}

(c) 6.0×10^{23} (d) 1.5×10^{22}

(e) 3.0×10^{22} (f) 3.0×10^{22}

5.71   (a) 1.20×10^{22}

(b) 1.51×10^{22}

(c) 6.02×10^{22}

(d) 3.01×10^{23}

(e) 1.20×10^{23}

(f) 6.02×10^{22}

5.73   (a) 2×10^{23} (b) 4×10^{22}

(c) 6×10^{23} (d) 6.0×10^{23}

(e) 6×10^{23}

5.75 CH_3O

5.77 CCl_2

5.79   GaP

5.81   (a) $CuCl_2$ (b) Mg_3N_2 (c) Fe_3O_4

(d) Na_2S

5.83   (a) CuBr   (b) $SnCl_2$ (c) $FeBr_2$

(d) Cr_2S_3

5.85 $C_6H_{10}O_4$

5.87 $C_{18}H_{22}O_2$

Chapter 6

6.3   Coefficients are given in order of appearance.

(a) 2,1,2   (b) 2,3,2   (c) 1,1,2   (d) 2,2,1

6.5   Coefficients are given in order of appearance.

(a) 1,1,1,2   (b) 4,1,4,3   (c) 2,4,3

(d) 1,1,2

6.7   Coefficients are given in order of appearance.

(a) 2,1,1   (b) 2,1,2   (c) 1,6,2,3

(d) 1,10,4

6.9   Coefficients are given in order of appearance.

(a) 2,4,1   (b) 2,2,1   (c) 2,3,2   (d) 1,1,1

6.11   Coefficients are given in order of appearance.

(a) 1,6,1,2   (b) 1,3,3,1   (c) 1,12,4,3

(d) 2,1,1,4

6.13   Coefficients are given in order of appearance.

(a) 2,7,4,6   (b) 1,8,5,6   (c) 2,15,10,10

(d) 1,7,5,4

6.15   Coefficients are given in order of appearance.

(a) 1,1,1,1,1   (b) 1,3,1,3   (c) 1,2,1,2

(d) 2,3,1,6

6.17   (a) $2\,Ba + O_2 \longrightarrow 2\,BaO$

(b) $H_2 + I_2 \longrightarrow 2\,HI$

(c) $2\,Li + S \longrightarrow Li_2S$

(d) $2\,Al + 3\,Cl_2 \longrightarrow 2\,AlCl_3$

6.19   (a) $Na_2O + H_2O \longrightarrow 2\,NaOH$

(b) $CaO + H_2O \longrightarrow Ca(OH)_2$

(c) $BaO + CO_2 \longrightarrow BaCO_3$

(d) $MgO + CO_2 \longrightarrow MgCO_3$

6.21   (a) $2\,HgO \longrightarrow 2\,Hg + O_2$

(b) $SrCO_3 \longrightarrow SrO + CO_2$

(c) $2\,Au_2O_3 \longrightarrow 4\,Au + 3\,O_2$

(d) $2\,Ag_2O \longrightarrow 4\,Ag + O_2$

6.23   (a) $2\,KBr + Cl_2 \longrightarrow 2\,KCl + Br_2$

(b) $SnO_2 + 2\,H_2 \longrightarrow Sn + 2\,H_2O$

(c) $4\,Mg + Fe_3O_4 \longrightarrow 3\,Fe + 4\,MgO$

(d) $2\,Al_2O_3 + 3\,C \longrightarrow 4\,Al + 3\,CO_2$

6.25   (a) $Zn + 2\,HCl \longrightarrow ZnCl_2 + H_2$

(b) $SiCl_4 + 2\,Mg \longrightarrow 2\,MgCl_2 + Si$

(c) $Br_2 + 2\,NaI \longrightarrow I_2 + 2\,NaBr$

(d) $2\,ZnO + C \longrightarrow CO_2 + 2\,Zn$

6.27   (a) $Pb(NO_3)_2 + 2\,HCl \longrightarrow PbCl_2 + 2\,HNO_3$

(b) $FeCl_2 + 2\,NaOH \longrightarrow Fe(OH)_2 + 2\,NaCl$

(c) $Cd(NO_3)_2 + H_2S \longrightarrow CdS + 2\,HNO_3$

(d) $AgNO_3 + CsCl \longrightarrow AgCl + CsNO_3$

6.29   (a) $MnSO_4 + (NH_4)_2S \longrightarrow MnS + (NH_4)_2SO_4$

(b) $CaCO_3 + H_2SO_4 \longrightarrow CaSO_4 + H_2O + CO_2$

(c) $CdSO_4 + 2\,NaOH \longrightarrow Cd(OH)_2 + Na_2SO_4$

(d) $3\,ZnCO_3 + 2\,H_3PO_4 \longrightarrow Zn_3(PO_4)_2 + 3\,H_2O + 3\,CO_2$

6.31   (a) $H_3PO_4 + 3\,KOH \longrightarrow 3\,H_2O + K_3PO_4$

(b) $2\,Fe(OH)_3 + 3\,H_2SO_4 \longrightarrow Fe_2(SO_4)_3 + 6\,H_2O$

(c) $2\,H_3PO_4 + 3\,Ca(OH)_2 \longrightarrow 6\,H_2O + Ca_3(PO_4)_2$

(d) $Zn(OH)_2 + 2\,HNO_3 \longrightarrow Zn(NO_3)_2 + 2\,H_2O$

6.33   (a) $U + 3\,F_2 \longrightarrow UF_6$

(b) $2\,P + 5\,Cl_2 \longrightarrow 2\,PCl_5$

(c) $2\,KCl + Br_2 \longrightarrow 2\,KBr + Cl_2$

(d) $SnO_2 + 2\,H_2 \longrightarrow Sn + 2\,H_2O$

7.3	(4 mole Al)/(3 mole MnO_2)
7.5	(2 mole AgCl)/(1 mole $BaNO_3$)
7.7	2.5
7.9	3.5
7.11	2.3
7.13	0.20
7.15	24
7.17	38
7.19	0.33
7.21	2.00
7.23	13
7.25	127
7.27	12.8
7.29	1271
7.31	19.4
7.33	1.11×10^3
7.35	0.0500
7.37	0.030
7.39	7.16
7.41	0.52 g S remains unreacted
7.43	94.9% yield
7.45	5.07

Chapter 8

8.3 (a) 8 (b) 18 (c) 4 (d) 6 (e) 15 (f) 3

8.5 (a) 19 (b) 35 (c) 24 (d) 29 (e) 32 (f) 36

8.7 (a) 2 (b) 1 (c) 4 (d) 3 (e) 5 (f) 6

8.9 $n = 1, 2n^2 = 2$ $n = 2, 2n^2 = 8$ $n = 3, 2n^2 = 18$ $n = 4, 2n^2 = 32$

8.11 (a) 2 in n = 1, 5 in n = 2
(b) 2 in n = 1, 8 in n = 2, 4 in n = 3
(c) 2 in n = 1, 3 in n = 2
(d) 2 in n = 1, 7 in n = 2
(e) 2 in n = 1, 8 in n = 2, 6 in n = 3
(f) 2 in n = 1, 8 in n = 2

8.13 (a) 2 in n = 1, 8 in n = 2, 8 in n = 3, 1 in n = 4
(b) 2 in n = 1, 8 in n = 2, 2 in n = 3
(c) 2 in n = 1, 8 in n = 2, 3 in n = 3
(d) 2 in n = 1, 8 in n = 2, 8 in n = 3, 2 in n = 4
(e) 2 in n = 1, 8 in n = 2, 18 in n = 3, 4 in n = 4
(f) 2 in n = 1, 8 in n = 2, 18 in n = 3, 3 in n = 4

8.15 (a) 5 (b) 4 (c) 3 (d) 7 (e) 6 (f) 2

8.17 (a) 1 (b) 2 (c) 3 (d) 2 (e) 4 (f) 3

8.19 (a) 2 (b) 4 (c) 3 (d) 5

8.21 4

8.23 (a) 2s,2p (b) 3s,3p (c) 2s (d) 2s,2p (e) 3s,3p (f) 2s

8.25 (a) 4s (b) 3s (c) 3s,3p (d) 4s (e) 4s,4p (f) 4s,4p

8.27 (a) 1 (b) 3 (c) 5 (d) 1 (e) 3 (f) 5

8.29 (a) 2 (b) 2 (c) 2 (d) 2

8.31 1p and 2d

8.33 2d, 1f, and 3g

8.35 (a) $1s^2 2s^2 2p^4$ (b) $1s^2 2s^2 2p^6 3s^2 3p^6$
(c) $1s^2 2s^2$ (d) $1s^2 2s^2 2p^2$
(e) $1s^2 2s^2 2p^6 3s^2 3p^3$ (f) $1s^2 2s^1$

8.37 (a) $1s^2 2s^2 2p^3$ (b) $1s^2 2s^2 2p^6 3s^2 3p^2$
(c) $1s^2 2s^2 2p^1$ (d) $1s^2 2s^2 2p^5$
(e) $1s^2 2s^2 2p^6 3s^2 3p^4$ (f) $1s^2$

8.39 $1s^2 2s^2 2p^3$ is correct

8.41 (a) 4 (b) 1 (c) 1 (d) 3 (e) 3 (f) 4

8.43 (a) 3 (b) 2 (c) 1 (d) 1 (e) 2 (f) 0

8.45 (a) 1 (b) 0 (c) 1 (d) 0 (e) 2 (f) 1

8.47 (a) 1 (b) 5 (c) 2 (d) 1 (e) 2 (f) 0

8.49 (a) $:\overset{\cdot\cdot}{\underset{\cdot}{O}}\cdot$ (b) $:\overset{\cdot\cdot}{\underset{\cdot\cdot}{Ar}}:$ (c) $:Be$
(d) $\cdot\overset{\cdot\cdot}{C}:$ (e) $\cdot\overset{\cdot\cdot}{P}:$ (f) $\cdot Li$

8.51 (a) $\cdot K$ (b) $:\overset{\cdot\cdot}{Br}\cdot$ (c) $Mg:$
(d) $Ca:$ (e) $\cdot Ge\cdot$ (f) $:\overset{\cdot\cdot}{\underset{\cdot\cdot}{Kr}}:$

Chapter 9

9.1 A period is a horizontal row in the periodic table, whereas a Group is a vertical column.

9.3 S and Te

9.5 Triads are in the same group.

9.7 The tides of the oceans or the phases of the moon.

9.9 technetium, Tc

9.11 (a) 2, IVA (b) 5, VIA (c) 5, IIA (d) 3, VA (e) 3, IVA (f) 5, VIIA

9.13 (a) 2, VIIA (b) 6, IA (c) 3, IIA (d) 4, VA (e) 3, VIIIA (f) 6, IVA

9.15 (a) P (b) Ga (c) Sn (d) Cs (e) Ra (f) 0

9.17 (a) Mg (b) Ge (c) I (d) Ba (e) B (f) F

9.19 (a) 2, VA (b) 4, IIA (c) 5, IIB (d) 4, VIIA (e) 2, IA (f) 3, IIIA

9.21 Ca, Se, Cl

9.23 Re, Ag, V

9.25 F, Cl

9.27 Ca, Rb, Ag, Ni

9.29 F

9.31 (a) nonmetal (b) metal (c) metal (d) metal (e) nonmetal (f) nonmetal

9.33 (a) Mg > Si (b) Ge > Br (c) Se > S (d) Sn > Si

9.35 Fr

9.37 (a) 5, IVB (b) 4, IVA (c) 6, VIB (d) 3, VA

9.39 (a) VA (b) IIIB (c) VIIA (d) IVA

9.41 (a) 4 (b) 6 (c) 5 (d) 8 (e) 1 (f) 5

9.43 (a) 2 (b) 3 (c) 1 (d) 2

9.45 (a) 0 (b) 1 (c) 2 (d) 0 (e) 1 (f) 2

9.47 (a) 0.89Å (b) 3.66 g/cm^3 (c) 87°C

9.51 (a) Cl (b) Si (c) K (d) C (e) Sb (f) Pb

9.53 He

9.55 (a) F (b) Li (c) Cl (d) S (e) Ge (f) Li

9.57 Ra

9.59 (a) F (b) S (c) Li (d) N (e) Cl (f) P

Chapter 10

10.3 M^{2+}

10.5 3 electrons from 4s^{2}3d^1

10.7 (a) 1s^0 (b) 1s^{2}2s^{2}2p^{6}3s^{2}3p^6 (c) 1s^2 (d) 1s^{2}2s^{2}2p^6 (e) 1s^{2}2s^{2}2p^{6}3s^{2}3p^6 (f) 1s^{2}2s^{2}2p^6

10.9 (a) 1s^2 (b) 1s^{2}2s^{2}2p^6 (c) 1s^{2}2s^{2}2p^{6}3s^{2}3p^6 (d) 1s^{2}2s^{2}2p^6 (e) 1s^{2}2s^{2}2p^{6}3s^{2}3p^6 (f) 1s^{2}2s^{2}2p^6

10.11 (a) H$^+$ (b) [Ar] (c) [He] (d) [Ne] (e) [Ar] (f) [Ne]

10.13 (a) 1s^2 (b) 2s^{2}2p^6 (c) 3s^{2}3p^6 (d) 2s^{2}2p^6 (e) 3s^{2}3p^6 (f) 2s^{2}2p^6

10.15 (a) H$^+$ (b) Ca^{2+} (c) Li$^+$ (d) Na$^+$ (e) K$^+$ (f) Mg^{2+}

10.17 (a) H : $^-$ (b) : F : $^-$ (c) : Cl : $^-$ (d) : N : $^{3-}$ (e) : S : $^{2-}$ (f) : O : $^{2-}$

10.19 (a) NaCl (b) CaCl$_2$ (c) MgO (d) MgS (e) AlF$_3$ (f) Na$_2$Se

10.21 (a) LiF (b) MgBr$_2$ (c) Li$_2$O (d) MgS (e) AlF$_3$ (f) Na$_2$Se

10.23 ScCl$_3$

10.25 anions are larger

10.27 (a) Mg (b) K (c) Al (d) Br (e) S (f) N

10.31 atom with higher electronegativity will be the more negative pole

10.33 (a) H (b) Br (c) H (d) O (e) Cl (f) Si

10.35 (a) H : O : Cl : (b) H : O : Cl : O :

(c) H : O : Br : O : (d) H : O : I : O :

(e) H : O : Se : O : H (f) H : O : P : O : H

10.37 Both Br atoms share one electron with each other.

10.39 (a) H—S—H (b) H—P—H with H below

(c) : F—C—F : with F above and F below (d) H—Se—H

(e) : Br—C—Br : with Br above and Br below (f) : Cl—Si—Cl : with Cl above and Cl below

10.41 (a) : S—H $^-$ (b) H—P—H $^+$ with H above and H below

(c) H—O—H $^+$ with H below (d) : C≡N : $^-$

(e) : O—S—O : $^{2-}$ with O above and O below (f) H—N—H $^+$ with H above and H below

10.43 SbH$_3$ H—Sb—H with H below

10.45 (a)

$:\!F\!:$
$:F\!-\!\overset{\displaystyle \ \ }{\underset{\displaystyle :\!F\!:}{C}}\!-\!F\!:$

(b)

$H\!-\!\overset{\displaystyle H}{\underset{\displaystyle H}{C}}\!-\!\ddot{O}\!:$

(c)

$H\!-\!\overset{\displaystyle H}{\underset{\displaystyle H}{C}}\!-\!\overset{\displaystyle H}{N}\!-\!H$

(d)

$\overset{\displaystyle Cl}{\underset{\displaystyle Cl}{C}}\!=\!\ddot{O}$

(e) $\quad H\!-\!C\!\equiv\!N\!:$

(f)

$\overset{\displaystyle H}{\underset{\displaystyle H}{C}}\!=\!\overset{\displaystyle :\ddot{O}\!-\!H}{N\!:}$

10.47 $\quad \ddot{O}\!=\!\overset{\ }{N}\!-\!\ddot{O}\!:^{-} \longleftrightarrow :\ddot{O}\!-\!\overset{\ }{N}\!=\!\ddot{O}^{-} \longleftrightarrow :\ddot{O}\!-\!\overset{\ }{N}\!-\!\ddot{O}\!:^{-}$

with $:\ddot{O}\!:$ below each

10.49 $\quad \ddot{O}\!=\!\overset{\ }{C}\!-\!\ddot{O}\!:^{2-} \longleftrightarrow :\ddot{O}\!-\!\overset{\ }{C}\!=\!\ddot{O}^{2-} \longleftrightarrow :\ddot{O}\!-\!\overset{\ }{C}\!-\!\ddot{O}\!:^{2-}$

with $:\ddot{O}\!:$ below each

10.51 (a) polar (b) polar (c) nonpolar
(d) polar (e) nonpolar (f) nonpolar

10.53 (a) 109.5° angles, tetrahedral
(b) 109.5° angles, tetrahedral
(c) 109.5° angles, tetrahedral
(d) 107° angles, trigonal pyramidal
(e) 105° angle, angular
(f) 105° angle, angular

10.55 (a) tetrahedral (b) trigonal planar
(c) tetrahedral (d) trigonal pyramidal
(e) tetrahedral (f) angular

Chapter 11

11.3 (a) chloride (b) fluoride (c) bromide
(d) iodide (e) oxide (f) sulfide

11.5 (a) magnesium ion (b) calcium ion
(c) potassium ion (d) aluminum ion
(e) mercury(II) ion (f) selenide

11.7 (a) hypochlorite (b) permanganate
(c) perbromate (d) chlorite
(e) sulfate (f) sulfite

11.9 (a) hypobromite (b) nitrite
(c) iodite (d) chromate
(e) nitrate (f) hydroxide

11.11 (a) copper(I) (b) iron(III)
(c) mercury(II) (d) mercury(I)
(e) iron(II) (f) copper(II)

11.13 (a) Na^{+} (b) Cu^{+} (c) Fe^{3+} (d) K^{+}
(e) Sn^{4+} (f) Hg_2^{2+}

11.15 (a) Li^{+} (b) Ag^{+} (c) Mg^{2+} (d) Fe^{2+}
(e) Ti^{4+} (f) Cs^{+}

11.17 (a) Br^{-} (b) SO_3^{2-} (c) ClO_3^{-}
(d) HCO_3^{-} (e) NO_3^{-} (f) ClO_4^{-}

11.19 (a) ClO_2^{-} (b) CrO_4^{2-} (c) F^{-}
(d) CO_3^{2-} (e) BrO^{-} (f) $Cr_2O_7^{2-}$

11.21 (a) -2 (b) $+3$ (c) $+4$ (d) $+4$
(e) -3 (f) $+2$

11.23 (a) -2 (b) $+3$ (c) $+4$ (d) $+8$
(e) $+2$ (f) $+2$

11.25 (a) $+5$ (b) $+3$ (c) $+3$ (d) $+6$
(e) $+7$ (f) $+6$

11.27 (a) $+4$ (b) $+6$ (c) $+3$ (d) $+5$
(e) $+7$ (f) $+3$

11.29 (a) sodium sulfide (b) barium oxide
(c) aluminum oxide (d) cobalt(II) bromide
(e) cadmium(II) fluoride (f) magnesium nitride
(g) potassium phosphide (h) zinc(II) selenide

11.31 (a) lead(II) iodide (b) mercury(II) fluoride
(c) iron(III) oxide (d) titanium(III) chloride
(e) iron(II) bromide (f) cerium(IV) oxide
(g) tin(II) oxide (h) gold(III) oxide

11.33 (a) lithium carbonate (b) silver iodate
(c) aluminum sulfate (d) zinc cyanide
(e) barium nitrate (f) sodium dichromate
(g) strontium fluoride (h) rubidium sulfite

11.35 (a) carbon tetrachloride
(b) tetraphosphorus decaoxide
(c) oxygen dichloride
(d) dinitrogen trichloride
(e) phosphorus trichloride
(f) carbon monoxide

11.37 (a) $+7$, manganese(VII) oxide
(b) $+4$, manganese(IV) oxide
(c) $+2$, manganese(II) oxide
(d) $+2$, manganese(II) chloride

11.39 (a) CaI_2 (b) Li_2S (c) Mg_3N_2
(d) Al_2S_3 (e) Na_2O (f) $BaSe$
(g) K_3P (h) $SrBr_2$

11.41 (a) Au_2O_3 (b) CuS (c) $SnBr_2$
(d) Fe_2Se_3 (e) FeO (f) $SnCl_4$
(g) $Cu(NO_3)_2$ (h) $HgBr_2$

11.43 (a) only $NaCl$ possible
(b) Fe_2O_3 and FeO possible
(c) $Cu(NO_3)_2$ or $Cu(NO_2)_2$ possible
(d) $HClO_4$ possible
(e) H_2SO_3 or H_2SO_4 possible
(f) $Ba_3(PO_4)_2$ possible
(g) NaH_2PO_4 or Na_2HPO_4 possible
(h) $Al(CN)_3$ possible

11.45 $LaSO_4$

11.46 $Zr(SO_4)_2$

11.47 Y_2S_3

11.49 (a) $LiOH$ (b) $Sr(OH)_2$ (c) $Al(OH)_3$
(d) $Zn(OH)_2$ (e) $Ni(OH)_2$ (f) $CsOH$

11.51 (a) nitrous acid (b) sulfuric acid
(c) phosphoric acid (d) carbonic acid
(e) hydrofluoric acid (f) bromous acid

11.53 $H_2C_2O_4$, oxalic acid

11.55 (a) $H_2SO_4 + 2\,KOH \longrightarrow K_2SO_4 + 2\,H_2O$
(b) $2\,H_3PO_4 + 3\,Ca(OH)_2 \longrightarrow Ca_3(PO_4)_2 + 6\,H_2O$
(c) $3\,HCl + Al(OH)_3 \longrightarrow AlCl_3 + 3\,H_2O$
(d) $3\,HNO_3 + Fe(OH)_3 \longrightarrow Fe(NO_3)_3 + 3\,H_2O$

11.57 (a) magnesium ammonium phosphate
(b) aluminum ammonium sulfate
(c) potassium aluminum sulfate
(d) calcium ammonium phosphate

11.59 $+3$

11.61 $x = 2$

Chapter 12

12.5 1 mm Hg

12.7 760 torr

12.11 (a) 0.500 atm (b) 38.0 cm Hg (c) 76.0 torr
(d) 19.0 cm Hg (e) 0.918 atm (f) 19.0 cm Hg

12.13 43.3 psi

12.17 225 mL

12.19 2 atm

12.21 1.4×10^3 L

12.23 3.0 L

12.25 375 K

12.27 112 mL

12.29 2 atm

12.31 400 K

12.33 179°C

12.35 750 mL

12.37 4 atm

12.39 300 K

12.41 equal numbers of H_2 molecules and He atoms

12.43 11 L H_2 and 11.2 L O_2

12.45 (a) 2.24 L; 0.100 mole
(b) 2.24 L; 0.100 mole
(c) 1.68 L; 0.075 mole
(d) 22.4 L; 1.00 mole

12.47 6.02×10^{23}; 2.7×10^{22}

12.49 44 amu

12.51 42 g/mole

12.53 64 g/mole

12.55 2.05

12.57 2.68 g/L

12.59 20 L H_2

12.61 6.72 L H_2

12.63 5.59 g Mg

12.65 200 torr

12.67 5.2 atm

12.69 310 mL

12.71 1.225

12.73 0.8289

12.75 1.054

12.77 232 g/mole

Chapter 13

13.5 temperature, humidity, amount of wind, and the surface area of the wet clothes

13.9 B particles have weaker attractive forces between them than A particles do.

13.11 No, mercury vapor would be present in the space.

13.13 $SbBr_3$

13.15 CSSe

13.17 120°C

13.19 SnI_4

13.21 Stronger intermolecular forces exist between $SiCl_4$ molecules.

13.23 5.4×10^3 cal

13.25 CCl_4, 8.26×10^3; CBr_4, 1.09×10^4 cal/mole

13.27 9.7×10^2 cal

13.29 -3°C

13.31 Mercury freezes at -39°C.

13.33 8.0×10^4 cal

13.35 224.4 cal/mole

13.37 1429 cal/mole

13.41 higher melting point under 100 atm pressure

13.43 London forces, dipole–dipole forces and hydrogen bonding

13.45 Butane is a heavier molecule with more electrons and has stronger London forces.

13.47 $SiCl_4$ has stronger London forces than SiF_4.

13.49 stronger London forces in BBr_3

13.51 CH_3CH_2F has dipole–dipole forces.

13.53 Both molecules are angular about the central atom. No hydrogen-bonding forces can exist in either compound, but CH_3OCH_3 should be somewhat more polar than CH_3SCH_3. The higher ΔH_v for CH_3SCH_3 is due to stronger London forces because S is larger and contains more electrons than O.

13.55 Ethyl alcohol should have a higher boiling point than ethyl mercaptan, because hydrogen bonding can occur between OH groups in ethyl alcohol, but not between SH groups in ethyl mercaptan.

13.57 lower boiling point in first compound because hydrogen bonding cannot exist

13.59 two sites where hydrogen may be donated and four sites where electron pairs may be supplied

13.61 Water molecule has 2 sites where H may be donated and 2 sites where electron pairs may be supplied for hydrogen bond formation.

13.63 Open cage-like network created by hydrogen bonding in the solid makes the solid less dense than the liquid.

13.65 Formation of a protective layer allows the water below to stay above freezing.

13.67 (a) $Na_2Cr_2O_7 \cdot 2H_2O$ (b) $SrBr_2 \cdot 6H_2O$
(c) $Al(BrO_3)_3 \cdot 9H_2O$ (d) $Ba(OH)_2 \cdot 8H_2O$
(e) $CaBr_2 \cdot 6H_2O$

13.69 (a) aluminum bromate nonahydrate
(b) barium bromide dihydrate

(c) aluminum iodide hexahydrate
(d) ammonium phosphate trihydrate
(e) ammonium sulfite monohydrate
(f) potassium sulfide pentahydrate

13.71 (a) strontium bromate monohydrate
(b) sodium bromide dihydrate
(c) potassium carbonate dihydrate
(d) nickel sulfite hexahydrate
(e) tin(IV) chloride tetrahydrate
(f) cobalt fluoride tetrahydrate

13.73 (a) 50.33% (b) 14.75% (c) 9.48%
(d) 13.85% (e) 43.22% (f) 37.35%

13.75 14.52%, 20.31%, 25.36%, 27.65%

13.77 32.17%, 37.22%, 45.36%

13.79 (a) $YBr_3 \cdot 9H_2O$ (b) $Zn_3(PO_4)_2 \cdot 8H_2O$
(c) $Zr(SO_4)_2 \cdot 4H_2O$

13.81 bonded to the metal cation by coordinate covalent bonds, bonded to the anion by hydrogen bonding, or entrapped in spaces in the crystal lattice

13.83 It is deliquescent and absorbs water from the air until a solution is formed.

Chapter 14

14.5 cobalt

14.7 unsaturated

14.9 (a) 1.00% (b) 4.00% (c) 3.0%
(d) 0.080%

14.11 (a) 20.0% (b) 5% (c) 0.3%

14.13 1 mL SO_2

14.15 9 parts per billion (ppb)

14.17 15 mL alcohol

14.19 (a) 0.20 M (b) 0.50 M (c) 4.22 M

14.21 (a) 45.0 g (b) 85.5 g (c) 15.0 g

14.23 (a) 0.40 L (b) 10.0 L

14.25 6.11×10^{-3} M

14.27 3.8×10^{-7} mole

14.29 (a) 0.60 M (b) 1.6 M (c) 4.5 M
(d) 1.5 M

14.31 (a) 150 mL (b) 2450 mL (c) 475 mL
(d) 3800 mL

14.33 3.1 m

14.35 1.1 m

14.37 11.4 m

14.39 $HBr + H_2O \longrightarrow H_3O^+ + Br^-$

14.43 unsaturated

14.45 Reduction of the pressure of CO_2 lowers the solubility of CO_2.

14.47 The oxygen content of river water is depleted when water used for cooling is returned to the river.

14.49 strong attractive forces between NH_3 and H_2O molecules due to hydrogen bonding

14.51 Both I_2 and CCl_4 are nonpolar.

14.53 Fresh water evaporates more rapidly because it has a higher vapor pressure.

14.55 Freezing point depression of NaCl should be double that of 0.10 M glucose.

14.57 150 g/mole

14.59 100.27°C

14.61 169 g/mole

14.63 −0.37°C

14.65 Water will pass into the cells by osmosis.

14.67 1.2 mm Hg

Chapter 15

15.1 exothermic

15.3 one that occurs without an outside source of energy

15.5 negative $\Delta G°$

15.7 an increase in the order of the system

15.9 (a) not spontaneous at any T (b) spontaneous at low T

15.11 One reaction is the reverse of the other in terms of the products and reactants.

15.13 The reaction will go faster at a higher temperature.

15.15 The chemical reaction associated with spoiling goes faster at higher temperatures, slower at lower temperatures.

15.17 an increase in concentration or contact between reactants, an increase in temperature, and an addition of a catalyst

15.19 (a) $K = \dfrac{[O_3]^2}{[O_2]^3}$ (b) $K = \dfrac{[NH_3]^2}{[N_2][H_2]^3}$

(c) $K = \dfrac{[CH_3Cl][HCl]}{[CH_4][Cl_2]}$ (d) $K = \dfrac{[CO_2]^2}{[CO]^2[O_2]}$

15.21 (a) $K = \dfrac{[CH_4][H_2S]^2}{[CS_2][H_2]^4}$ (b) $K = \dfrac{[HCl]^4[O_2]}{[Cl_2]^2[H_2O]^2}$

(c) $K = \dfrac{[NO_2]^2}{[N_2O_4]}$ (d) $K = \dfrac{[NO]^2}{[N_2][O_2]^2}$

15.23 They correspond to the exponents in the equilibrium constant expression.

15.25 61.0

15.27 4.65×10^{-3}

15.29 10.5

15.31 3.87×10^{-2} M

15.33 0.256 M

15.35 (a) shift right (b) shift left
(c) shift left (d) shift right

15.37 (a) shift right (b) shift left (c) shift left

15.39 (1) shift right (2) shift left (3) shift left

15.41 (1) shift right (2) no shift (3) shift left

15.43 (1) shift left (2) shift right (3) shift left

15.45 (1) Product formation is favored by high P, low T
(2) Product information is favored by low P, high T
(3) Product formation is favored by low T (P makes no difference)

15.47 (b) AgBr

15.49 (b) NH_4I and (f) Na_2CrO_4

15.51 (a) $NH_4I + AgNO_3 \longrightarrow AgI\downarrow + Na^+ + NO_3^-$
(b) $K_2SO_4 + Ba(NO_3)_2 \longrightarrow BaSO_4\downarrow + 2 K^+ + 2 NO_3^-$
(c) no precipitate

15.53 (a) no precipitate
(b) no precipitate
(c) $3 AgNO_3 + Na_3PO_4 \longrightarrow Ag_3PO_4\downarrow + 3 Na^+ + 3 NO_3^-$

15.55 a, b and c are all the same $Ag^+ + Br^- \longrightarrow AgBr\downarrow$

15.57 (a) $Ag^+ + I^- \longrightarrow AgI\downarrow$
(b) $Ba^{2+} + SO_4^{2-} \longrightarrow BaSO_4\downarrow$
(c) $Pb^{2+} + 2 Cl^- \longrightarrow PbCl_2\downarrow$
(d) $Hg^{2+} + S^{2-} \longrightarrow HgS\downarrow$
(e) $Ba^{2+} + CO_3^{2-} \longrightarrow BaCO_3\downarrow$

15.59 (a) $K_{sp} = [Ag^+][Br^-]$ (b) $K_{sp} = [Ba^{2+}][SO_4^{2-}]$
(c) $K_{sp} = [Pb^{2+}][I^-]^2$ (d) $K_{sp} = [Fe^{2+}][OH^-]^2$
(e) $K_{sp} = [Ag^+]^2[SO_4^{2-}]$ (f) $K_{sp} = [Cu^+][S^{2-}]$

15.61 (a) M^2 (b) M^2 (c) M^3
(d) M^5 (e) M^5 (f) M^5

15.63 6.4×10^{-9}

15.65 3.6×10^{-7}

15.67 1.1×10^{-2} g $PbSO_4$

15.69 6.3×10^{-3} M = $[Ba^{2+}]$

Chapter 16

16.1 (a) An acid is a proton donor; a base is a proton acceptor.
(b) Hydroxide ion is OH^-; hydronium ion is H_3O^+.

16.3 An acid has one more H^+ than its conjugate base.

16.5 It dissociates to produce H_3O^+ and Cl^- which conducts electricity.

16.7 Eye protection should be worn and an eye wash and safety shower should be available.

16.9 (a) H_3PO_4 (b) $HC_2H_3O_2$ (c) H_2SO_4
(d) $Ca(OH)_2$ (e) KOH (f) $Al(OH)_3$

16.11 add an indicator

16.13 H_2SO_4 and NH_3

16.15 (a) monoprotic (b) triprotic (c) monoprotic
(d) diprotic (e) diprotic (f) monoprotic

16.17 Yes, if it can lose or gain a proton such as in H_2O, HCO_3^-, and HSO_4^-

16.19 $H_2CO_3 + H_2O \rightleftharpoons H_3O^+ + HCO_3^+$
$HCO_3^- + H_2O \rightleftharpoons H_3O^+ + CO_3^{2-}$

16.21 (a) $HClO_4$ (b) HCl (c) HNO_3
(d) HCO_3^- (e) CH_3CO_2H (f) H_2SO_4

16.23 (a) OH^- (b) Cl^- (c) CO_3^{2-}
(d) HCO_3^- (e) NO_3^- (f) HSO_4^-

16.25

$$H-\underset{\underset{H}{|}}{\overset{\overset{H}{|}}{C}}-\overset{\overset{O}{\|}}{C}-\overset{\overset{O}{\|}}{C}-O^-$$ or $C_3H_3O_3^-$

16.27 $C_{18}H_{21}NO_3H^+$

16.29 The predominant species in solution is NH_3(aq)

16.31 (a) weak (b) weak

16.33 (a) weak (b) strong (c) strong
(d) weak (e) strong (f) weak

16.35 $K_a = \dfrac{[H_3O^+][CN^-]}{[HCN]}$

16.37 $K_{a_1} = \dfrac{[H_3O^+][H_2PO_4^-]}{[H_3PO_4]}$ $K_{a_2} = \dfrac{[H_3O^+][HPO_4^{2-}]}{[H_2PO_4^-]}$
$K_{a_3} = \dfrac{[H_3O^+][PO_4^{3-}]}{[HPO_4^{2-}]}$

16.39 6.6×10^{-4}

16.41 2.95×10^{-5} M

16.43 5.87×10^{-6}

16.45 1.7×10^{-5}

16.47 (a) $[H_3O^+] = 0.1$ M
$[OH^-] = (1 \times 10^{-14})/(0.1) = 1 \times 10^{-13}$ M
(b) $[H_3O^+] = 0.01$ M
$[OH^-] = (1 \times 10^{-14})/(10^{-2}) = 1 \times 10^{-12}$ M
(c) $[OH^-] = 10^{-3}$ M
$[H_3O^+] = (1 \times 10^{-14})/(10^{-3}) = 1 \times 10^{-11}$ M
(d) $[H_3O^+] = 0.1$ M
$[OH^-] = (1 \times 10^{-14})/(10^{-1}) = 1 \times 10^{-13}$ M
(e) $[OH^-] = 10^{-4}$ M
$[H_3O^+] = (1 \times 10^{-14})/(10^{-4}) = 1.0 \times 10^{-10}$ M
(f) $[OH^-] = 0.01$ M
$[H_3O^+] = (1 \times 10^{14})/(0.01) = 1 \times 10^{-12}$ M

16.49 The concentration of H_3O^+ will decrease.

16.51 (a) 3 (b) 2.7 (c) 12 (d) 11.5

16.53 (a) 3.7 (b) 6.7 (c) 11.7 (d) 1

16.55 CNO^- is a stronger base than F^- or NO_2^-.

16.57 5.0×10^{-4}

16.59 2.0×10^{-3} M $[OH^-]$

16.61 They contain a natural indicator which is one color in acid and another color in base.

16.63 red at pH 3; yellow at pH 7

16.67 $[H_3O^+] = K_a \dfrac{[A^-]}{[HA]}$

16.71 9.01

16.73 (a) 0.5 meq HCl (b) 0.5 meq HNO_3
(c) 10^3 meq HCl

16.75 (a) 0.20 N NaOH (b) 0.1 N HCl (c) 1.0 N H_2SO_4
(d) 1.0 N H_3PO_4 (e) 2.0×10^{-4} N $Ca(OH)_2$

16.77 0.2 g $C_3H_6O_3$

16.79 (a) 0.45 N (b) 0.40 N (c) 0.25 N
(d) 0.1 N (e) 0.40 N (f) 0.30 N

16.81 15 mL NaOH

16.83 0.0417 N $HClO_4$

Chapter 17

17.1 (a) Oxidation is loss of electrons.
(b) Reduction is gain of electrons.

(c) An oxidizing agent gains electrons and is reduced, allowing another substance to be oxidized.

(d) A reducing agent loses electrons and is oxidized, allowing another substance to be reduced.

17.3 The standard electrode potential of the hydrogen electrode was arbitrarily set at zero to give a reference point against which other half-reactions could be compared.

17.5 (a) $+1$ (b) -2 (c) $+7$
(d) $+4$ (e) -2 (f) $+6$

17.7 (a) $+2$ (b) $+3$ (c) $+6$
(d) $+5$ (e) $+5$ (f) $+3$

17.9 (a) Al oxidized, Co reduced
(b) Cu oxidized, Br reduced
(c) S oxidized, N reduced
(d) S oxidized, O reduced

17.11 (a) K oxidized, Br reduced
(b) Fe reduced, Al oxidized
(c) H reduced, C oxidized
(d) S oxidized, O reduced

17.13 (a) Hg^{2+} in HgS is the oxidizing agent; S^{2-} in HgS is the reducing agent.
(b) SnO_2 is the oxidizing agent; C in CO is the reducing agent.
(c) WO_3 is the oxidizing agent; H_2 is the reducing agent.
(d) NiS is the reducing agent; O_2 is the oxidizing agent.

17.15 (a) Sb_2O_3 is the oxidizing agent; Fe is the reducing agent.
(c) XeF_4 is the oxidizing agent; H_2O is the reducing agent.
(d) Cl_2 is the reducing agent; F_2 is the oxidizing agent.

17.17 Coefficients given in order of appearance.
(a) 2,3,2 (b) 2,3,2 (c) 2,3,2,2
(d) 3,2,3,1,3

17.19 Coefficients given in order of appearance.
(a) 3,2,2,3,2,1 (b) 1,4,1,1,2 (c) 1,1,1,1
(d) 1,2,1,1,2

17.21 Coefficients given in order of appearance.
(a) 1,7,8,8,4,4 (b) 2,2,3 (c) 3,6,1,5,3
(d) 8,5,4,4,1,4

17.23 Coefficients given in order of appearance.
(a) 2,1,2,1 (b) 1,2,2,1 (c) 3,8,2,3,2,4
(d) 1,6,14,2,6,7

17.25 Coefficients given in order of appearance.
(a) 2,2,2,1,2,2 (b) 3,2,8,3,2,4
(c) 3,1,5,3,2,4 (d) 3,2,6,3,2,3

17.27 Coefficients given in order of appearance.
(a) 3,2,4,3,2 (b) 3,4,2,3,4,4
(c) 3,5,3,6,5,6 (d) 1,1,2,4,2,1

17.29 (a) $Cu^{2+} > Zn^{2+}$ (b) $Zn^{2+} > Mg^{2+}$
(c) $Hg^{2+} > Cd^{2+}$ (d) $Hg^{2+} > Ag^+$
(e) $Br_2 > I_2$ (f) $Pb^{2+} > Sn^{2+}$

17.31 (a) $Li > K$ (b) $Na > Zn$ (c) $Zn > Cu$
(d) $Ni > Pb$ (e) $Cu > Hg$ (f) $I^- > Br^-$

17.33 a, b, and c will proceed as written.

17.35 (a) $Mg + Zn^{2+} \longrightarrow Mg^{2+} + Zn$ $E° = 1.61$ V
(b) $Mg + Fe^{2+} \longrightarrow Mg^{2+} + Fe$ $E° = 1.93$ V
(c) $Zn + Pb^{2+} \longrightarrow Zn^{2+} + Pb$ $E° = 0.63$ V
(d) $Cu + 2 Ag^+ \longrightarrow Cu^{2+} + 2 Ag$ $E° = 0.46$ V

17.37 The impure metal is used as the anode, and the cathode is a strip or bar of the same metal, ultra pure.

17.41 (a) $+4$ (b) $+5$ (c) $+5$

17.43 (a) $+6$ (b) $+4$ (c) $+5$

17.45 (a) Re is reduced; Sb is oxidized
(b) I is reduced; Fe is oxidized
(c) I is reduced; C is oxidized

17.47 (a) $KMnO_4$ is the oxidizing agent; HCl is the reducing agent
(b) $KMnO_4$ is the oxidizing agent; KI is the reducing agent
(c) HNO_3 is the oxidizing agent; Cu is the reducing agent

17.49 (a) $2 ReCl_5 + SbCl_3 \longrightarrow 2 ReCl_4 + SbCl_5$
(b) $2 K_2S_2O_3 + I_2 \longrightarrow K_2S_4O_6 + 2 KI$
(c) $3 Cu + 8 HNO_3 \longrightarrow 3 Cu(NO_3)_2 + 2 NO + 4 H_2O$

17.51 (a) $10 NO_3^- + 8 H^+ + I_2 \longrightarrow 10 NO_2 + 4 H_2O + 2 IO_3^-$
(b) $Br_2 + SO_2 + 2 H_2O \longrightarrow 4 H^+ + 2 Br^- + SO_4^{2-}$
(c) $MnO_4^{2-} + Se^{2-} + 2 H_2O \longrightarrow MnO_2 + Se + 4 OH^-$

17.53 (a) $Cu^{2+} > Cd^{2+}$ (b) $Hg^{2+} > Mg^{2+}$
(c) $Hg^{2+} > Zn^{2+}$ (d) $Ag^+ > Zn^{2+}$
(e) $Cl_2 > Br_2$ (f) $Pb^{2+} > Cd^{2+}$

17.55 (a) $K > Na$ (b) $Cd > Ni$ (c) $Cu > Hg$
(d) $K > Pb$ (e) $Zn > Hg$ (f) $Br^- > Cl^-$

17.57 a, b, and c will proceed as written.

17.59 (a) $Ni + 2 Ag^+ \longrightarrow Ni^{2+} + 2 Ag$ $E° = 1.05$ V
(b) $3Ba + 2 Al^{3+} \longrightarrow 3 Ba^{2+} + 2 Al$ $E° = 1.24$ V
(c) $Zn + Fe^{2+} \longrightarrow Zn^{2+} + Fe$ $E° = .32$ V
(d) $Cd + 2 Ag^+ \longrightarrow Cd^{2+} + 2 Ag$ $E° = 1.20$ V

Chapter 18

18.1 An alpha particle is ^4_2He.

18.3 (a) ^1_0n (b) ^1_1p (c) $^0_{-1}e$

18.5 (a) $^{14}_6C$ (b) $^{10}_5B$ (c) $^{238}_{92}U$
(d) $^{37}_{17}Cl$ (e) $^{65}_{30}Zn$ (f) $^{24}_{11}Na$

18.7 (a) 82 p, 124 n (b) 92 p, 143 n
(c) 53 p, 74 n (d) 7 p, 6 n
(e) 11 p, 14 n (f) 1 p, 2 n

18.9 (a) α or ^4_2He, $+2$, 4 amu (b) β or $^0_{-1}e$, -1, 0 amu (c) γ, 0, 0 amu

18.11 (a) $^{21}_9F \longrightarrow ^0_{-1}e + ^{21}_{10}Ne$
(b) $^{31}_{14}Si \longrightarrow ^0_{-1}e + ^{31}_{15}P$
(c) $^{28}_{12}Mg \longrightarrow ^0_{-1}e + ^{28}_{13}Al$

18.13 (a) $^{212}_{84}\text{Po} \longrightarrow {}^4_2\text{He} + {}^{208}_{82}\text{Pb}$

     (b) $^{240}_{96}\text{Cm} \longrightarrow {}^4_2\text{He} + {}^{236}_{94}\text{Pu}$

     (c) $^{252}_{99}\text{Es} \longrightarrow {}^4_2\text{He} + {}^{248}_{97}\text{Bk}$

18.15 (a) $^{243}_{96}\text{Cm} \longrightarrow {}^4_2\text{He} + {}^{239}_{94}\text{Pu}$

     (b) $^{222}_{86}\text{Rn} \longrightarrow {}^4_2\text{He} + {}^{218}_{84}\text{Po}$

     (c) $^{234}_{94}\text{Pu} \longrightarrow {}^4_2\text{He} + {}^{230}_{92}\text{U}$

     (d) $^{245}_{97}\text{Bk} \longrightarrow {}^4_2\text{He} + {}^{241}_{95}\text{Am}$

18.17 (a) 50%   (b) 25%   (c) 12.5%

18.19 17,190 years old

18.21 about 0.4% remaining

18.23 2 hours

18.25 Ionizing radiation strips electrons from stable molecules leaving highly reactive ions behind.

18.27 A Geiger counter detects ionizing radiation because the radiation produces ions and electrons in the low pressure gas chamber.

18.29 3.7×10^{10} Bk

18.33 The one with a short half-life will do more short-term damage, but the one with a long half-life will do more long-term damage.

18.35 16 m

18.37 emit gamma radiation so the radiation can be detected outside the body

18.39 $^{32}_{16}\text{S} + {}^1_0\text{n} \longrightarrow {}^{32}_{15}\text{P} + {}^1_1\text{H}$

18.41 Isotopes used in the therapy are alpha or beta emitters.

18.43 Phosphorus-32 is used in the form of phosphate ions to treat some forms of leukemia. The phosphate ions are incorporated in the bones, and the radiation emitted affects white blood cell formation in the bone marrow.

18.45 $^{58}_{28}\text{Ni} + {}^1_1\text{H} \longrightarrow {}^4_2\text{He} + {}^{55}_{27}\text{Co}$

18.47 $^{238}_{92}\text{U} + {}^{16}_{7}\text{N} \longrightarrow 5\,{}^1_0\text{n} + {}^{247}_{99}\text{Es}$

18.49 (a) $^{23}_{11}\text{Na} + {}^2_1\text{H} \longrightarrow {}^{24}_{11}\text{Na} + {}^1_1\text{H}$

     (b) $^{10}_{5}\text{B} + {}^4_2\text{He} \longrightarrow {}^{13}_{7}\text{N} + {}^1_0\text{n}$

     (c) $^{59}_{27}\text{Co} + {}^1_0\text{n} \longrightarrow {}^{56}_{25}\text{Mn} + {}^4_2\text{He}$

18.51 (a) $^{27}_{13}\text{Al} + {}^4_2\text{He} \longrightarrow {}^{30}_{15}\text{P} + {}^1_0\text{n}$

     (b) $^{15}_{7}\text{N} + {}^1_1\text{H} \longrightarrow {}^{12}_{6}\text{C} + {}^4_2\text{He}$

     (c) $^{242}_{96}\text{Cm} + {}^4_2\text{He} \longrightarrow {}^{245}_{98}\text{Cf} + {}^1_0\text{n}$

18.53 (a) $^{253}_{99}\text{Es} + {}^4_2\text{He} \longrightarrow {}^{256}_{101}\text{Md} + {}^1_0\text{n}$

     (b) $^{249}_{98}\text{Cf} + {}^{18}_{8}\text{O} \longrightarrow {}^{263}_{106}\text{Unh} + 4\,{}^1_0\text{n}$

     (c) $^{238}_{92}\text{U} + {}^{14}_{7}\text{N} \longrightarrow {}^{247}_{99}\text{Es} + 5\,{}^1_0\text{n}$

     (d) $^{238}_{92}\text{U} + {}^{16}_{8}\text{O} \longrightarrow {}^{249}_{100}\text{Fm} + 5\,{}^1_0\text{n}$

18.55 Uranium in nuclear reactors is not as pure as uranium used for bombs, and reactors have control rods of neutron-absorbing cadmium inserted between the uranium samples.

18.57 $^{235}_{92}\text{U} + {}^1_0\text{n} \longrightarrow {}^{137}_{56}\text{Ba} + {}^{94}_{36}\text{Kr} + 5\,{}^1_0\text{n}$

18.59 an atomic bomb "trigger"

Chapter 19

19.1 A molecular formula gives the number of atoms; the structural formula gives the arrangement of atoms.

19.3 Single bonds between carbon atoms

19.5 Structural isomers have the atoms arranged and bonded differently. Geometric isomers have identical atoms bonded to each other but the geometric arrangement is different about a double bond.

19.7 (a) C_5H_{12} (b) C_4H_{10} (c) C_6H_{14}

     (d) C_9H_{20}

19.9 (a) CH_3-CH_2-Br

     (b) CH_3-CH_2-Br

     (c) $Br-CH_2-CH_2-Br$

     (d) $CH_3-CH_2-CH_2-CH_2-CH_3$

19.11 (a) CH_3CH_2Br (b) CH_3CH_2Br

     (c) CH_2BrCH_2Br (d) $CH_3CH_2CH_2CH_2CH_3$

19.13 (a), (b), (c), (d)

19.15 (a), (b), (c), (d), (e), (f)

19.17 (a) conformers   (b) conformers

     (c) isomers   (d) isomers

19.19 (a) saturated   (b) unsaturated

     (c) unsaturated   (d) unsaturated

19.21 b, c, d are saturated compounds

19.23 (a) 22H (b) 18H (c) 12H

     (d) 32H (e) 14H (f) 10H

19.25 (a) C_6H_{12} (b) C_5H_8 (c) C_7H_{12}

     (d) C_4H_6

19.27 (a) methyl (b) isopropyl (c) t-butyl

     (d) sec-butyl

19.29 (a) heptane

     (b) 2,2-dimethylbutane

     (c) 2,2-dimethylpentane

     (d) 3-ethyl-2,5-dimethylhexane

19.31 (a) $CH_3-CH_2-\underset{\underset{\displaystyle CH_3}{|}}{CH}-CH_2-CH_3$

(b) $CH_3-CH_2-CH-CH-CH_2-CH_3$
         $\quad\quad\quad\quad\;\; | \quad\; |$
         $\quad\quad\quad\quad CH_3 \;\; CH_3$

(c)
$$CH_3-\underset{\underset{CH_3}{|}}{\overset{\overset{CH_3}{|}}{C}}-\underset{\underset{CH_3}{|}}{CH}-CH_2-CH_3$$

(d) $CH_3-CH_2-CH_2-CH-CH_2-CH_2-CH_3$
         $\quad\quad\quad\quad\quad\quad\quad |$
         $\quad\quad\quad\quad\quad\; CH_3-CH_2$

(e) $CH_3-CH-CH-CH-CH-CH_3$
         $\quad\quad\;\;\; | \quad\; | \quad\; | \quad\; |$
         $\quad\quad CH_3 \; CH_3 \; CH_3 \; CH_3$

19.33 **(a)** 2 **(b)** 2 **(c)** 4 **(d)** 3

19.35 **(a)** 2 **(b)** 3 **(c)** 0 **(d)** 0

19.37 **(a)** 0 **(b)** 0 **(c)** 0 **(d)** 1

19.39 **(a)** 6 **(b)** 6 **(c)** 12 **(d)** 9

19.41 **(a)** 2 **(b)** 2 **(c)** 2 **(d)** 3 **(e)** 4
(f) 3 **(g)** 2

19.43 **(a)** Cl on cyclopropane **(b)** cyclobutane with CH_3 and CH_3

(c) cycloheptane **(d)** cyclopentane with CH_3

19.45
$C=C$ (H, CH_2-CH_3 / H, H) ; $C=C$ (H, CH_3 / H, CH_3)

$C=C$ (CH_3, H / H, CH_3) ; $C=C$ (CH_3, CH_3 / H, H)

19.47
$C=C$ (H, $CH_2-CH_2-CH_3$ / H, H) ; $C=C$ (H, H / H, $CH-CH_3$ with CH_3)

1-pentene ; 3-methyl-1-butene

$C=C$ (CH_3, H / CH_2 CH_3, H) 2-methyl-1-butene ; $C=C$ (CH_3, CH_2-CH_3 / H, H) cis-2-pentene

$C=C$ (CH_3, H / H, CH_2-CH_3) trans-2-pentene ; $C=C$ (CH_3, CH_3 / H, CH_3) 2-methyl-2-butene

19.49 **(a)** 2-methylpropene
(b) 2,3-dimethyl-2-butene
(c) 2-methyl-2-butene

(d) 2,5-dimethyl-2-hexene
(e) 1-chloro-3-methyl-2-butene
(f) 3-bromopropene

19.51 **(a)** $C=C$ (CH_3, CH_2-CH_3 / CH_3, H)

(b) $C=C$ (H, $CH_2-CH_2-CH_2-CH_3$ / H, H)

(c) $C=C$ (H, H / CH_3-CH (with CH_3), CH_2-CH_3)

(d) $C=C$ (H, CH(with CH_3)$-CH_3$ / CH_3, H)

19.53 **(a)** 1-pentyne
(b) 2,2-dimethyl-3-hexyne
(c) 1-chloro-3-octyne
(d) 4,5-dibromo-2-hexyne
(e) 4-methyl-2-hexyne
(f) 2-chloro-6-methyl-3-octyne

19.55 Cyclohexane would not react with bromine.

19.57 **(a)** $CH_3CH_2CH=CH_2 + Br_2 \longrightarrow$
$\quad\quad CH_3CH_2CHBrCH_2Br$
(b) $CH_3CH_2CH=CH_2 + HBr \longrightarrow$
$\quad\quad CH_3CH_2CHBrCH_3$

19.59 **(a)** $CH_2ClCHClCH_3$
(b) CH_3CHCH_3
$\quad\quad\quad\;\; |$
$\quad\quad\quad OH$
(c) $CH_3-CH-CH_3$
$\quad\quad\quad\quad\; |$
$\quad\quad\quad\quad Br$
(d) $CH_2BrCH-CH_3$
$\quad\quad\quad\quad\quad |$
$\quad\quad\quad\quad\; Br$
(e) CH_3CH-CH_3
$\quad\quad\quad\quad |$
$\quad\quad\quad\; Cl$

19.61 **(a)** $CH_3C\equiv CH + 2\,H_2 \longrightarrow CH_3CH_2CH_3$
(b) $CH_3C\equiv CH + 2\,Br_2 \longrightarrow CH_3CBr_2CHBr_2$
(c) $CH_3C\equiv CH + HBr \longrightarrow$ $C=C$ (CH_3, H / Br, H)
(d) $CH_3C\equiv CH + 2\,HBr \longrightarrow CH_3CBr_2CH_3$

19.63 **(a)** Aniline **(b)** phenol **(c)** toluene
(d) chlorobenzene

19.65 **(a)** 1,2,4-trichlorobenzene **(b)** 2,4-dibromophenol
(c) m-nitroaniline **(d)** 3,4-dichlorotoluene

20.1 Alcohols are classified according to the number of carbon groups bonded to the carbon atom containing the hydroxyl group.

20.3 An aldehyde has a hydrogen atom bonded to the carbonyl carbon atom. A ketone has two carbon groups bonded to the carbonyl carbon atom. A carboxylic acid has a hydroxyl group bonded to the carbonyl carbon atom.

20.5 Amine have nitrogen bonded to a carbon atom other than a carbonyl carbon atom. Amides have nitrogen bonded to a carbonyl carbon atom.

20.7 (a)

(b)

(c)

(d)

(e)

(f)

20.9 (a) 2-methyl-2-butanol (b) 3-methyl-2-butanol
(c) 2-methyl-1-butanol (d) 3-methyl-1-butanol

20.11 c and d are primary; b is secondary; a is tertiary.

20.13 A primary alcohol at the top of molecule and a secondary alcohol attached to the ring.

20.15 There are two hydroxyl groups that can form intermolecular hydrogen bonds.

20.17 $2\ CH_3CH_2OH + 2\ K \longrightarrow 2\ CH_3CH_2OK + H_2$

20.19 The alcohol will react with sodium and release hydrogen gas whereas the alkane will not react.

20.21 (a) $CH_3CHClCH_3$ (b) $CH_3CH_2CH_2CH_2Br$

20.23 The rate of reactivity will be I > III > II

20.25 (a) $CH_2{=}CHCH_2CH_2CH_3$

(b) $CH_2{=}CHCH_2CH_2CH_3$ and cis and trans $CH_3CH{=}CHCH_2CH_3$

(c)

20.27 (a) CH_3CO_2H (b) no reaction
(c) $CH_3CH_2CH_2CO_2H$ (d) CH_3COCH_3
(e) CH_3CO_2H (f) $CH_3COCH_2CH_3$
(g) HCO_2H (h) $(CH_3)_2CHCO_2H$

20.29 $CH_3CH_2CH_2OCH_3$ $(CH_3)_2CHOCH_3$ $CH_3CH_2OCH_2CH_3$

20.31 They do not have an active hydrogen atom like alcohols do.

20.33 (a)

(b)

(c) CH_3CHO (d)

20.35 (a)

(b)

(c) $BrCH_2CH_2CH_2CH_2CHO$

(d)

20.37 (a) 2-methylbutanal (b) 2-methylpropanal
(c) 4-methylpentanal (d) 3-ethyl-3-methylpentanal

20.39 (a) 2-chloropropanal (b) 4-bromobutanal
(c) 4-chloro-3-methylbutanal (d) 3-iodopropanal

20.41 Aldehydes do not have a polar hydrogen atom with which to form hydrogen bonds to water.

20.43 A red precipitate of Cu_2O is formed.

20.45 (a) propionic acid (b) formic acid
(c) butyric acid (d) acetic acid

20.47 (a) valeric acid
(b) α-methylpropionic acid
(c) β-methylvaleric acid
(d) α,γ,-dimethylvaleric acid
(e) β-methylbutyric acid
(f) β,δ,δ-trimethylcaproic acid

20.49 butanal < 1-butanol < butanoic acid

20.51 (a) sodium butyrate, sodium butanoate
(b) potassium stearate, potassium hexadecanoate
(c) calcium propionate, calcium propanoate

20.53 (a) ethyl formate (b) methyl butyrate
(c) octyl acetate

20.55 (a) phenyl acetate (b) phenyl propionate

20.57 (a) $CH_3CO_2H + C_6H_5OH$
(b) $C_6H_5CO_2H + CH_3CH_2OH$

(c) $HCO_2H + CH_3CH_2OH$
(d) $CH_3CO_2H + CH_3CH_2CH_2OH$

20.59 (a) $CH_3CO_2Na + C_6H_5OH$
(b) $C_6H_5CO_2Na + CH_3CH_2OH$
(c) $HCO_2Na + CH_3CH_2OH$
(d) $CH_3CO_2Na + CH_3CH_2CH_2OH$

20.61 The nitrogen atom contained in the ring is secondary.

20.63 (a) secondary (b) tertiary (c) secondary
(d) primary (e) tertiary (f) secondary

20.65 (a) 3-aminohexane

(b) 1-(N,N-dimethylamino)butane
(c) 2-amino-4-methylpentane

20.67 (a) propionamide (b) N-ethylpropionamide
(c) α-methylbutyramide

20.69 (a) $CH_3CH_2CONHCH_2CH_3$
(b) $CH_3CH_2CH_2CON(CH_2CH_3)_2$
(c) $CH_3CONHC_6H_5$
(d) $CH_3CH_2CH_2CH_2CONH_2$

20.71 (a) $C_6H_5CO_2H + CH_3CH_2NH_2$
(b) $CH_3CH_2CO_2H + (CH_3)_2NH$

Index

Boldface page numbers indicate end-of-chapter definitions.

Helium, 411
  density, 42
  electron configuration, 164
  elemental symbol, 61
  Hund's rule and, 168
  Lewis octet rule and, 195
  mass, 74, 75
  nucleus (alpha particle), 396
  radius, 75
  solubility in water, 306
Hemoglobin, 194
  carbon monoxide and, 334
Henry's law, 306
Hepta, 227
Heterogeneous mixture, 58, **66**
Hexa, 227
Hibernation, 326
Homogeneous mixture, 58, **66**
Homologous series, 425, **443**
Hund's rule, 162, 167–68, **170**
Hydrate, 281, **286**
  formula, 282
Hydride ion, 223
Hydro-, 230, **233**
Hydrobromic acid, 230
Hydrocarbon, 424, **443**
  alkanes, 424, **443**
  alkenes, 425, 432, **443**
  alkynes, 425, 438, **443**
  aromatic, 425, 440, **443**
  cycloalkanes, 424, 431, **443**
  saturated, 424, **443**
  unsaturated, 424, **443**
Hydrochloric acid, 230, 354
  common name, 219
  concentrated solution, 347
  equivalent weight, 363
  in gastric juice, 347
  ionization, 350
  reaction with water, 348
Hydrofluoric acid, 230
Hydrogen
  classification, 427
  covalent bond, 200
  density, 42
  electrode potential, 382
  electron configuration, 164, 168
  elemental symbol, 61
  isotopes, 79
  mass, 74
  molecule, 82
  polyatomic ions containing, 233
Hydrogen bomb, 412
Hydrogen bond, 277, **286**
  in alcohols, 458
  in amides, 479
  in amines, 477
  in ammonia, 278
  in carbonyl compounds, 467
  in carboxylic acids, 471
  effect on boiling point, 278

in esters, 474
in ice, 282
in water, 278, 280–81
Hydrogen bromide, 230
Hydrogen chloride, 230
  bonding, 201
Hydrogen electrode, 383
Hydrogen fluoride, 230
  and hydrogen bonding, 278
Hydrogen iodide, 230
Hydrogen ion
  from acids, 348
  concentration in water, 353
  pH and, 355
  in water as hydronium ion, 348
Hydrogen peroxide, 110
Hydrogenation, 436
Hydroidic acid, 230
Hydrolysis, 359, **368**
  of amides, 480
  constant (K_h), 359
  of esters, 474
  of salts, 359
Hydrometer, 46
Hydronium ion, 228, 348, **368**
  buffers and, 361
  formation, 349
  structure, 349
Hydroxide ion, 86, 228, 353
Hydroxyl group, 455
Hygroscopic, 285, **286**
Hypo, 219
Hypo-, 86, 228, **233**
Hypochlorite ion, 86, 228
Hypochlorous acid, 203, 230
Hypothermia, 326
Hypothesis
  Avogadro's, 252, **262**
  definition, 10, **22**

Iatrochemist, 8
-ic, 225, **234**
Ice
  density, 280
  equilibrium with water, 280
-ide, 85, 227, **234**
Ideal gas, 244, **262**
  equation, 253
  law, 253, **262**
Improper fraction, 491
Indicator, 358
  colors, 358
Induced dipole, 275, **286**
Inert gas, 180
Inner transition element, 180, **188**
Intercept, 496
Intermolecular forces, 274, **286**
  in liquids, 267
International Union of Pure and Applied
  Chemistry, *see* IUPAC

Intramolecular forces, 274, **286**
Inverse proportionality, 495
Iodide ion, 85
  radius, 198
Iodine
  density, 176
  electrode potential, 382
  isotopes, 406, 408
  oxidation number, 221
  radius, 75, 198
Ion, 84, 198
  anion, 84, **90**
  cation, 84, **90**
  hydronium, 228, 348, **368**
  hydroxide, 86, 228
  naming, 85
  polyatomic, 85, 86, **90,** 233
  reaction with water (hydrolysis), 359
  sizes, 198
  symbols, 84
Ion product constant, 353
Ionic bond, 195, **214**
Ionic compound, 86, **90,** 195, **214**
  dissociation, 303, **314**
  names, 88, 89
  solubility, 135, 304, 336
  structure, 87, 198
Ionic equation
  net, 335, **339**
  in redox, 379
Ionic radius, 198, **214**
  table, 198
Ionization
  of acids, 348, **368**
  of atoms, 186, **188**
  of electrolytes, 303, **315**
  percent, 352
  of water, 353
Ionization constant, acid, 351, **367**
Ionization energy, 186, **188**
  table, 186
Ionizing radiation, 401, **413**
Iron
  abundance, 62
  density, 42
  electrode potential, 382
  elemental symbol, 61
  ions, 86, 226
  oxides, 226
  specific heat, 51
Isobutyl group, 428
Isomers, 422, **443**
  alkane, 422, 426
  alkene, 433
Isopropyl alcohol, 459
Isopropyl group, 428
Isotope
  definition, 79, **90**
  radio-, 396, 405, **413**
  table, 79, 396
-ite, 228, **234**

Radiation, *continued*
  doses, 404
  ionizing, 401, **413**
  sickness, 404
  units, 403
Radical, 401, **413**
Radius
  of atoms, 75, 184, **188**
  of ions, 198, **214**
Radioactive decay, 395
Radioactivity, 395
  detection, 402
  half-life, 399
  interaction with matter, 401
  measurement, 403
Radioisotope, 396, **413**
  use in diagnosis, 405
  use in therapy, 407
Radon, electron configuration, 165
Rare gas, *see* Noble gas
Rate of reaction, *see* Reaction rate
RBE (relative biological effectiveness), 403, **413**
Reactant, 7, **13**
  reaction rate and, 324
Reaction
  activation energy, 327
  addition, 436, **443**
  of alcohols, 460
  of alkanes, 430
  of alkenes, 436
  of alkynes, 439
  chain, 410, 411
  combination, 130, **137**
  decomposition, 131, 132, **137**
  definition, 7, **13**
  double replacement, 131, 134, **137**
  endothermic, 57, **66,** 322, 327, 332, **339**
  energy, 321
  enthalpy, 322
  entropy, 322
  of esters, 474
  exothermic, 57, **66,** 322, 327, 332, **339**
  half-, 378, **389**
  neutralization, 136, **137,** 232, 365
  nonspontaneous, 321, **339**
  nuclear, 397
  oxidation–reduction, 374
  precipitation, 335, **340**
  rate, *see* Reaction rate
  redox, 374, **389**
  single replacement, 131, 133, **137**
  spontaneous, 321, **340**
  substitution, 133, 440, 460
  types, 129
Reaction rate
  catalyst and, 326
  concentration and, 325
  reactant nature and, 324
  temperature and, 325
Real gas, 244, **262**

Redox reaction, 374, **389**
  equation balancing, 375, 378
Reducing agent, 375, **389**
Reduction
  of alkenes, 436
  definition, 374, **389**
Rem, 403, **413**
Representative element, 180, **188**
  atomic radius, 185
  electronegativity, 187
  ionization energy, 186
Resonance structure, 208, **214**
Roentgen, 403, **413**
Rolaids, 346
Roots, 493
Rounding off, 28
Rubbing alcohol, 459
Rubidium
  compounds, 184
  elemental symbol, 61
  radius, 75, 198
Rubidium ion, 86
  radius, 198

s orbital, 160
Salt
  acidic, 232, **233**
  basic, 232, **233**
  of carboxylic acids, 472
  formation from acid and base, 136, 232
  hydrolysis, 359, **368**
  mixed, 233, **234**
  solubility (table), 135, 336
  See also Sodium chloride
Saponification, 474, **481**
Saturated
  hydrocarbon, 424, **443**
  solution, 294, **315**
Science, 2, **13**
Scientific law, 9, **13**
Scientific method, 10, **13**
Scientific notation, 23, **36**
Scintillation counter, 403
Seaborg, Glenn, 9
Secondary
  alcohol, 457, **481**
  amine, 476, **481**
  carbon atom, 427, **443**
  hydrogen atom, 427, **443**
Selenide ion, 85
Selenium
  density, 176
  isotopes, 396
  melting point, 176
Self-ionization, water, 353
Semimetal, 65
  periodic table and, 183
Semipermeable membrane, 311, **315**
Shell, 157, **170**
  valence, 168, **170**

SI system, 30
Significant figures, 21
  addition and subtraction, 25
  multiplication and division, 26
  rounding, 27
Silicon
  abundance, 62
  electron configuration, 165
  elemental symbol, 61
  isotopes, 396
Silver
  electrode potential, 382
  elemental symbol, 61
  specific heat, 51
Silver ion, 86
Single replacement reaction, 113
Slaked lime, 219
Slope, 496
Sodium
  compounds, 184
  electrode potential, 382
  electron configuration, 164
  elemental symbol, 61
  radius, 75, 198
  reaction with alcohol, 460
  reaction with chlorine, 55, 196
  reaction with water, 55
Sodium bicarbonate, 132, 346
Sodium chloride (common salt), 194, 304
  boiling point, 54
  crystal structure, 87, 199
  dissolution, 305
  as electrolyte, 303
  ionic bonding, 196
  melting point, 54
  solubility, 294
Sodium hydroxide, 348
  equivalent weight, 363
Sodium ion, 84
  formation, 195
  radius, 198
Solid, 273
  characteristics, 53
  definition, 53, **66**
  density, 42
  heat of fusion, 273
  solution, 293
Solubility, 304, **315**
  of acids, 471
  of alcohols, 459
  of aldehydes, 467
  effect of pressure, 306
  effect of solvent, 307
  effect of temperature, 306
  of esters, 474
  of gases, 306
  of ionic compounds, 135, 336
  of oxygen, 293
  product constant, 337, **340**
  of sodium chloride, 294
Solute, 293, **315**

Atomic Numbers and Atomic Weights of the Elements

Based on $^{12}_{6}C$. Numbers in parentheses are the mass numbers of the most stable isotopes of radioactive elemen

Element	Symbol	Atomic Number	Atomic Weight	Element	Symbol	Atomic Number	Atomic Weight
Actinium	Ac	89	227.0278	Erbium	Er	68	167.26
Aluminum	Al	13	26.98154	Europium	Eu	63	151.96
Americium	Am	95	(243)	Fermium	Fm	100	(257)
Antimony	Sb	51	121.75	Fluorine	F	9	18.998403
Argon	Ar	18	39.948	Francium	Fr	87	(223)
Arsenic	As	33	74.9216	Gadolinium	Gd	64	157.25
Astatine	At	85	(210)	Gallium	Ga	31	69.72
Barium	Ba	56	137.33	Germanium	Ge	32	72.59
Berkelium	Bk	97	(247)	Gold	Au	79	196.9665
Beryllium	Be	4	9.01218	Hafnium	Hf	72	178.49
Bismuth	Bi	83	208.9804	Helium	He	2	4.00260
Boron	B	5	10.81	Holmium	Ho	67	164.9304
Bromine	Br	35	79.904	Hydrogen	H	1	1.0079
Cadmium	Cd	48	112.41	Indium	In	49	114.82
Calcium	Ca	20	40.08	Iodine	I	53	126.9045
Californium	Cf	98	(251)	Iridium	Ir	77	192.22
Carbon	C	6	12.011	Iron	Fe	26	55.847
Cerium	Ce	58	140.12	Krypton	Kr	36	83.80
Cesium	Cs	55	132.9054	Lanthanum	La	57	138.9055
Chlorine	Cl	17	35.453	Lawrencium	Lr	103	(260)
Chromium	Cr	24	51.996	Lead	Pb	82	207.2
Cobalt	Co	27	58.9332	Lithium	Li	3	6.941
Copper	Cu	29	63.546	Lutetium	Lu	71	174.967
Curium	Cm	96	(247)	Magnesium	Mg	12	24.305
Dysprosium	Dy	66	162.50	Manganese	Mn	25	54.9380
Einsteinium	Es	99	(252)	Mendelevium	Md	101	(258)